Handbook of Statistics
Volume 34

Data Gathering, Analysis and Protection of Privacy Through Randomized Response Techniques: Qualitative and Quantitative Human Traits

Handbook of Statistics

Series Editor

C.R. Rao
C.R. Rao AIMSCS, University of Hyderabad Campus,
Hyderabad, India

Handbook of Statistics

Volume 34

Data Gathering, Analysis and Protection of Privacy Through Randomized Response Techniques: Qualitative and Quantitative Human Traits

Edited by

Arijit Chaudhuri

Indian Statistical Institute, Kolkata, West Bengal, India

Tasos C. Christofides

University of Cyprus, Nicosia, Cyprus

C.R. Rao

C.R. Rao AIMSCS, University of Hyderabad Campus, Hyderabad, India

ELSEVIER

Amsterdam • Boston • Heidelberg • London • New York • Oxford
Paris • San Diego • San Francisco • Singapore • Sydney • Tokyo
North-Holland is an imprint of Elsevier

N·H

North-Holland is an imprint of Elsevier
Radarweg 29, PO Box 211, 1000 AE Amsterdam, Netherlands
The Boulevard, Langford Lane, Kidlington, Oxford OX5 1GB, UK

Notices

Knowledge and best practice in this field are constantly changing. As new research and experience broaden our understanding, changes in research methods, professional practices, or medical treatment may become necessary.

Practitioners and researchers must always rely on their own experience and knowledge in evaluating and using any information, methods, compounds, or experiments described herein. In using such information or methods they should be mindful of their own safety and the safety of others, including parties for whom they have a professional responsibility.

To the fullest extent of the law, neither the Publisher nor the authors, contributors, or editors, assume any liability for any injury and/or damage to persons or property as a matter of products liability, negligence or otherwise, or from any use or operation of any methods, products, instructions, or ideas contained in the material herein.

ISBN: 978-0-444-63570-9
ISSN: 0169-7161

Contents

Contributors

Numbers in Parentheses indicate the pages on which the author's contributions begin.

S. Abdelfatah (133), Cairo University, Giza, Egypt

S.A. Adeshiyan (273), US Energy Information Administration, Washington, DC, United States

A.K. Adhikary (191), Indian Statistical Institute, Kolkata, West Bengal, India

A. Arcos (155), University of Granada, Granada, Spain

R. Arnab (155, 169, 253), University of Botswana, Gaborone, Botswana; University of Kwazulu-Natal, Durban, South Africa

L. Barabesi (119), University of Siena, Siena, Italy

U. Böckenholt (287), Kellogg School of Management, Northwestern University, Evanston, IL, United States

M. Bose (341), Applied Statistics Unit, Indian Statistical Institute, Kolkata, India

C.N. Bouza-Herrera (209), Universidad de La Habana, La Habana, Cuba

A. Chaudhuri (13, 17), Indian Statistical Institute, Kolkata, India

T.C. Christofides (29), University of Cyprus, Nicosia, Cyprus

B. Cobo (155), University of Granada, Granada, Spain

M.J.L.F. Cruyff (287), Utrecht University, Utrecht, The Netherlands

G. Diana (119), University of Padova, Padova, Italy

K. Dihidar (427), Sampling and Official Statistics Unit, Indian Statistical Institute, Kolkata, West Bengal, India

L.E. Frank (287), Utrecht University, Utrecht, The Netherlands

M.L. Johnson (443), Texas A&M University-Kingsville, Kingsville, TX, United States

C.-S. Lee (353, 413), Coastal Bend College, Beeville, TX, United States

M. Marcovitz (317), Baylor University, Waco, TX, United States

R. Mazloum (133), Cairo University, Giza, Egypt

C. Mohamed (471), Texas A&M University-Kingsville, Kingsville, TX, United States

R. Mukerjee (331), Indian Institute of Management Calcutta, Kolkata, India

P. Mukhopadhyay (221), Indian Statistical Institute, Kolkata, India

K. Nandy (317), School of Nursing and Department of Biostatistics, Fielding School of Public Health, University of California, Los Angeles, CA, United States

T.K. Nayak (273), Center for Disclosure Avoidance Research, US Census Bureau; George Washington University, Washington, DC, United States

O. Odumade (169), Risk Analytics, Deloitte & Touche, Minneapolis, MN, United States

V.R. Padmawar (497), Stat-Math Unit, Indian Statistical Institute, Bangalore, Karnataka, India

S. Pal (405), West Bengal State University, Barasat, India

P.F. Perri (119), University of Calabria, Arcavacata di Rende (CS), Italy

A. Quatember (105), Johannes Kepler University, Linz, Austria

C.R. Rao (1), CRRao Advanced Institute of Mathematics, Statistics and Computer Science, Hyderabad, Telangana, India

T.J. Rao (1, 233), CRRao Advanced Institute of Mathematics, Statistics and Computer Science, Hyderabad, Telangana, India

S. Roy (405), Reserve Bank of India, Mumbai, India

M. Rueda (155, 253), University of Granada, Granada, Spain

J. Sarkar (233), Indiana University–Purdue University Indianapolis, IN, United States

S.A. Sedory (353, 413, 443, 471), Texas A&M University-Kingsville, Kingsville, TX, United States

S. Sengupta (91), University of Calcutta, Kolkata, India

P. Shaw (387), Reserve Bank of India, Mumbai, Maharashtra, India

S. Singh (43, 169, 353, 413, 443, 471), Texas A&M University-Kingsville, Kingsville, TX, United States

B.K. Sinha (233, 317), Retired Professor, Indian Statistical Institute, Kolkata, India

S.-C. Su (413), Freer Independent School District, Freer, TX, United States

P.G.M. van der Heijden (287), Utrecht University, Utrecht, The Netherlands; S3RI, University of Southampton, Southampton, United Kingdom

C. Zhang (273), George Washington University, Washington, DC, United States

Preface

Professor J.A. Fox, one of the two authors of the first monograph on "Randomized Response," made the initial move to bring about this Handbook. Although not contributing directly, he indicated his eagerness to celebrate in a befitting manner the fulfillment of the 50th year of the publication of the initial work on Randomized Response Technique. Some of the more enthusiastic among us duly responded and the outcome is this volume.

In the human society unpleasant phenomena happen to occur in plenty and abundance. We, as a part of it, are sometimes obliged to take serious notice of and propagate their occurrences among the conscientious public. This necessitates gathering of truthful and reliably adequate and accurate data. But social survey practices over the years were palpably found not to be sharp and incisive enough to elicit human responses through queries about sensitive and stigmatizing items of enquiry. Alcoholism, gambling habits, illegal drug use, tax evasion, rash driving of motorized vehicles, conjugal malpractices, and domestic violence are some of the features people like to hide from the human communities. Amounts gained or lost through gambling, duration of days behind bars on criminal prosecution, and expenses on treatment of venereal diseases are some of the facts people are usually reluctant to disclose on enquiries in direct manners in social surveys.

As a simple and easy way out, S.L. Warner developed and propagated a rather safe alternative of avoiding direct queries but going for indirect questioning techniques by dint of a probabilistic procedure of data gathering on this type of issues. This debut rapidly spawned a widespread volume of interesting literature on the subject. In the beginning, it attracted the social scientists. Sophisticated mathematical statisticians turned to this a little while afterward as a matter of fact. In course of time, doubtless, theoretical literature around this subject has accumulated in an admirable extent displaying qualitative scales attained as well. But applications of the developing and emerging techniques are not widespread enough to enthusiastically talk about among ourselves. Hoping an affectionate reception of this handbook among the community of readers this inadequacy may fruitfully be mitigated through official appreciation among government and private organizations in the corporate and other sectors.

Elsevier has shown an exemplary enthusiasm to propagate the study on the subject through a display of enterprising initiative and indulgent entrepreneurship, leaving the editors an opportunity to gather together established and budding contributors with a good deal of encouragement. Regrettably, there are established experts in the field who could not be persuaded to join. We sincerely believe that the contributors who have joined have done a highly praiseworthy service and we hope when the Handbook appears in the market it will be received with swift appreciation.

Arijit Chaudhuri
Tasos C. Christofides
C.R. Rao

Chapter 1

Review of Certain Recent Advances in Randomized Response Techniques

T.J. Rao and C.R. Rao[1]
CRRao Advanced Institute of Mathematics, Statistics and Computer Science, Hyderabad, Telangana, India
[1]*Corresponding author: e-mail: crr1@psu.edu*

ABSTRACT
During the past five decades, a number of research papers appeared in the literature on randomized response technique (RRT) introduced by Warner in 1965 for eliciting information on questions of a sensitive nature. In this chapter, we shall briefly review some of the recent advances relating to RRT.

Keywords: Warner's randomized response technique, Unrelated model, Cryptographic RRT, Stratified RRT, Reverse RRT, Cramér–Rao lower bound, Game theory, Meta analysis

1 INTRODUCTION

Nonresponse in sample surveys is a major problem. Two of the main sources of nonresponse are evasion of answers to questions of a sensitive nature, and "Not-at-Home's." We shall look at the randomized response technique (RRT) introduced by Warner for eliciting information on questions of a sensitive nature. During the past five decades, a number of important advances and reviews have appeared on this subject in sample survey literature. We refer to, among others, Chaudhuri and Mukerjee (1988); Chaudhuri and Mukerjee (1987), Chaudhuri (2011), Fox and Tracy (1986), and Chaudhuri and Christofides (2013) for comprehensive reviews on RRT.

In the next few paragraphs, we shall start with Warner's RRT and proceed to discuss some recent lines of research and comment on these. Since the volume contains background and reviews to various techniques, we shall briefly present certain recent theoretical and practical advances.

Handbook of Statistics, Vol. 34. http://dx.doi.org/10.1016/bs.host.2016.01.001

2 WARNER'S AND RELATED TECHNIQUES

Early sixties saw the growing menace of drug abuse in the United States initiated by the beats and hippies followed by college students. Though Warner (1965) begins his paper with the sentence: "for reasons of modesty, fear of being thought bigoted, or merely a reluctance to confide secrets to strangers, many individuals attempt to evade certain questions put to them by interviewers," we hazard a guess that Warner must have thought of the question of drug abuse which prompted him to build the alternative method for increasing cooperation while answering questions of sensitive nature.

We shall now briefly introduce Warner's method:

Based on a random sample of n individuals, let $\hat{\phi}$ be the sample proportion of "yes" answers to the following:

Statement 1: I belong to the sensitive category A [chosen with probability P under RRT];

Statement 2: I do not belong to the sensitive category A [chosen with probability $1-P$ under RRT].

If π denotes the population proportion of respondents who belong to the sensitive category and ϕ is the population proportion of yes responses under the above RRT, then we have the following relation

$$\phi = P\pi + (1-P)(1-\pi) = (2P-1)\pi + (1-P).$$

Assuming P to be known in advance to the investigator [interviewer], he estimates π by

$$\hat{\pi} = (\hat{\phi} - (1-P))/(2P-1);\, P \neq 0.5$$

which has a variance $V(\hat{\pi}) = \phi(1-\phi)/n(2P-1)^2$.

Being a social scientist himself and being more interested in practical aspects, Warner probably did not worry about the rigorous aspects of the parametric space as such and concluded that this $\hat{\pi}$ is an mle! Another interesting observation is that though Warner takes a spinner as a randomization device, several authors have changed it to a pack of cards or marbles of two different colors, etc. Though it is known that Warner's technique is less precise than the case when respondents respond truthfully to a direct question, one should appreciate that Warner's intention was to show that his technique provides an estimator with a smaller MSE than a direct answer with false claims or no answer at all.

Simmons argued that both statements of Warner relate to the sensitive nature of the question and modified Statement 2 to be unrelated to Statement 1 and a nonsensitive one (Greenberg et al., 1969; Horvitz et al., 1967). It is indeed true that in certain vernacular languages an answer to related Statement 2 in the form of yes/no might be confusing to the respondent. Simmons felt that the respondent might be more truthful if one statement is "completely innocuous and unrelated to the stigmatizing attribute." Under this unrelated model, we now have the relation $\phi = P\pi + (1-P)\pi'$ where π' is the [known] population proportion of individuals possessing the nonsensitive characteristic. Thus

$$\hat{\pi} = [\hat{\phi} - (1-P)\pi']/P$$

with

$$V(\hat{\pi}) = \phi(1-\phi)/nP^2.$$

The case of π' unknown and related inferential results followed in Greenberg et al. (1969). Chaudhuri and Mukerjee (1988) discuss several interesting RR devices and related results. At the end of the paper, Warner (1965) comments thus: "It should be noted that it is easy to extend the RRT to estimate distributions other than that appropriate to a simple dichotomous variable" and mentions the example of estimation of a five-class income distribution which arises in his field of economics. Subsequently, Warner (1971) discusses a class of RR models providing a theoretical framework. However, his BLUEs involve unknown parameters and Sen (1976) suggested a way out via a two-stage sequential sampling plan.

A simplified version of the unrelated RRT (see, for example, Rao, 1989, p. 35) is: Toss an unbiased coin. If Head appears, answer the sensitive statement truthfully and if Tail appears, answer yes / no to unrelated statement like: My telephone number ends with an even digit. If λ is the known proportion of telephone numbers in the population ending with an even digit, then $\pi/2 + \lambda/2 = \phi$, giving $\hat{\pi} = 2\hat{\phi} - \lambda$ with $V(\hat{\pi}) = 4\phi(1-\phi)/n$. This is an algebraically simple special case of the unrelated model and comparison of these estimators and related earlier results are discussed in sampling text books. See, for example, Cochran (1977).

During the recent past, another model which is a variant of Boruch's (1971) "forced response method" that gained popularity especially among cryptologists is as follows:

Toss a coin. If Tail appears, answer "Yes." If Head shows up, answer truthfully to the sensitive question. Consider the "No" group among the respondents. Half of them will get tails, giving an answer of "Yes," while the other half will answer truthfully and say "No." Thus, if ϕ per cent of the answers are "No," then among the respondents 2ϕ per cent belong to the "No" category and thus the desired estimate of the population proportion possessing the sensitive characteristic could be ascertained. We shall revert to this model in the next section.

3 CRYPTOGRAPHIC RRT

In Lecture 1 (Rao, 1989, pp. 1–12), it is noted that data analysis via encryption of messages and by eliciting information to sensitive questions are both related to the concepts of Randomness and Random Numbers. We observe that cryptographers now day are concerned with Cryptographic RRT based on these concepts which is briefly discussed in this section.

When a respondent maliciously deviates from the RRT to bias the survey outcome, cryptography is used to model a protocol which protects privacy as well as acts against such malicious behavior. This leads to the Cryptographic Randomized Response Technique (CRRT) which may be explained as follows.

The respondent has a private bit "b." Investigator wishes to obtain "b." Both of them use their strategies and exchange information and finally the investigator obtains a bit "c" such that $c=b$ with probability p, and $c=1-b$ with a probability $1-p$. Here, the respondent wants to make sure that the bit obtained by the investigator is not "b" with a certain probability, thereby protecting his privacy. This is called "plausible deniability." To guard against malicious behavior biassing the survey results, the investigator needs that $P[c=b] \leq p$, even if there is malicious behavior; thereby the respondent cannot bias the outcome unless he changes his bit "b." This is called "bounded bias." Ambainis et al. (2004) introduced CRRT and described these concepts followed by Moran and Naor (2006) who described two protocols to guard against respondent's malicious behavior and malicious investigator. Sarkar (2014) recently reviewed CRRT besides other connections between cryptography and statistics. With advances in cryptographic algorithms, RRT can assume a different role. Already a number of papers have appeared on CRRT and one sees a great opening for several applications along these lines.

4 REVERSE RRT

Consider the scenario where the investigator goes to a college and tells the students that he is eliciting information on their Cheating Habits "H" in the examinations. His randomization device consists of, say 65 black marbles and 35 white marbles. He then asks his sampled students to pick up Statement 1: I do belong to the Category H, if a black marble turns up when they randomly draw one, and pick up Statement 2: I do NOT belong to category H, if white marble turns up. Student body as a whole decide to reverse the choice of the two statements, thus twisting the two colors of the marbles. Investigator is, of course, unaware of this interim matter. His estimate as per Warner's method is: $\hat{\pi} = [\hat{\phi} - (1-P)]/(2P-1)$, with $P=0.65$ in the above illustration, whereas the real estimate is $\hat{\pi}^* = [\hat{\phi}^* - P]/(1-2P)$, where $\hat{\phi}$ and $\hat{\phi}^*$ are estimates of "yes" answers according to Warner's RRT and the "Reverse RRT." Note that, interestingly, both $\hat{\pi}$ and $\hat{\pi}^*$ have same variance but with different expectations.

We can also interpret this as two independent samples canvassed by two investigators who use P and P' for Statement 1 and $(1-P)$ and $(1-P')$ for Statement 2, respectively. [Clearly, $P+P'$ need not be unity]. If proportion of "Yes" answers are taken as ϕ_1 and ϕ_2, then $\phi_1 = \pi P + (1-\pi)(1-P)$ and

$\phi_2 = \pi P' + (1-\pi)(1-P')$. If $\phi_1 = \phi_2$, then either $P = P'$ or $\pi = 1/2$. For the unrelated model (Horvitz et al., 1967) $\hat{\pi} = [\hat{\phi}' - (1-P)\pi']/P$, whereas for the reverse randomization $\hat{\pi}^* = [\hat{\phi}' - P\pi']/(1-P)$. Here, π' is the known proportion of individuals possessing the unrelated characteristic. Here, the estimators have different variances unlike the related questions model and variance in the reverse Warner context is larger than in the reverse Horvitz et al context if $P < 1/2$. One is tempted to guess that there could be two-person kind of game-theoretic formulation here. Indeed, in a later section, we shall refer to Ljungqvist's (1993) concept and the more recent work of Blume et al. (2013).

5 CERTAIN RECENT THEORETICAL AND PRACTICAL RESULTS

5.1 Unified Theory

Suppose that the variate y takes value Y_i for the i-th person (unit) possessing the sensitive characteristic. To estimate $\bar{Y}$, Eriksson (1973) chose a fixed set of values $[(X_j; j = 1,2,\ldots,M)]$ likely to contain the true Y-values, or at least, broadly covering the Y-values. A unit (person) sampled according to a sampling design $p(s)$ is supplied with a RR device with $(q+1)$ options with probabilities C and $q_1, q_2, \ldots, q_M$ such that $C + \sum_j q_j = 1$. If the option with probability C is selected by the i-th respondent, he/she is supposed to report the true value $Z_i = Y_i$; otherwise, for j-th option, the reported value is $Z_i = X_j$ with probability q_j. For each selected unit i, one obtains randomized responses $[Z_{ir}; r = 1,2,\ldots,k]$ for k independent replicates using the same RR device. An unbiased estimator for Y_i is given by

$$\bar{Y}_i = (\bar{Z}_i - (1-C)\bar{X})/C$$

where $\bar{X} = \sum q_j X_j/(1-C)$. From these estimates, it is a routine task to estimate the population mean; also see, Bourke and Dalenius (1976).

Choosing $k = M = 1$ and assuming X_i known for all units in the sample, Godambe (1980) proves that there does not exist a UMV estimator and following his 1955 paper and the super population model approach, he obtains an optimal estimator with minimum expected variance under the model and it is given by

$$e = \sum_i [Z_i - (1-C)X_i]/\pi_i,$$

where π_i is the inclusion probability of the i-th unit.

Adhikary et al. (1984) extended the nonexistence result removing the linearity expression and showed how the optimality properties of classical Horvitz–Thompson Estimator [HTE] pass on to the RR-version given by e above. Extending Ramakrishnan's (1975) result on admissibility, Sengupta and Kundu (1989) proved a parallel result on admissibility of strategies based on a fixed sample size under a RR model for estimating the mean of a quantitative variable. Recently, however, Sengupta (2014) showed that neither result is generally true for estimating the population proportion, barring some restricted cases.

Continuing this work relating to the theoretical framework, Sengupta (2015a) identified a minimal sufficient statistic for this problem of estimating a sensitive proportion and obtained complete classes of unbiased and linear unbiased estimators. Sengupta (2015a) further proved the admissibility of two linear unbiased estimators and thereby the nonexistence of a best linear unbiased or a best unbiased estimator. He has also shown that for a given sampling design, for any linear unbiased estimator of a sensitive proportion P based on Warner's model with any given value of the model parameter, there exists a linear unbiased estimator of P based on Eriksson's model with a uniformly smaller variance for suitable choices of model parameters (vide Sengupta, 2015b). Further research along these theoretical lines would be a valuable addition to the literature on this subject.

5.2 Stratification and RRT

Stratification is known to have its own advantages. Researchers in the RR field automatically extended the available results on Randomized Response to stratified sampling and allocation of sample size, etc. However, we note that some of these extensions are of theoretical nature and it is difficult to envisage their adaptability in practical situations.

Hong et al. (1994) envisaged a straight forward stratified extension of RRT under proportional allocation of sample size to strata applying the same random device in each stratum. Kim and Warde (2004) maintain that Hong et al.'s proportionate sampling could lead to higher costs, though an expression for variance of the suggested estimator could be easily derived. They suggest Neyman's optimum allocation subject to $\sum n_i = n$, the total sample size. However, they do not derive the cost-optimum allocation due to Mahalanobis (1944) based on a simple linear cost constraint. It should be noted that all these extensions of allocations, being theoretical in nature, are difficult to use in practical applications. Kim and Warde (2004) cite certain situations but the optimum allocation depends on unknown quantities. One then needs to make model assumptions and derive user-friendly near-optimum allocations. Though this approach leads to meaningful results in classical situations (Rao, 2010), it is difficult to look for a related auxiliary information in the case of RR models to postulate a super population model. Extending the scrambled RR model proposed by Eichhorn and Hayre (1983), Mahajan et al. (1994), and Mahajan and Verma (2004) obtained optimum points of stratification on an auxiliary variable for SRSWR in each stratum using data on scrambled sensitive character. In the recent past, there have been several publications on these and related aspects but one feels that these extensions are more of an academic interest. Another recent line of theoretical research is to use scrambled responses in the presence of auxiliary information initiated by Singh et al. (2015). With a good choice of related auxiliary information, these results will add to the practical value.

5.3 Cramér–Rao Lower Bound

To draw inferences about a sensitive quantitative characteristic, Greenberg et al. (1971) devised a method by asking a randomly selected individual to report his true sensitive value with probability P and an unrelated innocuous value with a probability $1-P$. Using the sample mean of the Randomized Response values, they obtain an unbiased estimator of the mean of the sensitive characteristic. Furthermore they obtain a Cramér- Rao lower bound (CRLB) in the RR case and discuss the efficiency of estimator for certain situations. Researchers interested in statistical inference could probably proceed on these lines and think of extensions to Bhattacharyya's bounds and related inequalities. It should be pointed out that one has to make sure that the regularity conditions are satisfied in the RR situation as well. Recently, for the case of estimation of a sensitive proportion, Singh and Sedory (2011) obtain m.l.e. for the sensitive Proportion with a randomization device consisting of two decks of cards and also try to arrive at a CRLB. However, the m.l.e. so obtained has to be rigorously defined and the regularity conditions need to be properly verified.

5.4 Game Theory and RRT

Following Ljungqvist (1993) who studied RRT from the concept of maximizing utility, Blume et al. (2013) used a pay off function for the respondent that trades off lying aversion against stigmatization aversion and analyze RRT from a Game Theoretic point of view and test it in an experiment conducted in laboratory setting. Further research along the game theoretic lines could be a valuable theoretical contribution to this field.

5.5 Smart Phones and RRT

In the context of ever increasingly popular apps on the mobile phones for location based services, it is important not to disregard the privacy concerns of users' whereabouts. It has been observed that this privacy implication of commercial location based services is uneven at best and inadequate at worst according to a 2010 statement from the Center for Democracy and Technology. In view of the sensitivity of location information, the users of location-based services deserve a robust set of protections to tackle the privacy risks. Quercia et al. (2011) proposed an interesting way of aggregating location data that suits privacy-conscious individuals and, at the same time, results in a full-fledged location services. They apply a data obfuscation technique to the design of counting individuals in a location while at the same time guarding their privacy. The software called spotME is used to estimate such number of individuals. This particular software can report besides the actual location, certain erroneous locations as well selected by a RR algorithm. We shall next consider a similar new paradigm called Community Sharing.

This paradigm allows sharing of local conditions, community data, and mapping of physical phenomena. Data is aggregated from sensor equipped mobile phones. Applications of this are found in medical research relating to infectious diseases and chronic ailments. In this scenario, privacy of mobile phone users participating in the system needs to be strictly safe guarded. Since encryption is costly and time consuming for this type of application, data perturbation technique is used for preserving privacy. The sensitive data of users is protected when each sensing node creates perturbed data from the original data. However, earlier methods of perturbation by negative data(exactly opposite) gives clues to the malicious attackers. Aoki et al. (2013) and Aoki and Sezaki (2014) recently came up with multidimensional randomized response, where each sensor data is processed with duplicated perturbation. Malicious attackers have no way of tracking the sensitive data and the central server is programmed to construct the probability distribution of the original data preserving privacy.

5.6 Alternatives to RRT

Coutts and Jann (2011) in an experimental study to compare the use of RRT vs. the alternative Item Count Technique (ICT) for eliciting information on sensitive questions, find that ICT (also known as Unmatched Count Technique, cf. Raghavarao and Federer, 1979) is less confusing in the case of self-administered surveys. More research is needed to make simpler alternative RR techniques to be user friendly in the context of computer aided self-administered survey questionnaires.

5.7 Meta Analysis

With such a huge literature piled up on the subject, it is not surprising that Lensvelt-Mulders et al. (2005) attempted a very useful Meta Analysis of Randomized Response Research, considering 38 validation studies available during 35 year period of 1965–2000. With increasing applications as is evident from Conference Proceedings and Publications, it is perhaps time to have another Meta Analysis.

6 EPILOGUE

As early as in 1967, Greenberg et al. (1969) speculated that "the legendary Pandora's box has been opened with the Warner's technique." Chaudhuri and Mukerjee (1987) towards the end of their review mention thus: "yet a copious growth of literature through an uninterrupted flow of publications on the subject seems to promise a prosperous development and rapid progress in popular interest and creativity in RR techniques in the foreseeable future." And it has indeed been so.

A decade and half ago, there was a special invited article by Hoadley and Kettenring (1990) wherein serious problems that need resolution relating to statisticians and engineers are indicated and certain remedies are outlined. After five decades of Warners RRT, we note that very important and interesting applications of RRT are being researched, be it in conferences or publications or patents in communications engineering. For a useful interface between the subjects, the need of the day may be a better rapport between statisticians specializing in RRT and communication engineers and vice versa.

ACKNOWLEDGMENT

We are thankful to Professor Bikas K. Sinha for reading an earlier version of the manuscript and for making useful suggestions.

REFERENCES

Adhikary, A.K., Chaudhuri, A., Vijayan, K., 1984. Optimum sampling strategies for randomized response trials. Internat. Statist. Rev. 52 (2), 115–125 (With discussion by P. K. Sen, S. M. Tam and M. E. Thompson, and a reply by the authors).

Ambainis, A., Jakobsson, M., Lipmaa, H., 2004. Cryptographic randomized response techniques. In: Public Key Cryptography–PKC 2004, 425–438.

Aoki, S., Iwai, M., Sezaki, K., 2013. Privacy-aware community sensing using randomized response. In: 2013 IEEE 37th Annual Computer Software and Applications Conference Workshops (COMPSACW), pp. 127–132.

Aoki, S., Sezaki, K., 2014. Privacy-preserving community sensing for medical research with duplicated perturbation. In: 2014 IEEE International Conference on Communications (ICC), pp. 4252–4257.

Blume, A., Lai, E.K., Lim, W., 2013. Eliciting private information with noise: the case of randomized response. Institute of Mathematical Economics Working Paper No. 490.

Bourke, P.D., Dalenius, T., 1976. Some new ideas in the realm of randomized inquiries. Internat. Stat. Rev. 44, 219–221.

Boruch, R.F., 1971. Assuring confidentiality of responses in social research: a systematic analysis. Am. Psychol. 26, 413–430.

Chaudhuri, A., 2011. Randomized Response and Indirect Questioning Techniques in Surveys. CRC Press, Florida, USA.

Chaudhuri, A., Christofides, T.C., 2013. Indirect Questioning in Sample Surveys. Springer, Heidelberg.

Chaudhuri, A., Mukerjee, R., 1988. Randomized Response. Statistics: Textbooks and Monographs, vol. 85, Marcel Dekker, Inc., New York, NY.

Chaudhuri, A., Mukerjee, R., 1987. Randomized response techniques: a review. Statist. Neerlandica 41 (1), 27–44.

Cochran, W.G., 1977. Sampling Techniques, third ed. John Wiley & Sons, New York, NY.

Coutts, E., Jann, B., 2011. Sensitive questions in online surveys: experimental results for the randomized response technique (RRT) and the unmatched count technique (UCT). Sociol. Methods Res. 40 (1), 169–193.

Eichhorn, B.H., Hayre, L.S., 1983. Scrambled randomized response methods for obtaining sensitive quantitative data. J Stat. Plann. Inference 7 (4), 307–316.

Eriksson, S.A., 1973. A new model for randomized response. Internat. Stat. Rev. 41, 101–113.

Fox, J.A., Tracy, P.E., 1986. Randomized response: A method for sensitive surveys. Beverly Hills, CA.

Godambe, V.P., 1980. Estimation in randomised response trials. Internat. Stat. Rev. 48 (1), 29–32.

Greenberg, B.G., Abul-Ela, A.L.A., Simmons, W.R., Horvitz, D.G., 1969. The unrelated question randomized response model: Theoretical framework. J. Am. Stat. Assoc. 64, 520–539.

Greenberg, B.G., Kuebler Jr., R.R., Abernathy, J.R., Horvitz, D.G., 1971. Application of the randomized response technique in obtaining quantitative data. J. Am. Stat. Assoc. 66 (334), 243–250.

Hoadley, A.B., Kettenring, J.R., 1990. Communications between statisticians and engineers/physical scientists. Technometrics 32 (3), 243–247.

Hong, K., Yum, J., Lee, H., 1994. A stratified randomized response technique. Korean J. Appl. Stat. 7, 141–147.

Horvitz, D.G., Shah, B.V., Simmons, W.R., 1967. The unrelated randomized response model. Proc. Soc. Stat. Sect. Am. Stat. Assoc., 65–72.

Kim, J.M., Warde, W.D., 2004. A stratified Warner's randomized response model. J. Stat. Plann. Inference 120 (1-2), 155–165.

Lensvelt-Mulders, G.J.L.M., Hox, J.J., van der Heijden, P.G.M., Maas, C.J.M., 2005. Meta-analysis of randomized response research: thirty-five years of validation. Sociol. Methods Res. 33 (3), 319–348.

Ljungqvist, L., 1993. A unified approach to measures of privacy in randomized response models: a utilitarian perspective. J. Am. Stat. Assoc. 88 (421), 97–103.

Mahajan, P.K., Gupta, J.P., Singh, R., 1994. Determination of optimum strata boundaries for scrambled randomized response. Statistica (Bologna) 54 (3), 375–381. (1995).

Mahajan, P.K., Verma, M.R., 2004. An alternative method of optimum stratification for scrambled response. J. Indian Soc. Agric. Stat. 58 (3), 269–278.

Mahalanobis, P.C., 1944. On large-scale sample surveys. Phil. Trans. R. Soc. Lond. B Biol. Sci. 231 (584), 329–451.

Moran, T., Naor, M., 2006. Polling with physical envelopes: a rigorous analysis of a human-centric protocol. In: Advances in Cryptology-EUROCRYPT 2006. Springer, St. Petersburg, Russia, pp. 88–108.

Quercia, D., Leontiadis, I., McNamara, L., Mascolo, C., Crowcroft, J., 2011. SpotME if you can: randomized responses for location obfuscation on mobile phones. In: 2011 31st International Conference on Distributed Computing Systems (ICDCS). IEEE, Nara, Japan, pp. 363–372.

Raghavarao, D., Federer, W.T., 1979. Block total response as an alternative to the randomized response method in surveys. J. R. Stat. Soc. Ser. B 41 (1), 40–45.

Ramakrishnan, M.K., 1975. Choice of an optimum sampling strategy. I. Ann. Stat. 3, 669–679.

Rao, C.R., 1989. Statistics and Truth. International Co-operative Publishing House, Fairland, MD.

Rao, T.J., 2010. Near optimum allocations in stratified sampling. J. Stat. Theory Pract. 4 (1), 57–69.

Sarkar, P., 2014. On some connections between statistics and cryptology. J. Stat. Plann. Inference 148, 20–37.

Sen, P.K., 1976. Asymptotically optimal estimators of general parameters in randomized response models. Internat. Stat. Rev. 44, 223–224.

Sengupta, S., 2014. On the admissibility of Horvitz Thompson strategies for estimating finite population proportions in direct and randomized response surveys. Cal. Stat. Assoc. Bull. 66, 63–71.

Sengupta, S., 2015a. Estimation of finite population proportion in randomized response surveys using multiple responses. Sankhya B 77 (1), 75–83.

Sengupta, S., 2015b. On the comparison of Warner's and Eriksson's randomized response plans for estimating sensitive finite population proportions. Sankhya B 77 (1), 84–90.

Sengupta, S., Kundu, D., 1989. Estimation of finite population mean in randomized response surveys. J. Stat. Plann. Inference 23 (1), 117–125.

Singh, S., Sedory, S.A., 2011. Cramér-Rao lower bound of variance in randomized response sampling. Sociol. Methods Res. 40 (3), 536–546. ISSN 0049-1241.

Singh, S., Sedory, S.A., Arnab, R., 2015. Estimation of finite population variance using scrambled responses in the presence of auxiliary information. Commun. Stat. Simul. Comput. 44 (4), 1050–1065.

Warner, S.L., 1965. Randomized response: a survey technique for eliminating evasive answer bias. J. Am. Stat. Assoc. 60 (309), 63–69.

Warner, S.L., 1971. The linear randomized response model. J. Am. Stat. Assoc. 66 (336), 884–888.

Chapter 2

The Background and Genesis of Randomized Response Techniques

A. Chaudhuri*[,1]

*Indian Statistical Institute, Kolkata, India

[1]Corresponding author: e-mail: arijitchaudhuri1@rediffmail.com

ABSTRACT

The need for protecting respondent-privacy about stigmatizing personal features is usually appreciated. How an appropriate corrective device appeared and is being further pursued with is duly reported in brief.

Keywords: Sensitive issues, Randomized response, Warner's pioneering work and follow-ups

Half a century ago, Warner (1965) introduced an ingenious device to gather reliable data relating to issues people like to hide from their fellow human beings. Frequent drunken driving, abortions indiscreetly induced, misconduct with the spouse, finding HIV tests positive, under-reporting Income Tax returns, False Claims for social benefits, etc., are some of the features that may attach unethical stigmas to many delinquents. People generally dislike their revelations.

But in social studies capture of truthful and trustworthy incidences covering such sensitive features seems crucial often in practice.

Delicacy in an investigator in implementing direct queries in a social survey and possibility of facing reluctance to truthful reporting by the respondents if and when bravely addressed are commonly experienced. Warner (1965) devised a way to avoid venturing to procure direct responses (DR) and instead to request randomized responses (RRs) from each person sampled when a stigmatizing issue is under contemplation in a study.

Some details follow. Warner (1965) considered estimating the parameter θ which is the unknown proportion of people bearing a sensitive characteristic denoted by A, say, of which the complement is A^C. Being a social scientist with

Handbook of Statistics, Vol. 34. http://dx.doi.org/10.1016/bs.host.2016.01.002

a limited foundation in sophisticated theory of statistics, he permitted sample selection only by the primitive method of simple random sampling (SRS) with replacement (SRSWR). Thus, he ensured that on each successive draw of a sample person from the community of the people, θ remained the unaltered probability of a person bearing the stigmatizing characteristic A in the population. Warner's device requires the investigator to approach a sampled person with a box containing a large number of identical cards with a proportion $p(0<p<1, p\neq 1/2)$ of them marked A and the rest marked A^C. The sampled person is to shake and jerk the box, draw one of the cards, and return it to the box but truthfully to announce "Yes" if the card type matches his/her true characteristic A or A^C without divulging the card type noticed to the investigator. On getting such RRs from all the n (say) draws and counting the number n_Y (say) of "Yes" responses out of them an estimator for θ follows intuitively as:

$$\hat{\theta}=\frac{(n_Y/n)-(1-p)}{(2p-1)} \text{ because for}$$

every single draw the probability of finding a "Yes" response is

$$\begin{aligned}\text{Prob}(\text{Yes}) &= p\theta+(1-p)(1-\theta)\\ &= (1-p)+(2p-1)\theta\end{aligned}$$

Being a nonstatistician, Warner (1965) wrongly claimed $\hat{\theta}$ as the maximum likelihood estimator for θ but missed the point that $\hat{\theta}$ is an unbiased estimator for θ. Nevertheless, research proceeded quite vigorously from this rather shaky beginning. Flow was in various directions.

Some noticed A^C like A may be sensitive as well, and a fundamentally different device is then needed. To some neither A nor A^C may appear sensitive but to some others they may. Giving out a truthful RR may yet run the risk of revealing the secret. The question is how to protect the personal confidentiality. There may be interest in quantitative characteristics that are stigmatizing, for example, amount gained or lost in gambling, money spent in treatment of AIDS, number of days spent behind bars as punishment for criminal offenses detected, and such many others. Sample selection by SRS without replacement (SRSWOR) rather than SRSWR induces interesting problems for solution. A social survey exclusively covering sensitive attribute is hardly undertaken or paid for. Usually many items of enquiry need attention. Of them a few may be stigmatizing. A survey may be designed allowing stratified multistage sampling with unequal and equal probabilities at different stages. Then how to appropriately implement data analysis on gathering RR responses in one or more stages? RR technique (RRT) is one way of generating sensitive data in an indirect way. But a device employed may not be acceptable to a prospective respondent. So, alterative devices may be suitable to gather data in indirect manners on sensitive issues. Details galore are unfolded in this handbook by our colleagues as competent authors.

Fox and Tracy (1986) and Chaudhuri and Mukerjee (1988) published two books on the subject narrating prolific details on the subject RRT. Besides these classical ones recent monographs by Chaudhuri and Christofides (2013) and Chaudhuri (2011) have also enlightened the subject and so have some of the important comprehensive publications, for example, by Lensvelt-Mulders et al. (2005) and by Van der Heijden et al. (2000), among others.

Kuk (1990), Christofides (2003), Mangat and Singh (1990), Mangat (1994), and Singh and Joarder (1997) have introduced phenomenal improvements and modifications on Warner's (1965) pioneering work. Nayak and Adeshiyan (2009) have brought about critical unifications thereupon. A fundamental novelty of approach came from Horvitz et al. (1967) and Greenberg et al. (1969) who visualized the possibility that the complement A^C of a sensitive attribute A might itself be stigmatizing as well as calling for a dramatic overhaul of the Warner RRT itself.

Some recent and forthcoming volumes of MASA (Model-Assisted Statistical Applications) also are known to contain various aspects of the emerging novelty in RRTs. Chaudhuri (2001) showed how RRT related to attributes may be applied when samples are drawn following general sampling schemes rather than SRSWR alone. Eriksson (1973), Adhikary et al. (1984), Chaudhuri (1987, 1992), and Sengupta and Kundu (1989) were the principal early contributors to RRT covering quantitative characteristics and permitting unequal probability sampling.

Sensitive issues are now a days also being covered by techniques other than RRTs but by nonrandomized models, item counting techniques, nominative, and three-card method among others. These are mostly discussed by Chaudhuri and Christofides (2013). Some later developments are also discussed in this handbook as illustrated in its contents.

REFERENCES

Adhikary, A.K., Chaudhuri, A., Vijayan, K., 1984. Optimum sampling strategies for RR trials. Int. Stat. Rev. 52, 115–125.

Chaudhuri, A., 1987. Randomized response surveys of finite populations: a unified approach with quantitative data. J. Stat. Plan. Inference 15, 157–165.

Chaudhuri, A., 1992. Randomized response: estimating mean square errors of linear estimators and finding optimal unbiased strategies. Metrika 39, 341–357.

Chaudhuri, A., 2001. Using randomized response from a complex survey to estimate a sensitive proportion in a dichotomous finite population. J. Stat. Plan. Inference 94, 37–42.

Chaudhuri, A., 2011. Randomized Response and Indirect Questioning Techniques in Surveys. CRC Press, Taylor & Francis, Boca Raton, USA.

Chaudhuri, A., Christofides, T.C., 2013. Indirect Questioning in Sample Surveys. Springer-Verlag, Heidelberg.

Chaudhuri, A., Mukerjee, R., 1988. Randomized Response. Theory and Techniques. Marcel Dekker, Inc., New York, USA.

Christofides, T.C., 2003. A generalized randomized response techniques. Metrika 57, 195–200.

Eriksson, S.C., 1973. A new model for RR. Int. Stat. Rev. 41, 101–113.

Fox, J.A., Tracy, P.E., 1986. Randomized Response: a Method for Sensitive Surveys. Sage, London.

Greenberg, B.G., Abul-Ela, A.L., Simmons, W.R., Horvitz, D.G., 1969. The unrelated question RR model: theoretical framework. J. Am. Stat. Assoc. 64, 520–539.

Horvitz, D.G., Shah, B.V., Simmons, W.R., 1967. The unrelated question RR model. Proc. Soc. Stat. Sec. ASA, 65–72.

Kuk, A.Y.L., 1990. Asking sensitive questions indirectly. Biometrika 77 (2), 436–438.

Lensvelt-Mulders, G.J.L.M., Hox, J.J., van der Heijden, P.G.M., 2005. How to improve the efficiency of randomized response designs. Qual. Quant. 39, 253–265.

Mangat, N.S., 1994. An improved randomized response strategy. J. R. Stat. Soc. Ser. B 56 (1), 93–95.

Mangat, N.S., Singh, R., 1990. An alternative randomized response procedure. Biometrika 77 (2), 439–442.

Nayak, T.K., Adeshiyan, S.A., 2009. A unified framework, for analysis and comparison of randomized response survey of binary characteristics. J. Stat. Plan. Inference 139, 2757–2766.

Sengupta, S., Kundu, D., 1989. Estimation of finite population mean in randomized response surveys. J. Stat. Plan. Inference 23, 117–125.

Singh, S., Joarder, A.H., 1997. Unknown repeated trials in randomized response sampling. J. Indian Soc. Agric. Stat. 50, 70–74.

Van der Heijden, P.G.M., Van Gils, G., Bouts, J., Hox, J.J., 2000. A comparison of randomized response, computer-assisted self-interview and face-to-face direct questioning. Eliciting sensitive information in the context of welfare and unemployment benefit. Sociol. Methods Res. 28, 505–537.

Warner, S.L., 1965. RR: a survey technique for eliminating evasive answer bias. J. Am. Stat. Assoc. 60, 63–69.

Chapter 3

How Randomized Response Techniques Need not Be Confined to Simple Random Sampling but Liberally Applicable to General Sampling Schemes

A. Chaudhuri[1]
Indian Statistical Institute, Kolkata, India
[1]*Corresponding author: e-mail: arijitchaudhuri1@rediffmail.com*

ABSTRACT

S.L. Warner, the pioneer in research on randomized responses (RRs) to gather and analyze sensitive and stigmatizing data, started specifically with samples selected exclusively by the simplest scheme of simple random sampling (SRS) with replacement (SRSWR) and most of his successors followed it. But subsequent study remarkably in 2001 and onward has decisively shown that this restriction is uncalled for. Noteworthy aspects of RRs are unbiased and sometimes approximately unbiased estimation of the proportions of people bearing undesirable features in human communities and also totals/means of real values of sensitive variables of social interest. How to protect privacy in respondents' disclosures and also how to adjust data analysis with options allowed to respond directly if no sensitivity is perceived are important points of note covering qualitative as well as quantitative data. Indirect but nonrandomized data gathering is also attended to. Bayesian analysis is also permitted. In caring for such diverse aspects sampling procedures are freely permitted. How it is achieved is a story briefly unfolded here.

Keywords: Randomized response, Simple random sampling with replacement, General sampling schemes, Protection of privacy

1 INTRODUCTION

Warner Stanley (1965) inaugurated the journey of research program on randomized response (RR) surveys with a statistical approach. He considered how

Handbook of Statistics, Vol. 34. http://dx.doi.org/10.1016/bs.host.2016.01.003

to estimate the proportion of people bearing a qualitative characteristic A of dubious social approbation in a given community of people by dint of a face-to-face survey by sample selection. Though not clearly spelled out by him the complementary characteristic A^C need not invite any social condemnation after all. So, it is feared, on a direct query, one may not cheerfully like to give out that he/she may bear the characteristic A, though may not dislike to be identified as a person bearing A^C. If one intends to estimate the unknown proportion θ of people actually bearing the feature A, the effort may not be fruitful because correct information may not be forthcoming from a survey indulging only in direct enquiries from a sample of persons howsoever chosen. So, Warner Stanley (1965) prescribes a remedy allowing a sampled person an option to implement a random trial with a device that tells him or her only to say "Yes" or "No" if asked whether he/she bears A or in a random manner the alternative A^C.

If a sampled person is approached with a box of many identical cards marked either A or A^C in unequal proportions p and $(1-p)$, $(0<p\neq 1/2<1)$ and is requested to shake the box enough and blind-foldedly draw one of the cards without disclosing its type to the investigator and tell him/her truthfully either "Yes" or "No" according as the card type "matches" or "does not match" his/her true trait, then Warner Stanley (1965) enriches us with $\lambda = p\theta + (1-p)(1-\theta)$.

Here λ is the probability of a "Yes" response and the implicit claim is that the respondent need not hesitate to give the "truthful" response suspecting no possibility of his/her true trait to be thereby revealed to the investigator. The next obvious question is how to handle the situation if more samples are to be gathered and addressed. Warner Stanley's (1965) solution is to go on taking samples independently with equal probabilities from the population and repeat this exercise. As a result the above probability equation namely

$$\lambda = p\theta + (1-p)(1-\theta) \tag{1}$$

will continue to hold with respect to each person and is thus chosen and persuaded to go ahead with the same RR experiment to be implemented.

Thus, for an SRSWR sample selection and Warner's RR technique, an unbiased estimator for λ is given by the proportion of "Yes" responses in (say) n RRs thus gathered, and θ does not change over repeated draws with replacement. Hence an unbiased estimator of θ readily emerges as

$$\hat{\theta} = \frac{\hat{\lambda} - (1-p)}{2p-1}.$$

Here $\hat{\lambda}$ is the sample proportion of "Yes" responses gathered from the SRSWR by dint of Warner's RR technique. Even if one decides to choose a sample by SRSWOR (Simple Random Sampling Without Replacement), Eq. (1) will no longer hold for successively chosen sample persons as θ will change along with the successive draws by the sample persons in the same population using the same box of cards, each card being returned to the box each time.

Almost all the followers of Warner Stanley (1965) engaged in estimating population proportions adopting various alternative RR procedures continued over the decades to take only SRSWRs with marginally little exceptions. Consequently an impression went on gathering momentum in course of time that RRT (Randomized Response Technique) is not a survey sampling topic at all. Features of finite population like identifiability of population units and varying probability sample selection procedures were not being adopted at all over a long time, thus leading to avoidance of sophisticated estimation and variance and/or Mean Square Error (MSE) estimation as a matter of course.

Erikson (1973) was the first researcher on estimation of the total of a stigmatizing real variable like amount gained or lost in gambling and amount cheated in submitting income tax (IT) returns, by persons no matter chosen with what specified selection probabilities from a "finite population" U identified by labels $i=1, 2, 3, \ldots, N$.

Chaudhuri (2001) was the first to permit generation of RR observations implementing various RR devices from persons chosen no matter how demanding suitable probability sampling schemes only for statistical analysis of the qualitative RR data sophisticatedly generated as per mutual conveniences of the researcher and the respondent, which is essentially achieved as briefly noted.

2 TWO PROMINENT RR DEVICES REVISED FOR GENERAL APPLICATIONS

2.1 Warner Stanley's (1965) Device

Let $y_i=1$ if unit i bears A

$=0$ if i bears A^C

$I_i=1$ if i gets a match in card and trait

$=0$ if not

Writing E_R, V_R as expectation, variance operators generically over RRT one gets

$$E_R(I_i)=py_i+(1-p)(1-y_i)=(1-p)+(2p-1)y_i$$

Then,

$r_i=I_i-(1-p)/2p-1$ gives $E_R(r_i)=y_i$

$$V_R(I_i)=E_R(I_i)(1-E_R(I_i))=p(1-p)$$

because

$$y_i^2=y_i \text{ and}$$

$$I_i^2=I_i$$

$V_R(r_i)=p(1-p)/(2p-1)^2=V_i$, say, for simplicity.

V_P and V_R throughout will denote, respectively, design and RR-based variance operators.

2.2 Simmon's RR Device Revised

Unlike Warner Stanley (1965) some of the classical stalwarts (Greenberg et al., 1969; Horvitz et al., 1967) extended the scope of Warner Stanley's (1965) initial RRT by permitting A^C to be stigmatizing like A as well. For example, let A denote backing a political party and A^C as antagonism to it. Then, a second trait B denoting an innocuous one like preferring "sport" to "music" and B^C its contradiction is needed as a characteristic apparently unrelated to A and A^C. This unrelated model, called URL model, originated with Simmons as cited above. Generalizing Warner Stanley (1965) the above authors need two boxes containing cards marked A and B in respective proportions p_1: $(1-p_1)$ and p_2: $(1-p_2)$, $p_1 \neq p_2$, $(0<p_1, p_2<1)$.

Let $y_i=1(0)$ if i bears A (A^C), $x_i=1(0)$ if i bears B (B^C),
$I_i=1$ if i's trait A (A^C) or B (B^C) "matches" the type of the card drawn from the first box
$=0$ if it does not match;
$J_i=1$ if it matches when drawn from the second box
$=0$ if it does not match.

$$E_R(I_i)=p_1y_i+(1-p_1)x_i$$

Then,

$$E_R(J_i)=p_2y_i+(1-p_2)x_i$$

giving $r_i=(1-p_2)I_i-(1-p_1)J_i/(p_1-p_2)$ with $E_R(r_i)=y_i$

$$\text{Since } V_i - V_R(r_i) - E_R(r_i^2) - E_R^2(r_i)$$

$$=E_R(r_i^2)-y_i^2=E_R(r_i^2)-y_i=E_R[r_i(r_i-1)]$$

one gets

$$v_i=r_i(r_i-1)$$

as an unbiased estimator for V_i, $i \in U$,

With these backgrounds in Eqs. (2) and (3) let us appeal to well-established "Finite Population survey sampling theory" vide Chaudhuri (2010, 2014) giving us the following:

Using direct response (DR)-based survey data $(i, y_i|i\in s)$ from a sample s chosen with selection-probability $p(s)$ according to a design p admitting positive inclusion probabilities

$$\pi_i=\sum_{s\ni i}p(s)$$

of the first order and

$$\pi_{ij} = \sum_{s \ni i,j} p(s), i \neq j,$$

of the second order, one may employ for $Y = \sum_{i=1}^{N} y_i$ an unbiased estimator due to Horvitz and Thompson (1952),

$$t = \sum_{i \in s} \frac{y_i}{\pi_i}$$

having the variance vide Chaudhuri and Pal (2002)

$$V_P(t) = \sum_{i<}^{N} \sum_{j}^{N} (\pi_i \pi_j - \pi_{ij}) \left(\frac{y_i}{\pi_i} - \frac{y_j}{\pi_j} \right)^2 + \sum_{i=1}^{N} \frac{y_i^2}{\pi_i} \beta_i$$

$$\text{with } \beta_i = 1 + \frac{1}{\pi_I} \sum_{j \neq i}^{N} \pi_{ij} - \sum_{i=1}^{N} \pi_i = \frac{1}{\pi_i} \left(\sum_{s \ni i} v(s) p(s) \right) - \sum_{i=1}^{N} \pi_i$$

$v(s) \equiv$ the number of distinct units in s's.

This variance admits an unbiased estimator

$$v_P(t) = \sum_{i} \sum_{<j \in s} (\pi_i \pi_j - \pi_{ij}) \left(\frac{y_i}{\pi_i} - \frac{y_j}{\pi_j} \right)^2 \frac{I_{sij}}{\pi_{ij}} + \sum_{i \epsilon s} \frac{y_i^2}{\pi_i} \beta_i \frac{I_{si}}{\pi_i}$$

Writing $I_{si} = 1(0)$ if $s \ni i$ ($s \not\ni i$), $I_{sij} = 1(0)$ if $s \ni i,j$ ($s \not\ni i,j$).

We shall generically write

$$E = E_P E_R = E_R E_P \text{ and } V = E_P V_R + V_P E_R = E_R V_P + V_R E_P$$

Then, RR-based estimator for Y emerges as

$$e = \sum_{i \in s} \frac{r_i}{\pi_i}$$

with the variance

$$V(e) = V_R \left(\sum_{i=1}^{N} r_i \right) + E_R \left[\sum_{i<}^{N} \sum_{j}^{N} (\pi_i \pi_j - \pi_{ij}) \left(\frac{r_i}{\pi_i} - \frac{r_j}{\pi_j} \right)^2 + \sum_{i=1}^{N} \frac{r_i^2}{\pi_i} \beta_i \right]$$

$$= \sum_{i=1}^{N} V_i + \sum_{i<}^{N} \sum_{j}^{N} (\pi_i \pi_j - \pi_{ij}) \left[\left(\frac{y_i}{\pi_i} - \frac{y_j}{\pi_j} \right)^2 + \left(\frac{V_i}{\pi_i^2} - \frac{V_j}{\pi_j^2} \right) \right] + \sum_{i=1}^{N} \frac{y_i^2}{\pi_i} \beta_i + \sum_{i=1}^{N} \frac{V_i}{\pi_i} \beta_i \qquad (2)$$

We shall write W_i for V_i when, respectively, V_i is known or admits an unbiased estimator v_i for itself to get as an unbiased estimator for $V(e)$ which is

$$v(e) = \sum_{i \in s} \frac{W_i}{\pi_i} + \sum_{i} \sum_{<j \in s} \frac{(\pi_i \pi_j - \pi_{ij})}{\pi_{ij}} \left(\frac{r_i}{\pi_i} - \frac{r_j}{\pi_j} \right)^2 + \sum_{i \in s} \frac{r_i \beta_i}{\pi_i^2} + \sum_{i \in s} \frac{W_i}{\pi_i^2} \beta_i \text{ because } y_i^2 = y_i \tag{3}$$

These are generic formulae accommodating Warner Stanley's (1965) and Simmons' URL and other similar RR models addressed to estimate $Y = \sum_{i=1}^{N} y_i$ with $y_i = 1$ or 0 according as i bears A or A^C. Many details are described by Chaudhuri and Christofides (2013).

3 QUANTITATIVE RRs

Suppose for i in U the value y_i relates to a stigmatizing real variable y denoting, for example, number of days detained in jails for criminal delinquencies, amounts gained or lost in gambling, cheated in IT returns, paid against fines, treatments of AIDS, induced abortions, conjugal misconduct, offensive car-driving in speed, and the like.

A simple RR procedure is to approach a sample person with two boxes—one with cards marked numbers $a_1, a_2, a_3, \ldots, a_L$ with

$$\mu_1 = \frac{1}{L} \sum_{j=1}^{L} a_r \neq 0, \; \sigma_1^2 = \frac{1}{L} \sum_{1}^{L} (a_j - \mu_1)^2$$

and $b_1, b_2, b_3, \ldots, b_T$ with

$$\mu_2 = \frac{1}{T} \sum_{1}^{T} b_k, \; \sigma_2^2 = \frac{1}{T} \sum_{1}^{T} (b_k - \mu_2)^2$$

On a request to shake the boxes and to draw independently from the 2 boxes 2 cards numbered, say, a_j and b_k without revealing them to the investigator comes out the RR as

$$e_i = a_j y_i + b_k$$

$$\text{Then, } E_R(e_i) = y_i \mu_1 + \mu_2,$$

$V_R(e_i) = \sigma_1^2 y_i^2 + \sigma_2^2$.
Then,
$r_i = e_i - \mu_2 / \mu_1$, has $E_R(r_i) = y_i$
and $V_R(r_i) = V_i = y_i^2 (\sigma_1^2 / \mu_1^2) + (\sigma_2^2 / \mu_1^2)$.
Then, an unbiased estimator for V_i is

$$v_i = \frac{r_i^2 (\sigma_1 / \mu_1)^2 + (\sigma_2 / \mu_2)^2}{1 + (\sigma_1 / \mu_1)^2}$$

Then, for Y one may employ the unbiased estimator

$$e = \sum_{i \in s} \frac{r_i}{\pi_i}$$

Then, $E(e) = E_P E_R(e) = E_P\left(\sum_{i \in s} y_i/\pi_i\right) = Y$; also,

$$E(e) = E_R E_P(e) = E_R\left(\sum_1^N r_i\right) = Y \text{ and}$$

$$V(e) = E_R V_P(e) + V_R E_P(e)$$

$$= E_R\left[\sum_i \sum_{<j} (\pi_i \pi_j - \pi_{ij})\left(\frac{r_i}{\pi_i} - \frac{r_j}{\pi_j}\right)^2 + \sum \beta_i \frac{e_i^2}{\pi_i}\right] + \sum V_i;$$

also this immediately admits a simple unbiased estimator.

4 PROTECTION OF PRIVACY

An effective RR device should not only yield an effective estimator but should also ensure a belief in the respondent that his/her privacy is not compromised on giving out a truthful reply to the investigator's query.

4.1 When a Characteristic Is Qualitative and SRSWR Is Allowed

Chaudhuri and Christofides (2013) have reviewed the topic taking account of the early developments by Anderson (1975a,b,c, 1976, 1977), Lanke (1975, 1976), Leysieffer (1975), Leysieffer and Warner (1976) and more comprehensively by Ljungqvist (1993), Nayak (1994) as briefly reported by Chaudhuri and Mukerjee (1988) and also by Chaudhuri (2011). Guerriers and Sandri (2007), Hong et al. (2010), Nayak and Adeshiyan (2009), and Quatember (2009) are more recent contributors. Recalling θ as the proportion of people bearing a stigmatizing feature A and denoting $P(R|A)$, $P(R|A^C)$ the disclosure probabilities of an RR as R by a person bearing, respectively, A and A^C, by the Bayes theorem, one gets the posterior probability $P(A|R)$ that a respondent bears A when the truthful RR is given out as R. Similarly, $P(A^C|R)$ also like $P(A|R)$ is a revealing probability. The discrepancy between θ and $P(A|R)$ should measure how risky it may appear to a respondent to divulge an honest and truthful RR. Considering SRSWR of respondents the above authors have pushed appreciably the concerned theory.

4.2 When a General Sampling Design Is Allowed to Cover a Qualitative Characteristic

Chaudhuri and Saha (2004) followed by Chaudhuri et al. (2009) as reported by Chaudhuri (2011) and also by Chaudhuri and Christofides (2013) extended this topic covering general sampling schemes. Briefly, they refer to a prior probability L_i $(0 < L_i < 1)$ that a person labeled i bears a sensitive characteristic A, next introduce a posterior probability $L_i(R)$ that such a person yielding the RR as R bears A. Combining these one derives

$$J_i(R) = \frac{L_i(R)/L_i}{[1-L_i(r)]/(1-L_i)}, i \in U$$

as a "Jeopardy measure" indicating the risk of revelation of the ith person's bearing A on divulging the truthful response R.

They next take

$$\overline{J_i} = \frac{1}{|R|}\sum_R J_i(R)$$

as a corresponding average measure anticipating all the alternative responses. Further details we choose to omit as they are discussed plentifully in the literature cited. Here $|R|$ denotes the cardinality of the set of possible responses R, meaning the total number of possible responses.

4.3 Protection of Privacy Covering Quantitative Variables

Bose (2013) has given a procedure based exclusively on SRSWR and discrete-valued quantitative sensitive characteristics and presented a theory to examine how respondent privacy may be protected once RRs are truthfully given out. This is also succinctly reported by Chaudhuri and Christofides (2013) in their book.

Chaudhuri and Christofides (2013) have shown how privacy may be well protected when the quantitative RRs are gathered following the device described in the present section.

Let $L_i(y_i) = L_i$ be the unknown prior probability that y_i is the true but unknown value of a sensitive variable y applicable to the ith respondent selected according to a suitable sampling scheme. Let $P(e_i|y_i)$ be the conditional probability that the ith person reports the RR value as e_i when his/her true value is y_i.

Then, Bayes' theorem gives us the posterior probability that y_i is the true but unknown y-value of the ith person when his/her reported RR is e_i as

$$\begin{aligned} L(y_i|\, e_i) &= \frac{L_i P(e_i|\, y_i)}{P(e_i)} \\ &= \frac{L_i(1/LT)}{1/LT} = L_i \end{aligned}$$

Here L is the total number of numbers a_1, a_2, etc., in the first box and T is the total number of numbers b_1, b_2, etc., in the second box.

So, the RR reveals nothing untoward. Details are given by Chaudhuri and Christofides (2013).

5 OPTIONAL RR TECHNIQUES

Chaudhuri and Mukerjee (1985, 1988) illustrated a practical situation when the investigator presumes a query to be judged sensitive enough and approaches the respondents explaining how a randomization device is to be implemented to produce an RR presupposed to protect the respondent's privacy but some of the respondents may not feel embarrassed at all to directly reveal the truths. Experiencing such situations the device is revised allowing a sampled respondent an option to give out either a DR or an RR following the prescribed device. Theory is presented, allowing only SRSWR samples, by Chaudhuri and Mukerjee (1985, 1988), Gupta et al. (2002), Mangat and Singh (1994), and Singh and Joarder (1997). Later, Arnab (2004) and Chaudhuri and Saha (2005) permitted unequal probability sample selection to extend the relevant theory. Pal (2008) permitted every respondent to arbitrarily give either a DR or an RR without divulging which option has been exercised. Chaudhuri and Dihidar (2009) extended Pal's (2008) theory covering quantitative variables for the first time. We are tempted only to briefly recount this leaving the rest intact in the literature only drawing a curious reader's attention to theory. First, let us take, in the RR device in Section 3, the mean

$$\mu_1 = \frac{1}{L}\sum_{1}^{L} a_j = 1.$$

Further, let the device prescribe the use of one more box containing cards marked $b_1', b_2', b_3', \ldots, b_T'$ with $\mu_2' = 1/T\sum_{k=1}^{T} b_k' \neq \mu_2$.

A sampled person i is given an option either to use the first and second box giving a response

$$I_i = a_j y_i + b_k,$$

with probability $(1 - C_i)$ or to use the first and third boxes independently yielding the response

$$I_i' = a_j y_i + b_u',$$

with probability $(1 - C_i)$. But with probability $C_i (0 < C_i < 1)$ he/she may also give only the DR as y_i. The probability C_i is supposed to be unknown. The response is then either z_i with expectation

$$E_R(z_i) = C_i y_i + (1 - C_i)[y_i \mu_1 + \mu_2]$$

or z_i' with expectation

$$E_R\left(z_i'\right) = C_i y_i + (1 - C_i)\left[y_i \mu_1 + \mu_2'\right]$$

These yield

$$r_{1i} = \frac{\left(\mu_2' z_i - \mu_2 z_i'\right)}{\left(\mu_2' - \mu_2\right)}$$

Independently repeating this exercise once more one gets r_{2i} likewise observing that

$$r_i = \frac{1}{2}(r_{1i} + r_{2i}) \text{ has } E_R(r_i) = y_i;$$

one may unbiasedly estimate

$$V_R(r_i) \text{ by } \frac{1}{4}(r_{1i} - r_{2i})^2;$$

the rest is simple and clear.

REFERENCES

Anderson, H., 1975a. Efficiency versus protection in the RR designs for estimating proportions: Technical Report 9. University of Lund, Sweden.

Anderson, H., 1975b. Efficiency versus protection in a general RR model: Technical Report 10. University of Lund, Sweden.

Anderson, H., 1975c. Efficiency versus protection in RR designs. Mimco Notes. University of Lund, Lund, Sweden.

Anderson, H., 1976. Estimation of a proportion through RR. Int. Stat. Rev. 44, 213–217.

Anderson, H., 1977. Efficiency versus protection in a general RR model. Scand. J. Stat. 4, 11–19.

Arnab, R., 2004. Optional randomized response techniques for complex survey designs. Biom. J. 46, 114–124.

Bose, M., 2013. Respondent privacy and estimation efficiency in randomized response surveys for discrete-valued sensitive variables. arXiv 1303, 5172v1.

Chaudhuri, A., 2001. Using randomized response from a complex survey to estimate a sensitive proportion in a dichotomous finite population. J. Stat. Plan. Inference 94, 37–42.

Chaudhuri, A., 2010. Essentials of Survey Sampling. Prentice Hall of India, Delhi.

Chaudhuri, A., 2011. Randomized Response and Indirect Questioning Techniques in Surveys. CRC Press, Taylor and Francis, Boca Raton, FL.

Chaudhuri, A., 2014. Modern Survey Sampling. CRC Press Taylor and Francis, Boca Raton, FL.

Chaudhuri, A., Christofides, 2013. Indirect Questioning in Sample Surveys. Springer Verlag, Berlin, Germany.

Chaudhuri, A., Dihidar, K., 2009. Estimating means of stigmatizing qualitative and quantitative variables from discretionary responses randomized or direct. Sankhya B 71, 123–136.

Chaudhuri, A., Mukerjee, R., 1985. Optionally randomized responses techniques. CSAB 34, 225–229.

Chaudhuri, A., Mukerjee, R., 1988. Randomized Response Theory and Techniques. Marcel Dekker, New York.

Chaudhuri, A., Pal, S., 2002. Estimating proportions from unequal probability samples using randomized responses by Warner's and other devices. J. Ind. Soc. Agri. Stat. 55 (2), 174–183.

Chaudhuri, A., Saha, A., 2004. Utilizing covariates by logistic regression modelling in improved estimation of population proportions bearing stigmatizing features through randomized responses in complex surveys. J. Ind. Soc. Agri. Stat. 58 (2), 190–211.

Chaudhuri, A., Saha, A., 2005. Optional versus compulsory randomized response techniques in complex surveys. J. Stat. Plan. Inference 135, 516–527.

Chaudhuri, A., Christofides, T.C., Saha, A., 2009. Protection of privacy in efficient application of randomized response techniques. Stat. Math. Appl. 18, 389–418.

Erikson, S.C., 1973. A new model for RR. Int. Stat. Rev. 41, 101–113.

Greenberg, B.G., Abul-Ela, A.L., Simmons, W.R., Horvitz, D.G., 1969. The unrelated question RR model: theoretical framework. JASA 64, 520–539.

Guerriers, M., Sandri, M.E., 2007. A note on the comparison of some randomized response procedures. J. Stat. Plan. Inference 137, 2184–2190.

Gupta, S., Gupta, B., Singh, S., 2002. Estimation of sensitivity level of personal interview survey questions. J. Stat. Plan. Inference 100, 239–247.

Hong, Z., Yan, Z., Wei, L., 2010. A note of proposed privacy measures in randomized response models. Adv. Intel. Soft Comput. 77, 635–642.

Horvitz, D.G., Thompson, D.J., 1952. A generalization of sampling of sampling without replacement from finite universe. JASA 47, 663–685.

Horvitz, D.G., Shah, B.V., Simmons, W.R., 1967. One unrelated question RR model. Proc. Soc. Stat. Sec. ASA 65–72.

Lanke, J., 1975. Some contributions to the theory of survey sampling. Ph.D. Thesis. University of Lund, Lund, Sweden.

Lanke, J., 1976. On the degree of protection in randomized interviews. Int. Stat. Rev. 44, 197–203.

Leysieffer, R.W., 1975. Respondent Jeopardy in RR Procedures: Technical Report M338. Department of Statistics, Florida State University, Tallahassee, FL.

Leysieffer, R.W., Warner, S.L., 1976. Respondent jeopardy and optimal designs in randomized response models. J. Am. Stat. Assoc. 71, 649–656.

Ljungqvist, L., 1993. A unified approach to measures of privacy in randomized response models: a utilitarian perspective. J. Am. Stat. Assoc. 88, 97–103.

Mangat, N.S., Singh, S., 1994. Optional randomized response model. JISAS 32, 71–75.

Nayak, T.K., 1994. On randomized response surveys for estimating a proportion. Comm. Statist. Theory Method 23 (3), 3303–3321.

Nayak, T.K., Adeshiyan, S.A., 2009. A unified framework for analysis and comparison of randomized response survey of binary characteristics. J. Stat. Plan. Inference 139, 2757–2766.

Pal, S., 2008. Unbiasedly estimating the total of a stigmatizing variable from a complex survey on permitting options for direct or randomized responses. Stat. Pap. 49, 157–164.

Quatember, A., 2009. To copy with nonresponse and untruthful answering. Different questioning designs for different variables. In: Proceedings of the NTTS 2009—Conferences on New Techniques and Technologies for Statistics. www.ntts2009.eu.

Singh, S., Joarder, A.H., 1997. Optional randomized response technique for sensitive quantitative variable. Metron 55, 151–157.

Warner Stanley, L., 1965. Randomized response: a survey technique for eliminating evasive 264 answer bias. J. Am. Stat. Assoc. 60, 63–69.

Chapter 4

The Classical Randomized Response Techniques

Reading Warner (1965) and Greenberg et al. (1969) 50 Years Later

T.C. Christofides[*,1]
**University of Cyprus, Nicosia, Cyprus*
[1]Corresponding author: e-mail: tasos@ucy.ac.cy

ABSTRACT

In surveys of human populations, very often, a major source of nonsampling error is the presence of nonresponse. For various reasons, people refuse to respond. In case the issue under investigation is sensitive or stigmatizing, nonresponse seriously jeopardizes the validity of conclusions. It is very likely that some of the people agreeing to participate provide untruthful or misleading answers in order to reveal as little information as possible and thus to protect their privacy. Warner's (1965) innovative randomized response technique opened new ways in surveys of human populations and created a new area of research. In this review paper, we present the two classical randomized response models, Warner's pioneering technique and the unrelated question model as presented in Greenberg et al. (1969) and discuss very briefly some technical aspects of the two techniques. However, our emphasis in this chapter is not on the mathematics of the two classical techniques. Reading the two papers again, 50 years later, we offer some thoughts and stress the importance of these two influential publications to survey methodology.

Keywords: Indirect questioning, Randomized response, Protection of privacy, Warner's technique, The unrelated question model, Less than completely truthful reporting, Perceived protection of privacy

2000 MSC: 62D05

1 INTRODUCTION

The presence of nonresponse in surveys of human populations is a major issue. It has been, and still is, the object of scientific research. Various remedies, techniques, and estimation procedures have been proposed to reduce its negative effects. In case the issue under investigation is sensitive or even

Handbook of Statistics, Vol. 34. http://dx.doi.org/10.1016/bs.host.2016.01.028

worse is stigmatizing, nonresponse is a major threat to the validity of conclusions. Even if nonresponse is limited to a minimum, it is very likely that among the people agreeing to respond, a nonnegligible number of them will provide untruthful or misleading answers. It is very human that people will try to protect their privacy by revealing as little information as possible and thus conceal their true status. It is for these reasons that the pioneering work of Warner (1965) opened new avenues in survey sampling of human populations on issues of sensitive nature. His foundation work was only the beginning and a whole new area of research, the area of indirect questioning techniques started to be developed and, still is under very active development. In this paper we discuss Warner's randomized response technique as well as the unrelated question model, which followed soon after. The unrelated question model was at the time among the first attempts to extend Warner's work and to offer an alternative.

This paper is written for the sole purpose of stressing the importance of the two classical models. We will avoid unnecessary mathematical complications and details and we will attempt to read Warner (1965) and Greenberg et al. (1969) behind the lines.

In Section 2, we present in (reasonable) detail Warner's technique, initially for simple random sampling with replacement (SRSWR) and then under a general sampling design. In Section 3 we give the details of the unrelated question model of Greenberg et al. (1969), again for SRSWR and under a general sampling design. In Sections 4 and 5 we offer some comments by reading again, 50 years later the two original papers, Warner (1965) and Greenberg et al. (1969).

Throughout the paper, let A denote a stigmatizing category and θ its population prevalence, which is the target of our estimation procedures.

2 WARNER'S RANDOMIZED RESPONSE TECHNIQUE

Assume that we have a simple random sample of size n drawn from the population with replacement. Each person is provided with a randomization device, for example, as described in the original publication of Warner (1965) a spinner with a face marked so that the spinner points to A with probability P and to A^c with probability $1-P$. By using the randomization device (ie, spin the spinner) in the absence of the interviewer, the participant is instructed to respond truthfully (with a yes or no) whether or not the spinner points to the category to which the respondent belongs, ie, the participant responds with probability $P(P \neq 0.5)$ whether he/she belongs to the stigmatizing category A and with probability $1-P$ whether he/she belongs to the complementary category A^c. More formally, each sampled person is faced with the statements

(I) I belong to category A
(II) I belong to category A^c

but responds to either (I) or (II) with probability P or $1-P$.

Let λ be the probability that a person provides a "yes" response. Then

$$\lambda = P\theta + (1-P)(1-\theta)$$

from which we immediately have that

$$\theta = \frac{\lambda - (1-P)}{2P-1}. \tag{1}$$

Let $\hat{\lambda}$ be the sample proportion of people providing a "yes" response. Then Eq. (1) suggests as an estimator the quantity

$$\hat{\theta} = \frac{\hat{\lambda}}{(2P-1)} + \frac{P-1}{2P-1} \tag{2}$$

which is an unbiased estimator of the population proportion θ with variance

$$Var(\hat{\theta}) = \frac{\theta(1-\theta)}{n} + \frac{1}{n}\left\{\frac{1}{16(P-0.5)^2} - \frac{1}{4}\right\}. \tag{3}$$

Observe also that the RHS of (3) consists of two parts, the first being the variance due to random sampling and the second due to the use of the randomization device. The first part would be the variance of the estimator in case the responses are obtained by means of a direct response survey, and assuming (unrealistically) that the participants are completely honest.

The estimator provided by (2) is a moment estimator and has also been obtained as a maximum likelihood estimator by Warner (1965). However the estimator might as well fall outside the parameter space, ie, the interval [0, 1], and thus, strictly speaking it is not the MLE. This "mistake" has been identified by Singh (1976), Devore (1977), and Flinger et al. (1977). Let us denote the "corrected" MLE by $\tilde{\theta}$. Then, it is given as follows:

$$\tilde{\theta} = \begin{cases} 0 & \text{if } \hat{\lambda} \le 1-P < 0.5 \text{ or } 0.5 < 1-P \le \hat{\lambda} \\ \hat{\theta} & \text{if } 1-P < \hat{\lambda} < P,\ P > 0.5 \text{ or } P < \hat{\lambda} < 1-P,\ P < 0.5 \\ 1 & \text{if } 0.5 < P \le \hat{\lambda} \text{ or } \hat{\lambda} \le P < 0.5. \end{cases} \tag{4}$$

The "corrected" MLE given by (4) is obviously biased. However, since

$$|\tilde{\theta} - \theta| < |\hat{\theta} - \theta|,$$

then the "corrected" MLE $\tilde{\theta}$ has smaller mean squared error than the estimator $\hat{\theta}$.

Although the case of SRSWR is mathematically easier to present and understand, Chaudhuri (2001a,b, 2011) has made the case that for estimating θ, any sampling design will do. Based on his approach we proceed as follows:

For a finite population $U=(1,\ldots,i,\ldots,N)$ of N persons, let y_i take on the value 1 if the i member of the population has the stigmatizing characteristic A and the value 0 otherwise.

Assume now that the sample s is chosen according to a general sampling design p with positive first- and second-order probabilities of inclusion

$$\pi_i=\sum_{s\ni i} p(s),\quad i\in U \quad \text{and} \quad \pi_{ij}=\sum_{s\ni i,j} p(s),\quad i,j\in U,\ i\neq j.$$

Let E_p and E_R denote the expectation operators over the design p and the randomization device, respectively, and similarly let V_p and V_R denote the corresponding variance operators. Denoting by E and V the overall expectation and variance operators we will assume that

$$E=E_pE_R=E_RE_p$$

and

$$V=E_pV_R+V_pE_R=E_RV_p+V_RE_p.$$

Assume now that Warner's randomization technique with parameter $P(P\neq 0.5)$ is applied and let z_k take on two values, the value 1 if the person sampled on the kth draw responds "yes" and the value 0 if a "no" response is given. Then

$$E_R(z_k)=(2P-1)y_k+(1-P)$$

and thus

$$r_k=\frac{z_k-(1-P)}{2P-1}$$

is an unbiased estimator of y_k.

Consider now the Horvitz–Thompson estimator of θ

$$e=\frac{1}{N}\sum_{k\in s}\frac{r_k}{\pi_k}.$$

Then

$$E_p(e)=\frac{1}{N}\sum_{k=1}^{N} r_k,\quad E_R(e)=\frac{1}{N}\sum_{k\in s}\frac{y_k}{\pi_k}$$

and consequently,

$$E(e)=E_R\left[E_p(e)\right]=E_p[E_R(e)]=\theta.$$

In addition,

$$V_R(e) = \frac{1}{N^2}\sum_{k\in s}\frac{V_R(r_k)}{\pi_k^2} = \frac{P(1-P)}{N^2(2P-1)^2}\sum_{k\in s}\frac{1}{\pi_k^2},$$

$$V_p(e) = \frac{1}{N^2}\sum_{k}\sum_{l,l>k}(\pi_k\pi_l - \pi_{kl})\left(\frac{r_k}{\pi_k} - \frac{r_l}{\pi_l}\right)^2.$$

Therefore,

$$\begin{aligned} V(e) &= E_p V_R(e) + V_p E_R(e) \\ &= \frac{1}{N^2}\left[\frac{P(1-P)}{(2P-1)^2}\sum_{k=1}^{N}\frac{1}{\pi_k} + \sum_{k}\sum_{l,l>k}(\pi_k\pi_l - \pi_{kl})\left(\frac{y_k}{\pi_k} - \frac{y_l}{\pi_l}\right)^2\right] \\ &= E_R V_p(e) + V_R E_p(e) \\ &= \frac{1}{N^2}\left[\sum_{k}\sum_{l,l>k}(\pi_k\pi_l - \pi_{kl})E_R\left(\frac{r_k}{\pi_k} - \frac{r_l}{\pi_l}\right)^2 + \frac{NP(1-P)}{(2P-1)^2}\right]. \end{aligned} \tag{5}$$

An unbiased estimator of (5) is given by

$$v(e) = \frac{1}{N^2}\left[\sum_{k\in s}\sum_{l\in s,l>k}\left(\frac{\pi_k\pi_l - \pi_{kl}}{\pi_{kl}}\right)\left(\frac{r_k}{\pi_k} - \frac{r_l}{\pi_l}\right)^2 + \frac{P(1-P)}{(2P-1)^2}\sum_{k\in s}\frac{1}{\pi_k}\right].$$

Many details related to the mathematics of the technique, or to concepts such as the protection of privacy as well as numerous extensions and generalizations can be displayed. It would take many pages to describe very briefly all the extensions of the model that have been published. But the purpose of this paper is not to give all possible details and examine all aspects of the technique, and definitely not to list all its extensions and generalizations. Rather, the aim is to give the flavor of the model and stress its central role in survey methodology. The interested reader can find more detailed presentation of the technique elsewhere, for example, in monographs such as Chaudhuri and Mukerjee (1988), Chaudhuri (2011), and Chaudhuri and Christofides (2013).

3 THE UNRELATED QUESTION MODEL

The unrelated question model is based on an idea of Walt R. Simmons as presented in Horvitz et al. (1967). It is a modification of Warner's technique in which the second statement is replaced by a statement on an unrelated non-stigmatizing item, say B. A respondent is faced with two statements:

(I) I belong to category A

(II) I belong to category B.

Here, a randomization device instructs the interviewee to respond (with a yes or no) with probability P to statement (I) and with probability $1-P$ to statement (II). The main purpose for adopting such an approach, according to Greenberg et al. (1969), is to enhance the chances of participating in the survey. Walt R. Simmons, who suggested this modification, felt that the respondent might be more willing to provide a truthful response if one of the two questions is completely innocuous and unrelated to the stigmatizing attribute. In essence, Simmons thought that participants may feel that their privacy is better protected by implementing the new approach. We see here that early on, researchers realized that an important issue in indirect questioning methodology is the protection of privacy, as perceived by the people participating in the survey.

The prevalence P_B of the unrelated characteristic may be known or unknown. Assume for simplicity that it is known and also that our estimation will be based on a sample of size n drawn with replacement from the population.

Let λ be the probability of a "yes" response. Then

$$\lambda = P\theta + (1-P)P_B.$$

Solving for θ we obtain that

$$\theta = \frac{\lambda - P_B(1-P)}{P}$$

which immediately suggests that an estimator of θ is

$$\hat{\theta} = \frac{\hat{\lambda} - P_B(1-P)}{P},$$

where $\hat{\lambda}$ is the sample proportion of people providing a "yes" response. The estimator is unbiased with variance

$$\begin{aligned} Var(\hat{\theta}) &= \frac{\lambda(1-\lambda)}{nP^2} \\ &= \frac{1}{nP^2}\{P\theta + (1-P)P_B\}\{1 - P\theta - (1-P)P_B\}. \end{aligned}$$

The above variance can be estimated unbiasedly by the quantity

$$\widehat{Var}(\hat{\theta}) = \frac{\hat{\lambda}(1-\hat{\lambda})}{(n-1)P^2}.$$

Although the estimator can be obtained as the MLE using maximum likelihood, it may take values outside the appropriate parameter space [0, 1]. This pathological problem, similar to the one in the Warner's technique, can be remedied by a "corrected" version. In fact the correct maximum likelihood estimator $\tilde{\theta}$ (see Flinger et al., 1977) is given by

$$\tilde{\theta}=\begin{cases} 0 & \text{if } \hat{\lambda}<(1-P)P_B \\ \hat{\theta} & \text{if } (1-P)P_B<\hat{\lambda}<P+(1-P)P_B \\ 1 & \text{if } \hat{\lambda}\geq P+(1-P)P_B. \end{cases}$$

The "corrected" version of the estimator is obviously biased, but as in the case of Warner's technique it has smaller mean squared error than the uncorrected unbiased one.

The model with known P_B, although mathematically simple and practically easier to apply, has limited applicability and thus the more realistic scenario of unknown P_B should be considered. In this case, two independent random samples are needed. Again for simplicity assume that the two independent random samples of sizes n_1 and n_2, respectively, are drawn from the population with replacement. Each person from the first and second sample is instructed to follow the procedure described above with randomization parameters P_1 and P_2, respectively, with $P_1 \neq P_2$.

Let λ_1 and λ_2 be the probability of a "yes" response from a person of the first and second sample, respectively. Then

$$\lambda_1 = P_1\theta + (1-P_1)P_B \tag{6}$$

and

$$\lambda_2 = P_2\theta + (1-P_2)P_B. \tag{7}$$

Solving the system of Eqs. (6) and (7) we obtain that

$$\theta = \frac{\lambda_1(1-P_2)-\lambda_2(1-P_1)}{P_1-P_2}. \tag{8}$$

Eq. (8) suggests as an estimator for θ the quantity

$$\hat{\theta} = \frac{\hat{\lambda}_1(1-P_2)-\hat{\lambda}_2(1-P_1)}{P_1-P_2}, \tag{9}$$

where $\hat{\lambda}_1$ and $\hat{\lambda}_2$ are the sample proportions of people responding "yes" in the first and second sample, respectively. The estimator is unbiased with variance

$$Var(\hat{\theta}) = \frac{1}{(P_1-P_2)^2}\left[\frac{\lambda_1(1-\lambda_1)(1-P_2)^2}{n_1}+\frac{\lambda_2(1-\lambda_2)(1-P_1)^2}{n_2}\right].$$

An unbiased estimator of the variance is provided by the quantity

$$\widehat{Var}(\hat{\theta}) = \frac{1}{(P_1-P_2)^2}\left[\frac{\hat{\lambda}_1(1-\hat{\lambda}_1)(1-P_2)^2}{n_1-1}+\frac{\hat{\lambda}_2(1-\hat{\lambda}_2)(1-P_1)^2}{n_2-1}\right].$$

The method described above assumes samples drawn with replacement from the population. For a general sampling design p with positive first- and second-order inclusion probabilities, just as in the case of Warner's method we proceed as follows.

While y_k takes on the value 1 if the k member of the population has the stigmatizing characteristic A and the value 0 otherwise, let x_k take on the value 1 if the k member of the population has the innocuous characteristic B and the value 0 otherwise. Let also z_k take on the value 1 if the k sampled person from the first sample responds "yes" and the value 0 otherwise. Furthermore, let w_k take on the value 1 if the k sampled person from the second sample responds "yes" and the value 0 otherwise. Then

$$E_R(z_k) = P_1 y_k + (1 - P_1) x_k$$

and

$$E_R(w_k) = P_2 y_k + (1 - P_2) x_k.$$

Now let

$$r_k = \frac{z_k(1 - P_2) - w_k(1 - P_1)}{P_1 - P_2}.$$

Then r_k is unbiased for y_k and its variance can be expressed as

$$V_R(r_k) = \frac{(1 - P_1)(1 - P_2)(P_1 + P_2 - 2P_1 P_2)}{(P_1 - P_2)^2}(y_k - x_k)^2 \equiv V_k.$$

Consider now the Horvitz–Thompson estimator of θ

$$e = \frac{1}{N}\sum_{k \in s} \frac{r_k}{\pi_k}.$$

Its variance is given by

$$V(e) = \frac{1}{N^2}\left[\sum_k \sum_{l, l>k} (\pi_k \pi_l - \pi_{kl})\left(\frac{y_k}{\pi_k} - \frac{y_l}{\pi_l}\right)^2 + \sum_k \frac{V_k}{\pi_k}\right],$$

while an unbiased estimator of the variance is given by the quantity

$$v(e) = \frac{1}{N^2}\left[\sum_{k \in s} \sum_{l \in s, l>k} \left(\frac{\pi_k \pi_l - \pi_{kl}}{\pi_{kl}}\right)\left(\frac{r_k}{\pi_k} - \frac{r_l}{\pi_l}\right)^2 + \sum_{k \in s} \frac{r_k(r_k - 1)}{\pi_k}\right].$$

As in the case of Warner's technique, we choose not to present all technical aspects of the unrelated question model for the reasons described in Section 2. Of course, many details of the technique as well as extensions and generalizations can be found in monographs such as in Chaudhuri and Mukerjee (1988), Chaudhuri (2011), and Chaudhuri and Christofides (2013).

4 READING WARNER (1965) AND GREENBERG ET AL. (1969) 50 YEARS LATER

It is my humble opinion that strong and innovative ideas in statistics, very often, if not always, are very simple in nature and in many cases the mathematics involved are elementary and basic. Warner's idea fulfills this description 100%. The technical points and the mathematics displayed are so elementary that even undergraduate students would easily read the paper. I have the opinion that if a similar manuscript was to be submitted nowadays to the *Journal of the American Statistical Association* (and in fact to most probability and statistics journals), it is possible that the paper would not even make it to the refereeing process. That is why Warner's paper is so strong. Published in a top journal, a very short paper (six and a half pages long), it presents an innovative idea, using limited mathematics. By reading the paper many times I come to the conclusion that the simplicity of the paper was not accidental. I believe that Warner just wanted to put emphasis on his idea, and not on the technicalities and the mathematics. For example, he acknowledges the fact (in a footnote) that his estimator may take values outside the correct parameter space [0, 1], a problem pointed out in Section 2. But he chose not to put emphasis on that or to provide a corrected version of his estimator. Even the final section of his paper with his concluding remarks contains ideas for possible extensions, but he chose not to elaborate on that and leave it to others to pursue. Warner himself points out that his method "is only one of many possibilities" acknowledging the fact that other alternative approaches might emerge. Overall, I think that Warner just threw an idea and invited others to pursue it further. This is exactly what happened. Numerous extensions, improvements, generalizations, corrected versions, etc. have appeared. Of course, Warner's (1965) paper is always the point to start. It is no coincidence that it has been cited so far more than 1700 times in scientific journals alone, not counting books and conference proceedings. The speculation of Greenberg et al. (1969) that "the legendary Pandora's box has been opened with the Warner technique" has been proven to be a fact.

Warner (1965) introduces the concept of less than completely truthful reporting. But he assumes that this phenomenon takes place only in regular surveys, ie, in cases where the respondents are faced with the direct question and are asked to provide a direct answer. Of course, his assumption serves the purpose of making a case for the randomized response approach by comparing the mean square error of randomized response estimators and regular estimators (ie, estimators obtained by direct questioning) assuming constant probability of truthful reporting. But the values he chooses for the probability of less than truthful reporting are unreasonably high in my opinion and in fact, he does not do justice to his method. On the other hand, his numerical results assume unrealistically high values (0.5 and 0.6) for the prevalence of the stigmatizing characteristic. In addition, the tables with the numerical results are constructed assuming that the randomization parameter P takes on the values 0.6, 0.7, 0.8, and 0.9. Warner assumes that these values are low enough to "induce full cooperation in the randomized approach." But the last two values, in my opinion, are extremely high and unsuitable as randomization parameters.

The unrelated question randomized response model is considered by many people as a classical randomized response technique and along with Warner's model has been a starting point for research developments on randomized response. Although the technique officially is presented in the celebrated paper of Greenberg et al. (1969), the idea is credited (by the authors of that paper) to Walt R. Simmons and the technique initially is described in Horvitz et al. (1967). According to Greenberg et al. (1969) Simmons felt that by replacing the second statement in Warner's technique by a statement on an innocuous item, the respondents might be more truthful. Whether this is the case still remains to be verified. I do not know of any empirical evidence supporting such a claim.

Greenberg et al. (1969) is an easy to read paper just like Warner's (1965). The mathematics involved are elementary and easy to digest. I would make the same statement I made above for Warner's (1965) paper: It is possible that a similar paper submitted nowadays to the *Journal of the American Statistical Association* will not even make it to the refereeing process. But again, I feel that in many cases innovative ideas in statistics are simple and are sustained by basic mathematics. Although not as original as the technique described in Warner (1965), still it contains ideas for further developments and extensions on randomized response.

Greenberg et al. (1969) discuss the case of less than completely truthful reporting. Although they do not give any information as to how the probability of less than completely truthful reporting can be estimated, by considering this case, they recognize the possibility that participants might not respond truthfully even under randomized response.

A great part of the almost 20-page paper is dedicated (rightly so) to the optimal choice of the design randomization parameters P_1, P_2, the sample sizes n_1, n_2, and the prevalence of the unrelated innocuous characteristic. In fact, this is the strong advantageous point of the technique over Warner's. By having more parameters to fix and more flexibility, better estimates can be obtained. Although the technique is compared with Warner's model in detail, the comparison is based on the efficiencies of the two techniques. But another way of comparison could be carried out in terms of the protection of privacy. Of course it is not fair at all to expect the authors of Greenberg et al. (1969) to compare the unrelated question model with Warner's technique based on a concept such as the protection of privacy which, at that time, was not developed. However, the authors carried out a comparison between the unrelated question model and Warner's technique assuming less than completely truthful reporting for both methods. They assume that in the unrelated question model T_A is the probability that a person belonging to the stigmatizing category will tell the truth, while T_A' denotes the corresponding quantity for Warner's model. One may consider the parameters T_A and T_A' as implicitly related to the perceived protection of privacy, ie, not the protection of privacy as measured by various indexes developed in subsequent publications but to the protection of privacy as perceived by the participants, a concept not yet developed. It is reasonable to assume that a person will respond truthfully if he/she feels that his/her privacy is protected and not whether a mathematical quantity says so.

The unrelated question model has been devised because Simmons felt that the second statement on the innocuous item will make participants to be more truthful. Personally I have my reservations. In fact I believe (although I do not have any empirical evidence) that Warner's technique will generate more truthful responses. Given that a "no" response in the unrelated question model is not incriminating, I would expect at least some of the participants to opt for that response regardless of their true status about the sensitive characteristic. Of course, this disadvantage of the technique is recognized by the authors themselves who suggest that for that reason, as a remedial measure, the innocuous characteristic must be so chosen that it has high prevalence. Still, the problem is there.

Although the unrelated question model was devised as a competitor or improvement of the model developed by Warner, both models can be thought of as two special cases of a more general technique. Consider the following scenario:

Each participant is faced with the following statements:

(I) I belong to category A

(II) I belong to category A^c

(III) I belong to category B.

Here of course B denotes a nonstigmatizing category. With the use of a randomization device (unnoticed by the interviewer), the participant is instructed to respond to statements (I), (II), and (III) with probability P_1, P_2, and P_3, respectively, with $P_1+P_2+P_3=1$. To avoid trivialities, assume that $P_1 > 0$. Observe that the special case where $P_3=0$ gives Warner's procedure, whereas the case where $P_2=0$ gives the unrelated question model approach of Simmons. Depending on whether the prevalence of the innocuous characteristic is known or not, one or two independent samples are needed. It is quite possible that by choosing the randomization parameters P_1, P_2, and P_3 wisely as well as the unrelated characteristic B, one will end up with estimators which can be more efficient than either Warner's estimator or the estimator produced by the unrelated question model. In this model, just as in Warner's, a "no" response does not imply that the person questioned is not a member of the stigmatizing group, something which, in my opinion, is the Achilles' heel of the unrelated question model.

The model just described is a special case of an even more general model described by Quatember (2009). Quatember's approach is to describe various randomized response techniques in a unified manner, termed as standardization. The standardization, ie, the more general model, contains 16 different models. Two of them are Warner's technique and the unrelated question model.

5 EPILOGUE

Despite the minor weak points found in Warner (1965) and Greenberg et al. (1969), such as (unimportant) mathematical mistakes or poor choice of values for the design parameters in producing numerical results, these two milestone publications are among the most influential in survey methodology. In particular Warner's paper opened new avenues and established a whole new area of research. It is a fate of many (if not most) scientific papers that a few years after publication date will only be remembered and cited by very few. This is not the case for Warner (1965) and Greenberg et al. (1969). One can rarely find a manuscript on randomized response or indirect questioning techniques with these two publications absent from the list of references. Almost certainly, these two papers will continue to be the point of reference not only just for historical reasons, but because they contain strong innovative ideas, presented in a simple manner, using elementary mathematics in just a few pages.

REFERENCES

Chaudhuri, A., 2001a. Using randomized response from a complex survey to estimate a sensitive proportion in a dichotomous finite population. J. Stat. Plan. Inference 94, 37–42.

Chaudhuri, A., 2001b. Estimating sensitive proportions from unequal probability samples using randomized responses. Pak. J. Stat. 17, 259–270.

Chaudhuri, A., 2011. Randomized Response and Indirect Questioning Techniques in Surveys. Chapman and Hall, CRC Press, Taylor and Francis Group, Boca Raton.

Chaudhuri, A., Christofides, T.C., 2013. Indirect Questioning Techniques in Sample Surveys. Springer, Heidelberg.

Chaudhuri, A., Mukerjee, R., 1988. Randomized Response: Theory and Techniques. Marcel Dekker, New York.

Devore, J.L., 1977. A note on the randomized response technique. Commun. Stat. Theory Methods 6, 1525–1529.

Flinger, M.A., Policello, G.E., Singh, J., 1977. A comparison of two randomized response survey methods with consideration for the level of respondent protection. Commun. Stat. Theory Methods 6, 1511–1524.

Greenberg, B.G., Abul-Ela, A.L.A., Simmons, W.R., Horvitz, D.G., 1969. The unrelated question randomized response model: theoretical framework. J. Am. Stat. Assoc. 64, 520–539.

Horvitz, D.G., Shah, B.V., Simmons, W.R., 1967. The unrelated question randomized response model. In: Proceedings of the Social Statistics Section of the American Statistical Association, pp. 65–72.

Quatember, A., 2009. A standardization of randomized response strategies. Surv. Methodol. 35, 143–152.

Singh, J., 1976. A note on randomized response techniques. In: Proceedings of the Social Statistics Section of the American Statistical Association, p. 772.

Warner, S.L., 1965. Randomized response: a survey technique for eliminating evasive answer bias. J. Am. Stat. Assoc. 60, 63–69.

Chapter 5

On the Estimation of Correlation Coefficient Using Scrambled Responses

S. Singh[1]
Texas A&M University-Kingsville, Kingsville, TX, United States
[1]*Corresponding author: e-mail: sarjinder.singh@tamuk.edu*

ABSTRACT

The problem of estimating the correlation coefficient using scrambled responses based on the Eichhorn and Hayre (1983) model was considered by Singh (1991) who had studied the asymptotic behavior of the bias and variance expressions. As pointed out by Chaudhuri (2011, p. 185), Bellhouse (1995) has considered the problem of estimating the correlation coefficient, but the details are cumbersome and are not reported in his monograph. Chaudhuri (2011) further indicates that efforts are needed to refine developments to estimate correlation coefficient to switch over to the utilization of procedures when only randomized response survey data are available. He mentioned that no literature of relevance seems available yet. In this chapter, an attempt has been made to answer the question raised by Chaudhuri (2011).

Keywords: Sensitive variables, Randomized response techniques, Estimation of correlation coefficient

1 INTRODUCTION

The problem of estimating correlation coefficient between two variables in a finite population is well known in the field of survey sampling. Pearson (1896) was the first to define a very valuable parameter in the field of statistics and named it correlation coefficient. The problem of estimating this parameter has been widely discussed by Wakimoto (1971), Gupta et al. (1978, 1979), Rana (1989), Gupta and Singh (1990), Biradar and Singh (1992), Gupta et al. (1993), and Gupta (2002) under different survey sampling schemes. Singh et al. (2014) also suggested an empirical log-likelihood estimate of correlation coefficient. As pointed out by Chaudhuri (2011), a very limited effort has been made to estimate the value of correlation coefficient

Handbook of Statistics, Vol. 34. http://dx.doi.org/10.1016/bs.host.2016.01.004

between two sensitive variables which are observed through a randomization device. To our knowledge, Clickner and Iglewicz (1976) were the first to consider the problem of estimating correlation coefficient between two qualitative sensitive characteristics by following the Warner (1965) pioneer randomized response model technique. Recently, Lee et al. (2013) have also considered the problem of estimation of correlation coefficient between two qualitative sensitive characteristics with two different methods.

Horvitz et al. (1967) and Greenberg et al. (1971) extended the Warner (1965) model to the case where the responses to the sensitive question are quantitative rather than a simple "yes" or "no." The unrelated question model can also be used to estimate correlation between two sensitive characteristics. Fox and Tracy (1984) showed how the unrelated question model can be used to estimate correlation between two quantitative sensitive characteristics. In the unrelated question model, the respondent selects, by means of a randomization device, one of two questions. However, there are several difficulties which arise when using this unrelated question method. The main one is choosing the unrelated question. As Greenberg et al. (1971) note, it is essential that the mean and variance of the responses to the unrelated question be close to those for the sensitive question: otherwise, it will often be possible to recognize from the response which question was selected. However, the mean and variance of the responses to the sensitive question are unknown, making it difficult to choose good unrelated question. A second difficulty is that in some cases the answers to the unrelated question may be more rounded or regular, making it possible to recognize which question was answered. For example, Greenberg et al. (1971) considered the sensitive question: about how much money did the head of this household earn last year. This was paired with the question: about how much money do you think the average head of a household of your size earns in a year. An answer such as \$26,350 is more likely to be in response to the unrelated question, while an answer such as \$18,618 is almost certainly in response to the sensitive question. A third difficulty is that some people are hesitant in disclosing their answer to the sensitive question (even though they know that the interviewer cannot be sure that the sensitive question was selected). For example, some respondents may not want to reveal their income even though they know the interviewer can only be 0.75 certain, say, that the figure given is the respondent's income.

These difficulties are no longer present in the scrambled randomized response method introduced by Eichhorn and Hayre (1983). This method can be summarized as follows: each respondent scrambles in response X by multiplying it by a random scrambling variable S and only then reveals the scrambled result $Z=XS$ to the interviewer. The mean of the response, $E(X)$, can be estimated from a sample of Z values and the knowledge of the distribution of the scrambling variable S. This method may also be used to estimate the median or other parameters of the distribution function of X as reported by Ahsanullah and Eichhorn (1988). It will be worth mentioning that Bellhouse (1995) has also considered the problem of estimating the correlation coefficient, but his approach is too cumbersome to understand. The additive model due to Himmelfarb and Edgell (1980) has also been used to estimate correlation coefficient between two quantitative sensitive variables (see Fox, 2016).

In this chapter, we shall discuss randomized response techniques for estimating the correlation coefficient, introduced by Singh (1991), between the two sensitive variables X and Y. For example, X may stand for the respondents' income and Y may stand for the respondents' expenditure. The problem of estimating the correlation coefficient both between two sensitive variables, and between a sensitive and a nonsensitive variable is considered. Asymptotic properties of the proposed estimators are investigated through analytical expressions.

2 TWO SCRAMBLING VARIABLE RANDOMIZED RESPONSE TECHNIQUE

Suppose X denotes the response to the first sensitive question (eg, income), and Y denotes the response to the second sensitive question (eg, expenditure). Further, let S_1 and S_2 be the two scrambling random variables, each independent of X and Y and having finite means and variances. For simplicity, also assume that $X \geq 0, Y \geq 0, S_1 > 0$, and $S_2 > 0$. We now consider the following two cases:

(i) The respondent generates S_1 using some specified method, while S_2 is generated by using the linear relation, $S_2 = \alpha S_1 + \beta$ where α and β are known constants and, therefore, S_1 and S_2 are dependent random variables.

(ii) S_1 and S_2 are random variables following known distributions. The particular values of S_1 and S_2 to be used by any respondent are obtained from two separate randomization devices. This ways S_1 and S_2 become independent random variables (unsolved exercise 11.28 in Singh, 2003).

The interviewee multiplies his/her response X to the first sensitive question by S_1 and the response Y to the second sensitive question by S_2. The interviewer thus receives two scrambled answers $Z_1 = XS_1$ and $Z_2 = YS_2$. The particular values of S_1 and S_2 are not known to the interviewer, but their joint distribution is known. In this way the respondent's privacy is not violated.

Let $E(S_1) = \theta_1$, $E(S_2) = \theta_2$, $V(S_1) = \tau_{20}$, $V(S_2) = \tau_{02}$, $Cov(S_1, S_2) = \tau_{11}$, $E(X) = \mu_1$, $E(Y) = \mu_2$, $V(X) = \sigma_x^2 = m_{20}$, $V(Y) = \sigma_y^2 = m_{02}$, $\tau_{rs} = E[S_1 - \theta_1]^r [S_2 - \theta_2]^s$, and $m_{rs} = E[X - \mu_1]^r [Y - \mu_2]^s$, where θ_1, θ_2, τ_{20}, τ_{02}, τ_{11}, and τ_{rs} are known to the interviewer, but μ_1, μ_2, σ_x^2, σ_y^2, and m_{rs} are unknown. Also let $\sigma_{Z_1}^2$ and $\sigma_{Z_2}^2$ denote the variance of Z_1 and Z_2, respectively. We now have the following theorem.

Theorem 1. *The variance of the first sensitive variable X is given by:*

$$V(X) = \sigma_x^2 = \frac{\sigma_{Z_1}^2 - \tau_{20}\mu_1^2}{\tau_{20} + \theta_1^2} \tag{1}$$

Proof. We have $Z_1 = XS_1$. Since X and S_1 are independent, we have:

$$E(Z_1) = E(XS_1) = E(X)E(S_1)$$

or

$$E(X) = \frac{E(Z_1)}{E(S_1)} = \frac{E(Z_1)}{\theta_1} \tag{2}$$

Also, $E(Z_1^2) = E(S_1X)^2 = E(X^2S_1^2) = E(X^2)E(S_1^2)$
Thus,

$$E(X^2) = \frac{E(Z_1^2)}{E(S_1^2)} = \frac{E(Z_1^2)}{\tau_{20} + \theta_1^2} \tag{3}$$

By definition, we have

$$V(X) = \sigma_x^2 = E(X^2) - (E(X))^2$$

Using (2) and (3), we get

$$\begin{aligned} V(X) = \sigma_x^2 &= \frac{E(Z_1^2)}{\tau_{20} + \theta_1^2} - \frac{(E(Z_1))^2}{\theta_1^2} = \frac{\theta_1^2 E(Z_1^2) - (\tau_{20} + \theta_1^2)(E(Z_1))^2}{\theta_1^2(\tau_{20} + \theta_1^2)} \\ &= \frac{\theta_1^2\left[E(Z_1^2) - (E(Z_1))^2\right] - \tau_{20}(E(Z_1))^2}{\theta_1^2(\tau_{20} + \theta_1^2)} = \frac{\theta_1^2\left[E(Z_1^2) - (E(Z_1))^2\right] - \tau_{20}\theta_1^2\mu_1^2}{\theta_1^2(\tau_{20} + \theta_1^2)} \\ &= \frac{V(Z_1) - \tau_{20}\mu_1^2}{\tau_{20} + \theta_1^2} = \frac{\sigma_{Z_1}^2 - \tau_{20}\mu_1^2}{\tau_{20} + \theta_1^2} \end{aligned} \tag{4}$$

This proves the theorem. □

Corollary 1. *The variance of the sensitive variable Y is similarly obtained by replacing X by Y and S_1 by S_2 in Theorem 1 and is given by:*

$$V(Y) = \sigma_y^2 = \frac{\sigma_{Z_2}^2 - \tau_{02}\mu_2^2}{\tau_{02} + \theta_2^2} \tag{5}$$

3 SCRAMBLING VARIABLES ARE DEPENDENT

If S_1 and S_2 are dependent, then we have the following theorem.

Theorem 2. *The covariance between the two sensitive variables X and Y is given by:*

$$Cov(X,Y) = \frac{\sigma_{Z_1Z_2} - \tau_{11}\mu_1\mu_2}{\tau_{11} + \theta_1\theta_2} \tag{6}$$

Proof. We have

$$Z_1 = XS_1 \text{ and } Z_2 = YS_2$$

Thus

$$\begin{aligned} Cov(Z_1, Z_2) &= E(Z_1Z_2) - E(Z_1)E(Z_2) = E(XS_1YS_2) - E(XS_1)E(YS_2) \\ &= E(XYS_1S_2) - E(X)E(S_1)E(Y)E(S_2) \\ &= E(XY)E(S_1S_2) - E(X)E(Y)E(S_1)E(S_2) \\ &= E(XY)E(S_1S_2) - E(X)E(Y)E(S_1)E(S_2) \\ &= E(XY)(\tau_{11} + \theta_1\theta_2) - \mu_1\mu_2\theta_1\theta_2 \end{aligned}$$

or

$$E(XY) = \frac{Cov(Z_1, Z_2) + \mu_1\mu_2\theta_1\theta_2}{\tau_{11} + \theta_1\theta_2} = \frac{\sigma_{Z_1Z_2} + \mu_1\mu_2\theta_1\theta_2}{\tau_{11} + \theta_1\theta_2} \tag{7}$$

By the definition of covariance, we have

$$Cov(X,Y) = E(XY) - E(X)E(Y)$$

Using (7) we get

$$Cov(X,Y) = \frac{\sigma_{Z_1Z_2} + \mu_1\mu_2\theta_1\theta_2}{\tau_{11} + \theta_1\theta_2} - \mu_1\mu_2 = \frac{\sigma_{Z_1Z_2} - \mu_1\mu_2\tau_{11}}{\tau_{11} + \theta_1\theta_2} \tag{8}$$

This proves the theorem. □

Theorem 3. The correlation coefficient between two sensitive variables X and Y is then given by:

$$\rho_{xy} = \frac{(\sigma_{Z_1Z_2} - \tau_{11}\mu_1\mu_2)\sqrt{\tau_{20} + \theta_1^2}\sqrt{\tau_{02} + \theta_2^2}}{(\tau_{11} + \theta_1\theta_2)\sqrt{\sigma_{Z_1}^2 - \tau_{20}\mu_1^2}\sqrt{\sigma_{Z_2}^2 - \tau_{02}\mu_2^2}} \tag{9}$$

Proof. By definition of the usual correlation coefficient, we have

$$\rho_{xy} = \frac{Cov(X,Y)}{\sqrt{V(X)}\sqrt{V(Y)}} \tag{10}$$

Using relations (4)–(6) in (10), we have

$$\rho_{xy} = \frac{\dfrac{(\sigma_{Z_1Z_2} - \tau_{11}\mu_1\mu_2)}{(\tau_{11} + \theta_1\theta_2)}}{\sqrt{\dfrac{\sigma_{Z_1}^2 - \tau_{20}\mu_1^2}{\tau_{20} + \theta_1^2}}\sqrt{\dfrac{\sigma_{Z_2}^2 - \tau_{02}\mu_2^2}{\tau_{02} + \theta_2^2}}}$$

which on simplification gives (9). Hence the theorem. □

4 ESTIMATION OF THE CORRELATION COEFFICIENT ρ_{XY}

Suppose a sample of size n is drawn with simple random and with replacement sampling (SRSWR) from a population of size N. Let Z_{1i} and Z_{2i} denote the values of the scrambled variables Z_1 and Z_2, respectively, for the ith unit of the sample, $i = 1,2,\ldots,n$. We now define the following:

$$\bar{Z}_1 = n^{-1}\sum_{i=1}^{n} Z_{1i},\ \ \bar{Z}_2 = n^{-1}\sum_{i=1}^{n} Z_{2i},\ \ s_{Z_1}^2 = (n-1)^{-1}\sum_{i=1}^{n}(Z_{1i} - \bar{Z}_1)^2,$$

$$s_{Z_2}^2 = (n-1)^{-1}\sum_{i=1}^{n}(Z_{2i} - \bar{Z}_2)^2$$

$$s_{Z_1Z_2} = (n-1)^{-1}\sum_{i=1}^{n}(Z_{1i} - \bar{Z}_1)(Z_{2i} - \bar{Z}_2),\ \ \mu_{rs} = E[Z_1 - E(Z_1)]^r[Z_2 - E(Z_2)]^s,$$

$$Ars = \frac{\mu_{rs}}{{\mu_{20}}^{r/2}{\mu_{02}}^{s/2}}$$

$$C_{Z_1}^2 = \frac{\mu_{20}}{(\mu_1\theta_1)^2} = \frac{\sigma_{Z_1}^2}{(\mu_1\theta_1)^2},\ \ C_{Z_2}^2 = \frac{\mu_{02}}{(\mu_2\theta_2)^2} = \frac{\sigma_{Z_2}^2}{(\mu_2\theta_2)^2},\ \ \rho_{Z_1Z_2} = \frac{\mu_{11}}{\sqrt{\mu_{20}\mu_{02}}} = \frac{\sigma_{Z_1Z_2}}{\sigma_{Z_1}\sigma_{Z_2}}$$

Now we have the following theorem.

Theorem 4. *$s_{Z_1}^2$ is an unbiased estimator of $\sigma_{Z_1}^2$.*

Proof. We have

$$s_{Z_1}^2 = (n-1)^{-1}\sum_{i=1}^{n}(Z_{1i} - \bar{Z}_1)^2 = (n-1)^{-1}\left[\sum_{i=1}^{n} Z_{1i}^2 - n\bar{Z}_1^2\right]$$

Therefore,

$$E\left(s_{Z_1}^2\right) = (n-1)^{-1}\left[\sum_{i=1}^{n} E(Z_{1i}^2) - nE\left(\bar{Z}_1^2\right)\right] \tag{11}$$

Since

$$E(Z_{1i}^2) = E(X^2S_1^2) = E(X^2)E(S_1^2) = (\sigma_x^2 + \mu_1^2)(\tau_{20} + \theta_1^2) \tag{12}$$

and

$$V(\bar{Z}_1) = n^{-1}V(Z_1) = E\left[\bar{Z}_1^2\right] - [E(\bar{Z}_1)]^2$$

One gets

$$E\left(\bar{Z}_1^2\right) = V(\bar{Z}_1) + (E(\bar{Z}_1))^2 = \frac{\sigma_x^2(\tau_{20} + \theta_1^2) + \tau_{20}\mu_1^2}{n} + \mu_1^2\theta_1^2 \tag{13}$$

Using (12) and (13) in (11), we get

$$E(s_{Z_1}{}^2)$$
$$= (n-1)^{-1}\left[n\{\sigma_x^2(\tau_{20} + \theta_1^2) + \tau_{20}\mu_1^2 + \mu_1^2\theta_1^2\} - n\left\{\frac{\sigma_x^2(\tau_{20} + \theta_1^2) + \tau_{20}\mu_1^2}{n} + \mu_1^2\theta_1^2\right\}\right]$$
$$= \sigma_x^2(\tau_{20} + \theta_1^2) + \tau_{20}\mu_1^2 = \sigma_{Z_1}{}^2$$

which proves the theorem. □

Similarly, we have the following corollary.

Corollary 2. $s_{Z_2}^2 = (n-1)^{-1}\sum_{i=1}^{n}(Z_{2i} - \bar{Z}_2)^2$ *is an unbiased estimator of* $\sigma_{Z_2}^2$.

We have the following theorem.

Theorem 5. $s_{Z_1Z_2}$ *is an unbiased estimator of* $\sigma_{Z_1Z_2}$.

Proof. We have

$$s_{Z_1Z_2} = \frac{1}{(n-1)}\sum_{i=1}^{n}(Z_{1i} - \bar{Z}_1)(Z_{2i} - \bar{Z}_2) = \frac{1}{(n-1)}\left[\sum_{i=1}^{n}Z_{1i}Z_{2i} - n\bar{Z}_1\bar{Z}_2\right]$$

Thus

$$E(s_{Z_1Z_2}) = \frac{1}{(n-1)}\left[\sum_{i=1}^{n}E(Z_{1i}Z_{2i}) - nE(\bar{Z}_1\bar{Z}_2)\right] \tag{14}$$

Since

$$E(Z_{1i}Z_{2i}) = E[XS_1YS_2] = E(XY)E(S_1S_2) = (\sigma_{xy} + \mu_1\mu_2)(\tau_{11} + \theta_1\theta_2) \tag{15}$$

and

$$Cov(\bar{Z}_1, \bar{Z}_2) = \frac{Cov(Z_1, Z_2)}{n} = E(\bar{Z}_1\bar{Z}_2) - E(\bar{Z}_1)E(\bar{Z}_2)$$

We find that

$$E(\bar{Z}_1\bar{Z}_2)=\frac{Cov(Z_1,Z_2)}{n}+E(\bar{Z}_1)E(\bar{Z}_2)=\frac{\sigma_{xy}(\tau_{11}+\theta_1\theta_2)+\tau_{11}\mu_1\mu_2}{n}+\mu_1\mu_2\theta_1\theta_2 \tag{16}$$

Using (15) and (16) in (14), we get

$$E(s_{Z_1Z_2})$$
$$=\frac{1}{(n-1)}\left[n\{(\sigma_{xy}+\mu_1\mu_2)(\tau_{11}+\theta_1\theta_2)\}-n\left\{\frac{\sigma_{xy}(\tau_{11}+\theta_1\theta_2)+\tau_{11}\mu_1\mu_2}{n}+\mu_1\mu_2\theta_1\theta_2\right\}\right]$$
$$=\sigma_{xy}(\tau_{11}+\theta_1\theta_2)+\tau_{11}\mu_1\mu_2=\sigma_{Z_1Z_2}$$

which proves the theorem. □

Let

$$\varepsilon_1=\frac{\bar{Z}_1}{E(\bar{Z}_1)}-1,\ \ \varepsilon_2=\frac{\bar{Z}_2}{E(\bar{Z}_2)}-1,\ \ \varepsilon_3=\frac{s^2_{Z_1}}{\sigma^2_{Z_1}}-1,\ \ \varepsilon_4=\frac{s^2_{Z_2}}{\sigma^2_{Z_2}}-1,\ \text{ and }\ \varepsilon_5=\frac{s_{Z_1Z_2}}{\sigma_{Z_1Z_2}}-1$$

so that

$$E(\varepsilon_i)=0\ \text{ for all } i=1,2,3,4,5$$

We assume that the population size N is quite large as compared to sample size n so that the finite population correction factor may be ignored throughout. We then have

$$E(\varepsilon_1^2)=n^{-1}C^2_{Z_1},\ \ E(\varepsilon_2^2)=n^{-1}C^2_{Z_2},\ \ E(\varepsilon_1\varepsilon_2)=n^{-1}\rho_{Z_1Z_2}C_{Z_1}C_{Z_2},$$

$$E(\varepsilon_3^2)=n^{-1}(A_{40}-1),\ \ E(\varepsilon_4^2)=n^{-1}(A_{04}-1),\ \ E(\varepsilon_5^2)=n^{-1}\left(\frac{A_{22}}{\rho^2_{Z_1Z_2}}-1\right),$$

$$E(\varepsilon_1\varepsilon_3)=n^{-1}C_{Z_1}A_{30},\ \ E(\varepsilon_2\varepsilon_4)=n^{-1}C_{Z_2}A_{03},\ \ E(\varepsilon_1\varepsilon_4)=n^{-1}C_{Z_1}A_{12},$$

$$E(\varepsilon_2\varepsilon_3)=n^{-1}C_{Z_2}A_{21},\ \ E(\varepsilon_1\varepsilon_5)=n^{-1}C_{Z_1}\frac{A_{12}}{\rho_{Z_1Z_2}},\ \ E(\varepsilon_2\varepsilon_5)=n^{-1}C_{Z_2}\frac{A_{21}}{\rho_{Z_1Z_2}},$$

$$E(\varepsilon_3\varepsilon_5)=n^{-1}\left[\frac{A_{31}}{\rho_{Z_1Z_2}}-1\right],\ \ E(\varepsilon_4\varepsilon_5)=n^{-1}\left[\frac{A_{13}}{\rho_{Z_1Z_2}}-1\right],\ \text{ and}$$

$$E(\varepsilon_3\varepsilon_4)=n^{-1}(A_{22}-1).$$

These expected values may easily be obtained by following Sukhatme et al. (1984) or Srivastava and Jhajj (1981). For our purpose we need certain new results which we obtain in Lemmas 1–3.

Lemma 1. *The moments of order four or less of the joint distribution of* (X,Y) *are given by:*

$$E(X^2)=m_{20}+\mu_1^2=\alpha_{20};\ E(Y^2)=m_{02}+\mu_2^2=\alpha_{02};$$
$$E(X^3)=m_{30}+3\mu_1 m_{20}+\mu_1^3=\alpha_{30}$$
$$E(Y^3)=m_{03}+3\mu_2 m_{02}+\mu_2^3=\alpha_{03};\ E(X^4)=m_{40}+4\mu_1 m_{30}+6m_{20}\mu_1^2+\mu_1^4=\alpha_{40}$$
$$E(Y^4)=m_{04}+4\mu_2 m_{03}+6m_{02}\mu_2^2+\mu_2^4=\alpha_{04};\ E(XY)=m_{11}+\mu_1\mu_2=\alpha_{11}$$
$$E(XY^2)=m_{12}+\mu_1 m_{02}+\mu_1\mu_2^2+2\mu_2 m_{11}=\alpha_{12};$$
$$E(X^2Y)=m_{21}+\mu_2 m_{20}+\mu_2\mu_1^2+2\mu_1 m_{11}=\alpha_{21}$$
$$E(XY^3)=m_{13}+\mu_1 m_{03}+\mu_1\mu_2^3+3\mu_2 m_{12}+3\mu_1\mu_2 m_{02}+3\mu_2^2 m_{11}=\alpha_{13}$$
$$E(X^3Y)=m_{31}+\mu_2 m_{30}+\mu_2\mu_1^3+3\mu_1 m_{21}+3\mu_1\mu_2 m_{20}+3\mu_1^2 m_{11}=\alpha_{31}$$
$$E(X^2Y^2)=m_{22}+\mu_1^2 m_{02}+2\mu_1 m_{12}+\mu_2^2 m_{20}+\mu_1^2\mu_2^2+2\mu_2 m_{21}+4\mu_1\mu_2 m_{11}=\alpha_{22}$$

where we have defined, $\alpha_{rs}=E(X^rY^s)$, *r and s being nonnegative integers with* $(r+s)\le 4$.

Similarly one can obtain various moments for the joint distribution of (S_1,S_2). Fourth or lesser order moments are given in the following lemma.
Lemma 2. *The expression for* $E(S_1^rS_2^s)$ *with* $(r+s)\le 4$ *is given by:*

$$E(S_1^2)=\tau_{20}+\theta_1^2=e_{20},\ E(S_2^2)=\tau_{02}+\theta_2^2=e_{02},$$
$$E(S_1^3)=\tau_{30}+3\theta_1+\tau_{20}+\theta_1^3=e_{30}$$
$$E(S_2^3)=\tau_{03}+3\theta_2\tau_{02}+\theta_2^3=e_{03},\ E(S_1^4)=\tau_{40}+4\theta_1\tau_{30}+6\tau_{20}\theta_1^2+\theta_1^4=e_{40}$$
$$E(S_2^4)=\tau_{04}+4\theta_2\tau_{03}+6\tau_{02}\theta_2^2+\theta_2^4=e_{04},\ E(S_1S_2)=\tau_{11}+\theta_1\theta_2=e_{11}$$
$$E(S_1S_2^2)=\tau_{12}+\theta_1\tau_{02}+\theta_1\theta_2^2+2\theta_2\tau_{11}=e_{12},$$
$$E(S_1^2S_2)=\tau_{21}+\theta_2\tau_{20}+\theta_2\theta_1^2+2\theta_1\tau_{11}=e_{21}$$
$$E(S_1S_2^3)=\tau_{13}+\theta_1\tau_{03}+\theta_1\theta_2^3+3\theta_2\tau_{12}+3\theta_1\theta_2\tau_{02}+3\theta_2^2\tau_{11}=e_{13}$$
$$E(S_1^3S_2)=\tau_{31}+\theta_2\tau_{30}+\theta_2\theta_1^3+3\theta_1\tau_{21}+3\theta_1\theta_2\tau_{20}+3\theta_1^2\tau_{11}=e_{31},\ \text{and}$$
$$E(S_1^2S_2^2)=\tau_{22}+\theta_1^2\tau_{02}+2\theta_1\tau_{12}+\theta_2^2\tau_{20}+\theta_1^2\theta_2^2+2\theta_2\tau_{21}+4\theta_1\theta_2\tau_{11}=e_{22}$$

where $e_{rs}=E(S_1^rS_2^s)$. *Using the results obtained in Lemmas 1 and 2, one can easily get the expression for the fourth or lesser order central moments for the joint distribution of* (Z_1,Z_2) *as in Lemma 3.*

Lemma 3. *Fourth or lesser order central moments for the joint distribution of* (Z_1, Z_2) *are given by:*

$$\mu_{20} = \alpha_{20}e_{20} - \mu_1^2\theta_1^2,\quad \mu_{02} = \alpha_{02}e_{02} - \mu_2^2\theta_2^2,\quad \mu_{30} = \alpha_{30}e_{30} - 3\mu_1\theta_1\alpha_{20}e_{20} + 2\mu_1^3\theta_1^3$$

$$\mu_{03} = \alpha_{03}e_{03} - 3\mu_2\theta_2\alpha_{02}e_{02} + 2\mu_2^3\theta_2^3,$$
$$\mu_{40} = \alpha_{40}e_{40} - 4\mu_1\theta_1\alpha_{30}e_{30} + 6\mu_1^2\theta_1^2\alpha_{20}e_{20} - 3\mu_1^4\theta_1^4$$

$$\mu_{04} = \alpha_{04}e_{04} - 4\mu_2\theta_2\alpha_{03}e_{03} + 6\mu_2^2\theta_2^2\alpha_{02}e_{02} - 3\mu_2^4\theta_2^4,\quad \mu_{11} = \alpha_{11}e_{11} - \mu_1\theta_1\mu_2\theta_2$$

$$\mu_{12} = \alpha_{12}e_{12} - \mu_1\theta_1\alpha_{02}e_{02} - 2\mu_2\theta_2\alpha_{11}e_{11} + 2\mu_1\theta_1\mu_2^2\theta_2^2$$

$$\mu_{21} = \alpha_{21}e_{21} - \mu_2\theta_2\alpha_{20}e_{20} - 2\mu_1\theta_1\alpha_{11}e_{11} + 2\mu_2\theta_2\mu_1^2\theta_1^2$$

$$\mu_{13} = \alpha_{13}e_{13} - \mu_1\theta_1\alpha_{03}e_{03} - 3\mu_2\theta_2\alpha_{12}e_{12} + 3e_{02}\alpha_{02}\mu_1\theta_1\mu_2\theta_2 + 3\alpha_{11}e_{11}\mu_2^2\theta_2^2 - 3\mu_1\theta_1\mu_2^3\theta_2^3$$

$$\mu_{31} = \alpha_{31}e_{31} - \mu_2\theta_2\alpha_{30}e_{30} - 3\mu_1\theta_1\alpha_{21}e_{21} + 3e_{20}\alpha_{20}\mu_1\theta_1\mu_2\theta_2 + 3\alpha_{11}e_{11}\mu_1^2\theta_1^2 - 3\mu_2\theta_2\mu_1^3\theta_1^3, \text{ and}$$

$$\mu_{22} = \alpha_{22}e_{22} + \mu_1^2\theta_1^2\alpha_{02}e_{02} - 2\mu_1\theta_1\alpha_{12}e_{12} + \mu_2^2\theta_2^2e_{20}\alpha_{20} + 4e_{11}\alpha_{11}\mu_1\theta_1\mu_2\theta_2 - 2\alpha_{21}e_{21}\mu_2\theta_2 - 3\mu_1^2\theta_1^2\mu_2^2\theta_2^2$$

For obtaining the estimator of the correlation coefficient between the sensitive variables X and Y, we require the estimators for the variance and covariance terms for the variables X and Y. This we do in Theorems 6 and 7.
Theorem 6. *An unbiased estimator of $V(X)$ is given by:*

$$\hat{V}(X) = \frac{s_{Z_1}^2\left(1 + \dfrac{\tau_{20}}{n\theta_1^2}\right) - \dfrac{\tau_{20}\bar{Z}_1^2}{\theta_1^2}}{\tau_{20} + \theta_1^2} \tag{17}$$

Proof. Relation (17) in terms of ε_1 and ε_3 may be written as

$$\hat{V}(X) = \frac{\sigma_{Z_1}^2\left(1 + \dfrac{\tau_{20}}{n\theta_1^2}\right)(1+\varepsilon_3) - \tau_{20}\left(\dfrac{(\mu_1\theta_1)^2}{\theta_1^2}\right)(1+\varepsilon_1)^2}{\left(\tau_{20} + \theta_1^2\right)}$$
$$= \frac{\sigma_{Z_1}^2\left(1 + \dfrac{\tau_{20}}{n\theta_1^2}\right)(1+\varepsilon_3) - \tau_{20}\mu_1^2\left(1 + \varepsilon_1^2 + 2\varepsilon_1\right)}{\left(\tau_{20} + \theta_1^2\right)} \tag{18}$$

Taking expected value on both sides of (18), we get

$$E\left[\hat{V}(X)\right] = \frac{\sigma_{Z_1}^2\left(1+\dfrac{\tau_{20}}{n\theta_1^2}\right) - \tau_{20}\mu_1^2\left(1+n^{-1}C_{Z_1}^2\right)}{\tau_{20}+\theta_1^2}$$

$$= \frac{\sigma_{Z_1}^2\left(1+\dfrac{\tau_{20}}{n\theta_1^2}\right) - \tau_{20}\mu_1^2\left(1+\dfrac{\sigma_{Z_1}^2}{n\mu_1^2\theta_1^2}\right)}{\tau_{20}+\theta_1^2} = \frac{\sigma_{Z_1}^2 - \tau_{20}\mu_1^2}{\tau_{20}+\theta_1^2}$$

$$= \frac{\sigma_x^2\left(\tau_{20}+\theta_1^2\right)+\tau_{20}\mu_1^2-\tau_{20}\mu_1^2}{\tau_{20}+\theta_1^2} = \sigma_x^2 = V(X)$$

This completes the proof of the theorem. □

Corollary 3. *An unbiased estimator of variance V(Y) is similarly given by:*

$$\hat{V}(Y) = \frac{s_{Z_2}^2\left(1+\dfrac{\tau_{02}}{n\theta_2^2}\right) - \dfrac{\tau_{02}\bar{Z}_2^2}{\theta_2^2}}{\tau_{02}+\theta_2^2} \tag{19}$$

Theorem 7. *An unbiased estimator of the covariance Cov(X,Y) is given by:*

$$C\hat{o}v(X,Y) = \frac{s_{Z_1Z_2}\left(1+\dfrac{\tau_{11}}{n\theta_1\theta_2}\right) - \dfrac{\tau_{11}\bar{Z}_1\bar{Z}_2}{\theta_1\theta_2}}{\tau_{11}+\theta_1\theta_2} \tag{20}$$

Proof. Relation (20) in terms of $\varepsilon_1, \varepsilon_2$, and ε_5 can be written as

$$\begin{aligned} C\hat{o}v(X,Y) &= \frac{\sigma_{Z_1Z_2}\left(1+\dfrac{\tau_{11}}{n\theta_1\theta_2}\right)(1+\varepsilon_5) - \dfrac{\tau_{11}\mu_1\theta_1\mu_2\theta_2}{\theta_1\theta_2}(1+\varepsilon_1)(1+\varepsilon_2)}{\tau_{11}+\theta_1\theta_2} \\ &= \frac{\sigma_{Z_1Z_2}\left(1+\dfrac{\tau_{11}}{n\theta_1\theta_2}\right)(1+\varepsilon_5) - \tau_{11}\mu_1\mu_2(1+\varepsilon_1+\varepsilon_2+\varepsilon_1\varepsilon_2)}{\tau_{11}+\theta_1\theta_2} \end{aligned} \tag{21}$$

Taking expected value on both sides of (21), we get

$$E[C\hat{o}v(X,Y)] = \frac{\sigma_{Z_1Z_2}\left(1+\dfrac{\tau_{11}}{n\theta_1\theta_2}\right) - \tau_{11}\mu_1\mu_2\left(1+n^{-1}\rho_{Z_1Z_2}C_{Z_1}C_{Z_2}\right)}{\tau_{11}+\theta_1\theta_2}$$

$$= \frac{\sigma_{Z_1Z_2}\left(1+\dfrac{\tau_{11}}{n\theta_1\theta_2}\right) - \tau_{11}\mu_1\mu_2\left(1+\dfrac{\sigma_{Z_1Z_2}}{n\theta_1\theta_2\mu_1\mu_2}\right)}{\tau_{11}+\theta_1\theta_2} = \frac{\sigma_{Z_1Z_2} - \tau_{11}\mu_1\mu_2}{\tau_{11}+\theta_1\theta_2}$$

$$= \frac{\sigma_{xy}(\tau_{11}+\theta_1\theta_2)+\tau_{11}\mu_1\mu_2-\tau_{11}\mu_1\mu_2}{\tau_{11}+\theta_1\theta_2} = \sigma_{xy}$$

Hence the theorem. □

We now consider the problem of estimating the correlation coefficient between two sensitive variables information on which has been obtained by using two dependent scrambling devices. The usual estimator of the correlation coefficient ρ_{xy} is defined as:

$$r_{xy}=\frac{C\hat{o}v(X,Y)}{\sqrt{\hat{V}(X)}\sqrt{\hat{V}(Y)}} \tag{22}$$

Using relations (17), (19), and (20) in (22), we get an estimator of the correlation coefficient ρ_{xy} with scrambled responses as:

$$r_{xy}=\frac{\left[s_{Z_1Z_2}\left(1+\frac{\tau_{11}}{n\theta_1\theta_2}\right)-\frac{\tau_{11}\bar{Z}_1\bar{Z}_2}{\theta_1\theta_2}\right]\sqrt{\tau_{20}+\theta_1^2}\sqrt{\tau_{02}+\theta_2^2}}{(\tau_{11}+\theta_1\theta_2)\sqrt{s_{Z_1}^2\left(1+\frac{\tau_{20}}{n\theta_1^2}\right)-\frac{\tau_{20}\bar{Z}_1^2}{\theta_1^2}}\sqrt{s_{Z_2}^2\left(1+\frac{\tau_{02}}{n\theta_2^2}\right)-\frac{\tau_{02}\bar{Z}_2^2}{\theta_2^2}}} \tag{23}$$

5 BIAS AND MEAN SQUARED ERROR OF r_{xy}

To find the bias and mean squared error expressions of r_{xy}, we shall use certain results obtained in Lemmas 4–9. For this let us define

$$\delta_0=\frac{s_{Z_1}^2\left(1+\frac{\tau_{20}}{n\theta_1^2}\right)-\frac{\tau_{20}\bar{Z}_1^2}{\theta_1^2}}{V(X)\left(\tau_{20}+\theta_1^2\right)}-1 \tag{24}$$

$$\delta_1=\frac{s_{Z_2}^2\left(1+\frac{\tau_{02}}{n\theta_2^2}\right)-\frac{\tau_{02}\bar{Z}_2^2}{\theta_2^2}}{V(Y)\left(\tau_{02}+\theta_2^2\right)}-1 \tag{25}$$

and

$$\delta_2=\frac{s_{Z_1Z_2}\left(1+\frac{\tau_{11}}{n\theta_1\theta_2}\right)-\frac{\tau_{11}\bar{Z}_1\bar{Z}_2}{\theta_1\theta_2}}{Cov(X,Y)(\tau_{11}+\theta_1\theta_2)}-1 \tag{26}$$

so that

$$E(\delta_i)=0 \quad \text{for} \, i=0,1,2.$$

We then have the following lemmas. Proofs for these lemmas are straight forward and are omitted.

Lemma 4. *Expected value of δ_0^2 is given by*

$$E\left(\delta_0^2\right)=n^{-1}A_1 \tag{27}$$

where

$$A_1 = \frac{\left[\sigma_{Z_1}^2 (A_{40}-1) + \frac{\tau_{20}^2 \sigma_{Z_1}^4}{n\theta_1^4}\left\{1+n^{-1}(A_{40}-1)\right\} + 4\tau_{20}^2 \mu_1^4 C_{Z_1}^2 - 4\tau_{20}\mu_1^2 \sigma_{Z_1}^2 C_{Z_1} A_{30} - \frac{2\tau_{20}^2 \mu_1^2 \sigma_{Z_1}^2}{n\theta_1^2} C_{Z_1}^2\right]}{\sigma_x^4 \left(\tau_{20}+\theta_1^2\right)^2}$$

Lemma 5. *Expected value of* δ_1^2 *is given by*

$$E\left(\delta_1^2\right) = n^{-1} B_1 \tag{28}$$

where

$$B_1 = \frac{\left[\sigma_{Z_2}^2 (A_{04}-1) + \frac{\tau_{02}^2 \sigma_{Z_2}^4}{n\theta_2^4}\left\{1+n^{-1}(A_{04}-1)\right\} + 4\tau_{02}^2 \mu_2^4 C_{Z_2}^2 - 4\tau_{02}\mu_2^2 \sigma_{Z_2}^2 C_{Z_2} A_{03} - \frac{2\tau_{02}^2 \mu_2^2 \sigma_{Z_2}^2}{n\theta_2^2} C_{Z_2}^2\right]}{\sigma_y^4 \left(\tau_{02}+\theta_2^2\right)^2}$$

Lemma 6. *Expected value of* δ_2^2 *is similarly obtained as*

$$E\left(\delta_2^2\right) = n^{-1} C_1 \tag{29}$$

where

$$C_1 = \frac{\begin{aligned}&\left[\sigma_{Z_1Z_2}^2 \left(1+\frac{\tau_{11}}{n\theta_1\theta_2}\right)^2 \left(\frac{A_{22}}{\rho_{Z_1Z_2}^2} - 1\right) + \tau_{11}^2 \mu_1^2 \mu_2^2 \left(C_{Z_1}^2 + C_{Z_2}^2 + \rho_{Z_1Z_2} C_{Z_1} C_{Z_2}\right) + \frac{\tau_{11}^2 \sigma_{Z_1Z_2}^2}{n\theta_1^2\theta_2^2}\right. \\ &\left. -2\tau_{11}\mu_1\mu_2\sigma_{Z_1Z_2}\left(1+\frac{\tau_{11}}{n\theta_1\theta_2}\right)\left(C_{Z_1}\frac{A_{12}}{\rho_{Z_1Z_2}} + C_{Z_2}\frac{A_{21}}{\rho_{Z_1Z_2}}\right) - \frac{2\tau_{11}^2 \sigma_{Z_1Z_2}\mu_1\mu_2}{n\theta_1\theta_2}\rho_{Z_1Z_2} C_{Z_1} C_{Z_2}\right]\end{aligned}}{\sigma_{xy}^2 (\tau_{11}+\theta_1\theta_2)^2}$$

Lemma 7. *The expected value of the product* $\delta_0\delta_2$ *is given by*

$$E(\delta_0\delta_2) = n^{-1} D_1 \tag{30}$$

where

$$D_1 = \frac{\begin{aligned}&\left[\frac{\sigma_{Z_1}^2 \tau_{20} \sigma_{Z_1Z_2} \tau_{11}}{n\theta_1^3 \theta_2} - \frac{\sigma_{Z_1Z_2}\tau_{11}\tau_{20}\mu_1^2}{n\theta_1\theta_2} C_{Z_1}^2 + \sigma_{Z_1Z_2}\sigma_{Z_1}^2 \left(1+\frac{\tau_{20}}{n\theta_1^2}\right)\left(1+\frac{\tau_{11}}{n\theta_1\theta_2}\right)\left(\frac{A_{31}}{\rho_{Z_1Z_2}} - 1\right)\right. \\ &-2\sigma_{Z_1Z_2}\tau_{20}\mu_1^2 \left(1+\frac{\tau_{12}}{n\theta_1\theta_2}\right) C_{Z_1}\frac{A_{12}}{\rho_{Z_1Z_2}} - \frac{\sigma_{Z_1}^2 \tau_{20}\mu_1\mu_2\tau_{11}}{n\theta_1^2}\rho_{Z_1Z_2} C_{Z_1} C_{Z_2} - \sigma_{Z_1}^2 \tau_{11}\mu_1\mu_2\left(1+\frac{\tau_{20}}{n\theta_1^2}\right) \\ &\left. \times\left\{C_{Z_1}A_{30} + C_{Z_2}A_{21}\right\} + 2\tau_{20}\mu_1^2\tau_{11}\mu_1\mu_2\left\{C_{Z_1}^2 + \rho_{Z_1Z_2} C_{Z_1} C_{Z_2}\right\}\right]\end{aligned}}{\left\{\sigma_x^2 \sigma_{xy}\left(\tau_{20}+\theta_1^2\right)(\tau_{11}+\theta_1\theta_2)\right\}}$$

Lemma 8. *The expected value of the product* $\delta_1\delta_2$ *is given by*

$$E(\delta_1\delta_2) = n^{-1}F_1 \tag{31}$$

where

$$F_1 = \frac{\begin{array}{l}\left[\dfrac{\sigma_{Z_2}^2\tau_{02}\sigma_{Z_1Z_2}\tau_{11}}{n\theta_2^3\theta_1} - \dfrac{\sigma_{Z_1Z_2}\tau_{11}\tau_{02}\mu_2^2}{n\theta_1\theta_2}C_{z_2}^2 + \sigma_{Z_1Z_2}\sigma_{Z_2}^2\left(1+\dfrac{\tau_{02}}{n\theta_2^2}\right)\left(1+\dfrac{\tau_{11}}{n\theta_1\theta_2}\right)\left(\dfrac{A_{13}}{\rho_{Z_1Z_2}}-1\right)\right.\\ -2\sigma_{z_1z_2}\tau_{02}\mu_2^2\left(1+\dfrac{\tau_{11}}{n\theta_1\theta_2}\right)C_{Z_2}\dfrac{A_{21}}{\rho_{Z_1Z_2}} - \dfrac{\sigma_{Z_2}^2\tau_{02}\mu_1\mu_2\tau_{11}}{n\theta_2^2}\rho_{Z_1Z_2}C_{Z_1}C_{Z_2} - \sigma_{Z_2}^2\tau_{11}\mu_1\mu_2\left(1+\dfrac{\tau_{02}}{n\theta_2^2}\right)\\ \left.\times\left\{C_{Z_2}A_{03}+C_{Z_1}A_{12}\right\} + 2\tau_{02}\mu_2^2\tau_{11}\mu_1\mu_2\left\{C_{Z_2}^2+\rho_{z_1z_2}C_{Z_1}C_{Z_2}\right\}\right]\end{array}}{\left\{\sigma_y^2\sigma_{xy}\left(\tau_{02}+\theta_2^2\right)(\tau_{11}+\theta_1\theta_2)\right\}}$$

Lemma 9. *We have*

$$E(\delta_1\delta_0) = n^{-1}G_1 \tag{32}$$

where

$$G_1 = \frac{\begin{array}{l}\left[\sigma_{Z_1}^2\sigma_{Z_2}^2(A_{22}-1) + \dfrac{\sigma_{Z_1}^2\sigma_{Z_2}^2\tau_{20}}{n\theta_1^2}(A_{22}-1) - 2\tau_{20}\mu_1^2\sigma_{Z_2}^2C_{Z_1}A_{12} + \dfrac{\sigma_{Z_1}^2\sigma_{Z_2}^2\tau_{02}}{n\theta_2^2}(A_{22}-1)\right.\\ +\dfrac{\sigma_{Z_1}^2\sigma_{Z_2}^2\tau_{20}\tau_{02}}{n\theta_1^2\theta_2^2}\left\{1+n^{-1}(A_{22}-1)\right\} - \dfrac{\sigma_{Z_2}^2\tau_{02}\tau_{20}\mu_1^2}{n\theta_2^2}\left\{C_{Z_1}^2+2C_{Z_1}A_{12}\right\} - 2\sigma_{Z_1}^2\tau_{02}\mu_2^2C_{Z_2}A_{21}\\ \left.-\dfrac{\sigma_{Z_1}^2\tau_{20}\tau_{02}\mu_2^2}{n\theta_1^2}\left\{C_{Z_2}^2+2C_{Z_2}A_{21}\right\} + 4\tau_{20}\tau_{02}\mu_1^2\mu_2^2\rho_{Z_1Z_2}C_{Z_1}C_{Z_2}\right]\end{array}}{\left\{\sigma_x^2\sigma_y^2\left(\tau_{20}+\theta_1^2\right)\left(\tau_{02}+\theta_2^2\right)\right\}}$$

Theorem 8. *The bias of the estimator* r_{xy} *defined at (22) of* ρ_{xy} *is seen to be approximately*

$$B(r_{XY}) = n^{-1}\rho_{XY}\left[\frac{3}{8}(A_1+B_1) - \frac{1}{2}(D_1+F_1) + \frac{1}{4}G_1\right] \tag{33}$$

where A_1, B_1, D_1, F_1, *and* G_1 *have been defined in Lemmas 4–9.*
Proof. We have

$$r_{xy} = \frac{\left[s_{Z_1Z_2}\left(1+\dfrac{\tau_{11}}{n\theta_1\theta_2}\right) - \dfrac{\tau_{11}\bar{Z}_1\bar{Z}_2}{\theta_1\theta_2}\right]\sqrt{\tau_{20}+\theta_1^2}\sqrt{\tau_{02}+\theta_2^2}}{(\tau_{11}+\theta_1\theta_2)\sqrt{s_{Z_1}^2\left(1+\dfrac{\tau_{20}}{n\theta_1^2}\right) - \dfrac{\tau_{20}\bar{Z}_1^2}{\theta_1^2}}\sqrt{s_{Z_2}^2\left(1+\dfrac{\tau_{02}}{n\theta_2^2}\right) - \dfrac{\tau_{02}\bar{Z}_2^2}{\theta_2^2}}} \tag{34}$$

Relation (34) in terms of δ_0, δ_1, and δ_2 may be written as

$$r_{xy} = \frac{\sigma_{xy}(1+\delta_2)}{\sqrt{\sigma_x^2(1+\delta_0)}\sqrt{\sigma_y^2(1+\delta_1)}} = \rho_{xy}(1+\delta_2)(1+\delta_0)^{-1/2}(1+\delta_1)^{-1/2} \quad (35)$$

Assuming that $|\delta_0| < 1$ and $|\delta_1| < 1$ and using the binomial theorem to expand the right hand side of (35), we get

$$r_{xy} = \rho_{xy}(1+\delta_2)\left[1 - \frac{1}{2}\delta_0 + \frac{3}{8}\delta_0^2 + -\right]\left[1 - \frac{1}{2}\delta_1 + \frac{3}{8}\delta_1^2 + -\right]$$

$$= \rho_{xy}\left[1 + \delta_2 - \frac{1}{2}\delta_0 - \frac{1}{2}\delta_1 + \frac{3}{8}\delta_0^2 + \frac{3}{8}\delta_1^2 - \frac{1}{2}\delta_0\delta_2 - \frac{1}{2}\delta_1\delta_2 + \frac{1}{4}\delta_0\delta_1 + O(\delta^2)\right]$$

Taking expected value on both sides, we get

$$E(r_{xy}) = \rho_{xy}\left[1 + n^{-1}\left\{\frac{3}{8}(A_1 + B_1) - \frac{1}{2}(D_1 + F_1) + \frac{1}{4}G_1\right\}\right]$$

Therefore, the expression for the bias in r_{XY} is obtained as

$$B(r_{xy}) = E(r_{xy}) - \rho_{xy} = n^{-1}\rho_{xy}\left[\frac{3}{8}(A_1 + B_1) - \frac{1}{2}(D_1 + F_1) + \frac{1}{4}G_1\right] \quad (36)$$

Relation (36) shows that the bias in the estimator r_{XY} of ρ_{XY} is of order $O(n^{-1})$. It will, therefore, be reasonably small for large sample sizes. We now find the expression for the mean square error of r_{XY} in Theorem 9. □

Theorem 9. *The mean square error of the estimator r_{XY} up to terms of order $O(n^{-1})$ is given by*

$$\text{MSE}(r_{xy}) = n^{-1}\rho_{xy}^2\left[C_1 + \frac{1}{4}(A_1 + B_1) + \frac{1}{2}G_1 - D_1 - F_1\right] \quad (37)$$

where terms A_1, B_1, C_1, D_1, F_1, and G_1 have been defined in Lemmas 4–9.

Proof. We have

$$\text{MSE}(r_{xy}) = E\left(r_{xy} - \rho_{xy}\right)^2 \approx E\left[\rho_{xy}\{1 + \delta_2 - \tfrac{1}{2}\delta_0 - \tfrac{1}{2}\delta_1 + O(\delta^2)\} - \rho_{xy}\right]^2$$

$$\approx \rho_{xy}^2 E\left[\delta_2^2 + \frac{1}{4}\delta_0^2 + \frac{1}{4}\delta_1^2 - \delta_0\delta_2 - \delta_1\delta_2 + \frac{1}{2}\delta_0\delta_1\right]$$

$$= n^{-1}\rho_{xy}^2\left[C_1 - D_1 - F_1 + \frac{1}{4}(A_1 + B_1) + \frac{1}{2}G_1\right]$$

which proves the theorem. □

6 SCRAMBLING VARIABLES ARE INDEPENDENT

We now consider the case when S_1 and S_2 are independent random variables. They may be two numbers drawn from two different decks of cards. The numbers for which follow known two distributions. For this case we first have the Theorem 10.

Theorem 10. *The covariance between the two sensitive variables X and Y is given by*

$$Cov(X,Y)=\frac{\sigma_{Z_1Z_2}}{\theta_1\theta_2} \tag{38}$$

Proof. We have

$$\begin{aligned}Cov(Z_1,Z_2) &= E(Z_1Z_2)-E(Z_1)E(Z_2)=E(XS_1YS_2)-E(XS_1)E(YS_2)\\ &= E(XY)E(S_1S_2)-E(XS_1)E(YS_2)\\ &= E(XY)E(S_1)E(S_2)-E(X)E(S_1)E(Y)E(S_2)\\ &= E(XY)\theta_1\theta_2-\mu_1\mu_2\theta_1\theta_2\end{aligned}$$

or,

$$E(XY)=\frac{Cov(Z_1,Z_2)+\mu_1\mu_2\theta_1\theta_2}{\theta_1\theta_2}=\frac{\sigma_{Z_1Z_2}+\mu_1\mu_2\theta_1\theta_2}{\theta_1\theta_2}$$

By definition, we have

$$Cov(X,Y)=E(XY)-E(X)E(Y)=\frac{\sigma_{Z_1Z_2}+\mu_1\mu_2\theta_1\theta_2}{\theta_1\theta_2}-\mu_1\mu_2=\frac{\sigma_{Z_1Z_2}}{\theta_1\theta_2}$$

which proves the theorem. □

From relations (1), (5), and (38), the correlation coefficient between the two sensitive variables X and Y is given by

$$\rho_{xy}=\frac{\sigma_{Z_1Z_2}\sqrt{\tau_{20}+\theta_1^2}\sqrt{\tau_{02}+\theta_2^2}}{\theta_1\theta_2\sqrt{\sigma_{Z_1}^2-\tau_{20}\mu_1^2}\sqrt{\sigma_{Z_2}^2-\tau_{02}\mu_2^2}} \tag{39}$$

For developing an estimator for ρ_{xy} we shall need the estimator of $Cov(X,Y)$, since estimators for $V(X)$ and $V(Y)$ are already available. For this we have the following theorem.

Theorem 11. *An unbiased estimator of $Cov(X,Y)$ is given by*

$$\hat{Cov}(X,Y)=\frac{s_{Z_1Z_2}}{\theta_1\theta_2} \tag{40}$$

Proof. Relation (40) in terms of ε_5 may be written as

$$\hat{Cov}(X,Y)=\frac{\sigma_{Z_1Z_2}(1+\varepsilon_5)}{\theta_1\theta_2} \tag{41}$$

Taking expected value on both sides of (41), we get

$$E[C\hat{o}v(X,Y)] = \frac{\sigma_{Z_1Z_2}}{\theta_1\theta_2} = \frac{\theta_1\theta_2 Cov(X,Y)}{\theta_1\theta_2} = Cov(X,Y) = \sigma_{xy}$$

This completes the proof. □

An estimator $\hat{r}_1$ of ρ_{xy} is now straight forward and is given by

$$\hat{r}_1 = \frac{s_{Z_1Z_2}\sqrt{\tau_{20}+\theta_1^2}\sqrt{\tau_{02}+\theta_2^2}}{\theta_1\theta_2\sqrt{s_{Z_1}^2\left(1+\frac{\tau_{20}}{n\theta_1^2}\right)-\frac{\tau_{20}\bar{Z}_1^2}{\theta_1^2}}\sqrt{s_{Z_2}^2\left(1+\frac{\tau_{02}}{n\theta_2^2}\right)-\frac{\tau_{02}\bar{Z}_2^2}{\theta_2^2}}} \tag{42}$$

7 BIAS AND MEAN SQUARE ERROR OF $\hat{r}_1$

To find the bias and mean squared error expressions of $\hat{r}_1$, we shall use certain results which are obtained in Lemmas 10–12. The proofs of the lemmas are straightforward and are therefore omitted. Here since S_1 and S_2 are independent, we have

$$E\left[S_1^r S_2^s\right] = E\left(S_1^r\right)E\left(S_2^s\right)$$

We thus have

Lemma 10. *For independent S_1 and S_2, we have*

$$E\left(S_1S_2^2\right) = \theta_1 e_{02},\ E\left(S_1^2S_2\right) = \theta_2 e_{20},\ E\left(S_1S_2^3\right) = \theta_1 e_{03},$$
$$E\left(S_1^3S_2\right) = \theta_2 e_{30},\ \text{and}\ E\left(S_1^2S_2^2\right) = e_{20}e_{02}$$

The values of μ_{rs} so obtained will be used to find the bias and mean squared error of the proposed estimator $\hat{r}_1$ of ρ_{xy}.

Lemma 11. *The expected value of the product $\delta_0\varepsilon_5$ is given by*

$$E(\delta_0\varepsilon_5) = n^{-1}H_1 \tag{43}$$

where

$$H_1 = \frac{\left[\sigma_{Z_1}^2\left(1+\frac{\tau_{20}}{n\theta_1^2}\right)\left(\frac{A_{31}}{\rho_{Z_1Z_2}}-1\right) - 2\tau_{20}\mu_1^2 C_{Z_1}\frac{A_{12}}{\rho_{Z_1Z_2}}\right]}{\sigma_x^2\left(\tau_{20}+\theta_1^2\right)}$$

Lemma 12. *The expected value of the product $\delta_1\varepsilon_5$ is given by*

$$E(\delta_1\varepsilon_5) = n^{-1}I_1 \tag{44}$$

where

$$I_1 = \frac{\left[\sigma_{Z_2}^2\left(1+\frac{\tau_{02}}{n\theta_2^2}\right)\left(\frac{A_{13}}{\rho_{Z_1Z_2}}-1\right)-2\tau_{02}\mu_2^2C_{Z_2}\frac{A_{21}}{\rho_{Z_1Z_2}}\right]}{\sigma_y^2\left(\tau_{02}+\theta_2^2\right)}$$

In Theorems 12 and 13, we obtain expressions for the bias and mean squared error of the estimator $\hat{r}_1$.

Theorem 12. *The bias of the estimator $\hat{r}_1$ defined in (42) is approximately*

$$B(\hat{r}_1) = n^{-1}\rho_{xy}\left[\frac{3}{8}(A_1+B_1)-\frac{1}{2}(H_1+I_1)+\frac{1}{4}G_1\right] \tag{45}$$

Proof. Relation (42) in terms of ε_5, δ_0, and δ_1 may be written as

$$\hat{r}_1 = \frac{\sigma_{xy}(1+\varepsilon_5)}{\sqrt{\sigma_x^2(1+\delta_0)}\sqrt{\sigma_y^2(1+\delta_1)}} = \rho_{xy}(1+\varepsilon_5)(1+\delta_0)^{-1/2}(1+\delta_1)^{-1/2} \tag{46}$$

Assuming that $|\delta_0|<1$ and $|\delta_1|<1$, and on using the binomial theorem to expand the right hand side of (46), we get

$$\begin{aligned}\hat{r}_1 &\approx \rho_{xy}(1+\varepsilon_5)\left(1-\frac{1}{2}\delta_0+\frac{3}{8}\delta_0^2+\cdots\right)\left(1-\frac{1}{2}\delta_1+\frac{3}{8}\delta_1^2+\cdots\right)\\ &= \rho_{xy}\left[1+\varepsilon_5-\frac{1}{2}\delta_0-\frac{1}{2}\delta_1+\frac{3}{8}\delta_0^2+\frac{3}{8}\delta_1^2-\frac{1}{2}\delta_0\varepsilon_5-\frac{1}{2}\delta_1\varepsilon_5+\frac{1}{4}\delta_0\delta_1+\cdots\right]\end{aligned}$$

Taking expected value on both sides, one finds

$$E(\hat{r}_1) = \rho_{xy}\left[1+n^{-1}\left\{\frac{3}{8}(A_1+B_1)-\frac{1}{2}(H_1+I_1)+\frac{1}{4}G_1\right\}\right]$$

Therefore, the expression for bias in $\hat{r}_1$ is obtained as

$$B(\hat{r}_1) = E(\hat{r}_1)-\rho_{xy} = n^{-1}\rho_{xy}\left[\frac{3}{8}(A_1+B_1)-\frac{1}{2}(H_1+I_1)+\frac{1}{4}G_1\right]$$

which proves the theorem. □

Theorem 13. *The mean squared error of the estimator $\hat{r}_1$ (up to terms of order n^{-1}) is given by*

$$\mathrm{MSE}(\hat{r}_1) = n^{-1}\rho_{xy}^2\left[\left(\frac{A_{22}}{\rho_{Z_1Z_2}^2}-1\right)+\frac{1}{4}(A_1+B_1)-(H_1+I_1)+\frac{1}{2}G_1\right] \tag{47}$$

Proof. We have

$$\text{MSE}(\hat{r}_1) = E\left(\hat{r}_1 - \rho_{xy}\right)^2 \approx E\left[\rho_{xy}\left\{1 + \varepsilon_5 - \tfrac{1}{2}\delta_0 + \tfrac{1}{2}\delta_1 + O\left(\delta^2\right)\right\} - \rho_{xy}\right]^2$$

$$\approx \rho_{xy}^2 E\left[\varepsilon_5^2 + \frac{1}{4}\delta_0^2 + \frac{1}{4}\delta_1^2 - \delta_0\delta_5 + \frac{1}{2}\delta_0\delta_1 - \delta_1\delta_5\right]$$

$$= n^{-1}\rho_{xy}^2\left[\left(\frac{A_{22}}{\rho_{Z_1Z_2}^2} - 1\right) + \frac{1}{4}(A_1 + B_1) - (H_1 + I_1) + \frac{1}{2}G_1\right.$$

Hence the theorem. □

Remark 1. The estimator r_{xy} defined at (23) reduces to the estimator $\hat{r}_1$ defined at (42) if $\tau_{11} = 0$

8 SINGLE SCRAMBLING VARIABLE RANDOMIZED RESPONSE TECHNIQUE

Suppose X denotes the response to the first sensitive question (say, income), and Y denotes the response to the second sensitive question (eg, expenditure, or amount of alcohol used last year). Further let S_1 be a random variable, independent of X and Y, and having a finite mean and variance. For simplicity, also assume that $X \geq 0$ and $Y \geq 0$. Assume a respondent obtains a value of S_1 using some specified method and then multiplies his/her sensitive answer X by S_1 and Y by S_1. The interviewer thus receives two scrambled answers $Z_1 = XS_1$ and $Z_2 = YS_1$. The particular value of S_1 drawn by different respondents is unknown to the interviewer, but its distribution is known. In this way, the respondents' privacy is also not violated. An example is given in Table 1.

In other words, Table 1 shows that the scrambled income and scrambled expenditure reported for two different respondents may be same, although actual income and actual expenditure for these respondents are not identical. Since the values of the scrambling variable S_1 are not known to the interviewer, the interviewer cannot detect the actual income and actual expenditure of the interviewee. Thus we have the following corollary from Theorem 1.

Corollary 4. *The variance of the sensitive variables X and Y is, respectively, given by*

$$V(X) = \frac{\sigma_{Z_1}^2 - \tau_{20}\mu_1^2}{\tau_{20} + \theta_1^2} \tag{48}$$

$$V(Y) = \frac{\sigma_{Z_2}^2 - \tau_{20}\mu_2^2}{\tau_{20} + \theta_1^2} \tag{49}$$

We also have the following theorem.

TABLE 1 True and Scrambled Responses

Respondents	X	Y	S_1	Z_1	Z_2
1	400	90	10	4,000	900
2	5,000	700	1	5,000	700
3	2,500	350	2	5,000	700
4	20,000	6,000	0.1	2,000	600
5	3,200	720	1.25	4,000	900
6	2,200	300	2.5	5,500	750

Theorem 14. *The covariance between the variables X and Y is obtained as*

$$Cov(X, Y) = \frac{\sigma_{Z_1 Z_2} - \mu_1 \mu_2 \tau_{20}}{\tau_{20} + \theta_1^2} \tag{50}$$

Proof. We have $Z_1 = XS_1$ and $Z_2 = YS_1$. Thus

$$\begin{aligned} Cov(Z_1, Z_2) &= E(Z_1 Z_2) - E(Z_1)E(Z_2) \\ &= E(XS_1 YS_1) - E(XS_1)E(YS_1) \\ &= E(XYS_1^2) - E(XS_1)E(YS_1) \\ &= E(XY)E(S_1^2) - E(X)E(S_1)E(Y)E(S_1) \\ &= E(XY)(\tau_{20} + \theta_1^2) - \mu_1 \mu_2 \theta_1^2 \end{aligned}$$

or

$$E(XY) = \frac{Cov(Z_1, Z_2) + \mu_1 \mu_2 \theta_1^2}{\tau_{20} + \theta_1^2} = \frac{\sigma_{Z_1 Z_2} + \mu_1 \mu_2 \theta_1^2}{\tau_{20} + \theta_1^2} \tag{51}$$

By definition, we have

$$Cov(X, Y) = E(XY) - E(X)E(Y)$$

which from (51) yields

$$Cov(X, Y) = \sigma_{xy} = \frac{\sigma_{Z_1 Z_2} + \mu_1 \mu_2 \theta_1^2}{\tau_{20} + \theta_1^2} - \mu_1 \mu_2 = \frac{\sigma_{Z_1 Z_2} - \mu_1 \mu_2 \tau_{20}}{\tau_{20} + \theta_1^2}$$

This proves the theorem. □

We now have Corollaries 5 and 6 from earlier results, which result in estimating ρ_{xy} in the present case.

Corollary 5. *Using (48)–(50), it can be easily seen that the correlation coefficient between the two sensitive variables X and Y is given by*

$$\rho_{xy} = \frac{\sigma_{Z_1Z_2} - \tau_{20}\mu_1\mu_2}{\sqrt{\sigma_{Z_1}^2 - \tau_{20}\mu_1^2}\sqrt{\sigma_{Z_2}^2 - \tau_{20}\mu_2^2}} \tag{52}$$

Corollary 6. *An unbiased estimator of covariance Cov(X,Y) is given by*

$$C\hat{o}v(X,Y) = \frac{s_{Z_1Z_2}\left(1 + \frac{\tau_{20}}{n\theta_1^2}\right) - \frac{\tau_{20}\bar{Z}_1\bar{Z}_2}{\theta_1^2}}{\tau_{20} + \theta_1^2} \tag{53}$$

Using (17), (19), and (53), one may easily get an estimator $\hat{r}_2$ *of* ρ_{xy} *as*

$$\hat{r}_2 = \frac{s_{Z_1Z_2}\left(1 + \frac{\tau_{20}}{n\theta_1^2}\right) - \frac{\tau_{20}\bar{Z}_1\bar{Z}_2}{\theta_1^2}}{\sqrt{s_{Z_1}^2\left(1 + \frac{\tau_{20}}{n\theta_1^2}\right) - \frac{\tau_{20}\bar{Z}_1^2}{\theta_1^2}}\sqrt{s_{Z_2}^2\left(1 + \frac{\tau_{20}}{n\theta_1^2}\right) - \frac{\tau_{20}\bar{Z}_2^2}{\theta_1^2}}} \tag{54}$$

In this case since a single scrambling variable S_1 *is being used, the values of* e_{rs} *used in Lemma 2 should be replaced by* e_{r+s}*, for all r and s. In particular, we have* $e_{12} = e_{1+2,0} = e_{30} = E(S_1^3)$; $e_{21} = e_{2+1,0} = e_{30} = E(S_1^3)$; $e_{13} = e_{1+3,0} = e_{40} = E(S_1^4)$; $e_{31} = e_{3+1,0} = e_{40} = E(S_1^4)$; *and* $e_{22} = e_{2+2,0} = e_{40} = E(S_1^4)$. *Also, we have* $\theta_2 = \theta_1$. *The values of* μ_{rs} *so obtained for the present case will be used to find the bias and mean squared error of the estimator* $\hat{r}_2$.

9 BIAS AND MEAN SQUARED ERROR OF $\hat{r}_2$

To find the bias and mean squared error of $\hat{r}_2$, we shall use certain results given below. For this let us define

$$\delta_{01} = \frac{s_{Z_1}^2\left(1 + \frac{\tau_{20}}{n\theta_1^2}\right) - \frac{\tau_{20}\bar{Z}_1^2}{\theta_1^2}}{V(X)\left(\tau_{20} + \theta_1^2\right)} - 1 \tag{55}$$

$$\delta_{11} = \frac{s_{Z_2}^2\left(1 + \frac{\tau_{20}}{n\theta_1^2}\right) - \frac{\tau_{20}\bar{Z}_2^2}{\theta_1^2}}{V(Y)\left(\tau_{20} + \theta_1^2\right)} - 1 \tag{56}$$

$$\delta_{21} = \frac{s_{Z_1Z_2}\left(1 + \frac{\tau_{20}}{n\theta_1^2}\right) - \frac{\tau_{20}\bar{Z}_1\bar{Z}_2}{\theta_1^2}}{Cov(X,Y)\left(\tau_{20} + \theta_1^2\right)} - 1 \tag{57}$$

so that

$$E(\delta_{i1}) = 0 \text{ for } i = 0,1,2$$

Thus we have the following corollaries from the corresponding earlier results of this chapter.

Corollary 7. *Expected value of δ_{01}^2 is given by*

$$E\left(\delta_{01}^2\right) = n^{-1}A \tag{58}$$

where

$$A = \frac{\left[\sigma_{Z_1}^4(A_{40}-1) + \frac{\tau_{20}^2\sigma_{Z_1}^4}{n\theta_1^4}\{1+n^{-1}(A_{40}-1)\} + 4\tau_{20}^2\mu_1^4C_{Z_1}^2 - 4\tau_{20}\mu_1^2\sigma_{Z_1}^2C_{Z_1}A_{30} - \frac{2\tau_{20}^2\mu_1^2\sigma_{Z_1}^2}{n\theta_1^2}C_{Z_1}^2\right]}{\sigma_x^4\left(\tau_{20}+\theta_1^2\right)^2}$$

Corollary 8. *Expected value of δ_{11}^2 is given by*

$$E\left(\delta_{11}^2\right) = n^{-1}B \tag{59}$$

where

$$B = \frac{\left[\sigma_{Z_2}^2(A_{04}-1) + \frac{\tau_{20}^2\sigma_{Z_2}^4}{n\theta_1^4}\{1+n^{-1}(A_{04}-1)\} + 4\tau_{20}^2\mu_2^4C_{Z_2}^2 - 4\tau_{20}\mu_2^2\sigma_{Z_2}^2C_{Z_2}A_{03} - \frac{2\tau_{20}^2\mu_2^2\sigma_{Z_2}^2}{n\theta_1^2}C_{Z_2}^2\right]}{\sigma_y^4\left(\tau_{20}+\theta_1^2\right)^2}$$

Corollary 9. *We have*

$$E\left(\delta_{21}^2\right) = n^{-1}C \tag{60}$$

where

$$C = \frac{\begin{aligned}&\left[\sigma_{Z_1Z_2}^2\left(\frac{A_{22}}{\rho_{Z_1Z_2}^2}-1\right) + \frac{\tau_{20}^2}{n\theta_1^4}\left\{1+n^{-1}\left(\frac{A_{22}}{\rho_{Z_1Z_2}^2}-1\right)\right\} + \tau_{20}^2\mu_1^2\mu_2^2\right.\\ &\times\left\{C_{Z_1}^2 + C_{Z_2}^2 + 2\rho_{Z_1Z_2}C_{Z_1}C_{Z_2}\right\} + 2\frac{\sigma_{Z_1Z_2}\tau_{20}}{n\theta_1^2}\left(\frac{A_{22}}{\rho_{Z_1Z_2}^2}-1\right) - 2\tau_{20}\mu_1\mu_2\sigma_{Z_1Z_2}\\ &\left.\times\left\{C_{Z_1}\frac{A_{12}}{\rho_{Z_1Z_2}} + C_{Z_2}\frac{A_{21}}{\rho_{Z_1Z_2}}\right\} - 2\frac{\tau_{20}^2\mu_1\mu_2}{n\theta_1^2}\left\{\rho_{Z_1Z_2}C_{Z_1}C_{Z_2} + C_{Z_1}\frac{A_{12}}{\rho_{Z_1Z_2}} + C_{Z_2}\frac{A_{21}}{\rho_{Z_1Z_2}}\right\}\right]\end{aligned}}{\sigma_{xy}^2\left(\tau_{20}+\theta_1^2\right)^2}$$

Corollary 10. *The expected value of the product $\delta_{01}\delta_{21}$ is given by*

$$E(\delta_{01}\delta_{21}) = n^{-1}D \tag{61}$$

where

$$D = \frac{\left[\begin{aligned}&\sigma_{Z_1}^2\sigma_{Z_1Z_2}\left(\frac{A_{31}}{\rho_{Z_1Z_2}}-1\right) - 2\tau_{20}\mu_1^2\sigma_{Z_1Z_2}C_{Z_1}\frac{A_{12}}{\rho_{Z_1Z_2}} + \frac{\sigma_{Z_1}^2\tau_{20}\sigma_{Z_1Z_2}}{n\theta_1^2}\left(\frac{A_{31}}{\rho_{Z_1Z_2}}-1\right) - \tau_{20}\mu_1\mu_2\sigma_{Z_1}^2\\ &\times\left\{C_{Z_1}A_{30}+C_{Z_2}A_{21}\right\} + 2\tau_{20}^2\mu_1^3\mu_2\left\{C_{Z_1}^2+\rho_{Z_1Z_2}C_{Z_1}C_{Z_2}\right\} - \frac{\sigma_{Z_1}^2\tau_{20}^2\mu_1\mu_2}{n\theta_1^2}\\ &\times\left\{\rho_{Z_1Z_2}C_{Z_1}C_{Z_2}+C_{Z_1}A_{30}+C_{Z_2}A_{21}\right\} + \frac{\sigma_{Z_1}^2\sigma_{Z_1Z_2}\tau_{20}}{n\theta_1^2}\left(\frac{A_{31}}{\rho_{Z_1Z_2}}-1\right) - \frac{\sigma_{Z_1Z_2}\tau_{20}\mu_1^2}{n\theta_1^2}\\ &\times\left\{C_{Z_1}^2+2C_{Z_1}\frac{A_{12}}{\rho_{Z_1Z_2}}\right\} + \frac{\sigma_{Z_1}^2\sigma_{Z_1Z_2}\tau_{20}^2}{n\theta_1^4}\left\{1+n^{-1}\left(\frac{A_{31}}{\rho_{Z_1Z_2}}-1\right)\right\}\end{aligned}\right]}{\sigma_x^2\sigma_{xy}\left(\tau_{20}+\theta_1^2\right)^2}$$

Corollary 11. *It can be seen that*

$$E(\delta_{11}\delta_{21}) = n^{-1}F \tag{62}$$

where

$$F = \frac{\left[\begin{aligned}&\sigma_{Z_2}^2\sigma_{Z_1Z_2}\left(\frac{A_{13}}{\rho_{Z_1Z_2}}-1\right) - 2\tau_{20}\mu_2^2\sigma_{Z_1Z_2}C_{Z_2}\frac{A_{21}}{\rho_{Z_1Z_2}} + \frac{\sigma_{Z_1}^2\tau_{20}\sigma_{Z_1Z_2}}{n\theta_1^2}\left(\frac{A_{13}}{\rho_{Z_1Z_2}}-1\right) - \tau_{20}\mu_1\mu_2\sigma_{Z_2}^2\\ &\times\left\{C_{Z_2}A_{03}+C_{Z_1}A_{12}\right\} + 2\tau_{20}^2\mu_2^3\mu_1\left\{C_{Z_1}^2+\rho_{Z_1Z_2}C_{Z_1}C_{Z_2}\right\} - \frac{\sigma_{Z_2}^2\tau_{20}^2\mu_1\mu_2}{n\theta_1^2}\\ &\times\left\{\rho_{Z_1Z_2}C_{Z_1}C_{Z_2}+C_{Z_2}A_{03}+C_{Z_2}A_{12}\right\} + \frac{\sigma_{Z_2}^2\sigma_{Z_1Z_2}\tau_{20}}{n\theta_1^2}\left(\frac{A_{13}}{\rho_{Z_1Z_2}}-1\right) - \frac{\sigma_{Z_1Z_2}\tau_{20}\mu_2^2}{n\theta_1^2}\\ &\times\left\{C_{Z_2}^2+2C_{Z_2}\frac{A_{21}}{\rho_{Z_1Z_2}}\right\} + \frac{\sigma_{Z_1}^2\sigma_{Z_1Z_2}\tau_{20}^2}{n\theta_1^4}\left\{1+n^{-1}\left(\frac{A_{13}}{\rho_{Z_1Z_2}}-1\right)\right\}\end{aligned}\right]}{\sigma_y^2\sigma_{xy}\left(\tau_{20}+\theta_1^2\right)^2}$$

Corollary 12. *The expected value of the product $\delta_{01}\delta_{11}$ is given by*

$$E(\delta_{01}\delta_{11}) = n^{-1}G \tag{63}$$

where

$$G=\frac{\left[\begin{aligned}&\sigma_{Z_1}^2\sigma_{Z_2}^2(A_{22}-1)-2\tau_{20}\mu_1^2C_{Z_1}A_{12}+\frac{\sigma_{Z_1}^2\tau_{20}}{n\theta_1^2}(A_{22}-1)+\frac{\sigma_{Z_1}^2\sigma_{Z_2}^2\tau_{20}}{n\theta_1^2}(A_{22}-1)\\&+\frac{\sigma_{Z_1}^2\sigma_{Z_2}^2\tau_{20}}{n\theta_1^4}\{1+n^{-1}(A_{22}-1)\}-\frac{\sigma_{Z_2}^2\tau_{20}^2\mu_1^2}{n\theta_1^2}\left\{C_{Z_1}^2+2C_{Z_1}A_{12}\right\}-2\sigma_{Z_1}^2\tau_{20}\mu_2^2C_{Z_2}A_{21}\\&-\frac{\sigma_{Z_1}^2\tau_{20}^2\mu_2^2}{n\theta_1^2}\left\{C_{Z_2}^2+2C_{Z_2}A_{21}\right\}+4\tau_{20}^2\mu_1^2\mu_2^2\rho_{Z_1Z_2}C_{Z_1}C_{Z_2}\end{aligned}\right]}{\sigma_x^2\sigma_y^2\left(\tau_{20}+\theta_1^2\right)^2}$$

We now obtain expressions for the bias and mean squared error of the estimator $\hat{r}_2$ in Theorems 15 and 16.

Theorem 15. *The bias of the estimator $\hat{r}_2$ defined at (54) is approximately*

$$B(\hat{r}_2)=n^{-1}\rho_{xy}\left[\frac{3}{8}(A+B)-\frac{1}{2}(D+F)+\frac{1}{4}G\right] \tag{64}$$

Proof. We have

$$\hat{r}_2=\frac{s_{Z_1Z_2}\left(1+\frac{\tau_{20}}{n\theta_1^2}\right)-\frac{\tau_{20}\bar{Z}_1\bar{Z}_2}{\theta_1^2}}{\sqrt{s_{Z_1}^2\left(1+\frac{\tau_{20}}{n\theta_1^2}\right)-\frac{\tau_{20}\bar{Z}_1^2}{\theta_1^2}}\sqrt{s_{Z_2}^2\left(1+\frac{\tau_{20}}{n\theta_1^2}\right)-\frac{\tau_{20}\bar{Z}_2^2}{\theta_2^2}}} \tag{65}$$

Relation (65) in terms of δ_{01},δ_{11}, and δ_{21} may be written as

$$\hat{r}_2=\frac{\sigma_{xy}(1+\delta_{21})}{\sqrt{\sigma_x^2(1+\delta_{01})}\sqrt{\sigma_y^2(1+\delta_{11})}}=\rho_{xy}(1+\delta_{21})(1+\delta_{01})^{-1/2}(1+\delta_{11})^{-1/2} \tag{66}$$

Again assuming that $|\delta_{01}|<1$ and $|\delta_{11}|<1$, and using the binomial theorem to expand right hand side of (66), we get

$$\begin{aligned}\hat{r}_2&=\rho_{xy}(1+\delta_{21})\left(1-\frac{1}{2}\delta_{01}+\frac{3}{2}\delta_{01}^2+\cdots\right)\left(1-\frac{1}{2}\delta_{11}+\frac{3}{8}\delta_{11}^2+\cdots\right)\\&=\rho_{xy}\left(1+\delta_{21}-\frac{1}{2}\delta_{01}-\frac{1}{2}\delta_{11}+\frac{3}{8}\delta_{01}^2+\frac{3}{8}\delta_{11}^2-\frac{1}{2}\delta_{01}\delta_{21}-\frac{1}{2}\delta_{11}\delta_{21}+\frac{1}{4}\delta_{01}\delta_{11}+O(\delta^2)\right)\end{aligned}$$

Taking expected value on both sides, we get, to the first order of approximation,

$$E(\hat{r}_2) = \rho_{xy}\left[1 + n^{-1}\left\{\frac{3}{8}(A+B) - \frac{1}{2}(D+F) + \frac{1}{4}G\right\}\right]$$

Therefore, the expression for the bias in r_2 is obtained as

$$B(\hat{r}_2) = E(\hat{r}_2) - \rho_{xy} = n^{-1}\rho_{xy}\left[\frac{3}{8}(A+B) - \frac{1}{2}(D+F) + \frac{1}{4}G\right]$$ □

Theorem 16. *The mean squared error of the estimator $\hat{r}_2$ (up to terms of order n^{-1}) is given by*

$$\mathrm{MSE}(\hat{r}_2) = n^{-1}\rho_{xy}^2\left[C + \frac{1}{4}(A+B) + \frac{1}{2}G - D - F\right] \tag{67}$$

Proof. We have

$$\begin{aligned}\mathrm{MSE}(\hat{r}_2) &= E\left(\hat{r}_2 - \rho_{xy}\right)^2 = E\left[\rho_{xy}\left\{1 + \delta_{21} - \tfrac{1}{2}\delta_{01} - \tfrac{1}{2}\delta_{11} + O(\delta^2)\right\} - \rho_{xy}\right]^2 \\ &= n^{-1}\rho_{xy}^2\left[C + \frac{1}{4}(A+B) + \frac{1}{2}G - D - F\right]\end{aligned}$$

which proves the theorem. □

Remark 2. The estimator r_{XY} defined at (23) reduces to the estimator $\hat{r}_2$ defined at (54) if we choose $\alpha = 1$ and $\beta = 0$. Moreover, the value of α and β is known to both the interviewer and interviewee.

10 CORRELATION BETWEEN SENSITIVE AND NONSENSITIVE VARIABLE

Suppose X denotes the response to the sensitive question (eg, abortion cases) and Y denotes the response to the nonsensitive question (eg, # of children). Further, let S_1 be a random variable, independent of X and having finite mean and variance. For simplicity, also assume that $X \geq 0$ and $S_1 > 0$. The respondent generates S_1 using some specified method and multiplies his/her sensitive answer X by S_1. The nonsensitive question Y is asked directly without using any randomization device. The interviewer thus receives one scrambled response $Z_1 = XS_1$ and other direct response $Z_2 = Y$. Since the particular value of S_1 is unknown to the interviewer, the respondents privacy cannot be violated. Singh et al. (1996) have used such a method to fit a regression model. Again suppose, $E(S_1) = \theta_1$, $V(S_1) = \tau_{20}$, $E(X) = \mu_1$, $V(X) = \sigma_x^2$, $E(Z_2) = \mu_2$, and $V(Z_2) = \sigma_{Z_2}^2$, where θ_1 and τ_{20} are known but μ_1, σ_x^2, μ_2, and σ_y^2 are unknown. The expressions for $V(X)$, $V(Y)$, and $Cov(X,Y)$ in terms of the parameters of observed variables Z_1 and Z_2 and those of the distribution of S_1 are then given by the following corollaries and Theorem 17.

Corollary 13. *Following Theorem 1, the variance of the sensitive variable X is given by:*

$$V(X) = \sigma_x^2 = \frac{\sigma_{Z_1}^2 - \tau_{20}\mu_1^2}{\tau_{20} + \theta_1^2} \tag{68}$$

Corollary 14. *The variance of the nonsensitive variable Y is seen to be*

$$V(Y) = V(Z_2) = \sigma_{Z_2}^2 \tag{69}$$

Theorem 17. *The covariance between the sensitive variable X and nonsensitive variable Y is obtained as*

$$Cov(X, Y) = \frac{\sigma_{Z_1Z_2}}{\theta_1} \tag{70}$$

Proof. Using the independence of X and Y from S_1, we have

$$\begin{aligned} Cov(Z_1, Z_2) &= E(Z_1Z_2) - E(Z_1)E(Z_2) = E(XS_1Y) - E(XS_1)E(Y) \\ &= E(XY)E(S_1) - E(X)E(S_1)E(Y) = \theta_1 E(XY) - \mu_1\mu_2\theta_1 \end{aligned}$$

or

$$E(XY) = \frac{\sigma_{Z_1Z_2} + \mu_1\mu_2\theta_1}{\theta_1}$$

By the definition, we have

$$Cov(X, Y) = E(XY) - E(X)E(Y) = \frac{\sigma_{Z_1Z_2} + \mu_1\mu_2\theta_1}{\theta_1} - \mu_1\mu_2 = \frac{\sigma_{Z_1Z_2}}{\theta_1}$$

For the case under consideration, the correlation coefficient between the sensitive variable X and a nonsensitive variable Y is given by

$$\rho_{xy} = \frac{\sigma_{Z_1Z_2}\sqrt{\tau_{20} + \theta_1^2}}{\theta_1 \sigma_{Z_2}\sqrt{\sigma_{Z_1}^2 - \tau_{20}\mu_1^2}} \tag{71}$$

For developing an estimator of ρ_{xy} we need estimators for $V(X), V(Y)$, and $Cov(X,Y)$. The unbiased estimator of $V(X)$ has been obtained in Theorem 6, while $s_{Z_2}^2$ and $s_{Z_1Z_2}$ are unbiased estimators of $\sigma_{Z_2}^2$ and $\sigma_{Z_1Z_2}$.

An estimator $\hat{r}_3$ of ρ_{xy} is, therefore, given by

$$\hat{r}_3 = \frac{s_{Z_1Z_2}\sqrt{\tau_{20} + \theta_1^2}}{\theta_1 s_{Z_2}\sqrt{s_{Z_1}^2\left(1 + \frac{\tau_{20}}{n\theta_1^2}\right) - \frac{\tau_{20}\bar{Z}_1^2}{\theta_1^2}}} \quad \square \tag{72}$$

11 BIAS AND MEAN SQUARE ERROR OF $\hat{r}_3$

The bias and mean squared error expressions of $\hat{r}_3$ can be obtained by proceeding along the lines of previous sections. Using the notations introduced in these sections we have certain results presented in Corollaries 15–17.

Corollary 15. *The expected value of the product* $\delta_0\varepsilon_4$ *is seen to be*

$$E(\delta_0\varepsilon_4) = n^{-1}J_1 \tag{73}$$

where

$$J_1 = \frac{\left[\sigma_{Z_1}^2\left(1+\frac{\tau_{20}}{n\theta_1^2}\right)(A_{22}-1) - 2\tau_{20}\mu_1^2 C_{Z_1}A_{12}\right]}{\sigma_x^2\left(\tau_{20}+\theta_1^2\right)}$$

Using the binomial theorem the explicit form of the estimator $\hat{r}_3$ *of* ρ_{xy} *can be approximately put in terms of* ε*'s and* δ*'s*

$$\hat{r}_3 = \rho_{xy}\left[1+\varepsilon_5 - \frac{1}{2}\delta_0 - \frac{1}{2}\varepsilon_4 + \frac{3}{8}\delta_0^2 + \frac{3}{8}\varepsilon_4^2 - \frac{1}{2}\delta_0\varepsilon_5 - \frac{1}{2}\varepsilon_4\varepsilon_5 + \frac{1}{4}\delta_0\varepsilon_4 + O\left(\varepsilon^2\right)\right] \tag{74}$$

Corollary 16. *The bias of the estimator* $\hat{r}_3$ *of* ρ_{xy} *to the order* $O\left(n^{-1}\right)$ *is given by*

$$B(\hat{r}_3) = n^{-1}\rho_{xy}\left[\frac{3}{8}(A_1 + A_{04} - 1) - \frac{1}{2}\left(H_1 + \frac{A_{13}}{\rho_{Z_1Z_2}} - 1\right) + \frac{1}{4}J_1\right] \tag{75}$$

Corollary 17. *The mean squared error* (up to terms of order $O\left(n^{-1}\right)$) *of the estimator* $\hat{r}_3$ *is obtained as*

$$\mathrm{MSE}(\hat{r}_3) = n^{-1}\rho_{xy}^2\left[\left(\frac{A_{22}}{\rho_{Z_1Z_2}^2} - 1\right) + \frac{1}{4}(A_1 + A_{04} - 1) - \left(H_1 + \frac{A_{13}}{\rho_{Z_1Z_2}} - 1\right) + \frac{1}{2}J_1\right] \tag{76}$$

Remark 3. The estimator r_{xy} proposed at (22) reduces to the estimator $\hat{r}_3$ of ρ_{xy} at (72) if $\alpha = 0$ and $\beta = 1$.

In the next section, we did a simulation study to investigate the performance of the estimators of the finite population correlation coefficient when the variables get scrambled.

12 SIMULATION STUDY

Following Singh and Horn (1998), we generated a bivariate population of $N = 10{,}000$ units with two variables Y and X having a desired correlation coefficient ρ_{xy}, whose values are given by:

$$y_i = 2.8 + \left(\sqrt{1-\rho_{xy}^2}\right)y_i^* + \rho_{xy}\frac{S_y}{S_x}x_i^* \tag{77}$$

and

$$x_i = 2.4 + x_i^* \tag{78}$$

where $x_i^* \sim G(a_x, b_x)$ and $y_i^* \sim G(a_y, b_y)$ follow independent gamma distributions. In particular, we chose $a_x = 1.5$, $b_x = 1.2$, $a_y = 2.1$, and $b_y = 2.0$, and $\rho_{xy} \in [-0.90, +0.90]$ with a step of 0.2. For each value of the correlation coefficient ρ_{xy}, we generated a population. From the given population of $N = 10{,}000$ units, using a SRSWOR scheme, we selected NITR = 5000 samples each of size n in the range 50–200 with a step of 50 units. From a given sample of n units, we calculate the value of the sample correlation coefficient $r_{xy|(k)}$, $k = 1,2,\ldots,$NITR. We calculate the percent relative bias and mean squared error of the estimator of the finite population correlation coefficient as:

$$\mathrm{RB}(1) = \frac{\frac{1}{\mathrm{NITR}} \sum_{k=1}^{\mathrm{NITR}} r_{xy(k)} - \rho_{xy}}{\rho_{xy}} \times 100\% \tag{79}$$

and

$$\mathrm{MSE}(1) = \frac{1}{\mathrm{NITR}} \sum_{k=1}^{\mathrm{NITR}} \left[r_{xy(k)} - \rho_{xy} \right]^2 \tag{80}$$

We also generated a bivariate population of $N = 10{,}000$ units of two scrambling variables S_1 and S_2 whose values are given by:

$$S_{1i} = 2.8 + \left(\sqrt{1 - \rho_{s_1 s_2}^2} \right) s_{1i}^* + \rho_{s_1 s_2} \frac{\sqrt{\tau_{20}}}{\sqrt{\tau_{02}}} s_{2i}^* \tag{81}$$

and

$$S_{2i} = 2.4 + s_{2i}^* \tag{82}$$

where $s_{1i}^* \sim G(a_{s_1}, b_{s_1})$ and $s_{2i}^* \sim G(a_{s_2}, b_{s_2})$ follow independent gamma distributions. In particular, we choose $a_{s_1} = 0.9$, $b_{s_1} = 0.1$, $a_{s_2} = 1.2$, and $b_{s_2} = 0.2$, and $\rho_{s_1 s_2} = \tau_{11}/\sqrt{\tau_{20}\tau_{02}} \in [-0.90, +0.90]$ with a step of 0.3. We obtained two scrambled data values in the entire population as $Z_{1i} = x_i S_{1i}$ and $Z_{21i} = y_i S_{2i}$, $i = 1,2,\ldots N$. From the given population $N = 10{,}000$ scrambled responses, using SRSWOR scheme we selected NITR = 5000 samples each of size n in the range 50–200 with a step of 50 units. From a given sample of n units, we calculate the value of the sample correlation coefficient $r_{xy|(k)}^s$, $k = 1,2,\ldots,$NITR, where suffix stands for scrambled responses and is given in Eq. (23). We calculate the percent relative bias and mean squared error of the estimator of the finite population correlation coefficient obtained from the scrambled responses

$$\mathrm{RB}(2) = \frac{\frac{1}{\mathrm{NITR}} \sum_{k=1}^{\mathrm{NITR}} r_{xy(k)}^s - \rho_{xy}}{\rho_{xy}} \times 100\% \tag{83}$$

and

$$\text{MSE}(2) = \frac{1}{\text{NITR}} \sum_{k=1}^{\text{NITR}} \left[r^s_{xy(k)} - \rho_{xy} \right]^2 \tag{84}$$

It is to be expected that there will be an increase in the value of the MSE when using the scrambled responses compared to using the actual X, Y values, that is, MSE(2) is expected to be larger than MSE(1), but the percent relative bias may increase or decrease. Thus we define a measure of relative loss in percent relative efficiency due to scrambling the variables as:

$$\text{RLoss} = \frac{\text{MSE}(2) - \text{MSE}(1)}{\text{MSE}(2)} \times 100\% \tag{85}$$

The results obtained are presented in Table 2. The results are very encouraging toward the use of scrambled responses when estimating correlation coefficient between two sensitive variables. The values of RB(1) and RB(2) remain negligible in almost all cases. The value of RLoss lies between 10% and 21% in the entire simulation study.

To have a clear view of the results displayed in Table 2, we use a graphical visualization of the effect of the three parameters, viz., ρ_{xy}, $\rho_{s_1 s_2}$, and sample size n. From Fig. 1 we see that the value of RLoss remains between 10% to 21%; if the value of ρ_{xy} is −0.90, then RLoss value decreases until the value of ρ_{xy} reaches zero value and again its value starts increasing up to almost 21% as the value of ρ_{xy} becomes 0.90. Thus the RLoss depends on the value of the population correlation coefficient between the two variables being estimated.

Fig. 2 shows that the RB(1) value is free from the value of the correlation coefficient $\rho_{s_1 s_2}$ between the two scrambling variables, which should be the case because no scrambling has been applied while computing RB(1) values. The RB(2) values are function of $\rho_{s_1 s_2}$ and it seems that if one makes use of highly correlated scrambling variables, then the value of the relative bias is more sensitive to the value of ρ_{xy} than if one uses less correlated scrambling variables. The variation in the value of RLoss seems free from the value of the correlation coefficient between the scrambling variables.

From Fig. 3 we see that the variation in the value or RLoss seems free from the sample size. The value of RB(1) is negligible in the range of −1.3% to 0.9% as the sample size changes between 50 and 200. The value RB(2) remains higher in case of small sample sizes, but again it becomes less than 5% as the sample size is increased to 100. In case of scrambled responses, a large sample size is expected to be needed to obtain trustworthy estimates. Table 3 gives the three different values of the population correlation coefficient between the two variables. In the first column ρ_{xy} is the value of the desired correlation coefficient supplied in the transformations (77) and (78); ρ^o_{xy} is the value of the observed correlation coefficient between the 10,000 values of X and Y; and ρ^s_{xy} is the value of the population correlation

TABLE 2 RB(1), RB(2), and RLoss for Different Values of Correlation Coefficients and Sample Sizes

ρ_{xy}	$\rho_{s_1s_2}$	n	RB(1)	RB(2)	RLoss	ρ_{xy}	$\rho_{s_1s_2}$	n	RB(1)	RB(2)	RLoss
−0.901	−0.903	50	−0.695	−0.515	15.93	0.097	−0.903	50	−1.051	4.952	12.85
−0.901	−0.903	100	−0.334	−0.297	16.73	0.097	−0.903	100	0.573	3.024	12.97
−0.901	−0.903	150	−0.206	−0.204	17.47	0.097	−0.903	150	0.259	1.530	13.05
−0.901	−0.903	200	−0.175	−0.228	17.95	0.097	−0.903	200	0.772	1.431	13.11
−0.901	−0.606	50	−0.695	−0.403	16.02	0.097	−0.606	50	−1.051	1.176	13.30
−0.901	−0.606	100	−0.334	−0.176	15.72	0.097	−0.606	100	0.573	0.722	13.39
−0.901	−0.606	150	−0.206	−0.079	15.68	0.097	−0.606	150	0.259	−0.605	13.45
−0.901	−0.606	200	−0.175	−0.090	15.68	0.097	−0.606	200	0.772	−0.755	13.49
−0.901	−0.302	50	−0.695	−0.349	15.11	0.097	−0.302	50	−1.051	−1.623	13.63
−0.901	−0.302	100	−0.334	−0.104	15.05	0.097	−0.302	100	0.573	−0.715	13.70
−0.901	−0.302	150	−0.206	−0.006	15.03	0.097	−0.302	150	0.259	−1.918	13.75
−0.901	−0.302	200	−0.175	−0.009	15.04	0.097	−0.302	200	0.772	−1.984	13.78
−0.901	0.005	50	−0.695	−0.333	15.16	0.097	0.005	50	−1.051	−3.987	13.89
−0.901	0.005	100	−0.334	−0.069	15.33	0.097	0.005	100	0.573	−1.770	13.96
−0.901	0.005	150	−0.206	0.032	15.40	0.097	0.005	150	0.259	−2.852	13.99
−0.901	0.005	200	−0.175	0.039	15.47	0.097	0.005	200	0.772	−2.770	14.03
−0.901	0.311	50	−0.695	−0.346	16.02	0.097	0.311	50	−1.051	−6.047	14.12

−0.901	0.311	100	−0.334	−0.063	16.34	0.097	0.311	100	0.573	−2.546	14.19
−0.901	0.311	150	−0.206	0.041	16.50	0.097	0.311	150	0.259	−3.499	14.22
−0.901	0.311	200	−0.175	0.059	16.64	0.097	0.311	200	0.772	−3.216	14.25
−0.901	0.611	50	−0.695	−0.384	17.48	0.097	0.611	50	−1.051	−7.799	14.35
−0.901	0.611	100	−0.334	−0.085	17.93	0.097	0.611	100	0.573	−3.023	14.41
−0.901	0.611	150	−0.206	0.023	18.16	0.097	0.611	150	0.259	−3.833	14.45
−0.901	0.611	200	−0.175	0.051	18.34	0.097	0.611	200	0.772	−3.291	14.48
−0.901	0.904	50	−0.695	−0.448	19.47	0.097	0.904	50	−1.051	−8.976	14.58
−0.901	0.904	100	−0.334	−0.143	20.04	0.097	0.904	100	0.573	−2.907	14.66
−0.901	0.904	150	−0.206	−0.034	20.34	0.097	0.904	150	0.259	−3.558	14.70
−0.901	0.904	200	−0.175	0.002	20.58	0.097	0.904	200	0.772	−2.636	14.73
−0.702	−0.903	50	−1.067	−0.728	16.94	0.298	−0.903	50	−0.801	2.227	15.05
−0.702	−0.903	100	−0.494	−0.178	15.65	0.298	−0.903	100	0.046	1.453	15.20
−0.702	−0.903	150	−0.294	−0.041	14.96	0.298	−0.903	150	0.004	0.965	15.30
−0.702	−0.903	200	−0.289	−0.062	14.46	0.298	−0.903	200	0.136	0.833	15.37
−0.702	−0.606	50	−1.067	−0.444	12.98	0.298	−0.606	50	−0.801	0.983	15.60
−0.702	−0.606	100	−0.494	0.057	12.46	0.298	−0.606	100	0.046	0.733	15.71
−0.702	−0.606	150	−0.294	0.190	12.15	0.298	−0.606	150	0.004	0.306	15.78
−0.702	−0.606	200	−0.289	0.188	11.93	0.298	−0.606	200	0.136	0.169	15.84

Continued

TABLE 2 RB(1), RB(2), and RLoss for Different Values of Correlation Coefficients and Sample Sizes—Cont'd

ρ_{xy}	$\rho_{s_1s_2}$	n	RB(1)	RB(2)	RLoss	ρ_{xy}	$\rho_{s_1s_2}$	n	RB(1)	RB(2)	RLoss
−0.702	−0.302	50	−1.067	−0.269	11.30	0.298	−0.302	50	−0.801	0.042	16.01
−0.702	−0.302	100	−0.494	0.196	11.12	0.298	−0.302	100	0.046	0.279	16.10
−0.702	−0.302	150	−0.294	0.326	11.01	0.298	−0.302	150	0.004	−0.104	16.15
−0.702	−0.302	200	−0.289	0.332	10.93	0.298	−0.302	200	0.136	−0.202	16.20
−0.702	0.005	50	−1.067	−0.157	10.74	0.298	0.005	50	−0.801	−0.775	16.33
−0.702	0.005	100	−0.494	0.277	10.76	0.298	0.005	100	0.046	−0.067	16.40
−0.702	0.005	150	−0.294	0.406	10.76	0.298	0.005	150	0.004	−0.405	16.44
−0.702	0.005	200	−0.289	0.418	10.77	0.298	0.005	200	0.136	−0.443	16.48
−0.702	0.311	50	−1.067	−0.092	10.85	0.298	0.311	50	−0.801	−1.504	16.59
−0.702	0.311	100	−0.494	0.312	11.00	0.298	0.311	100	0.046	−0.336	16.65
−0.702	0.311	150	−0.294	0.441	11.08	0.298	0.311	150	0.004	−0.626	16.69
−0.702	0.311	200	−0.289	0.456	11.14	0.298	0.311	200	0.136	−0.586	16.72
−0.702	0.611	50	−1.067	−0.068	11.43	0.298	0.611	50	−0.801	−2.143	16.81
−0.702	0.611	100	−0.494	0.303	11.67	0.298	0.611	100	0.046	−0.521	16.86
−0.702	0.611	150	−0.294	0.431	11.80	0.298	0.611	150	0.004	−0.757	16.90
−0.702	0.611	200	−0.289	0.446	11.91	0.298	0.611	200	0.136	−0.619	16.93
−0.702	0.904	50	−1.067	−0.104	12.37	0.298	0.904	50	−0.801	−2.604	17.01

−0.702	0.904	100	−0.494	0.225	12.69	0.298	0.904	100	0.046	−0.526	17.06
−0.702	0.904	150	−0.294	0.349	12.87	0.298	0.904	150	0.004	−0.703	17.10
−0.702	0.904	200	−0.289	0.356	13.02	0.298	0.904	200	0.136	−0.428	17.13
−0.503	−0.903	50	−0.767	−0.257	12.42	0.499	−0.903	50	−1.134	1.085	17.42
−0.503	−0.903	100	−0.327	0.260	12.10	0.499	−0.903	100	−0.321	0.696	17.56
−0.503	−0.903	150	−0.175	0.346	11.90	0.499	−0.903	150	−0.212	0.562	17.65
−0.503	−0.903	200	−0.253	0.264	11.74	0.499	−0.903	200	−0.124	0.451	17.73
−0.503	−0.606	50	−0.767	0.270	11.33	0.499	−0.606	50	−1.134	0.369	17.94
−0.503	−0.606	100	−0.327	0.651	11.12	0.499	−0.606	100	−0.321	0.305	18.04
−0.503	−0.606	150	−0.175	0.724	10.99	0.499	−0.606	150	−0.212	0.208	18.11
−0.503	−0.606	200	−0.253	0.667	10.89	0.499	−0.606	200	−0.124	0.104	18.16
−0.503	−0.302	50	−0.767	0.621	10.64	0.499	−0.302	50	−1.134	−0.185	18.32
−0.503	−0.302	100	−0.327	0.886	10.54	0.499	−0.302	100	−0.321	0.056	18.40
−0.503	−0.302	150	−0.175	0.948	10.46	0.499	−0.302	150	−0.212	−0.014	18.45
−0.503	−0.302	200	−0.253	0.896	10.41	0.499	−0.302	200	−0.124	−0.088	18.49
−0.503	0.005	50	−0.767	0.878	10.31	0.499	0.005	50	−1.134	−0.675	18.61
−0.503	0.005	100	−0.327	1.035	10.28	0.499	0.005	100	−0.321	−0.142	18.67
−0.503	0.005	150	−0.175	1.089	10.26	0.499	0.005	150	−0.212	−0.182	18.71
−0.503	0.005	200	−0.253	1.035	10.24	0.499	0.005	200	−0.124	−0.215	18.74

Continued

TABLE 2 RB(1), RB(2), and RLoss for Different Values of Correlation Coefficients and Sample Sizes—Cont'd

ρ_{xy}	$\rho_{s_1s_2}$	n	RB(1)	RB(2)	RLoss	ρ_{xy}	$\rho_{s_1s_2}$	n	RB(1)	RB(2)	RLoss
−0.503	0.311	50	−0.767	1.066	10.27	0.499	0.311	50	−1.134	−1.124	18.83
−0.503	0.311	100	−0.327	1.118	10.32	0.499	0.311	100	−0.321	−0.304	18.88
−0.503	0.311	150	−0.175	1.164	10.34	0.499	0.311	150	−0.212	−0.313	18.91
−0.503	0.311	200	−0.253	1.102	10.36	0.499	0.311	200	−0.124	−0.294	18.94
−0.503	0.611	50	−0.767	1.190	10.51	0.499	0.611	50	−1.134	−1.527	19.01
−0.503	0.611	100	−0.327	1.133	10.61	0.499	0.611	100	−0.321	−0.425	19.05
−0.503	0.611	150	−0.175	1.171	10.68	0.499	0.611	150	−0.212	−0.400	19.08
−0.503	0.611	200	−0.253	1.094	10.73	0.499	0.611	200	−0.124	−0.318	19.10
−0.503	0.904	50	−0.767	1.214	10.99	0.499	0.904	50	−1.134	−1.833	19.15
−0.503	0.904	100	−0.327	1.038	11.16	0.499	0.904	100	−0.321	−0.453	19.19
−0.503	0.904	150	−0.175	1.064	11.26	0.499	0.904	150	−0.212	−0.390	19.21
−0.503	0.904	200	−0.253	0.955	11.34	0.499	0.904	200	−0.124	−0.223	19.23
−0.304	−0.903	50	−0.212	0.127	11.25	0.700	−0.903	50	−1.299	0.434	19.48
−0.304	−0.903	100	−0.119	0.558	11.17	0.700	−0.903	100	−0.539	0.189	19.60
−0.304	−0.903	150	−0.027	0.679	11.10	0.700	−0.903	150	−0.343	0.248	19.68
−0.304	−0.903	200	−0.239	0.522	11.06	0.700	−0.903	200	−0.261	0.150	19.74
−0.304	−0.606	50	−0.212	1.163	10.97	0.700	−0.606	50	−1.299	−0.029	19.92

−0.304	−0.606	100	−0.119	1.269	10.90	0.700	−0.606	100	−0.539	−0.043	20.00
−0.304	−0.606	150	−0.027	1.355	10.84	0.700	−0.606	150	−0.343	0.043	20.06
−0.304	−0.606	200	−0.239	1.233	10.81	0.700	−0.606	200	−0.261	−0.044	20.10
−0.304	−0.302	50	−0.212	1.885	10.75	0.700	−0.302	50	−1.299	−0.396	20.23
−0.304	−0.302	100	−0.119	1.702	10.71	0.700	−0.302	100	−0.539	−0.194	20.29
−0.304	−0.302	150	−0.027	1.761	10.67	0.700	−0.302	150	−0.343	−0.088	20.33
−0.304	−0.302	200	−0.239	1.637	10.65	0.700	−0.302	200	−0.261	−0.150	20.37
−0.304	0.005	50	−0.212	2.452	10.63	0.700	0.005	50	−1.299	−0.730	20.46
−0.304	0.005	100	−0.119	1.993	10.63	0.700	0.005	100	−0.539	−0.319	20.50
−0.304	0.005	150	−0.027	2.029	10.61	0.700	0.005	150	−0.343	−0.193	20.53
−0.304	0.005	200	−0.239	1.886	10.60	0.700	0.005	200	−0.261	−0.222	20.56
−0.304	0.311	50	−0.212	2.905	10.65	0.700	0.311	50	−1.299	−1.043	20.63
−0.304	0.311	100	−0.119	2.177	10.67	0.700	0.311	100	−0.539	−0.429	20.66
−0.304	0.311	150	−0.027	2.190	10.68	0.700	0.311	150	−0.343	−0.280	20.68
−0.304	0.311	200	−0.239	2.015	10.69	0.700	0.311	200	−0.261	−0.271	20.70
−0.304	0.611	50	−0.212	3.250	10.80	0.700	0.611	50	−1.299	−1.331	20.75
−0.304	0.611	100	−0.119	2.249	10.86	0.700	0.611	100	−0.539	−0.519	20.77
−0.304	0.611	150	−0.027	2.237	10.90	0.700	0.611	150	−0.343	−0.346	20.79
−0.304	0.611	200	−0.239	2.015	10.92	0.700	0.611	200	−0.261	−0.290	20.80

Continued

TABLE 2 RB(1), RB(2), and RLoss for Different Values of Correlation Coefficients and Sample Sizes—Cont'd

ρ_{xy}	$\rho_{s_1s_2}$	n	RB(1)	RB(2)	RLoss	ρ_{xy}	$\rho_{s_1s_2}$	n	RB(1)	RB(2)	RLoss
−0.304	0.904	50	−0.212	3.413	11.10	0.700	0.904	50	−1.299	−1.562	20.84
−0.304	0.904	100	−0.119	2.127	11.21	0.700	0.904	100	−0.539	−0.557	20.85
−0.304	0.904	150	−0.027	2.084	11.26	0.700	0.904	150	−0.343	−0.357	20.86
−0.304	0.904	200	−0.239	1.781	11.31	0.700	0.904	200	−0.261	−0.241	20.87
−0.103	−0.903	50	0.644	−1.394	11.46	0.900	−0.903	50	−0.770	0.659	21.06
−0.103	−0.903	100	−0.233	−0.264	11.51	0.900	−0.903	100	−0.359	0.157	21.15
−0.103	−0.903	150	−0.031	0.592	11.53	0.900	−0.903	150	−0.224	0.190	21.21
−0.103	−0.903	200	−0.653	0.342	11.55	0.900	−0.903	200	−0.177	0.054	21.26
−0.103	−0.606	50	0.644	1.983	11.64	0.900	−0.606	50	−0.770	0.376	21.39
−0.103	−0.606	100	−0.233	1.912	11.67	0.900	−0.606	100	−0.359	0.040	21.45
−0.103	−0.606	150	−0.031	2.634	11.67	0.900	−0.606	150	−0.224	0.094	21.49
−0.103	−0.606	200	−0.653	2.463	11.69	0.900	−0.606	200	−0.177	−0.028	21.52
−0.103	−0.302	50	0.644	4.423	11.75	0.900	−0.302	50	−0.770	0.140	21.61
−0.103	−0.302	100	−0.233	3.253	11.77	0.900	−0.302	100	−0.359	−0.040	21.65
−0.103	−0.302	150	−0.031	3.875	11.78	0.900	−0.302	150	−0.224	0.029	21.68
−0.103	−0.302	200	−0.653	3.661	11.79	0.900	−0.302	200	−0.177	−0.071	21.70
−0.103	0.005	50	0.644	6.419	11.85	0.900	0.005	50	−0.770	−0.086	21.75

−0.103	0.005	100	−0.233	4.200	11.88	0.900	0.005	100	−0.359	−0.115	21.78
−0.103	0.005	150	−0.031	4.729	11.89	0.900	0.005	150	−0.224	−0.031	21.80
−0.103	0.005	200	−0.653	4.415	11.90	0.900	0.005	200	−0.177	−0.104	21.81
−0.103	0.311	50	0.644	8.099	11.98	0.900	0.311	50	−0.770	−0.305	21.85
−0.103	0.311	100	−0.233	4.850	12.02	0.900	0.311	100	−0.359	−0.189	21.87
−0.103	0.311	150	−0.031	5.283	12.04	0.900	0.311	150	−0.224	−0.088	21.88
−0.103	0.311	200	−0.653	4.824	12.06	0.900	0.311	200	−0.177	−0.131	21.89
−0.103	0.611	50	0.644	9.469	12.15	0.900	0.611	50	−0.770	−0.516	21.91
−0.103	0.611	100	−0.233	5.192	12.21	0.900	0.611	100	−0.359	−0.259	21.92
−0.103	0.611	150	−0.031	5.518	12.25	0.900	0.611	150	−0.224	−0.140	21.93
−0.103	0.611	200	−0.653	4.861	12.27	0.900	0.611	200	−0.177	−0.149	21.94
−0.103	0.904	50	0.644	10.291	12.40	0.900	0.904	50	−0.770	−0.699	21.95
−0.103	0.904	100	−0.233	4.955	12.49	0.900	0.904	100	−0.359	−0.308	21.95
−0.103	0.904	150	−0.031	5.158	12.53	0.900	0.904	150	−0.224	−0.171	21.96
−0.103	0.904	200	−0.653	4.197	12.57	0.900	0.904	200	−0.177	−0.138	21.96

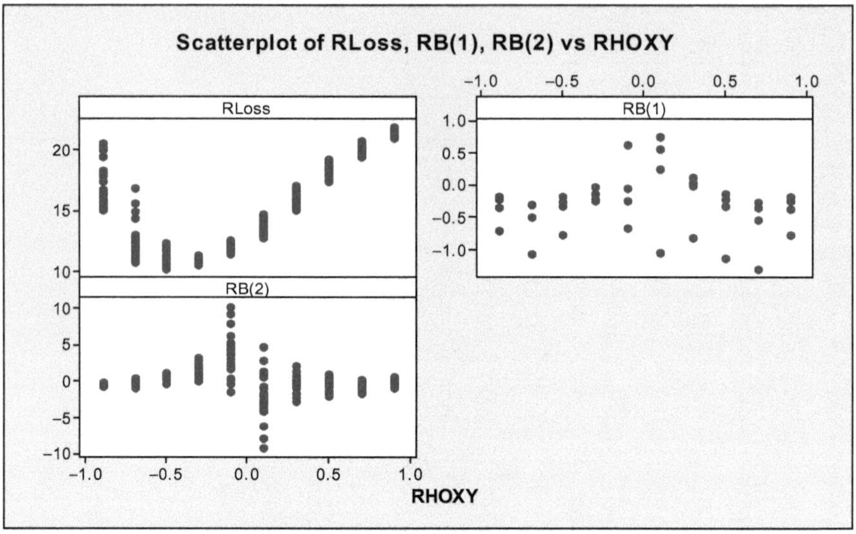

FIG. 1 RB(1), RB(2), and RLoss values as a function of ρ_{xy}.

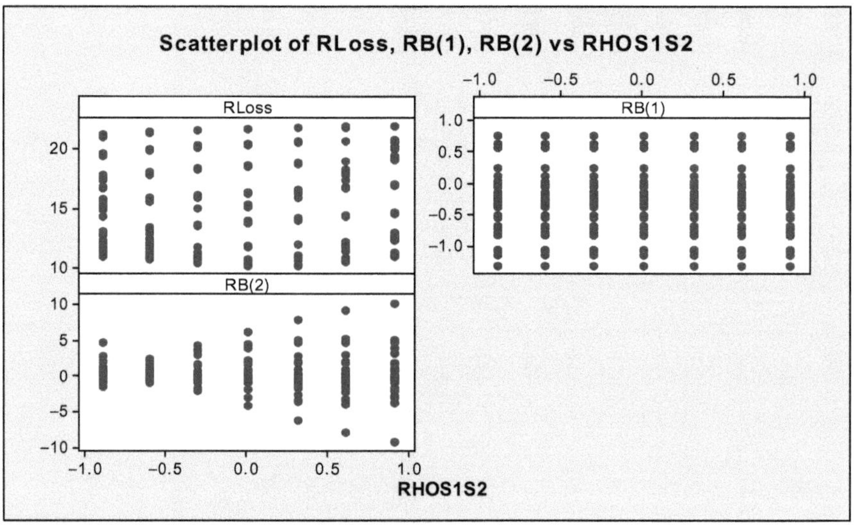

FIG. 2 RB(1), RB(2), and RLoss values as a function of $\rho_{s_1s_2}$.

coefficient obtained by using formula in Eq. (9). One can observe that there is not much difference between all the three values. Thus, the value of the correlation coefficient between the two scrambling variables can be transformed back to the original value of the correlation coefficient using information on the parameters of the scrambling variables.

The FORTRAN code used in producing these results is given in the Appendix.

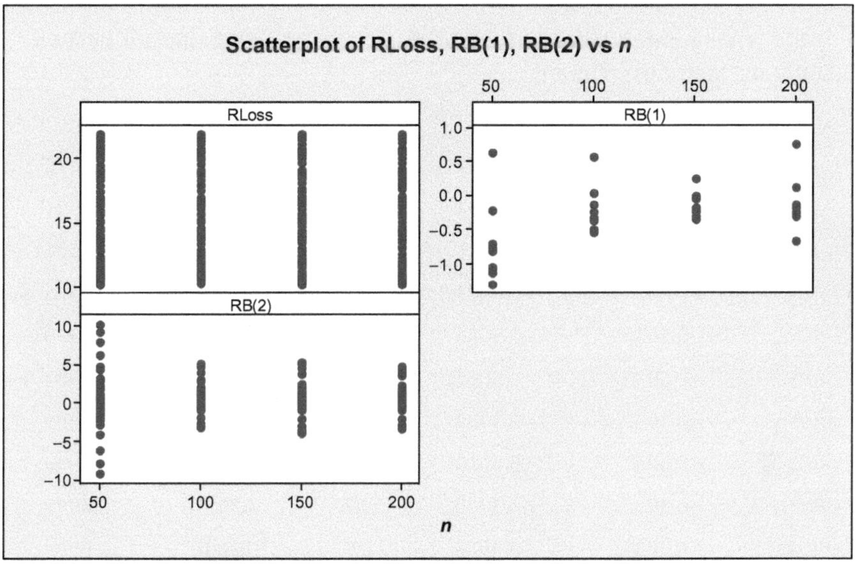

FIG. 3 RB(1), RB(2), and RLoss values as a function of n.

TABLE 3 Computed Values of Population Correlation Coefficient by Two Different Methods

ρ_{xy}	ρ^o_{xy}	ρ^s_{xy}	ρ_{xy}	ρ^o_{xy}	ρ^s_{xy}
−0.9	−0.9008	−0.8995	0.1	0.0974	0.0960
−0.9	−0.9008	−0.9009	0.1	0.0974	0.0945
−0.9	−0.9008	−0.9017	0.1	0.0974	0.0939
−0.9	−0.9008	−0.9023	0.1	0.0974	0.0937
−0.9	−0.9008	−0.9026	0.1	0.0974	0.0937
−0.9	−0.9008	−0.9027	0.1	0.0974	0.0942
−0.9	−0.9008	−0.9023	0.1	0.0974	0.0953
−0.7	−0.7023	−0.7033	0.3	0.2984	0.2977
−0.7	−0.7023	−0.7049	0.3	0.2984	0.2964
−0.7	−0.7023	−0.7059	0.3	0.2984	0.2959
−0.7	−0.7023	−0.7064	0.3	0.2984	0.2958
−0.7	−0.7023	−0.7067	0.3	0.2984	0.2960
−0.7	−0.7023	−0.7066	0.3	0.2984	0.2964

Continued

TABLE 3 Computed Values of Population Correlation Coefficient by Two Different Methods—Cont'd

ρ_{xy}	ρ_{xy}^{o}	ρ_{xy}^{s}	ρ_{xy}	ρ_{xy}^{o}	ρ_{xy}^{s}
−0.7	−0.7023	−0.7059	0.3	0.2984	0.2976
−0.5	−0.5032	−0.5054	0.5	0.4993	0.4991
−0.5	−0.5032	−0.5071	0.5	0.4993	0.4981
−0.5	−0.5032	−0.5081	0.5	0.4993	0.4978
−0.5	−0.5032	−0.5086	0.5	0.4993	0.4978
−0.5	−0.5032	−0.5087	0.5	0.4993	0.4981
−0.5	−0.5032	−0.5085	0.5	0.4993	0.4985
−0.5	−0.5032	−0.5076	0.5	0.4993	0.4996
−0.3	−0.3035	−0.3060	0.7	0.7001	0.6995
−0.3	−0.3035	−0.3077	0.7	0.7001	0.6989
−0.3	−0.3035	−0.3086	0.7	0.7001	0.6989
−0.3	−0.3035	−0.3090	0.7	0.7001	0.6991
−0.3	−0.3035	−0.3091	0.7	0.7001	0.6994
−0.3	−0.3035	−0.3088	0.7	0.7001	0.6998
−0.3	−0.3035	−0.3078	0.7	0.7001	0.7008
−0.1	−0.1032	−0.1054	0.9	0.9003	0.8983
−0.1	−0.1032	−0.1070	0.9	0.9003	0.8983
−0.1	−0.1032	−0.1078	0.9	0.9003	0.8985
−0.1	−0.1032	−0.1081	0.9	0.9003	0.8989
−0.1	−0.1032	−0.1081	0.9	0.9003	0.8993
−0.1	−0.1032	−0.1077	0.9	0.9003	0.8997
−0.1	−0.1032	−0.1066	0.9	0.9003	0.9003

ACKNOWLEDGMENTS

This chapter is from the dissertation of the author Sarjinder Singh which was completed under the supervision of late Dr. Ravindra Singh from Punjab Agricultural University during 1991. The author would also like to thank Sneha Sunkara, MS student in Electrical Engineering at TAMUK, for retyping the portion of this chapter related to estimation of correlation coefficient. The author is also thankful to Prof. Arijit Chaudhuri, Prof. Stephen A. Sedory, Purnima Shaw, and a referee for their valuable comments on the original version of this chapter.

APPENDIX

```
!! FORTRAN CODE USED IN THE SIMULATION: FILE NAME SAR15.F95
        USE NUMERICAL_LIBRARIES
        IMPLICIT NONE
        INTEGER NP,I,ISEED0,ISEED1,ISEED2,ISEED3,ISEED6, NS,IR1(10000)
        INTEGER NITR, IIII, NS1
        REAL Y(10000), X(10000),RHO, YP(10000),XP(10000)
        REAL AS1, BS1, AS2, BS2, VS1, VS2, S1(10000), S2(10000)
        REAL S1P(10000), S2P(10000),RHOS1S2,Z1P(10000),Z2P(10000)
        REAL AY, BY, AX, BX, VY, VX, AMU1,AMU2, SUMX, SUMY, SUMXY
        REAL SUMX2,SUMY2, VARX,VARY,COVXY,RHOXY,ANP
        REAL TH1, SUMS1, TH2, SUMS2,VARS1,VARS2,COVS1S2
        REAL TAU20,TAU02,TAU11, VARZ1,SUMZ1, SUMZ2,VARZ2,COVZ1Z2
        REAL RHOXYN,RHOXYD,RHOXYT,Z1M,Z2M,SUMZ12,SUMZ22,
SUMZ1Z2
        REAL FACT1, FACT2, Z1S(10000),Z2S(10000),SUMZ1S,SUMZ2S
        REAL Z1MS, Z2MS, SUMZ1S2,SUMZ2S2,SUMZ1Z2S
        REAL SUMXS, SUMYS, XS(1000), YS(1000), XSM, YSM
        REAL SUMX2S, SUMY2S, SUMXYS,VARXS,VARYS,COVXYS,RXYS,SZ1Z2
        REAL SZ12,SZ22,T1,T2,T3,T4,ANS,D1,F1,F2,D2,D3,RXYSC
        REAL BRXYS,BRXYSC, AMSE1,AMSE2, RE, RHO1, RHO2
        REAL SRXYS,SRXYSC
        CHARACTER*20 OUT_FILE
        WRITE(*,'(A)') 'NAME OF THE OUTPUT FILE'
        READ(*,'(A20)') OUT_FILE
        OPEN(42, FILE = OUT_FILE, STATUS = 'UNKNOWN')
        NP = 10000
        ANP = NP
        AY = 2.1
        BY = 2.0
        AX = 1.5
        BX = 1.2
        VY = AY*BY**2
        VX = AX*BX**2
!       WRITE(42, 121)AY,BY,AX,BX
!121     FORMAT(2X,'AY = ',F9.3,2X,'BY = ',F9.3,2X,'AX = ',F9.3,
!      12X,'BX = ',F9.3/)
```

Continued

```
      ISEED0 = 130131963
      CALL RNSET(ISEED0)
      CALL RNGAM(NP,AY,YP)
      CALL SSCAL(NP,BY,YP,1)
      ISEED1 = 123457
      CALL RNSET(ISEED1)
      CALL RNGAM(NP,AX,XP)
      CALL SSCAL(NP,BX,XP,1)
      DO 6666 RHO1 = -0.90, 0.91, 0.20
      DO 3333 RHO2 = -0.90, 0.91, 0.30
      RHO = RHO1
      DO 111 I =1, NP
      Y(I) = 2.8 + SQRT(1.-RHO**2)*YP(I) + RHO*SQRT(VY)*XP(I)/SQRT(VX)
      X(I) = 3.4 + XP(I)
111   CONTINUE
      SUMX = 0.0
      SUMY = 0.0
      DO 51 I = 1, NP
      SUMX = SUMX + X(I)
51    SUMY = SUMY + Y(I)
!     WRITE(42, 126)SUMX, SUMY
!126  FORMAT(2X,'SUMX = ',F9.1,2X,'SUMY = ',F9.1)
      AMU1 = SUMX/ANP
      AMU2 = SUMY/ANP
      SUMX2 = 0.0
      SUMY2 = 0.0
      SUMXY = 0.0
      DO 52 I =1, NP
      SUMX2 = SUMX2 + (X(I)-AMU1)**2
      SUMY2 = SUMY2 + (Y(I)-AMU2)**2
52    SUMXY = SUMXY + (X(I)-AMU1)*(Y(I)-AMU2)
      VARX = SUMX2/(ANP-1)
      VARY = SUMY2/(ANP-1)
      COVXY = SUMXY/(ANP-1)
      RHOXY = COVXY/SQRT(VARX*VARY)
      AS1 = 0.9
```

Continued

```
      BS1 = 0.1
      AS2 = 1.2
      BS2 = 0.2
      VS1 = AS1*BS1**2
      VS2 = AS2*BS2**2
!     WRITE(42, 122)AS1,BS1,AS2,BS2
!122  FORMAT(2X,'AS1 = ',F9.3,2X,'BS1 = ',F9.3,2X,'AS2 = ',F9.3,
!    1 2X,'BS2 = ',F9.3/)
      ISEED2 = 130131963
      CALL RNSET(ISEED2)
      CALL RNGAM(NP,AS1,S1)
      CALL SSCAL(NP,BS1,S1,1)
      ISEED3 = 123457
      CALL RNSET(ISEED3)
      CALL RNGAM(NP,AS2,S2)
      CALL SSCAL(NP,BS2,S2,1)
      RHOS1S2 = RHO2
      DO 115 I =1, NP
      S1P(I) = 1.8 + SQRT(1.-RHOS1S2**2)*S1(I)
     1 + RHOS1S2*SQRT(VS1)*S2(I)/SQRT(VS2)
      S2P(I) = 1.4 + S2(I)
115   CONTINUE
      SUMS1 = 0.0
      SUMS2 = 0.0
      DO 116 I =1, NP
      SUMS1 = SUMS1 + S1P(I)
116   SUMS2 = SUMS2 + S2P(I)
      TH1 = SUMS1/ANP
      TH2 = SUMS2/ANP
      VARS1 = 0.0
      VARS2 = 0.0
      COVS1S2 = 0.0
      DO 117 I =1, NP
      VARS1 = VARS1 + (S1P(I)-TH1)**2
      VARS2 = VARS2 + (S2P(I)-TH2)**2
```

Continued

```
117   COVS1S2 = COVS1S2 + (S1P(I)-TH1)*(S2P(I)-TH2)
      TAU20 = VARS1/(ANP-1)
      TAU02 = VARS2/(ANP-1)
      TAU11 = COVS1S2/(ANP-1)
      RHOS1S2 = TAU11/SQRT(TAU20*TAU02)
!     WRITE(42, 129)RHO, RHOXY, RHOS1S2
!129  FORMAT(2X,'RHO = ',F9.3,2X,'RHOXY = ',F9.3,2X,'RHOS1S2 = ', F9.4)
      DO 118 I =1, NP
      Z1P(I) = X(I) * S1P(I)
      Z2P(I) = Y(I) * S2P(I)
!     WRITE(*,555) X(I), Y(I), S1P(I),S2P(I)
!555  FORMAT(2X,4(F9.3,2X))
118   CONTINUE
      SUMZ1 = 0.0
      SUMZ2 = 0.0
      DO 119 I =1, NP
      SUMZ1 = SUMZ1 + Z1P(I)
119   SUMZ2 = SUMZ2 + Z2P(I)
      Z1M = SUMZ1/ANP
      Z2M = SUMZ2/ANP
      SUMZ12 = 0.0
      SUMZ22 = 0.0
      SUMZ1Z2 = 0.0
      DO 226 I =1,NP
      SUMZ12 = SUMZ12 + (Z1P(I)-Z1M)**2
      SUMZ22 = SUMZ22 + (Z2P(I)-Z2M)**2
226   SUMZ1Z2 = SUMZ1Z2 + (Z1P(I)-Z1M)*(Z2P(I)-Z2M)
      VARZ1 = SUMZ12/(ANP-1)
      VARZ2 = SUMZ22/(ANP-1)
      COVZ1Z2 = SUMZ1Z2/(ANP-1)
      FACT1 = VARZ1-TAU20*AMU1**2
      FACT2 = VARZ2-TAU02*AMU2**2
!     WRITE(42, 215)FACT1, FACT2
!215  FORMAT(2X,2(F9.5,2X))
      RHOXYN = (COVZ1Z2-TAU11*AMU1*AMU2)
     1 *SQRT(TAU20 + TH1**2)*SQRT(TAU02 + TH2**2)
      RHOXYD = (TAU11 + TH1*TH2)*SQRT(FACT1)*SQRT(FACT2)
```

Continued

```
        RHOXYT = RHOXYN/RHOXYD
!       WRITE(*, 112)NP, RHO, RHOXY, RHOXYT
!        WRITE(42, 112)NP, RHO, RHOXY, RHOXYT
!112     FORMAT(2X,I7,2X, 3(F9.5,2X))
        NITR = 5000
        DO 7777 NS1 = 50, 200, 50
        SRXYS = 0.0
        SRXYSC = 0.0
        DO 9999 IIII = 1, NITR
        ISEED6 = IIII
        CALL RNSET(ISEED6)
        NS = NS1
        ANS = NS
        CALL RNSRI(NS, NP, IR1)
        DO 18 I = 1, NS
        XS(I) = X(IR1(I))
        YS(I) = Y(IR1(I))
        Z1S(I) = Z1P(IR1(I))
        Z2S(I) = Z2P(IR1(I))
18      CONTINUE
        SUMZ1S = 0.0
        SUMZ2S = 0.0
        SUMXS = 0.0
        SUMYS = 0.0
        DO 19 I =1,NS
        SUMXS = SUMXS + XS(I)
        SUMYS = SUMYS + YS(I)
        SUMZ1S = SUMZ1S + Z1S(I)
19      SUMZ2S = SUMZ2S + Z2S(I)
        Z1MS = SUMZ1S/ANS
        Z2MS = SUMZ2S/ANS
        YSM = SUMYS/ANS
        XSM = SUMXS/ANS
        SUMX2S = 0.0
        SUMY2S = 0.0
        SUMXYS = 0.0
```

Continued

```
      SUMZ1S2 = 0.0
      SUMZ2S2 = 0.0
      SUMZ1Z2S = 0.0
      DO 20 I = 1, NS
      SUMX2S = SUMX2S + (XS(I)-XSM)**2
      SUMY2S = SUMY2S + (YS(I)-YSM)**2
      SUMXYS = SUMXYS + (XS(I)-XSM)*(YS(I)-YSM)
      SUMZ1S2 = SUMZ1S2 + (Z1S(I)-Z1MS)**2
      SUMZ2S2 = SUMZ2S2 + (Z2S(I)-Z2MS)**2
20    SUMZ1Z2S = SUMZ1Z2S + (Z1S(I)-Z1MS)*(Z2S(I)-Z2MS)
      VARXS = SUMX2S/(ANS-1)
      VARYS = SUMY2S/(ANS-1)
      COVXYS = SUMXYS/(ANS-1)
      RXYS = COVXYS/SQRT(VARXS*VARYS)
      SZ1Z2 = SUMZ1Z2S/(ANS-1)
      SZ12 = SUMZ1S2/(ANS-1)
      SZ22 = SUMZ2S2/(ANS-1)
      T1 = SZ1Z2*(1 + TAU11/(ANS*TH1*TH2))
      T2 = TAU11*Z1MS*Z2MS/(TH1*TH2)
      T3 = SQRT(TAU20 + TH1**2)
      T4 = SQRT(TAU02 + TH2**2)
      D1 = (TAU11 + TH1*TH2)
      F1 = TAU20*Z1MS**2/TH1**2
      F2 = TAU02*Z2MS**2/TH2**2
      D2 = SQRT(SZ12*(1 + TAU20/(ANS*TH1**2))-F1)
      D3 = SQRT(SZ22*(1 + TAU02/(ANS*TH2**2))-F2)
      RXYSC = (T1-T2)*T3*T4/(D1*D2*D3)
      SRXYS = SRXYS + RXYS
      SRXYSC = SRXYSC + RXYSC
      AMSE1 = AMSE1 + (RXYS - RHOXY)**2
      AMSE2 = AMSE2 + (RXYSC - RHOXY)**2
!     WRITE(42,234)NS,RXYS, RXYSC
!234  FORMAT(2X,I5,2X, F9.4,2X,F9.4)
9999  CONTINUE
      BRXYS = (SRXYS/DBLE(NITR)-RHOXY)*100/RHOXY
      BRXYSC = (SRXYSC/DBLE(NITR)-RHOXY)*100/RHOXY
```

Continued

```
      RE = (AMSE2-AMSE1)*100/AMSE2
      WRITE(*,235)NP,RHOXY,RHOS1S2,NS,BRXYS,BRXYSC, RE
      WRITE(42,235)NP,RHOXY,RHOS1S2,NS,BRXYS,BRXYSC, RE
235   FORMAT(1X,I6,2X,F7.4,2X,F7.4,2X,I4,2X,F9.4,2X,F9.4,2X,F7.2)
7777  CONTINUE
3333  CONTINUE
6666  CONTINUE
      STOP
      END
```

REFERENCES

Ahsanullah, M., Eichhorn, B.H., 1988. On estimation of response from scrambled quantitative data. Pak. J. Stat. 4A (2), 83–91.

Bellhouse, D.R., 1995. Estimation of correlation in randomized response. Surv. Methodol. 21, 13–19.

Biradar, R.S., Singh, H.P., 1992. A class of estimators for finite population correlation coefficient using auxiliary information. J. Indian Soc. Agric. Stat. 44, 271–285.

Chaudhuri, A., 2011. Randomized Response and Indirect Questioning Techniques in Surveys. Chapman & Hall/CRC, Boca Raton, FL.

Clickner, R.P., Iglewicz, B., 1976. Warner's Randomized Response Technique: The Two Sensitive Question Case. In: Social Statistics Section, Proceedings of the American Statistical Association, pp. 260–263.

Eichhorn, B.H., Hayre, L.S., 1983. Scrambled randomized response methods for obtaining sensitive quantitative data. J. Stat. Plan. Inference 7, 307–316.

Fox, J.A., 2016. Randomized Response and Related Methods. Second Edition. Sage, Los Angeles.

Fox, J.A., Tracy, P.E., 1984. Measuring associations with randomized response. Soc. Sci. Res. 13, 188–197.

Greenberg, B.G., Kuebler, R.R., Abernathy, J.R., Horvitz, D.G., 1971. Application of the randomized response technique in obtaining quantitative data. J. Am. Stat. Assoc. 66, 243–250.

Gupta, J.P., 2002. Estimation of the correlation coefficient in probability proportional to size with replacement sampling. Stat. Pap. 43 (4), 525–536.

Gupta, J.P., Singh, R., 1990. A note on usual correlation coefficient in systematic sampling. Statistica 50, 255–259.

Gupta, J.P., Singh, R., Lal, B., 1978. On the estimation of the finite population correlation coefficient-I. Sankhyā C 41, 38–59.

Gupta, J.P., Singh, R., Lal, B., 1979. On the estimation of the finite population correlation coefficient-II. Sankhyā C 42, 1–39.

Gupta, J.P., Singh, R., Kashani, H.B., 1993. An estimator of the correlation coefficient in probability proportional to size with replacement sampling. Metron 51 (3–4), 165–177.

Himmelfarb, S., Edgell, S.E., 1980. Additive constants model: a randomized response technique for eliminating evasiveness to quantitative response questions. Psychol. Bull. 87, 525–530.

Horvitz, D.G., Shah, B.V., Simmons, W.R., 1967. The Unrelated Question Randomized Response Model. In: Proceedings of Social Statistics Section, American Statistical Association, pp. 65–72.

Lee, C.S., Sedory, S.A., Singh, S., 2013. Estimating at least seven measures of qualitative variables from a single sample using randomized response technique. Stat. Probab. Lett. 83, 399–409.

Pearson, K., 1896. Mathematical contribution to the theory of evolution. III. Regression: heredity and panmixia. Philos. Trans. A R. Soc. Lond. 187, 253–318.

Rana, R.S., 1989. Concise estimator of bias and variance of the finite population correlation coefficient. J. Indian Soc. Agric. Stat. 41, 69–76.

Singh, S., 1991. On improved strategies in survey sampling. Unpublished dissertation submitted to the Department of Mathematics and Statistics, Punjab Agricultural University, Ludhiana.

Singh, S., 2003. Advanced Sampling Theory with Applications: How Michael Selected Amy, vols. 1 and 2. Kluwer Academic Publisher, The Netherlands.

Singh, S., Horn, S., 1998. An alternative estimator in multi-character surveys. Metrika, 48, 99–107.

Singh, S., Joarder, A.H., King, M.L., 1996. Regression analysis using scrambled responses. Aust. J. Stat. 38 (2), 201–211.

Singh, S., Sedory, S.A., Kim, J.M., 2014. An empirical likelihood estimate of the finite population correlation coefficient. Commun. Stat. Simul. Comput. 43 (6), 1430–1441.

Srivastava, S.K., Jhajj, H.S., 1981. A class of estimators of the population mean in survey sampling using auxiliary information. Biometrika 68, 341–343.

Sukhatme, P.V., Sukhatme, B.V., Sukhatme, S., Ashok, C., 1984. Sampling Theory of Surveys with Applications. Iowa State University Press and Indian Society of Agricultural Statistics, New Delhi.

Wakimoto, K., 1971. Stratified random sampling (III): estimation of the correlation coefficient. Ann. Inst. Stat. Math. 23, 339–355.

Warner, S.L., 1965. Randomized response: a survey technique for eliminating evasive answer bias. J. Am. Stat. Assoc. 60, 63–69.

Chapter 6

Admissible and Optimal Estimation in Finite Population Sampling Under Randomized Response Models

S. Sengupta[1]
University of Calcutta, Kolkata, India
[1]Corresponding author: e-mail: samindras@yahoo.co.in

ABSTRACT
We consider the problem of unbiased estimation of finite population mean (or proportion) related to a sensitive character under some randomized response models covering different randomized response plans and present a comprehensive review of various nonexistence, admissibility, and optimality results on the problem paralleling those for direct surveys.

Keywords: Admissibility, Optimality, Population mean, Population proportion, Randomized response, Sampling strategy, Super-population model, Unbiased estimator, Variance estimation

AMS (2000) Classification: 62D05, 62C15

1 INTRODUCTION

Consider a finite population of labeled units and suppose that the problem is to estimate certain population parameters on surveying a random sample of units. In an open setup it is assumed that an exact response can be obtained from each sampled unit through a direct survey. However, if the character of interest is sensitive or stigmatizing such as drinking alcohol or gambling habit, drug addiction, tax evasion, and history of induced abortions, a direct survey is likely to yield unreliable responses and an alternative technique, introduced by Warner (1965), is to obtain responses through a *randomized response* (RR) survey wherein every sampled unit is asked to give a response

Handbook of Statistics, Vol. 34. http://dx.doi.org/10.1016/bs.host.2016.01.005

through an RR device as per instructions from the investigator. We refer to Chaudhuri and Mukerjee (1988), Chaudhuri (2011), and Chaudhuri and Christofides (2013) for a comprehensive review of such RR procedures.

Bellhouse (1980), Adhikary et al. (1984), Chaudhuri (1987, 1992), Sengupta and Kundu (1989), Arnab (1995a, 1998, 2002), and Padmawar and Vijayan (2000) considered the problem of unbiased estimation of a finite population mean of a sensitive quantitative variable (or the proportion bearing a sensitive attribute) under some RR models and obtained various nonexistence, admissibility, and optimality results paralleling those based on direct responses assuming tacitly that a single RR is obtained from every sampled unit even if selected more than once in the sample. However, for with replacement (WR) sampling, a population unit may be selected more than once and independent RRs may be obtained from it as many times as it is selected in the sample. In fact, for several RR devices suggested in the literature including that of Warner (1965), an infinite population setup (or equivalently simple random WR sampling) had been assumed, and the proposed estimators of population parameters were based on independent repeated responses from the sampled units. Estimators based on independent repeated responses in the finite population setup were also discussed, eg, in Arnab (1999) and Chaudhuri et al. (2011a,b). Arnab (1995b) and Sengupta (2015a,b) considered the problem of estimation of the population mean (or proportion) related to a sensitive character based on such independent multiple RRs and obtained some similar nonexistence, admissibility, and optimality results.

In this paper, we present a comprehensive review of these nonexistence, admissibility, and optimality results on the problem of unbiased estimation of finite population mean (or proportion) related to sensitive characters under some RR models. In Section 2, we discuss some preliminary notions and describe different RR models covering various RR plans. Estimation based on a single response is discussed in Section 3, and estimation based on independent multiple responses is discussed in Section 4.

2 NOTATIONS AND PRELIMINARIES

Let $U=\{1, 2, \ldots, i, \ldots, N\}$ be a finite population of N labeled units and Y be a real variable with unknown value y_i for the population unit i, $1 \leq i \leq N$. For a dichotomous population, Y is considered to be an indicator variable where y_i is 1 or 0 according as the unit i does or does not possess a certain attribute. A sequence of units of U with or without repetitions is considered as a sample s which is selected with a known probability $p(s)$ (>0) from a set of possible samples. It is assumed that $p(s)$ is independent of $\mathbf{y}=(y_1, y_2, \ldots, y_N)$. The collection of all such $p(s)$ is called a sampling design p. The problem of interest is to estimate unbiasedly the unknown population mean (or proportion) $\theta = \frac{1}{N}\sum_{i=1}^{N} y_i$ on surveying a sample of units selected according to a given sampling design p. It is assumed that for p, the inclusion probability $\pi_i = \pi_i\ (p)$ of population unit i in a sample, defined by $\sum_{s \supset i} p(s)$, is positive for every $i = 1, \ldots, N$. Any p is said to be a without replacement (WOR) sampling design if $f_{si} = 1$ for each $i \in s$ and for each s and a WR sampling design otherwise, where f_{si} is the number of times the population unit i is selected in s. We shall write $n(s)$ as the number of units (including repetitions) in s and $v(s)$ as the number of distinct population units in s. A sampling design p will be denoted by p_n if $E_p(n(s)) \leq n$ and by p_n^* if $E_p(v(s)) \leq n$ for a fixed positive integer n, where the suffix p is used on E (or V) to denote the expectation (or variance) with respect to the sampling design p. A WOR sampling design p_n will be denoted as p_n^+ when $n(s) = n$ in every s and as $p_n^+(\boldsymbol{w})$ when in addition $\pi_i = (nw_i/W),\ 1 \leq i \leq N$ for a fixed vector $\boldsymbol{w} = (w_1, \ldots, w_N)$ of positive real numbers, $W = \sum_{i=1}^{N} w_i$.

We consider the character Y to be sensitive and suppose that some RR device R is employed to produce an RR z_i on the population unit i when included in s. We assume that under R, z_is are independently distributed with

$$E_R(z_i) = a_i y_i + b_i \text{ and } V_R(z_i) = \phi_i, \quad 1 \leq i \leq N \tag{1}$$

for some known constants a_i ($\neq 0$) and b_i and for some ϕ_i depending only on characteristics of the population unit i. The suffixes R and both p and R on E (or V) will be used to denote the expectations (or variances) with respect to R and both p and R.

The model (1) can be equivalently written as

$$E_R(r_i) = y_i \text{ and } V_R(r_i) = \frac{\phi_i}{a_i^2} = \psi_i, \quad 1 \leq i \leq N \tag{2}$$

in terms of transformed RR $r_i = (z_i - b_i)/a_i$. An RR plan satisfying (2) will be denoted by R_1 when $\psi_i = \psi\ (y_i)$ for some known function ψ and by R_2 when in addition $\psi(y_i) = c_1 y_i^2 + c_2 y_i + c_3$ for known constants c_1, c_2, and c_3 with $c_1 = 0$ for a dichotomous population.

For a dichotomous population with binary response z_i, a special case of R_2 will be denoted by R_0 under which

$$P_R[z_i = 1] = ay_i + b, \quad P_R[z_i = 0] = 1 - P_R[z_i = 1], \quad 1 \leq i \leq N \tag{3}$$

for some known constants a ($\neq 0$) and b independent of i, $0 < a + b$, $b < 1$. The RR model (3) holds for many RR devices, eg, for that due to Warner (1965) which consists of asking a population unit i when included in a sample to report y_i or $1 - y_i$ with probability q or $1 - q$ ($0 < q < 1$, $q \neq 1/2$) which are known to the investigator. The RR device due to Eriksson (1973) is also a special case of (3) in which a population unit i is asked to report y_i, 1 or 0 with given probabilities q_0, q_1 or $1 - q_0 - q_1$ ($q_0, q_1 > 0$, $q_0 + q_1 < 1$). As illustrated in Chaudhuri and Christofides (2013, Chapter 4) the model (3) is also true for several other RR plans suggested in the literature, eg, for those suggested in Boruch (1972), Mangat and Singh (1990), and Mangat (1994).

For a dichotomous population, some RR procedures also use polychotomous responses with possible responses $k_1,\ldots,k_m$, $m \geq 2$ satisfying

$$P_R[z_i = k_j] = a_j y_i + b_j, \quad 1 \leq j \leq m, \quad 1 \leq i \leq N \tag{4}$$

for some known constants a_j and b_j independent of i, $\sum_{j=1}^{m} a_j k_j \neq 0$ (see Nayak and Adeshiyan, 2009). The RR plans due to Chow (1973) discussed in Leysieffer and Warner (1976), Kuk (1990), and Christofides (2003) are of this type and are again special cases of R_2.

The model R_2 holds for several RR plans for a quantitative character as well (see Chaudhuri, 1987; Chaudhuri and Christofides, 2013, Chapter 5). As for example, this is true for the RR plan due to Eriksson (1973) which consists of asking a sampled unit i to report its true value y_i with probability q_o and to report a value x_j out of M known values $x_1,\ldots,x_M$ covering the anticipated range of Y with probability q_j, where q_o and q_js are known to the investigator, q_o, $q_j > 0 \forall j$, $q_0 + \sum_{j=1}^{M} q_j = 1$. Another RR plan for a quantitative character satisfying R_2 is a scrambled RR plan mentioned in Warner (1971) and extensively discussed in Pollock and Bek (1976) and Eichhorn and Hayre (1983) wherein a population unit i is asked to report $Ay_i + B$, where A and B are two independent random variables with known probability distributions, $E(A) \neq 0$. This may be implemented, eg, by asking a sampled unit i to independently choose at random a pair of numbers A_j, B_k out of two given sets of numbers $A_1,\ldots,A_L$ and $B_1,\ldots,B_M$, L, $M > 1$, $\sum_{j=1}^{L} A_j \neq 0$ and to report the value $z_i = A_j y_i + B_k$. Bellhouse (1980) and Arnab (1998) also discussed a modified scrambled RR plan wherein a population unit i is asked to report $Ay_i + Bw_i$ for a fixed vector $\boldsymbol{w} = (w_1, \ldots, w_N)$ of positive real numbers which again satisfies (1).

For some RR procedures suggested in the literature the RR is vector valued but can be fitted into the model (2) for some known transformed response r_i. As, for example, for the unrelated question plan due to Horvitz et al. (1967) wherein a population unit i when included in a sample is asked to report twice independently y_i or x_i with given probability q_1 or $1-q_1$ in the first trial and q_2 or $1-q_2$ in the second trial ($0 < q_1, q_2 < 1, q_1 \neq q_2$), where x_i is the value of the unit i for an innocuous character X completely unrelated to Y, one may consider $r_i = \left((1-q_2)z_i^{(1)} - (1-q_1)z_i^{(2)}\right)/(q_1 - q_2)$ with $z_i^{(k)}$ as the RR obtained from population unit i in the kth trial, $k=1, 2$ so that (2) holds with $\psi_i = \frac{(1-q_1)(1-q_2)(q_1(1-q_2)+q_2(1-q_1))}{(q_1-q_2)^2}(y_i - x_i)^2$.

For some RR plans satisfying (2), the following completeness property also holds:

$$E_R[h(\boldsymbol{r})] = 0 \forall \boldsymbol{y} \text{ implies } h(\boldsymbol{r}) = 0 \forall \mathbf{r} \text{ for any real-valued function } h(\boldsymbol{r}) \tag{5}$$

where $\boldsymbol{r} = (r_1, \ldots, r_N)$. The condition (5) is satisfied, eg, for an RR plan R_0 and also for the RR plan due to Eriksson (1973) for a quantitative variable (see

Adhikary et al., 1984; Sengupta, 2014). An R (or R_1 or R_2) satisfying (5) will be denoted as R_c (or R_{1c} or R_{2c}).

For an RR survey under an RR plan R, an estimator e_R is a function of survey data and the pair $H_R = (p, e_R)$ is called a sampling strategy for estimating θ. An estimator e_R for a given p or a strategy $H_R = (p, e_R)$ is said to be unbiased if $E_{pR}(e_R) = \theta \forall \mathbf{y}$. For a given p, an unbiased estimator e_R is considered to be better than another unbiased estimator e'_R if

$$V_{pR}(e_R) \leq V_{pR}(e'_R) \ \forall \ \mathbf{y} \ \text{ with strict inequality for at least one } \mathbf{y} \tag{6}$$

An estimator e_{0R} € a class C of unbiased estimators is said to be admissible in C if there does not exist any estimator e_R€C better than e_{0R} and e_{0R} is said to be uniformly best in C if $V_{pR}(e_{0R}) \leq V_{pR}(e_R) \ \forall \ \mathbf{y} \ \forall \ e_R$ € C.

An unbiased strategy $H_R = (p, e_R)$ is considered to be better than another unbiased strategy $H'_R = (p', e'_R)$ if

$$\begin{aligned} V(H_R) &= V_{pR}(e_R) \leq V(H'_R) \\ &= V_{p/R}(e'_R) \ \forall \ \mathbf{y} \ \text{ with strict inequality for at least one } y. \end{aligned} \tag{7}$$

An unbiased strategy $H_{0R} = (p_0, e_{0R})$ belonging to a class of unbiased strategies $\mathcal{H}$ is said to be admissible in $\mathcal{H}$ if there does not exist any $H_R \in \mathcal{H}$ which is better than H_{0R}.

In what follows, as in the open setup, we will postulate some super-population model M, ie, a class of super-population distributions ζ of $\mathbf{y}$ in search of an optimal strategy. An unbiased strategy $H_{0R} = (p_0, e_{0R})$ belonging to a class of unbiased strategies $\mathcal{H}$ will be said to be optimal in $\mathcal{H}$ under the super-population model M if

$$E_\zeta V(H_{0R}) \leq E_\zeta V(H_R) \forall \zeta \in M \forall H_R \in \mathcal{H} \tag{8}$$

where we use the suffix ζ on E to denote the expectation with respect to ζ. Writing $Y_i^* = (y_i/w_i)$ for a fixed vector $\mathbf{w} = (w_1, \ldots, w_N)$ of positive real numbers we will specifically consider the following super-population models for a quantitative variable Y: (i) M_1 ($\mathbf{w}$) consisting of ζs such that Y_i^*s are uncorrelated with common unknown mean and variance, (ii) M_2 ($\mathbf{w}$) consisting of ζs such that Y_i^*s have common unknown mean, variance, and covariance, (iii) M_3 ($\mathbf{w}$) consisting of ζs such that Y_i^*s are independently distributed with common unknown mean and variance, (iv) M_4 ($\mathbf{w}$) consisting of ζs such that for any fixed numbers $y_1^*, \ldots, y_N^*$, $\Pr\left[Y_1^* = y_{i_1}^*, \ldots, Y_N^* = y_{i_N}^*\right] = 1/N!$ for each permutation $i_1, \ldots, i_N$ of $1, \ldots, N$, (v) M_5 ($\mathbf{w}$) consisting of ζs such that Y_i^*s have an exchangeable absolutely continuous distribution; and (vi) M_6 ($\mathbf{w}$) consisting of ζs such that Y_i^*s have common unknown mean and variance and the correlation between Y_i^* and Y_j^* is $\rho(i-j)$, $1 \leq i, j \leq N$ where the correlation function $\rho(u)$ is convex (see Cochran, 1946; Hajek, 1959). For an indicator variable Y, the above super models are not generally tenable and we will consider only M_i ($\mathbf{1}$), $i = 1, 2, 3, 4, 6$.

3 ESTIMATION BASED ON SINGLE RR

If a single response is obtained from each distinct population unit in s, one obtains the data $d=\{r_i: i€s\}$, where $\{i: i€s\}$ is the set of distinct population units in s. Let A_u be the class of all unbiased estimators $e_R=e_R(d)$ based on d and A_{uo} be the subclass of unbiased estimators satisfying $E_p(e_R)=\frac{1}{N}\sum_{i=1}^{N} r_i \forall \mathbf{r}$. For an R_c, A_{uo} and A_u are, however, identical. Let L_u be the subclass of all linear unbiased estimators (lue's) of A_{uo} of the form

$$e_R=\alpha_s+\sum_{i\in s} b_{si}r_i, \quad \sum_{s}\alpha_s p(s)=0, \quad \sum_{s\supset i} b_{si}p(s)=1/N\forall i \tag{9}$$

The subclass of homogeneous lue's (hlue's) of the form (9) with $\alpha_s=0$ will be denoted by L_{uo}.

For an estimator $e_R€A_{uo}$, the estimator e obtained by replacing r_i by y_i is an unbiased estimator of θ in the open setup and vice versa. We shall call e_R to be the derived estimator from e and the strategy $H_R=(p, e_R)$ to be the derived strategy from $H=(p, e)$ in the open setup. The estimator $e_{HTR}=\frac{1}{N}\sum_{i\in s}\frac{r_i}{\pi_i}$ is thus the derived estimator from the Horvitz and Thompson (1952) estimator e_{HT} in the open setup. We note that for $p_n^+(\mathbf{1}), e_{HTR}=\frac{1}{n}\sum_{i\in s} r_i$.

3.1 Nonexistence of a Best Estimator

It is known (see Basu, 1971; Godambe, 1955; Hanurav, 1966) that in the open setup, there does not exist an uniformly minimum variance unbiased estimator (umvue) or an umvlue for a given p unless p is a census, ie, every s includes all the population units and there exists an umvhlue if and only if p is a *uni-cluster sapling design* (UCSD), ie, any two samples either include the same set of distinct population units or do not include any common population unit. Using a result of Rao (1952), Sengupta and Kundu (1989) proved similar results for the RR setup (see also Adhikary et al., 1984). The results are stated in the following theorem.

Theorem 1. *Under an RR plan R and for a given p, (a) there does not exist a uniformly best estimator in A_u, A_{uo}, and L_u if p is noncensus and (b) there exists a uniformly best estimator in L_{uo} if and only if p is an UCSD.*

3.2 Admissibility Results

Under an RR plan R, the derived estimator from an inadmissible unbiased estimator for a given p or the derived strategy from an inadmissible unbiased strategy in the open setup will generally be inadmissible. This follows from Theorem 2 proved in Sengupta and Kundu (1989).

Theorem 2. *Let* $H=(p,\ e)$ *and* $H'=(p',\ e')$ *be two unbiased strategies in the open setup such that* $V_p(e) \leq V_{p'}(e')\forall \mathbf{y} \in R_N$. *Then for the derived strategies* H_R *and* H'_R *under an RR plan* R, $V(H_R) \leq V\left(H'_R\right)\forall \mathbf{y} \in R_N$. *If, further,* Ω_0 *is the subset of* R_N *for which strict inequality holds in the open setup, then under* R, *strict inequality holds for those* $\mathbf{y}$ *for which* $P_R[\mathbf{r} \in \Omega_0|y] > 0$.

Examples 1 and 2 in Sengupta and Kundu (1989) and Sengupta (2014) illustrate that the derived estimator from an admissible unbiased estimator for a given p or the derived strategy from an admissible unbiased strategy in the open setup may also be inadmissible.

Example 1. Consider a dichotomous population with $N=3$ and let $p(s)=1/6$ $\forall s=\{1\},\ \{2\},\ \{3\},\ \{1,\ 2\},\ \{1,3\},\ \{2,\ 3\}$. Consider the unbiased estimator $e=\dfrac{1}{n(s)}\sum_{i\in s} y_i$ in the open setup. It can be shown that under the Warner's (1965) RR plan R, the estimator e_{HTR} is better than the derived estimator e_R for certain values of q so that e_R is inadmissible in L_{uo} although e is admissible in the class of all unbiased estimators in the open setup (see, as for example Sengupta, 1983).

Example 2. Consider a dichotomous population with $N=3$ and let $H_{1R}=(p_1, e_{HTR}), H_{2R}=(p_2, e_{HTR})$, where $p_1(s)=\dfrac{1}{3}\forall s=\{1,2\},\{1,3\},\{2,3\}$, $p_2(s)=\dfrac{1}{2}$ for $s=\{1,2\}$ and $p_2(s)=\dfrac{1}{4}$ for $s=\{1,3\},\ \{2,3\}$. It can be verified that under the Warner's (1965) RR plan R, H_{1R} is better than H_{2R} for certain values of q and thus H_{2R} is inadmissible in the class of all unbiased strategies based on a WOR sampling design of fixed size two. However, $H_2=(p_2, e_{HT})$ is known to be admissible in the corresponding class of all unbiased strategies in the open setup (see Ramakrishnan, 1975; Sengupta, 2014).

One can, however, prove (see Sengupta, 2014; Sengupta and Kundu, 1989) the admissibility of the estimator e_{HTR} and a strategy $\left(p_n^+(\mathbf{1}), e_{HTR}\right)$ using the corresponding known results in the open setup (see Godambe and Joshi, 1965; Ramakrishnan, 1975; Sengupta, 2014) and the following lemma.

Lemma 1. *Sengupta and Kundu, 1989 For an* $H_R=(p, e_R)$ *under an RR plan* R, $e_R \in A_{uo}$, $V(H_R) \geq V_p E_R(e_R)+\dfrac{1}{N^2}\sum_{i=1}^{N}\dfrac{\psi_i}{\pi_i}$ *with equality if and only if* $V_R(e_R-e_{HTR})=0\forall s$.

These admissibility results are stated in the following theorem.

Theorem 3. *(a) Under an RR plan* R (R_c), *the estimator* e_{HTR} *is admissible in* A_{uo} (A_u) *for a given* p. *(b) Under an RR plan* R_1 (R_{1c}), *a strategy* $\left(p_n^+(\mathbf{1}), e_{HTR}\right)$ *is admissible in the class* $\mathcal{H}_{no}^*(\mathcal{H}_n^*)$ *consisting of strategies* (p_n^*, e_R), $e_R \in A_{uo}$ (A_u).

For a quantitative variable Y, Sengupta and Kundu (1989) also proved the result stated below establishing the admissibility of a strategy $\left(p_n^+, e_{HTR}\right)$ in the class $\mathcal{H}_{no}^*(\mathcal{H}_n^*)$ defined above under an RR plan R_2 (R_{2c}).

Theorem 4. *For estimating the population mean of a quantitative variable under an RR plan R_2 (R_{2c}), a strategy $\left(p_n^+, e_{HTR}\right)$ is admissible in $\mathcal{H}_{no}^*(\mathcal{H}_n^*)$.*

It may be noted that Theorem 4 is not generally true for a qualitative variable (see Example 2).

3.3 Optimality Results

In an open setup a strategy $H_o = \left(p_n^+(\boldsymbol{w}), e_{HT}\right)$ is known to be optimum in different comparable classes of unbiased strategies under the super-population model M_i ($\boldsymbol{w}$), $i=1, 2, 3, 4, 5$ (see Cassel et al., 1976, 1977; Godambe, 1955; Godambe and Joshi, 1965; Godambe and Thompson, 1973; Rao, 1971; Thompson, 1971). Examples can be provided (see Sengupta and Kundu, 1989) to demonstrate that such optimality results are not generally true for the derived strategy $H_{oR} = \left(p_n^+(\boldsymbol{w}), e_{HTR}\right)$ in the corresponding classes of unbiased strategies in the RR setup. However, using Lemma 1 it is easy to verify Theorem 5.

Theorem 5. *Sengupta and Kundu, 1989 Let a strategy $H = (p_o, e_{HT})$ be optimum under a super-population model M in a class of unbiased strategies in the open setup. Then under an RR plan R (R_c), the derived strategy H_R is optimum under M in the corresponding subclass of unbiased strategies (p, e_R), $e_R € A_{uo}(A_u)$ with $\pi_i(p) = \pi_i(p_o)$, $1 \leq i \leq N$.*

Thus from the known optimality results in the open setup one can establish (see also Chaudhuri, 1987) the optimality of a strategy H_{oR} in the subclass of $\mathcal{H}_{no}^*(\text{or}\,\mathcal{H}_n^*)$ under $M_3(\boldsymbol{w})$ or in the subclass of $\mathcal{H}_{nLo}^*$ under $M_1(\boldsymbol{w})$ with $\pi_i(p) = nw_i/W$, where $\mathcal{H}_{no}^*(\text{or}\,\mathcal{H}_n^*)$ is defined in Theorem 3 and $\mathcal{H}_{nLo}^*$ is the class of strategies $(p_n^*, e_R) e_R € L_{uo}$. The optimality of a strategy H_{oR} in the subclass of $\mathcal{H}_{no}^+\left(\text{or}\,\mathcal{H}_n^+\right)$ under $M_i(\boldsymbol{w})$, $i=4, 5$ or in the subclass of $\mathcal{H}_{nL}^+$ under $M_2(\boldsymbol{w})$ with $\pi_i(p) = nw_i/W$ similarly follows, where $\mathcal{H}_{no}^+\left(\text{or}\,\mathcal{H}_n^+\right)$ is the class of strategies $\left(p_n^+, e_R\right)$, $e_R € A_{uo}(A_u)$ and $\mathcal{H}_{nL}^+$ is the class of strategies $\left(p_n^+, e_R\right), e_R € L_u$. It further follows from a result of Hajek (1959) in the open setup that under $M_6(\boldsymbol{w})$, the strategy $\left(p_{no}^+(\mathbf{w}), e_{HTR}\right)$ is optimum in the class of strategies $\left(p_n^+(\mathbf{w}), e_{HTR}\right)$ where $p_{no}^+(\mathbf{w})$ is the pps systematic sampling design based on size measures w_is suggested in Madow (1949).

For an RR plan R_2 (or R_{2c}), a strategy $\left(p_n^+(\mathbf{1}), e_{HTR}\right)$ is optimal under $M_i(\mathbf{1})$, $i=1, 2, 3, 4, 5$ in such classes of unbiased strategies without any restriction on π_i (p). This follows from the following theorem which again can be proved using Lemma 1.

Theorem 6. *Sengupta and Kundu, 1989 Let a strategy $\left(p_n^+(\mathbf{1}), e_{HT}\right)$ be optimum under a super-population model M in a class of unbiased strategies in the open setup. Then under an RR plan R (R_c) for which $E_\xi \psi_i$ is constant $\forall \xi \in M$, the derived strategy $\left(p_n^+(\mathbf{1}), e_{HTR}\right)$ is optimum under M in the corresponding class of unbiased strategies (p, e_R), $e_R € A_{uo}(A_u)$.*

Using Lemma 1, Theorem 6 can also be generalized as follows (see also Arnab, 1995a, 1998).

Theorem 7. *Let a strategy* $\left(p_n^+(\boldsymbol{w}), e_{HT}\right)$ *be optimum under a super-population model M in a class of unbiased strategies in the open setup. Then under an RR plan R for which* $E_\xi \psi_i$ *is proportional to* $w_i^2 \forall \xi \in M$, *the derived strategy* $\left(p_n^+(\boldsymbol{w}), e_{HTR}\right)$ *is optimum under M in the corresponding class of unbiased strategies* $(p, e_R), e_R \in A_{uo}$.

The condition in Theorem 7 on R under $M_i(\boldsymbol{w})$, $i=1, 2, 3, 4, 5$, is satisfied, eg, for the scrambled RR plan when the distribution of B is degenerate and for the modified scrambled RR plan described in Section 2. Hence, for these RR plans, a strategy H_{oR} is optimal under $M_i(\boldsymbol{w})$, $i=1, 2, 3, 4, 5$ in different classes of unbiased strategies described earlier without the restriction, viz., $\pi_i(p) = nw_i/W, \ \ 1 \le i \le N$.

4 ESTIMATION BASED ON INDEPENDENT MULTIPLE RESPONSES

If independent RRs are obtained from each population unit as many times as it is selected in the sample, one obtains the data $d^* = \{r_{ij}, j=1, \ldots, f_{si}: i \in s\}$, where r_{ij} is the transformed response from unit i in its jth selection. Let A_u^* be the class of all unbiased estimators $e_R^* = e_R^*(d^*)$ based on d^* and L_u^* be the subclass of all linear unbiased estimators of the form

$$e_R^* = \alpha_s + \sum_{i \in s} \sum_{j=1}^{f_{si}} b_{sij} r_{ij}, \ \ \sum_s \alpha_s p(s) = 0, \ \ \sum_{s \supset i} b'_{si} p(s) = 1/N \forall i \tag{10}$$

where $b'_{si} = \sum_{j=1}^{f_{si}} b_{sij}$. The subclass of estimators of the form (10) with $\alpha_s = 0$ will be denoted by L_{u0}^*. Clearly the classes A_u, L_u, and L_{uo} are, respectively, subclasses of A_u^*, L_u^*, and L_{u0}^* since d is equivalent to $\{r_{i1} : i \in s\}$. Writing $d_0^* = \{r'_i : i \in s\}$ with $r'_i = \sum_{j=1}^{f_{si}} r_{ij}$, we also denote by A_u^{**} the subclass of A_u^* consisting of all unbiased estimators $e_R^{**} = e_R^{**}\left(d_0^*\right)$ based on d_0^* and by L_u^{**} the subclass of L_u^* consisting of estimators of the form

$$e_R^{**} = \alpha_s + \sum_{i \in s} b_{si}^* r'_i, \ \ \sum_s \alpha_s p(s) = 0, \ \ \sum_{s \supset i} b_{si}^* f_{si} p(s) = 1/N \forall i. \tag{11}$$

The subclass of L_{u0}^* consisting of estimators of the form (11) with $\alpha_s = 0$ will be denoted by L_{u0}^{**}. For a WOR sampling design, the classes A_u, A_u^*, and A_u^{**} (or L_u, L_u^*, and L_u^{**} or L_{uo}, L_{u0}^*, and L_{u0}^{**}) are all identical.

Sengupta (2015a,b) had shown that for a given p, any e_R^* of the form (10) but not of the form (11) can be improved upon by an e_R^{**} of the form (11) with $b_{si}^* = b'_{si}/f_{si}$ and thus had obtained the following theorem.

Theorem 8. *Under an RR plan R and for a given p, the class of estimators* L_u^{**}(*or* L_{u0}^{**}) *is complete in* L_u^*(*or* L_{u0}^*).

Remark 1. Under an RR plan R_0 and for a given p, d_0^* is, in fact, a minimal sufficient statistic and Theorem 8 follows by an application of Rao–Blackwell theorem (see Sengupta, 2015a). More generally it follows that in this case A_u^{**} is also complete in A_u^*.

It follows from Theorem 8 (see also Arnab, 1999) that for a WR sampling design p, an estimator e_R of the form (9) can be improved upon by an e_R^{**} of the form (11) with $b_{si}^* = b_{si}/f_{si}$. Such an e_R^{**} may be called the derived estimator from an lue e and the strategy $H_R^* = (p, e_R^{**})$ may be called the derived strategy from a linear unbiased strategy $H = (p, e)$ in the open setup based on multiple responses. The derived estimators e_R^{**} and e_R corresponding to an e or the derived strategies H_R^* and H_R corresponding to an H in the open setup based on a WOR sampling design are clearly identical.

4.1 Nonexistence of a Best Estimator

As in Theorem 1, it can be proved that there does not exist a uniformly best estimator in A_u^* and L_u^* under an RR plan R for a noncensus p. The following result in Sengupta (2015a) shows that there also generally does not exist a uniformly best estimator in L_{u0}^* under an RR plan R_1 for a given p.

Theorem 9. *Under an RR plan R_1 and for a given p, there exists a uniformly best estimator in L_{u0}^* if and only if p is a UCSD with $f_{si} = f_i \forall s \supset i \forall i$.*

4.2 Admissibility Results

Theorem 2 is not true for derived strategies from linear unbiased strategies in the open setup based on multiple responses. In fact, for a given p, the derived estimator e_R^{**} from an inadmissible lue e in the open setup may turn out to be admissible. Also the derived estimator e_R^{**} from an admissible lue e and even the derived estimator e_{HTR}^{**} from e_{HT} may be inadmissible. These may be demonstrated through the following example from Sengupta (2015a).

Example 3. Consider Warner's (1965) RR plan with $q = 0.4$ and let p be the SRSWR design involving n draws. Let e_{1R}^{**} and e_{2R}^{**} be the derived estimators from the sample means $\bar{y}_n$ and $\bar{y}_v$ in the open setup based, respectively, on all n units and on v distinct population units in the sample. It can be shown that e_{1R}^{**} is better than e_{2R}^{**} for $N = 4$, $n = 3$ and e_{1R}^{**} is better than e_{HTR}^{**} for $N = 3$, $n = 2$. Thus both e_{2R}^{**} and e_{HTR}^{**} are inadmissible for these two sets of values of N and n although $\bar{y}_v$ and e_{HT} are admissible lue's in the open setup. Further, the estimator e_{1R}^{**} can be shown to be admissible in L_u^* (see Theorem 10) whatever be N and n although in the open setup $\bar{y}_v$ is known to be better than $\bar{y}_n$ for $n > 2$ (see Basu, 1958; Raj and Khamis, 1958).

Under an RR plan R_1 and for a given p, one can prove (see Arnab, 1995b; Sengupta, 2015a) the admissibility of two lue's based on multiple responses, viz.,

$$e_{1R}^{**} = \sum_{i \in s} b_i^* r_i',\ b_i^* = \frac{1}{NE_p(f_{si})}$$

and

$$e_{2R}^{**} = \sum_{i \in s} b_{si}^* r_i',\ b_{si}^* = \frac{1}{Nf_{si}} \frac{\left(1 + \frac{\psi(1)+(N-1)\psi(0)}{f_{si}}\right)^{-1}}{E_p\left(1 + \frac{\psi(1)+(N-1)\psi(0)}{f_{si}}\right)^{-1}}$$

The results are given in the following theorem.

Theorem 10. *Under an RR plan R_1 and for a given p, (i) e_{1R}^{**} is admissible in L_u^* and (ii) e_{2R}^{**} is admissible in L_{u0}^*.*

For a WOR sampling design, both the estimators e_{1R}^{**} and e_{2R}^{**} reduce to e_{HTR}. However, for a WR sampling design, the two estimators are different unless for each $i = 1, \ldots, N, f_{si} = f_i \forall s \supset i$. It may also be noted that for SRSWR involving n draws, e_{1R}^{**} reduces to Warner's (1965) type estimator $e_{WR}^{**} = \frac{1}{n}\sum_{i \in s} r_i'$.

Sengupta (2015b) also proved the following result on the admissibility of a strategy $H_{0R} = \left(p_n^+(\mathbf{1}), e_{HTR}\right)$ under an RR plan R_1 in the class $\mathcal{H}_n$ consisting of strategies $(p_n, e_R^*), e_R^* \in L_u^*$.

Theorem 11. *Under an RR plan R_1, an H_{OR} is admissible in $\mathcal{H}_n$.*

4.3 Optimality Results

For an RR plan R, the following theorem on the optimality of a strategy H_{OR} defined earlier in a class of strategies based on multiple responses follows from Sengupta (2015b).

Theorem 12. *Let a strategy $\left(p_n^+(\mathbf{1}) e_{HT}\right)$ be optimum under a super-population model M in the class of linear (or homogeneous linear) unbiased strategies based on sampling designs p_n^* in the open setup. Then under an RR plan R for which $E_\xi \psi_i$ is constant $\forall \xi \in M$, the strategy $H_{OR} = \left(p_n^+(\mathbf{1}), e_{HTR}\right)$ is optimum under M in the class of strategies $\mathcal{H}_n$(or $\mathcal{H}_{n0}$), where $\mathcal{H}_n$ is defined earlier and $\mathcal{H}_{n0}$ is the subclass of $\mathcal{H}_n$ with $e_R^* \in L_{u0}^*$.*

It thus follows that under an RR plan R_2, a strategy $H_{OR} = \left(p_n^+(\mathbf{1}) e_{HTR}\right)$ is optimal in $\mathcal{H}_n$ under M_3 (**1**) and in $\mathcal{H}_{n0}$ under M_1 (**1**).

5 CONCLUDING REMARKS

Along with estimating the population mean (or proportion) unbiasedly, one is generally also interested in the problem of unbiased estimation of the variance of the unbiased estimator of the parameter. The problem of unbiased estimation of the population variance is also of considerable practical interests. The problem of unbiased variance estimation for a dichotomous population based on single and independent multiple RRs had been studied in Sengupta and Kundu (1991) and Sengupta (2015c,d) and some admissibility and optimality results were established. We refer to these papers for details.

REFERENCES

Adhikary, A.K., Chaudhuri, A., Vijayan, K., 1984. Optimum sampling strategies for randomized response trials. Int. Stat. Rev. 52, 115–125.

Arnab, R., 1995a. Optimal estimation of a finite population total under randomized response surveys. Statistics 27, 175–180.

Arnab, R., 1995b. On admissibility and optimality of sampling strategies in randomized response surveys. Sankhya B 57, 385–390.

Arnab, R., 1998. Randomized response surveys: optimum estimation of a finite population total. Stat. Pap. 39, 405–408.

Arnab, R., 1999. On use of distinct respondents in RR surveys. Biom. J. 41, 507–513.

Arnab, R., 2002. Optimum sampling strategies under randomized response surveys. Biom. J. 44, 490–495.

Basu, D., 1958. On sampling with and without replacement. Sankhya 20, 287–294.

Basu, D., 1971. An essay on the logical foundations of survey sampling I. In: Godambe, V.P., Sprott, D.A. (Eds.), Foundations of Statistical Inference. Hold, Rinehart and Winston, Toronto, pp. 203–242.

Bellhouse, D.R., 1980. Linear models for randomized response designs. J. Am. Stat. Assoc. 75, 1001–1004.

Boruch, R.F., 1972. Relations among statistical methods for assuring confidentiality of social research data. Soc. Sci. Res. 1, 403–414.

Cassel, C.M., Sarndal, C.E., Wretman, J.H., 1976. Some results on generalized difference estimation and generalized regression estimation for finite populations. Biometrika 63, 615–620.

Cassel, C.M., Sarndal, C.E., Wretman, J.H., 1977. Foundations of Inference in Survey Sampling. John Wiley, New York.

Chaudhuri, A., 1987. Randomized response surveys of finite populations: a unified approach with quantitative data. J. Stat. Plann. Inference 15, 157–165.

Chaudhuri, A., 1992. Randomized response: estimating mean square errors of linear estimators and finding optimal unbiased strategies. Metrika 39, 341–357.

Chaudhuri, A., 2011. Randomized Response and Indirect Questioning Techniques in Surveys. CRC Press, Chapman and Hall, Taylor & Francis Group, Boca Raton, FL.

Chaudhuri, A., Bose, M., Dihidar, K., 2011a. Estimating sensitive proportions by Warner's randomized response technique using multiple responses from distinct persons. Stat. Pap. 52, 111–124.

Chaudhuri, A., Bose, M., Dihidar, K., 2011b. Estimation of a sensitive proportion by Warner's randomized response data through inverse sampling. Stat. Pap. 52, 343–354.

Chaudhuri, A., Christofides, T.C., 2013. Indirect Questioning in Sample Surveys. Springer-Verlag, Berlin, Heidelberg, Germany.

Chaudhuri, A., Mukerjee, R., 1988. Randomized Response: Theory and Techniques. Marcel Dekker, New York.

Chow, L.P., 1973. Memorandum on a New Randomized Response Technique: Multiple Answer Model. Department of Population Dynamics, The Johns Hopkins University, Baltimore, MD.

Christofides, T.C., 2003. A generalized randomized response technique. Metrika 57, 195–200.

Cochran, W.G., 1946. Relative accuracy of systematic and stratified random samples for a certain class of populations. Ann. Math. Stat. 17, 164–177.

Eichhorn, B.H., Hayre, L.S., 1983. Scrambled randomized response method for obtaining sensitive quantitative data. J. Stat. Plann. Inference 7, 307–316.

Eriksson, S.A., 1973. A new model for randomized response. Int. Stat. Rev. 41, 101–113.

Godambe, V.P., 1955. A unified theory of sampling from finite populations. J. R. Stat. Soc. Ser. B 17, 269–278.

Godambe, V.P., Joshi, V.M., 1965. Admissibility and Bayes estimation in sampling finite populations I. Ann. Math. Stat. 36, 1707–1722.

Godambe, V.P., Thompson, M.E., 1973. Estimation in sampling theory with exchangeable prior distributions. Ann. Stat. 1, 1212–1221.

Hajek, J., 1959. Optimum strategy and other problems in probability sampling. Casopis Pro Pestovani Matemtiky 84, 387–423.

Hanurav, T.V., 1966. Some aspects of unified sampling theory. Sankhya A 28, 175–204.

Horvitz, D.G., Shah, B.V., Simmons, W.R., 1967. The unrelated question RR model. Proc. ASA Soc. Stat. Sec. 65–72.

Horvitz, D.G., Thompson, D.J., 1952. A generalization of sampling without replacement. J. Am. Stat. Assoc. 47, 663–685.

Kuk, A.Y.C., 1990. Asking sensitive questions indirectly. Biometrika 77, 436–438.

Leysieffer, R.W., Warner, S.L., 1976. Respondent jeopardy and optimal design in RR models. J. Am. Stat. Assoc. 71, 649–656.

Madow, W.G., 1949. On the theory of systematic sampling II. Ann. Math. Stat. 20, 333–354.

Mangat, N.S., 1994. An improved randomized response strategy. J. R. Stat. Soc. Ser. B 56, 93–95.

Mangat, N.S., Singh, R., 1990. An alternative randomized response procedure. Biometrika 77, 439–442.

Nayak, T.K., Adeshiyan, S.A., 2009. A unified framework for analysis and comparison of randomized response surveys of binary characteristics. J. Stat. Plann. Inference 139, 2757–2766.

Padmawar, V.R., Vijayan, K., 2000. Randomized response revisited. J. Stat. Plann. Inference 90, 293–304.

Pollock, K.H., Bek, Y., 1976. A comparison of three randomized response models for quantitative data. J. Am. Stat. Assoc. 71, 884–886.

Raj, D., Khamis, S.H., 1958. Some remarks on sampling with replacement. Ann. Math. Stat. 39, 550–557.

Ramakrishnan, M.K., 1975. Choice of an optimum sampling strategy I. Ann. Stat. 3, 669–679.

Rao, C.R., 1952. Some theorems on minimum variance estimation. Sankhya 12, 27–42.

Rao, C.R., 1971. Some aspects of statistical inference in problems of sampling from finite populations. In: Godambe, V.P., Sprott, D.A. (Eds.), Foundations of Statistical Inference. Hold, Rinehart and Winston, Toronto, pp. 177–202.

Sengupta, S., 1983. Admissibility of unbiased estimators in finite population sampling for samples of size at most two. Cal. Stat. Assoc. Bull. 32, 91–102.

Sengupta, S., 2014. On the admissibility of Horvitz Thompson strategies for estimating finite population proportions in direct and randomized response surveys. Cal. Stat. Assoc. Bull. 66, 63–71.

Sengupta, S., 2015a. Estimation of finite population proportion in randomized response surveys using multiple responses. Sankhya B 77, 75–83.

Sengupta, S., 2015b. Admissible and optimal sampling strategy for estimating finite population mean in randomized response surveys with multiple responses. Comm. Stat. Theor. Meth., accepted for publication. http://dx.doi.org/10.1080/03610926.2014.957859.

Sengupta, S., 2015c. Admissibility of a variance estimator in finite population sampling under Warner's randomized response plan with multiple responses. Comm. Stat. Theor. Meth., accepted for publication http://dx.doi.org/10.1080/03610926.2014.948208.

Sengupta, S., 2015d. Optimal design unbiased strategy for estimating dichotomous finite population variance under Warner's randomized response plan. J. Stat. Theor. Pract., accepted for publication. http://dx.doi.org/10.1080/15598608.2016.1141128.

Sengupta, S., Kundu, D., 1989. Estimation of finite population mean in randomized response surveys. J. Stat. Plann. Inference 23, 117–125.

Sengupta, S., Kundu, D., 1991. Admissible unbiased variance estimation in finite population sampling under randomized response. Metrika 38, 71–82.

Thompson, M.E., 1971. Discussion of paper by C. R. Rao. In: Godambe, V.P., Sprott, D.A. (Eds.), Foundations of Statistical Inference. Hold, Rinehart and Winston, Toronto, pp. 196–198.

Warner, S.L., 1965. Randomized response—a survey technique for eliminating evasive answer bias. J. Am. Stat. Assoc. 60, 63–69.

Warner, S.L., 1971. The linear randomized response model. J. Am. Stat. Assoc. 66, 884–888.

Chapter 7

A Mixture of True and Randomized Responses in the Estimation of the Number of People Having a Certain Attribute

A. Quatember[1]
Johannes Kepler University, Linz, Austria
[1]*Corresponding author: e-mail: Andreas.Quatember@jku.at*

ABSTRACT

In practice, when sensitive information is asked from survey units by an indirect questioning technique, some of the respondents may be willing to divulge their true values of the sensitive variable even without being asked. For instance, while explaining a randomized response (RR) method, the interviewee may say "Stop the explanations—I don't mind answering this question directly!" With respect to estimation accuracy, such an offer must not be ignored. For the estimation of a group size, an RR questioning design allowing such a mixture of true and masked answers is discussed for general probability sampling applying a Horvitz–Thompson type estimator. This estimation process, including compensation for nonresponse, can be vividly illustrated by the concept of pseudo-populations. In the author's teaching experience, this can improve the comprehension of the estimation process especially for those with only little knowledge of probability theory.

Keywords: Randomized response, Horvitz–Thompson estimator, General probability sampling, Weighting adjustment, Data imputation, Nonresponse, Pseudo-population

1 INTRODUCTION

Nonsampling errors can be caused, for example, by nonresponse and untruthful answering of respondents. As the term implies, such errors occur not only in sample surveys but also in censuses.

Handbook of Statistics, Vol. 34. http://dx.doi.org/10.1016/bs.host.2016.01.006

For the theoretical discussion, let U denote a population of size N with $U=\{1,2,\ldots,N\}$ and U_a the subpopulation of N_a units ($U_a \subseteq U$) bearing a certain attribute a in which we are interested. Moreover, let the value of a binary variable y for population element k be

$$y_k = \begin{cases} 1 & \ldots \text{ if } k \in U_a \\ 0 & \ldots \text{ otherwise.} \end{cases}$$

With the unbiased Horvitz–Thompson estimator for a total of a variable, in a general probability sample s of size n with $s=\{1, 2, \ldots, n\}$, the group size $N_a = \sum_U y_k$ is estimated by

$$\widehat{N}_{a,HT} = \sum_s y_k \cdot d_k \tag{1}$$

(Σ_U and Σ_s denote the sum overall population and sample units, respectively). Here, d_k denotes the design weight of element k, defined as the reciprocal of the element's first-order inclusion probability π_k (cf., for instance, Horvitz and Thompson, 1952 or Särndal et al., 1992, p. 43ff). The theoretical variance of $\widehat{N}_{a,HT}$ is given by

$$V\left(\widehat{N}_{a,HT}\right) = \sum\sum_U \Delta_{kl} \cdot \frac{y_k}{\pi_k} \cdot \frac{y_l}{\pi_l} \tag{2}$$

with Δ_{kl} as the covariance of the sample inclusion indicators I_k and I_l where $I_k=1$ if a population element k is included in the sample and zero otherwise (k, $l=1, 2, \ldots, N$). Expectation, variance, and covariance are given by $E(I_k)=\pi_k$, $V(I_k)\equiv\Delta_{kk}=\pi_k\cdot(1-\pi_k)$, and $C(I_k,I_l)\equiv\Delta_{kl}=\pi_{kl}-\pi_k\cdot\pi_l$. Here, π_{kl} denotes the second-order inclusion probability that both elements k and l of U are in s. Provided that $\pi_{kl}>0$ for k, $l \in U$, an unbiased estimator of $V\left(\widehat{N}_{a,HT}\right)$ is given by

$$\hat{V}\left(\widehat{N}_{a,HT}\right) = \sum\sum_s \frac{\Delta_{kl}}{\pi_{kl}} \cdot \frac{y_k}{\pi_k} \cdot \frac{y_l}{\pi_l}. \tag{3}$$

In the presence of missing or untruthful values of y, the sample s is divided into three pairwise disjoint sets: a set s_t of truthful answering sample elements, a set s_u of untruthful answering sample elements, and a set s_m of sample units, of whom the y-values are missing ($s=s_t \cup s_u \cup s_m$). Hence, the Horvitz–Thompson estimator of size N_a of population group U_a according to (1) is decomposed into

$$\widehat{N}_{a,HT} = \sum_s y_k \cdot d_k = \sum_{s_t} y_k \cdot d_k + \sum_{s_u} y_k \cdot d_k + \sum_{s_m} y_k \cdot d_k. \tag{4}$$

Only for $s_u = s_m = \varnothing$, this equation reduces to the original estimator presented in (1). For s_u nonempty, false y-values are observed for the sample units belonging to this group without recognizing this fact. For s_m nonempty, no y-values are observed in s_m. Hence, such respondents' behavior may cause serious problems when estimating the parameters under study using only the available cases.

Assuming that at least $s_u = \varnothing$, two different approaches exist for the estimation of N_a by $\widehat{N}_{a,HT}$ in the presence of nonresponse: either one tries to estimate the parameter of interest using the observations in $s_t = s - s_m$ only, or one tries to estimate the sum $\sum_{s_m} y_k \cdot d_k$ itself. In the first case, the design weights d_k of the responding units have to be adjusted according to the increased respondents' burden. This is done by multiplying each d_k with the reciprocal of the estimated response probability of element k leading to the adjusted weight w_k. This weighting adjustment (W) yields the estimator

$$\widehat{N}_{a,W} = \sum_{s_t} y_k \cdot w_k.$$

of N_a. In the second, in the set s_m, reasonable substitute values $y_k{}^I$ for the missing y-values have to be imputed. Hence, the imputation (I) estimator for N_a is given by

$$\widehat{N}_{a,I} = \sum_{s_t} y_k \cdot d_k + \sum_{s_m} y_k{}^I \cdot d_k.$$

In both cases, the nonresponse mechanism that lies behind the occurrence of the missing y-values has to be modeled to provide valid estimates. The selection of the response set from a given sample s follows a mechanism that determines the probabilities of certain sets s_t. Little and Rubin (2002) distinguish between different types of nonresponse mechanisms (cf., ibid., p. 11ff), which are called "missing completely at random," "missing at random," and "not missing at random." In the first two cases, nonresponse can be compensated because the necessary auxiliary information for its modeling is available for the sampling units. For weighting adjustment, this information determines the modeling of the response probabilities to increase the original design weights d_k. For data imputation, this modeling is incorporated in the estimation process by the choice of a reasonable imputation method (cf., Lohr, 2012, section 8.6).

In fact, weighting adjustment and data imputation are two supplementing methods to compensate for nonresponse. Usually, weighting adjustment is applied to compensate for unit nonresponse, whereas imputation is used in the presence of item nonresponse. In practice, the two techniques can be applied in succession with the compensation of item nonresponse by data imputation followed by that of unit nonresponse by weighting adjustment.

Anyway, before these methods of weighting adjustment and data imputation have to be applied to try to compensate for a nonresponse that already occurred, an attempt should be made to avoid this situation. Empirical social researchers have investigated survey design features that affect the quantity of responses starting from motivation letters to financial incentives (cf., for instance, Groves et al., 2004, chapter 6).

Nonresponse clearly increases when questions on sensitive subjects such as harassment at work, domestic violence, or drug usage are asked in a survey. With respect to the negative effect of nonresponse on the quality of survey results, interviewers should be encouraged to consider unusual alternatives in observing and collecting information. Such an alternative is offered by indirect questioning techniques such as randomized response (RR) designs that avoid the direct questions on sensitive subjects (cf., for an overview, for instance, Chaudhuri and Christofides, 2013). Instead, with these methods, the question actually asked from the respondent is randomly drawn from different possible questions according to assigned design probabilities. The idea is that according to this, the interviewer will not be able to assign the given answer of any survey element directly to the sensitive variable. This technique reduces the respondents' fear of an embarrassing outcome and thus assures the cooperation of the survey units.

2 A GENERAL RR TECHNIQUE FOR THE ESTIMATION OF GROUP SIZE

For the given problem of estimating N_a, at first, the RR questioning design, naturally enough, has to include the direct question (Q1) on the sensitive subject of interest: "Do you belong to group U_a?" A second question (Q2) that may be part of the RR design was suggested by Warner (1965) in his pioneering work on RR techniques. It is the question: "Do you belong to the complementary group $U_a{}^C$?" Herein, $U_a{}^C$ denotes the set of elements of U, for which $k \notin U_a$ applies. A third possible question (Q3) that can be used in the RR design is the question on membership of a group U_b. The N_b members of U_b are characterized by the possession of a completely innocuous attribute b (for instance, this attribute could be "having birthday in the first two months of a year"). Moreover, the membership of U_b should not be related to the membership of U_a (cf., Greenberg et al., 1969; Horvitz et al., 1967). As further possible alternatives to question Q1, survey units may be instructed just to say "yes" (I4), or "no" (I5) (cf., for instance, Fidler and Kleinknecht, 1977 or Singh et al., 1995).

In the RR technique, R discussed herein, question Q1 is assigned a design probability of P_{1R} of being chosen, Q2 has one of P_{2R}, Q3 has a probability of P_{3R}. The instruction I4 just to say "yes" is given a design probability of P_{4R}, and eventually, the last alternative I5 just to say "no" has the remaining probability of P_{5R} ($P_{1R} > 0$, $0 \leq P_{iR} < 1$ for $i = 2, 3, 4, 5$, and $\sum_{i=1}^{5} P_{iR} = 1$). Questioning design R is very general and one can choose the design probabilities $P_{1R}, \ldots, P_{5R}$ according to one's own preferences. A number of already existing RR procedures such as Warner's technique with $0 < P_{iR} < 1$ for $i = 1, 2$, and $P_{1R} + P_{2R} = 1$ are included in this "standardized" design (cf., Quatember, 2009, 2012). Note that this is clearly not an invitation to use all five questions with nonzero design probabilities. Rather, the objective is to provide a theoretical framework that can be applied with all possible combinations of these questions.

In practice though, when information is asked from survey units by an indirect questioning technique such as strategy R, some of the respondents may be willing to divulge their true values y_k of variable y. Of course, the interviewee should not be asked if he or she wants to disclose the true status with respect to y or to mask it. To opt for the privacy-protecting questioning design might be interpreted as an admission that one bears the sensitive attribute. However, for instance, during a telephone or face-to-face interview, while explaining the indirect questioning design to the survey unit, the interviewee may say something like "Stop the explanations—I don't mind answering this question directly!" With respect to the efficiency of the estimator of interest, such an offer from a respondent to deliver the true answer on the direct question on membership of U_a in any case instead of answering a randomly selected question must not be ignored. In fact, such a behavior affects the accuracy of the survey results and has to be included in the formula of the estimator, the theoretical variance, and the variance estimator to be able to provide valid $(1-\alpha)$-confidence intervals and correct statistical hypothesis tests at the desired α-level of significance (cf., for a discussion of categorical and quantitative variables, respectively, Quatember, 2015, chapter 6).

3 COMBINING TRUE AND RANDOMIZED RESPONSES

Extending the general questioning design R of Quatember (2009) to the questioning design Q, which also allows for true answers in the sense presented at the end of Section 2, let the answer z of unit k be defined by

$$z_k = \begin{cases} 1 & \ldots \text{ if unit } k \text{ answers "yes"} \\ 0 & \ldots \text{ otherwise.} \end{cases}$$

Further, let population U be divided into a subpopulation U_R of N_R elements, who will not voluntarily provide the desired answer on the sensitive subject on their own but use the offered RR questioning design R, and a complementary group U_D of N_D elements, who, without being asked, are willing to answer the direct question on possession of attribute a not using the RR device ($U_R \cap U_D = \varnothing$, $U = U_R \cup U_D$, and $N = N_R + N_D$). This possible behavior of the respondents leads to the following statistical properties regarding questioning design Q: The probability of the answer $z_k = 1$ is given by

$$P_Q(z_k = 1) = P_{1k} \cdot y_k + P_{2k} \cdot (1 - y_k) + P_{3k} \cdot \frac{N_b}{N} + P_{4k}.$$

Let $u_k \equiv P_{2k} + P_{3k} \cdot \frac{N_b}{N} + P_{4k}$ and $v_k = P_{1k} - P_{2k}$ with

$$P_{1k} = \begin{cases} 1 & \ldots \text{ if } k \in U_D \\ P_{1R} & \ldots \text{ otherwise} \end{cases}$$

and

$$P_{ik} = \begin{cases} P_{iR} & \ldots \text{ if } k \in U_R \\ 0 & \ldots \text{ otherwise} \end{cases}$$

(for $i=2, 3, 4, 5$). Hence, for questioning design R, the expression

$$y'_k = \frac{z_k - u_k}{v_k} \tag{5}$$

is unbiased for the true y_k ($k \in U$). For $u_R \equiv P_{2R} + P_{3R} \cdot \frac{N_b}{N} + P_{4R}$ and $v_R = P_{1R} - P_{2R}$, this leads to

$$u_k = \begin{cases} u_R & \ldots \text{ if } k \in U^R \\ 0 & \ldots \text{ otherwise} \end{cases}$$

and

$$v_k = \begin{cases} 1 & \ldots \text{ if } k \in U^D \\ v_R & \ldots \text{ otherwise.} \end{cases}$$

Using these y'_k-values in (5) as substitutes of the values y_k in (1), the following theorem applies.

Theorem. *For questioning design Q, combining values of the variable y under study masked by the RR method R and true values delivered by respondents answering the direct question on y without request,*

$$\widehat{N}_{a,Q} = \sum\nolimits_s y'_k \cdot d_k \tag{6}$$

is an unbiased estimator of N_a for any probability sampling scheme P with inclusion probabilities π_k ($k=1, 2, \ldots, N$). The estimator $\widehat{N}_{a,Q}$ has a theoretical variance of

$$V\left(\widehat{N}_{a,Q}\right) = \sum\sum\nolimits_U \Delta_{kl} \cdot \frac{y_k}{\pi_k} \cdot \frac{y_l}{\pi_l} + \frac{u_R(1-u_R)}{v_R^2} \cdot \sum\nolimits_{U_R} d_k + \frac{1 - v_R - 2u_R}{v_R} \cdot \sum\nolimits_{U_R} y_k \cdot d_k. \tag{7}$$

In (7), the first summand is the variance of the Horvitz–Thompson estimator for full response. This means that with the two other summands of the variance $V\left(\widehat{N}_{a,Q}\right)$, the interviewer has to pay the price in terms of accuracy for the higher protection of privacy of the respondents, which in turn shall lead to higher response rates. For $P_{1R}=1$ for all $k \in U$, $V\left(\widehat{N}_{a,Q}\right)$ reduces to the variance (2) of the Horvitz–Thompson estimator.

Provided that $\pi_{kl}>0$ for $k, l \in U$, an unbiased estimator of $V\left(\widehat{N}_{a,Q}\right)$ is given by

$$\widehat{V}\left(\widehat{N}_{a,Q}\right) = \sum\sum\nolimits_s \frac{\Delta_{kl}}{\pi_{kl}} \cdot \frac{y'_k}{\pi_k} \cdot \frac{y'_l}{\pi_l} + \frac{u_R(1-u_R)}{v_R^2} \cdot \sum\nolimits_{s \cap U_R} d_k + \frac{1 - v_R - 2u_R}{v_R} \cdot \sum\nolimits_{s \cap U_R} y'_k \cdot d_k. \tag{8}$$

Proof. E_P and E_Q denote the expectations over the probability sampling scheme P and questioning design Q, respectively. Because of $E_Q(y_k')=y_k$, the expected value of (6) is

$$E\left(\widehat{N}_{a,Q}\right)=E_P\left[E_Q\left(\sum_s y_k'\cdot d_k|s\right)\right]=E_P\left[\sum_s y_k\cdot d_k\right]=\sum_U y_k=N_a,$$

that proves its unbiasedness for N_a.

For (7), the theoretical variance of $\widehat{N}_{a,Q}$ can be decomposed through:

$$V\left(\widehat{N}_{a,Q}\right)=V_P\left(E_Q\left(\widehat{N}_{a,Q}|s\right)\right)+E_P\left(V_Q\left(\widehat{N}_{a,Q}|s\right)\right)$$

with V_P and V_Q denoting the variances over P and Q, respectively. The first summand on the right-hand side of this equation leads to

$$V_P\left(E_Q\left(\widehat{N}_{a,Q}|s\right)\right)=V_P\left(\sum_s y_k\cdot d_k\right)=\sum\sum_U \Delta_{kl}\cdot\frac{y_k}{\pi_k}\cdot\frac{y_l}{\pi_l},$$

the variance of the common Horvitz–Thompson estimator of N_a. Moreover, for $V_Q\left(\widehat{N}_{a,Q}|s\right)$ in $E_P\left(V_Q\left(\widehat{N}_{a,Q}|s\right)\right)$ from the variance decomposition above,

$$\begin{aligned}V_Q\left(\widehat{N}_{a,Q}|s\right)&=V_Q\left(\sum_s y_k'\cdot d_k|s\right)=V_Q\left(\sum_s y_k'\cdot d_k\cdot I_k|s\right)\\&=V_Q\left(\sum_{U_D} y_k\cdot d_k\cdot I_k|s+\sum_{U_R}\frac{z_k-u_R}{v_R}\cdot d_k\cdot I_k|s\right)\\&=\sum_{U_R} d_k\cdot V_Q\left(\frac{z_k-u_R}{v_R}\right)\end{aligned}$$

applies. The variance $V_Q\left(\frac{z_k-u_R}{v_R}\right)$ over the questioning design Q among the population elements $k\in U_R$ results in

$$V_Q\left(\frac{z_k-u_R}{v_R}\right)=\frac{1}{v_R{}^2}\cdot V_Q(z_k).$$

Hence,

$$\begin{aligned}V_Q(z_k)&=E_Q\left(z_k{}^2\right)-E_Q{}^2(z_k)\\&=v_R\cdot E_Q\left(y_k'\right)+u_R-\left(v_R\cdot E_Q\left(y_k'\right)+u_R\right)^2\\&=u_R\cdot(1-u_R)+v_R\cdot(1-v_R-2\cdot u_R)\cdot y_k,\end{aligned}$$

which results in

$$V_Q\left(\frac{z_k-u_R}{v_R}\right)=\frac{u_R(1-u_R)}{v_R{}^2}+\frac{1-v_R-2\cdot u_R}{v_R}\cdot y_k.$$

Hence, the expected value $E_P\left(V_Q\left(\widehat{N}_{a,Q}|s\right)\right)$ is given by

$$E_P\left(V_Q\left(\widehat{N}_{a,Q}|s\right)\right) = \frac{u_R(1-u_R)}{v_R^2}\cdot\sum\nolimits_{U_R} d_k + \frac{1-v_R-2u_R}{v_R}\cdot\sum\nolimits_{U_R} d_k\cdot y_k,$$

which completes the proof of the expression of the theoretical variance $V\left(\widehat{N}_{a,Q}\right)$ as presented in (7).

For the proof of the unbiasedness of $\hat{V}\left(\widehat{N}_{a,Q}\right)$ for $V\left(\widehat{N}_{a,Q}\right)$, the expectation of the first summand on the right-hand side of Eq. (8) over the questioning design Q is given by

$$\begin{aligned} E_Q\left(\sum\sum\nolimits_s \frac{\Delta_{kl}}{\pi_{kl}}\cdot\frac{y_k'}{\pi_k}\cdot\frac{y_l'}{\pi_l}\right) &= \sum\sum\nolimits_U \frac{\Delta_{kl}}{\pi_{kl}}\cdot\frac{E_Q(y_k'y_l')}{\pi_k\pi_l} I_k\cdot I_l \\ &= \sum\nolimits_U \frac{\Delta_{kk}}{\pi_k}\cdot\frac{E_Q\left[(y_k')^2\right]}{\pi_k^2}\cdot I_k^2 \\ &\quad + \sum\sum\nolimits_{U(k\neq l)} \frac{\Delta_{kl}}{\pi_{kl}}\cdot\frac{E_Q(y_k'y_l')}{\pi_k\pi_l} I_k\cdot I_l \end{aligned}$$

For $k\in U_R$, the expectation is given by

$$\begin{aligned} E_Q\left[(y_k')^2\right] &= \frac{1}{v_R^2}\cdot E_Q\left(z_k^2 - 2\cdot u_R\cdot z_k + u_R^2\right) \\ &= \frac{1}{v_R^2}\cdot\left[v_R\cdot y_k + u_R - 2\cdot u_R\cdot(v_R\cdot y_k + u_R) + u_R^2\right] \\ &= \frac{1-2u_R}{v_R}\cdot y_k + \frac{u_R(1-u_R)}{v_R^2}. \end{aligned}$$

This results in

$$E_Q\left[(y_k')^2\right] = \begin{cases} y_k & \ldots \text{ if } k\in U_D \\ \dfrac{1-2\cdot u_R}{v_R}\cdot y_k + \dfrac{u_R(1-u_R)}{v_R^2} & \ldots \text{ otherwise.} \end{cases}$$

Continuing the development from the above, with the variance $\Delta_{kk} = \pi_k\cdot(1-\pi_k)$ of the sample membership indicators, the expectation is expressed by

$$\begin{aligned} E_Q\left(\sum\sum\nolimits_s \frac{\Delta_{kl}}{\pi_{kl}}\cdot\frac{y_k'}{\pi_k}\cdot\frac{y_l'}{\pi_l}\right) &= \sum\sum\nolimits_{U(k\neq l)} \frac{\Delta_{kl}}{\pi_{kl}}\cdot\frac{y_k}{\pi_k}\cdot\frac{y_l}{\pi_l}\cdot I_k\cdot I_l \\ &\quad + \sum\nolimits_{U_D}(1-\pi_k)\cdot\frac{y_k}{\pi_k^2}\cdot I_k^2 \\ &\quad + \frac{1}{v_R^2}\cdot\sum\nolimits_{U_R}(1-\pi_k)\cdot\frac{v_R(1-2u_R)y_k + u_R(1-u_R)}{\pi_k^2}\cdot I_k^2 \end{aligned}$$

Furthermore, the expected value over sampling scheme P is given by

$$\begin{aligned}
&E_P\left[E_Q\left(\sum\sum_s \frac{\Delta_{kl}}{\pi_{kl}} \cdot \frac{y'_k}{\pi_k} \cdot \frac{y'_l}{\pi_l}\right)\right] \\
&= \sum\sum_{U(k \neq l)} \Delta_{kl} \cdot \frac{y_k}{\pi_k} \cdot \frac{y_l}{\pi_l} \\
&\quad + \sum_{U_D} (1-\pi_k) \cdot \frac{y_k}{\pi_k} + \frac{1}{v_R^2} \cdot \sum_{U_R} (1-\pi_k) \cdot \frac{v_R(1-2u_R)y_k + u_R(1-u_R)}{\pi_k} \\
&= \sum\sum_U \Delta_{kl} \cdot \frac{y_k}{\pi_k} \cdot \frac{y_l}{\pi_l} - \sum_U (1-\pi_k) \cdot \frac{y_k}{\pi_k} + \sum_{U_D} (1-\pi_k) \cdot \frac{y_k}{\pi_k} \\
&\quad + \frac{1}{v_R^2} \cdot \sum_{U_R} (1-\pi_k) \cdot \frac{v_R(1-2u_R)y_k + u_R(1-u_R)}{\pi_k} \\
&= \sum\sum_U \Delta_{kl} \cdot \frac{y_k}{\pi_k} \cdot \frac{y_l}{\pi_l} - \sum_{U_R} (1-\pi_k) \cdot \frac{y_k}{\pi_k} + \frac{1-2u_R}{v_R} \cdot \sum_{U_R} (1-\pi_k) \cdot \frac{y_k}{\pi_k} \\
&\quad + \frac{u_R(1-u_R)}{v_R^2} \cdot \sum_{U_R} \frac{1-\pi_k}{\pi_k} \\
&= \sum\sum_U \Delta_{kl} \cdot \frac{y_k}{\pi_k} \cdot \frac{y_l}{\pi_l} + \frac{1-v_R-2u_R}{v_R} \cdot \sum_{U_R} (1-\pi_k) \cdot \frac{y_k}{\pi_k} \\
&\quad + \frac{u_R(1-u_R)}{v_R^2} \cdot \sum_{U_R} \frac{1-\pi_k}{\pi_k}
\end{aligned}$$

Consequently, for $\hat{V}\left(\widehat{N}_{a,Q}\right)$ to be an unbiased estimator of $V\left(\widehat{N}_{a,Q}\right)$, two terms have to be added to $\sum\sum_s \frac{\Delta_{kl}}{\pi_{kl}} \cdot \frac{y'_k}{\pi_k} \cdot \frac{y'_l}{\pi_l}$. The first term is $\frac{u_R(1-u_R)}{v_R^2} \cdot \sum_{s \cap U_R} d_k$ with $s \cap U_R$ denoting the part of s that consists of elements belonging to the response behavior group U_R. For this term, $E_P\left(\sum_{s \cap U_R} d_k\right) = N_R$ applies. The second term to be added is $\frac{1-v_R-2u_R}{v_R} \cdot \sum_{s \cap U_R} y'_k \cdot d_k$ with $E_P\left[E_Q\left(\sum_{s \cap U_R} y'_k \cdot d_k\right)\right] = \sum_{U_R} y_k$. This completes the proof of the unbiasedness of variance estimator (8).

Applying these results to simple random sampling without replacement (SI), for instance, estimator (5) is presented by

$$\widehat{N}^{SI}_{a,Q} = N \cdot \frac{1}{n} \cdot \sum_s y'_k = N \cdot \overline{y'}_s$$

with $\overline{y'}_s$, the sample mean of all y_k'-values in s. According to (7), for the SI case, the theoretical variance is given by

$$\begin{aligned}
V\left(\widehat{N}^{SI}_{a,Q}\right) = {} & N^2 \cdot \frac{N-n}{N-1} \cdot \frac{\frac{N_a}{N}\left(1-\frac{N_a}{N}\right)}{n} + \frac{N}{n} \cdot N_R \cdot \frac{u_R(1-u_R)}{v_R^2} \\
& + \frac{N}{n} \cdot \frac{1-v_R-2u_R}{v_R} \cdot \sum_{U_R} y_k,
\end{aligned}$$

where $\sum_{U_R} y_k$ denotes the number of units of group U_R bearing attribute a.

Eventually, for SI sampling, the estimator (8) of this theoretical variance is given by

$$\hat{V}\left(\widehat{N}_{a,Q}^{\mathrm{SI}}\right) = N^2 \cdot \frac{N-n}{N} \cdot \frac{S^2_{y',s}}{n} + \frac{N}{n} \cdot n_{s\cap U_R} \cdot \frac{u_R(1-u_R)}{{v_R}^2} + N \cdot \frac{1-v_R-2u_R}{v_R} \cdot \overline{y'}_{s\cap U_R}$$

with $S^2_{y',s}$, the sample variance of the observed y_k'-values. Moreover, the term $n_{s\cap U_R}$ denotes the number of sample elements using the RR strategy R, and $\overline{y'}_{s\cap U_R}$ denotes the mean value of the y_k'-values over these sample units.■

4 A VIVID ILLUSTRATION OF THIS STRATEGY INCLUDING TRUE AND MASKED RESPONSES

In this contribution, the parameter of interest is the total $N_a = \sum_U y_k$ of the binary variable y indicating whether a population unit k possesses a (sensitive) attribute a or not. Regardless of the chosen probability sampling method P with arbitrary sample inclusion probabilities π_k $(k \in U)$, the Horvitz–Thompson estimator (1),

$$\widehat{N}_{a,HT} = \sum_s y_k \cdot d_k,$$

is an unbiased estimator of N_a. Looking at this expression, the rationale behind this estimation approach can be described by the generation of an artificial population estimating the original population with respect to variable y under study, or to be more precise: with respect to the total of y.

As shown in Fig. 1, the process starts with population U. Each element k of U is assigned a fixed value y_k of variable y, and the (unknown) parameter N_a is the sum of all of these y-values. In the next step, one of all possible samples, which can be drawn according to a given probability sampling scheme P with individual inclusion probabilities π_k $(k = 1, \ldots, N)$, is realized. A certain

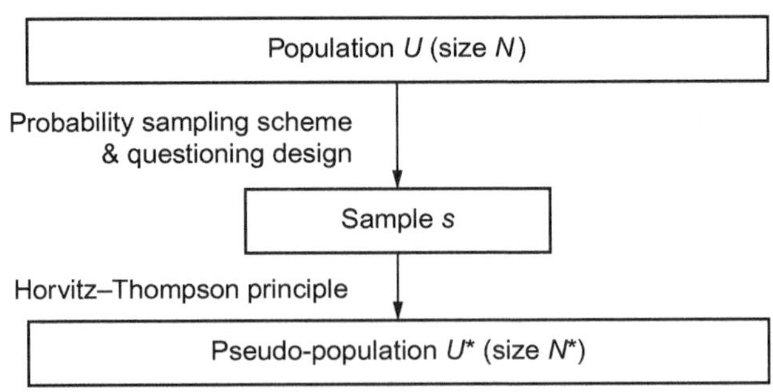

FIG. 1 The Horvitz–Thompson principle of estimating a total under a certain questioning design.

questioning design such as direct questioning, or the RR questioning design R or Q, is applied to this sample s of n elements to observe relevant information about y.

For the direct questioning, based on the true y-values, the Horvitz–Thompson estimation principle as shown in formula (1) estimates the original population U of size N with respect to the total N_a of variable y by a pseudo-population U^*. For the actual generation of U^*, according to (1), the variable value y_1 of the first element in the sample is replicated d_1 times delivering d_1 "clones" of y_1 for the pseudo-population; the value y_2 of the second sample element is replicated d_2 times delivering d_2 such replicates of y_2, and so on. Hence, the design weights $d_k = \frac{1}{\pi_k}$ can be seen as the replication factors of the pseudo-population concept. Consequently, pseudo-population U^* has $N^* = \sum_s d_k$ elements. The expectation of size N^* is N, the size of the original population U:

$$E(N^*) = E\left(\sum_s d_k\right) = E\left(\sum_U d_k \cdot I_k\right) = N$$

In U^*, the total N_a of y in U is estimated by the total $\widehat{N}_{a,HT}$ of the replications of y in U^*. Assuming that all design weights were integers, with these replications denoted as y^*, the Horvitz–Thompson estimator of the size

$$N_a = \sum_U y_k$$

is presented by

$$\widehat{N}_{a,HT} = \sum_{U^*} y_k^*.$$

This means that the sum of the y-values in U is estimated by the sum of the replicated y-values in U^*.

Applying the RR technique R discussed in Section 2 of this contribution as a questioning design to this process instead of direct questioning (Fig. 1), in obtaining information on y in the sample s, the true y-values are substituted by masked values y' ($k=1, \ldots, n$). Hence, in the concept of pseudo-populations, these values are replicated according to their design weights as replication factors to generate pseudo-population U^* in Fig. 1. This adds to the inefficiency of the estimation process, when we sum up all the y'-values in U^* to estimate the sum of the true y-values in U. The added inaccuracy due to this masking process can be calculated according to (7) when $U_R = U$ for the last two summands.

Eventually, applying the questioning design Q instead of R, which was theoretically developed in the Theorem in Section 3, some of the units in s deliver their true values of y without being asked to do so. Therefore, in the part of s that belongs to the subgroup U_D of U ($s \cap U_D$), to generate U^*, these true values are replicated according to the design weights assigned to these sample units, whereas in the remaining group of s ($s \cap U_R$), the masked values y_k' are cloned according to the design weights of these elements. Therefore, the resulting pseudo-population U^* (of the same size as the pseudo-population for the direct questioning design or questioning design R) consists of a mixture of replicated true and masked values of y. Also in this case, the total N_a of y in U is estimated by the total $\widehat{N}_{a,Q}$ of the replications y^* in U^*.

In the presence of nonresponse, Fig. 1 changes because the true and masked values of y are not observed for all units of s (Fig. 2). As discussed in Section 1, data imputation is frequently used to compensate for item nonresponse and then followed by weighting adjustment to compensate also for the unit nonresponse that has occurred. When these methods are applied to data observed by the use of the very general questioning design Q, this means that a substitute value has to be imputed for each unobserved true or masked value y_k' in the item nonresponse subset of the set s_m consisting of all sample elements with missing observations on y. This is done according to a model of the nonresponse mechanism as discussed in Section 1.

For this purpose, for each unit k belonging to the item nonresponse subset of s_m, a presumption of the membership of U_D or U_R, respectively, might be used to build imputation classes (cf., for instance, Särndal et al., 1992, section 15.7).

The result is an "imputed sample" s_I as part of s consisting of actually observed (true or masked) and imputed values. Subsequently, to compensate also for the unit nonresponse that has occurred, the design weights of the elements of s_I have to be adjusted. This is done with the help of a model with

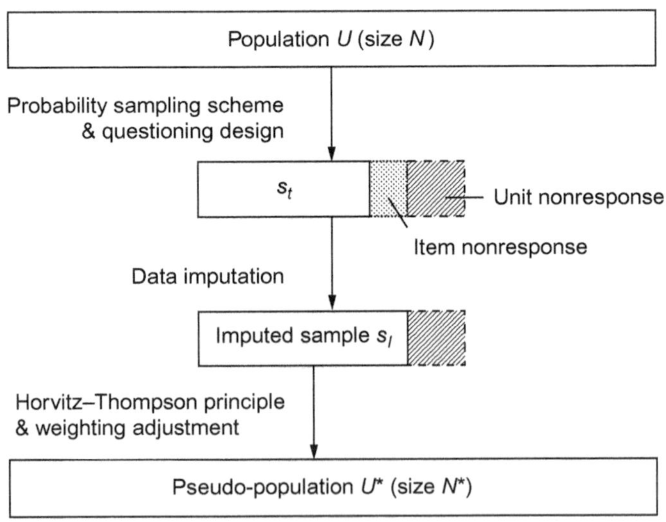

FIG. 2 The Horvitz–Thompson principle of estimating a total under a certain questioning design in the presence of nonresponse using imputation for item nonresponse and weighting adjustment for unit nonresponse.

respect to the response probabilities for the elements of s_I. For $s_I \neq s$, the individual sample elements have to bear a heavier representative burden because of the occurring unit nonresponse.

With the picture of the pseudo-population concept, this two-stage process can be described in the following way: For the generation of U^* based on the sample s (Fig. 1), in the presence of item and unit nonresponse, after data imputation, the values only of the imputed sample s_I are replicated to build U^*. For this purpose, the design weights would be too small as replication factors. The adjustment of the design weights assigns an increased number of replicates to each element of s_I to form a pseudo-population U^* that can be reasonably used to estimate the interesting parameter N_a.

REFERENCES

Chaudhuri, A., Christofides, T.C., 2013. Indirect Questioning in Sample Surveys. Springer, Heidelberg.

Fidler, D.S., Kleinknecht, R.E., 1977. Randomized response versus direct questioning: two data collection methods for sensitive information. Psychol. Bull. 84 (5), 1045–1049.

Greenberg, B.G., Abul-Ela, A.-L.A., Simmons, W.R., Horvitz, D.G., 1969. The unrelated question randomized response model: theoretical framework. J. Am. Stat. Assoc. 64, 520–539.

Groves, R.M., Fowler, F.J., Couper, M.P., Lepkowski, G.M., Singer, E., Tourangeau, R., 2004. Survey Methodology. John Wiley & Sons, Hoboken.

Horvitz, D.G., Thompson, D.J., 1952. A generalization of sampling without replacement from a finite universe. J. Am. Stat. Assoc. 47, 663–685.

Horvitz, D.G., Shah, B.V., Simmons, W.R., 1967. The unrelated question randomized response model. In: Proceedings of the Social Statistics Section of the American Statistical Association, pp. 65–72.

Little, R.J.A., Rubin, D.B., 2002. Statistical Analysis with Missing Data. John Wiley & Sons, Hoboken.

Lohr, S.L., 2012. Sampling: Design and Analysis. Brooks/Cole Cengage Learning, Boston.

Quatember, A., 2009. A standardization of randomized response strategies. Surv. Method. 35 (2), 143–152.

Quatember, A., 2012. An extension of the standardized randomized response technique to a multi-stage setup. Stat. Methods Appl. 21 (4), 475–484.

Quatember, A., 2015. Pseudo-Populations-A Basic Concept in Statistical Surveys. Springer, Heidelberg.

Särndal, C.-E., Swensson, B., Wretman, J., 1992. Model Assisted Survey Sampling. Springer, New York.

Singh, R., Singh, S., Mangat, N.S., Tracy, D.S., 1995. An improved two stage randomized response strategy. Stat. Pap. 36, 265–271.

Warner, S.L., 1965. Randomized response: a survey technique for eliminating evasive answer bias. J. Am. Stat. Assoc. 60, 63–69.

Chapter 8

Estimation of Complex Population Parameters Under the Randomized Response Theory

L. Barabesi*,[1], G. Diana† and P.F. Perri‡

**University of Siena, Siena, Italy*
†University of Padova, Padova, Italy
‡University of Calabria, Arcavacata di Rende (CS), Italy
[1]Corresponding author: e-mail: lucio.barabesi@unisi.it

ABSTRACT

When a sensitive quantitative variable is under study and the Randomized Response Theory is adopted, a great deal of literature has been devoted to the estimation of the population mean (or total) or – at most – simple functions of population totals. However, in many real surveys the main interest might rely on the estimation of a complex parameter, usually a nonlinear combination of population totals. Hence, in order to face with this problem, we suppose to collect data by means of the well-known unrelated question method proposed by Greenberg et al. (1971), and under the design-based framework, we propose to handle such a complex parameter as a population functional by suitably extending the linearization approach proposed by Deville (1999). The considered strategy permits to obtain parameter estimation by means of the substitution method based on the empirical functional, and to achieve the corresponding variance estimator. Some selected illustrative examples are provided mostly concerning the estimation of two inequality indices, namely the Gini concentration index and the Atkinson index, widely discussed in the social and economic literature.

Keywords: Design-based approach, Unrelated Question Model, Functional linearization, Influence function, Gini index, Atkinson index

1 INTRODUCTION

Since Warner's (1965) seminal work on the Randomized Response (RR) Theory (RRT), the literature on the topic has been devoted to conceive new randomization devices aimed at increasing respondents' cooperation and

Handbook of Statistics, Vol. 34. http://dx.doi.org/10.1016/bs.host.2016.01.007

obtaining reliable estimates of unknown target parameters. In the course of time, researchers have mostly focused their attention on the estimation of "simple" population parameters, mainly the prevalence of individuals bearing one or more stigmatizing attributes, or the mean of a quantitative sensitive variable. A full description of the methods may be found, among others, in the monographs by Fox and Tracy (1986), Chaudhuri and Mukerjee (1988), Chaudhuri (2011), and Chaudhuri and Christofides (2013). Nonetheless, researchers, survey practitioners, and final users of sensitive data could be interested in other population characteristics, usually "complex" parameters, or their concerns could rely on nonstandard situations. For instance, Singh et al. (1996) and Strachan et al. (1998) discussed regression analysis under an RR setup; Perri and van der Heijden (2012) investigated the use of the RRT for classification purposes; Singh et al. (2014) employed auxiliary information to obtain more efficient estimates of the variance of a quantitative sensitive variable; Poole (1974), Duffy and Waterton (1984), and Barabesi et al. (2013) considered the estimation of the distribution function of a quantitative sensitive variable; Barabesi et al. (2015) directed their attention to the estimation of the Gini concentration index; Christofides (2005) and Lee et al. (2013) introduced some RR devices to elicit information about two sensitive attributes for the same population at the same time; Barabesi et al. (2014) analyzed the nonstandard situation where the RRT is applied in a nonresponse setup; Singh et al. (1998), Arnab and Singh (2002), and Saha (2007) discussed the simultaneous estimation of a hidden group size and the mean value (or total) of a sensitive quantitative variable within the group; Barabesi et al. (2013) focused on the problem of estimating simultaneously the size of a hidden gang and the distribution function of a sensitive variable for the members of the group.

It is at once apparent that, despite sporadic attempts, the use of the RRT in nonstandard situations is still very limited and this might preclude a complete awareness of the effectiveness of the approach in real analyses and slow down its use in large-scale surveys. This chapter would, therefore, be a contribution to fill this gap. Hence, motivated by the more and more increasing interest in the RRT, and aimed at spreading new ideas for applied researches, we provide, under a generic sampling design, some results for estimating a complex population parameter—rephrased in terms of a statistical functional—and for obtaining variance estimation. In particular, moving from Deville's (1999) linearization approach based on the concept of influence function in the design-based framework and conceived for data collected by direct questioning survey modes, we extend the approach to the case of quantitative sensitive data collected by means of the Unrelated Question Model (UQM) proposed by Greenberg et al. (1971). The discussed results turn out to be particularly useful for the estimation of inequality indices generally computed on data concerning sensitive variables as income or wealth. We hope, therefore, that the content of this chapter may be valuable not only for researchers in the field of the RRT but also for economists, sociologists, and survey practitioners.

The remaining part of this chapter is organized as follows. In Section 2, we introduce the basics of the linearization approach with a focus on the technique proposed by Deville (1999) and on a computational rule for deriving the linearized quantities of a complex population functional. In Section 3, we formalize an extension of Deville's approach to the case where data are supposed to be collected in the RR setting by means of the UQM. Two simple examples concerning the estimation of a population total and the population variance are illustrated in Section 4, while Section 5 is devoted to the study of two inequality indices, namely the Gini index and the Atkinson index. Some final remarks are given in Section 6.

2 FOUNDATIONS OF FUNCTIONAL LINEARIZATION

Under the usual design-based approach, let $U=\{1,\ldots,N\}$ be a fixed population of identifiable individuals labeled by the first N integers and let y_i be the variable value on the i-th individual. Moreover, let $\theta=\theta(y_1,\ldots,y_N)$ be an unknown population parameter to be estimated from a random sample S of size n selected from U on the basis of a given sampling design. In addition, let π_i and π_{ij} denote the first- and second-order positive inclusion probabilities, respectively, and let $w_i=\pi_i^{-1}$ be the design weight for the i-th population unit.

The linearization approach provides an approximation of the parameter θ, as well as a "stochastic" approximation of its estimator $\hat{\theta}$. More precisely, the rationale underlying linearization consists in obtaining a linearized quantity v_i for each observation y_i in such a way that

$$\hat{\theta}-\theta=\sum_{i\in S} w_i v_i-\sum_{i\in U} v_i+R_p,$$

where R_p represents a stochastically negligible remainder term. Hence, the variance of the random variable $\sum_{i\in S} w_i v_i$ may be used to approximate the variance of the estimator $\hat{\theta}$.

Several techniques have been proposed to provide the v_i's. In this chapter, we consider Deville's (1999) approach stemming from the concept of influence function in the design-based framework. In this setting, a discrete measure on $\mathbb{R}$ is considered, say

$$M=\sum_{i\in U}\delta_{y_i},$$

where δ_y represents the Dirac mass at y, and the target population parameter θ is reformulated as a functional F with respect to M, namely $\theta=F(M)$. In addition, let $\hat{M}$ be the empirical measure corresponding to M and given by

$$\hat{M}=\sum_{i\in S} w_i\delta_{y_i}.$$

Hence, an estimator of θ based on the sample S can be achieved by means of the so-called substitution principle which consists in plugging $\hat{M}$ in $\theta = F(M)$ and obtaining $\hat{\theta} = F(\hat{M})$. As an easy example, if the population total is considered, then $\theta = \int y dM(y) = \sum_{i \in U} y_i$, while $\hat{\theta} = \int y d\hat{M}(y) = \sum_{i \in S} w_i y_i$.

By requiring some assumptions on the sampling design, if F is homogeneous of degree α (i.e., there exists $\alpha > 0$ such that $F(rM) = r^{\alpha} F(M)$ for each $r > 0$) and is Fréchet differentiable, Deville (1999) has shown that the following linearization holds

$$N^{-\alpha}(\hat{\theta} - \theta) = N^{-\alpha} \int IF_F(u;M) d(\hat{M} - M)(u) + o(\rho(N^{-1}\hat{M}, N^{-1}M)), \quad (1)$$

where

$$IF_F(u;M) = \lim_{\epsilon \to 0} \frac{F(M + \epsilon \delta_u) - F(M)}{\epsilon} \quad (2)$$

represents the influence function in the design-based approach, while ρ is a suitable metric such as the total variation norm or the sup norm. Goga and Ruiz-Gazen (2014) remarked that the strong assumption of Fréchet differentiability can be weakened to Hadamard differentiability if the sup norm is assumed. From a mathematical perspective, $IF_F(u;M)$ is actually the Gâteaux differential of $F(M)$ in the direction of the Dirac mass at u. The role of $IF_F(u;M)$ is central since (1) may be trivially rewritten as

$$N^{-\alpha}(\hat{\theta} - \theta) = N^{-\alpha} \left(\sum_{i \in S} w_i v_{F,i} - \sum_{i \in U} v_{F,i} \right) + o(\rho(N^{-1}\hat{M}, N^{-1}M)),$$

where $v_{F,i} = IF_F(y_i;M)$ is the i-th linearized quantity. Since $v_{F,i}$ is not known in practical cases, its substitution estimator is given by

$$\hat{v}_{F,i} = IF_F(y_i;\hat{M}). \quad (3)$$

Accordingly, the Horvitz–Thompson-type estimator for $Var[\hat{\theta}]$ turns out to be

$$\widehat{Var}[\hat{\theta}] = \sum_{i,j \in S} \frac{\pi_{ij} - \pi_i \pi_j}{\pi_i \pi_j \pi_{ij}} \hat{v}_{F,i} \hat{v}_{F,j}.$$

The direct application of (2) to obtain the influence function may be somewhat tedious. In order to simplify the task, Deville (1999) provided basic rules for deriving the influence function for some selected functionals, while, recently, Barabesi et al. (2016) formalized a user-friendly rule for linearizing a quite general family of complex population functionals. Indeed, let F be the functional

$$F(M) = \int \psi_y(L_y(M)) dM(y),$$

where $L_y(M)=(L_{1,y}(M),\ldots,L_{q,y}(M))^t$ is a vector of functionals (eventually) indexed by y and $\psi_y:\mathbb{R}^q\mapsto\mathbb{R}$ is a function family assumed to be differentiable and regularly indexed by y. If $L_{j,y}(M)$ is Gâteaux differentiable for each j, the influence function of F is given by

$$IF_F(u;M)=\psi_u(L_u(M))+\int\nabla\psi_y(L_y(M))^t IF_{L_y}(u;M)dM(y),$$

where $IF_{L_y}(u;M)=(IF_{L_{1,y}}(u;M),\ldots,IF_{L_{q,y}}(u;M))^t$. Readers interested in the proof of this result are referred to Barabesi et al. (2016). Additionally, if the functional of interest is expressed in the form $\varphi(F(M))=(\varphi\circ(M))$, with $\varphi:\mathbb{R}\mapsto\mathbb{R}$ a smooth and differentiable function, the influence function of $(\varphi\circ F)$ is easily obtained as

$$IF_{\varphi\circ F}(u;M)=\varphi'(F)IF_F(u;M).$$

In Deville's approach, the starting point for obtaining the linearized quantity is represented by the population parameter to estimate. Alternatively, Demnati and Rao (2004) focused on the estimator of the parameter – expressed as a function of sample totals – and not on the parameter. Under this proposal, called Deville-Demnati-Rao approach by Graf and Tillé (2014), the estimated influence function promptly stems from the definition of partial derivative as

$$\hat{v}_{F,i}=\frac{\partial F(\hat{M})}{\partial w_i}. \tag{4}$$

It is worth noting that under this approach, the computation of $\hat{v}_{F,i}$ by means of (4) solely requires ordinary partial derivatives, while (3) involves Gâteaux derivatives which may be rather calculus-demanding in some cases. In any case, we observe that the influence function $IF_F(u;M)$ has its intrinsic interest and it should be computed in order to assess the structure of the parameter to be estimated. Indeed, roughly speaking, a parameter with a "smooth" influence function is easier to estimate than a parameter with a "coarse" influence function.

3 FUNCTIONAL LINEARIZATION WITH THE RR TECHNIQUE

By assuming the RR framework, let us suppose that the UQM is considered. Under this model, the i-th individual reports her/his true value y_i with probability q, while with probability $(1-q)$ she/he generates a value from a suitably chosen random variable X with probability law P_X. Let us denote by $\mu_{X,s}=\int x^s dP_X(x)$ the s-th moment of X–by letting for ease of notation that $\mu_X=\mu_{X,1}$–and assume that $\mu_{X,2}<\infty$. In addition, let us assume that the sampling design and the randomization procedure are independently carried out, as well as the randomization step is independently performed on each individual. According to this scheme, the i-th individual produces a realization of a random variable, say Z_i, with probability law given by the mixture $q\delta_{y_i} + (1-q)P_X$. Thus, in order to extend Deville's approach to the RR setting, we consider the measure on $\mathbb{R}$

$$M_R = \sum_{i \in U} (q\delta_{y_i} + (1-q)P_X) = qM + (1-q)NP_X,$$

in such a way that the empirical measure corresponding to M_R is given by

$$\hat{M}_R = \sum_{i \in S} w_i \delta_{Z_i}.$$

It is worth remarking that $E[\delta_{Z_i}] = q\delta_{y_i} + (1-q)P_X$ and hence $E[\hat{M}_R] = M_R$ for each Borel set in $\mathbb{R}$. Moreover, by suitably reformulating the original functional, i.e. by determining the functional $F_R(M_R) = F(M)$ with a change in measure, the substitution estimator for $\theta = F_R(M_R)$ is obviously given by $\hat{\theta}_R = F_R(\hat{M}_R)$.

Since the Z_i's are assumed to be i.i.d. and independent of the sampling design, by considering a suitable framework similar to that described in Section 2, the following linearization may be achieved

$$N^{-\alpha}(\hat{\theta}_R - \theta) = N^{-\alpha} \int IF_{F_R}(u; M_R) d(\hat{M}_R - M_R)(u) + o(\rho(N^{-1}\hat{M}_R, N^{-1}M_R)),$$

where the influence function is in turn given by

$$IF_{F_R}(u; M_R) = \lim_{\epsilon \to 0} \frac{F_R(M_R + \epsilon\delta_u) - F_R(M_R)}{\epsilon}.$$

Obviously, the Gâteaux derivative rules given by Deville and the results formalized in Section 2 may be also adopted for the computation of $IF_{F_R}(u; M_R)$ once that $F_R(M_R)$ is explicitly achieved. Moreover, the comments given in the previous section also apply in this case, in the sense that $v_{F_R,i} = IF_{F_R}(y_i; M_R)$ is the i-th linearized quantity and the corresponding substitution estimator is given by

$$\hat{v}_{F_R,i} = IF_{F_R}(Z_i; \hat{M}_R). \tag{5}$$

Accordingly, it follows that

$$\widehat{Var}[\hat{\theta}_R] = \sum_{i,j \in S} \frac{\pi_{ij} - \pi_i \pi_j}{\pi_i \pi_j \pi_{ij}} \hat{v}_{F_R,i} \hat{v}_{F_R,j}.$$

Finally, the Deville-Demnati-Rao approach produces

$$\hat{v}_{F_R,i} = \frac{\partial F_R(\hat{M}_R)}{\partial w_i} \tag{6}$$

in the RR framework.

4 SOME SIMPLE EXAMPLES

We consider some easy settings in order to illustrate the linearization technique. For the sake of simplicity, even with a slight abuse in notation, we will use the same symbol for the parameter and the corresponding functional. First, let us consider the functional

$$T_s = T_s(M) = \int y^s dM(y),$$

which encompasses as special cases the population total for $s=1$, which is simply denoted by $T=T(M)=T_1(M)$, and the population size for $s=0$, i.e., $N=N(M)=T_0(M)$. Here, it is assumed that N is unknown and has to be estimated. Equivalently, the parameter T_s may be reformulated as

$$T_s = T_{R,s}(M_R) = \frac{1}{q}\int (y^s - (1-q)\mu_{X,s})dM_R(y),$$

which is simply denoted by $T=T_R(M_R)$ for $s=1$. Hence, the substitution estimator for T_s is given by

$$\hat{T}_{R,s} = T_{R,s}(\hat{M}_R) = \frac{1}{q}\int (y^s - (1-q)\mu_{X,s})d\hat{M}_R(y) = \frac{1}{q}\sum_{i\in S} w_i Z_i^s - \frac{(1-q)\mu_{X,s}}{q}\hat{N},$$

where, since $N(M)=N(M_R)$,

$$\hat{N} = N(\hat{M}_R) = \sum_{i\in S} w_i,$$

while $\hat{T}_R = T_R(\hat{M}_R)$ for $s=1$. Now, on the basis of Example 1 in Deville (1999), it holds

$$IF_{T_{R,s}}(u;M_R) = \frac{1}{q}u^s - \frac{(1-q)\mu_{X,s}}{q},$$

and, by means of (5), the i-th estimated linearized quantity turns out to be

$$\hat{v}_{T_{R,s},i} = \frac{1}{q}Z_i^s - \frac{(1-q)\mu_{X,s}}{q}.$$

The same result can be promptly achieved through (6).

As a second example, let us consider the population variance, which may be written as the functional

$$V = V(M) = \frac{1}{N}T_2 - \frac{1}{N^2}T^2.$$

Equivalently, the parameter may be reformulated as

$$V = V_R(M_R) = \frac{1}{N(M_R)}T_{R,2}(M_R) - \frac{1}{N(M_R)^2}T_R(M_R)^2.$$

Hence, the substitution estimator for V is given by

$$\hat{V}_R = V_R(\hat{M}_R) = \frac{1}{\hat{N}}\hat{T}_{R,2} - \frac{1}{\hat{N}^2}\hat{T}_R^2.$$

By considering the rules given by Deville (1999), it holds

$$IF_{V_R}(u;M_R) = \frac{1}{qN}(u^2 - (1-q)\mu_{X,2}) - \frac{2T}{qN^2}(u - (1-q)\mu_X) - \frac{T_2}{N^2} + \frac{2T^2}{N^3}$$

and from (5) the i-th estimated linearized quantity is given by

$$\hat{v}_{V_R,i} = \frac{1}{q\hat{N}}(Z_i^2 - (1-q)\mu_{X,2}) - \frac{2\hat{T}_R}{q\hat{N}^2}(Z_i - (1-q)\mu_X) - \frac{\hat{T}_{R,2}}{\hat{N}^2} + \frac{2\hat{T}_R^2}{\hat{N}^3}.$$

It is promptly checked that the use of (6) produces the same estimated linearized quantity.

The two previous examples may be differently (and more simply) managed if the population size is assumed to be known, even if this issue implies a different reformulation of the functionals T_s and V. In this case, the influence functions of the two functionals are, respectively, given by

$$IF_{T_{R,s}}(u;M_R) = \frac{1}{q}u^s$$

and

$$IF_{V_R}(u;M_R) = \frac{1}{qN}u^2 - \frac{2T}{qN^2}u$$

and the derivation of the corresponding estimated linearized quantities easily follows.

Finally, on the basis of these two simple examples, it is at once apparent that it is a simple matter to obtain the linearization of parameters given by "smooth" functions of population totals, such as the correlation coefficient. More involved examples will be considered in the next section.

5 FURTHER EXAMPLES WITH SOME INEQUALITY INDICES

We now consider the linearization of two well-known and widely adopted inequality indices, namely the Gini concentration index and the Atkinson index (see, e.g., Cowell, 2011). The estimation of these indices under the RR setup seems to be worthwhile since income is notoriously considered a hard sensitive variable to be surveyed. In fact, it is a well-established fact that surveys based on direct questioning on income have to contend with untruthful reporting or noncooperation mostly in the case of income from self-employment, property, and financial assets. This aspect is generally underrated and data for the analysis are treated as they were released by people who are completely honest in their responses. As stated in Greenberg et al. (1971, p. 247) "*There appears to be considerable evidence in the literature on economic surveys that respondents to direct questions tend to understate income.*" Indeed, while the rich tend to understate income so as to reduce their tax liability, the poor tend to overstate their income for sense of shame or, to a lesser extent, it may happen that taxpayers declare slightly more than they earn to avoid controls. Consequently, nonsampling errors arising from refusal to respond (nonresponse bias) or deliberate falsification of answers (response bias) may make it impossible to achieve estimation of inequality indices at a desired level of accuracy. Hence, the need to investigate the feasibility of the RRT in this setting.

By using the same notation introduced in Section 4, the Gini index may be expressed as (see, e.g., Barabesi et al., 2015)

$$G = G(M) = \frac{2C}{NT} - 1,$$

where

$$C = C(M) = \int K_y(M)dM(y).$$

In addition, let $K_y(\lambda) = \int x I_{[y,\infty)}(x) d\lambda(x)$ be, in general, a functional indexed by y and depending on a given measure λ, with I_B denoting the usual indicator function of a set B. In the following, the functionals $H_y(\lambda) = \int I_{[x,\infty)}(y) d\lambda(x)$ and $L_y(\lambda) = K_y(\lambda) + yH_y(\lambda)$ will be also adopted. As an example, $K_y(M)$ practically represents the total of the population values which are greater or equal to y, while $H_y(M)$ represents the number of population values which are less or equal to y. Hence, the following parameter reformulation holds (Barabesi et al., 2015)

$$G = G_R(M_R) = \frac{2C_R(M_R)}{N(M_R)T_R(M_R)} - 1,$$

where

$$C = C_R(M_R) = \frac{1}{q^2}\alpha - \frac{1-q}{q^2}N\beta + \frac{(1-q)^2\gamma_X}{q^2}N^2,$$

with

$$\alpha = \alpha(M_R) = \int K_y(M_R)dM_R(y),$$

$$\beta = \beta(M_R) = \int L_y(P_X)dM_R(y),$$

and

$$\gamma_X = \int K_y(P_X)dP_X(y).$$

Obviously, γ_X and $L_y(P_X)$ are, respectively, a known constant and a known functional since they are determined by the adopted RR device. Hence, the substitution estimator of the concentration index is given by

$$\hat{G}_R = G_R(\hat{M}_R) = \frac{2\hat{C}_R}{\hat{N}\hat{T}_R} - 1,$$

where

$$\hat{C}_R = C_R(\hat{M}_R) = \frac{1}{q^2}\hat{\alpha} - \frac{1-q}{q^2}\hat{N}\hat{\beta} + \frac{(1-q)^2\gamma_X}{q^2}\hat{N}^2,$$

$$\hat{\alpha} = \alpha(\hat{M}_R) = \sum_{i\in S} w_i K_{Z_i}(\hat{M}_R),$$

$$\hat{\beta} = \beta(\hat{M}_R) = \sum_{i\in S} w_i L_{Z_i}(\hat{M}_R),$$

while

$$K_{Z_i}(\hat{M}_R) = \sum_{j\in S} w_j Z_j I_{[Z_i,\infty)}(Z_j)$$

and

$$L_{Z_i}(\hat{M}_R) = \sum_{j\in S} w_j (Z_j I_{[Z_i,\infty)}(Z_j) + Z_i I_{[Z_j,\infty)}(Z_i)).$$

Hence, on the basis of the results given in Section 2 (for more details, see Barabesi et al., 2015, 2016), it holds that

$$\begin{aligned} IF_{G_R}(u;M_R) = {} & \frac{2}{q^2NT}L_u(M_R) - \frac{2(1-q)}{q^2T}L_u(P_X) - \frac{G+1}{qT}u \\ & + \frac{2(1-q)\beta}{q^2NT} + \frac{4(1-q)^2\gamma_X}{q^2T} - (G+1)\left(\frac{1}{N} - \frac{(1-q)\mu_X}{qT}\right), \end{aligned}$$

which provides

$$\begin{aligned} \hat{v}_{G_R,i} = {} & \frac{2}{q^2\hat{N}\hat{T}_R}L_{Z_i}(\hat{M}_R) - \frac{2(1-q)}{q^2\hat{T}_R}L_{Z_i}(P_X) - \frac{\hat{G}_R+1}{q\hat{T}_R}Z_i \\ & + \frac{2(1-q)\hat{\beta}}{q^2\hat{N}\hat{T}_R} + \frac{4(1-q)^2\gamma_X}{q^2\hat{T}_R} - (\hat{G}_R+1)\left(\frac{1}{\hat{N}} - \frac{(1-q)\mu_X}{q\hat{T}_R}\right). \end{aligned}$$

After some tedious algebra, the same result may be achieved through (6).

With regard to the Atkinson index, using the same notation of the previous section, it may be rewritten as (see, e.g., Barabesi et al., 2016)

$$A_\epsilon = A_\epsilon(M) = 1 - \left(\frac{T_{1-\epsilon}}{N^\epsilon T^{1-\epsilon}}\right)^{1/(1-\epsilon)},$$

where $\epsilon \in [0, 1)$. In this case, the following parameter reformulation holds

$$A_\epsilon = A_{R,\epsilon}(M_R) = 1 - \left(\frac{T_{R,1-\epsilon}(M_R)}{N(M_R)^\epsilon T_R(M_R)^{1-\epsilon}}\right)^{1/(1-\epsilon)}.$$

Hence, the substitution estimator of the index is given by

$$\hat{A}_{R,\epsilon} = A_{R,\epsilon}(\hat{M}_R) = 1 - \left(\frac{\hat{T}_{R,1-\epsilon}}{\hat{N}^\epsilon \hat{T}_R^{1-\epsilon}}\right)^{1/(1-\epsilon)}.$$

By applying the results given in Section 2 (for more details, see Barabesi et al., 2016), it also holds

$$IF_{A_{R,\epsilon}}(u;M_R) = -\frac{(1-A_\epsilon)^\epsilon}{(1-\epsilon)qN^\epsilon T^{1-\epsilon}}u^{1-\epsilon} + \frac{(1-A_\epsilon)^\epsilon T_{1-\epsilon}}{qN^\epsilon T^{2-\epsilon}}u + \frac{(1-q)(1-A_\epsilon)^\epsilon \mu_{X,1-\epsilon}}{(1-\epsilon)qN^\epsilon T^{1-\epsilon}}$$
$$+\frac{\epsilon(1-A_\epsilon)^\epsilon T_{1-\epsilon}}{(1-\epsilon)N^{\epsilon+1}T^{1-\epsilon}} - \frac{(1-q)(1-A_\epsilon)^\epsilon T_{1-\epsilon}\mu_X}{qN^\epsilon T^{2-\epsilon}}$$

which gives rise to

$$\hat{v}_{A_{R,\epsilon},i} = -\frac{(1-\hat{A}_{R,\epsilon})^\epsilon}{(1-\epsilon)q\hat{N}^\epsilon \hat{T}_R^{1-\epsilon}}Z_i^{1-\epsilon} + \frac{(1-\hat{A}_{R,\epsilon})^\epsilon \hat{T}_{R,1-\epsilon}}{q\hat{N}^\epsilon \hat{T}_R^{2-\epsilon}}Z_i + \frac{(1-q)(1-\hat{A}_{R,\epsilon})^\epsilon \mu_{X,1-\epsilon}}{(1-\epsilon)q\hat{N}^\epsilon \hat{T}_R^{1-\epsilon}}$$
$$+\frac{\epsilon(1-\hat{A}_{R,\epsilon})^\epsilon \hat{T}_{R,1-\epsilon}}{(1-\epsilon)\hat{N}^{\epsilon+1}\hat{T}_R^{1-\epsilon}} - \frac{(1-q)(1-\hat{A}_{R,\epsilon})^\epsilon \hat{T}_{R,1-\epsilon}\mu_X}{q\hat{N}^\epsilon \hat{T}_R^{2-\epsilon}}.$$

In turn, the same result may be tediously achieved on the basis of (6).

6 FINAL REMARKS

In this chapter we have intended to draw researchers' attention on the estimation of complex population parameters when data are assumed to be collected through the UQM proposed by Greenberg et al. (1971). We have shown how Deville's (1999) linearization method—widely discussed in the sampling theory—can be extended to the RR setting. Incidentally, we have also discussed the so-called Deville-Demnati-Rao linearization approach. Obviously, other RR mechanisms than the UQM could be considered. Conceptually, no differences emerge but the algebra for deriving the linearized quantities may be somewhat tedious. As final outcome of this chapter, we expect that our contribution can produce some advance in the RRT by inspiring researchers to issue new challenges in nonstandard estimation problems.

ACKNOWLEDGMENTS

The work of P.F. Perri was supported by the Italian project PRIN-2012F42NS8 "Household wealth and youth unemployment: new survey methods to meet current challenges."

REFERENCES

Arnab, R., Singh, S., 2002. Estimation of the size and the mean values of a stigmatized characteristic of a hidden gang in a finite population: a unified approach. Ann. Inst. Stat. Math. 54 (3), 659–666.

Barabesi, L., Diana, G., Perri, P.F., 2013. Design-based distribution function estimation for stigmatized populations. Metrika 76 (7), 919–935.

Barabesi, L., Diana, G., Perri, P.F., 2014. Horvitz-Thompson estimation with randomized response and nonresponse. Model Assist. Stat. Appl. 9 (1), 3–10.

Barabesi, L., Diana, G., Perri, P.F., 2015. Gini concentration index estimation under randomized response sampling. Adv. Stat. Anal. 99 (1), 45–62.

Barabesi, L., Diana, G., Perri, P.F., 2016. Linearization of inequality indices in the design-based framework. Stat. J. Theor. Appl. Stat. http://dx.doi.org/10.1080/02331888.2015.1135924 (in press).

Chaudhuri, A., 2011. Randomized Response and Indirect Questioning Techniques in Surveys. Chapman and Hall/CRC, Boca Raton, FL.

Chaudhuri, A., Christofides, T.C., 2013. Indirect Questioning in Sample Surveys. Springer, Berlin.

Chaudhuri, A., Mukerjee, R., 1988. Randomized Response: Theory and Techniques. Marcel Dekker, Inc., New York.

Christofides, T.C., 2005. Randomized response technique for two sensitive characteristics at the same time. Metrika 62 (1), 53–63.

Cowell, F.A., 2011. Measuring Inequality. Oxford University Press, New York.

Demnati, A., Rao, J.N.K., 2004. Linearization variance estimators for surveys data (with discussion). Surv. Methodol. 30 (1), 17–26.

Deville, J.C., 1999. Variance estimation for complex statistics and estimators: linearization and residual techniques. Surv. Methodol. 25 (2), 193–203.

Duffy, J.C., Waterton, J.J., 1984. Randomized response models for estimating the distribution function of a quantitative character. Int. Stat. Rev. 52 (2), 165–171.

Fox, J.A., Tracy, P.E., 1986. Randomized Response: A Method for Sensitive Survey. Sage Publication, Inc., Newbury Park.

Goga, C., Ruiz-Gazen, A., 2014. Efficient estimation of non-linear finite population parameters by using non-parametrics. J. R. Stat. Soc. B 76 (1), 113–140.

Graf, E., Tillé, Y., 2014. Variance estimation using linearization for poverty and social exclusion indicators. Surv. Methodol. 40 (1), 61–79.

Greenberg, B.G., Kubler, R.R., Abernathy, J.R., Horvitz, D.G., 1971. Applications of the randomized response technique in obtaining quantitative data. J. Am. Stat. Assoc. 66 (334), 243–250.

Lee, C.-S., Sedory, S.A., Singh, S., 2013. Estimating at least seven measures of qualitative variables from a single sample using randomized response technique. Stat. Probab. Lett. 83 (1), 399–409.

Perri, P.F., van der Heijden, P.G.M., 2012. A property of the CHAID partitioning method for dichotomous randomized response data and categorical predictors. J. Classif. 29 (1), 76–90.

Poole, W.K., 1974. Estimation of the distribution function of a continuous type random variable through randomized response. J. Am. Stat. Assoc. 69 (348), 1002–1005.

Saha, A., 2007. A simple randomized response technique in complex surveys. Metron LXV (1), 59–66.

Singh, S., Joarder, A.H., King, M.L., 1996. Regression analysis using scrambled responses. Aust. J. Stat. 38 (2), 201–211.

Singh, S., Horn, S., Chowdhury, S., 1998. Estimation of stigmatized characteristics of a hidden gang in finite population. Aust. N. Z. J. Stat. 40 (3), 291–297.

Singh, S., Sedory, S.A., Arnab, R., 2014. Estimation of finite population variance using scrambled responses in the presence of auxiliary information. Commun. Stat. Simul. Comput. 44 (4), 1050–1065.

Strachan, R., King, M., Singh, S., 1998. Likelihood-based estimation of regression model with scrambled responses. Aust. N. Z. J. Stat. 40 (3), 279–290.

Warner, S.L., 1965. Randomized response: a survey technique for eliminating evasive answer bias. J. Am. Stat. Assoc. 60 (309), 63–69.

Chapter 9

An Efficient Randomized Response Model Using Two Decks of Cards Under Simple and Stratified Random Sampling

S. Abdelfatah[1] and R. Mazloum

Cairo University, Giza, Egypt

[1]*Corresponding author: e-mail: sally_abdelfatah@feps.edu.eg*

ABSTRACT

In an attempt to obtain a more efficient estimator of the population proportion possessing a sensitive characteristic, a randomized response model is proposed as a modification to Abdelfatah et al. (2011) model. In the proposed model, one of the decks includes the same statements used in the first deck of Abdelfatah et al. (2011) model, while the second deck includes the same statements used in the forced response model. Under simple random sampling, the proposed randomized response model has turned out to be more efficient than the Abdelfatah et al. (2011) model. A simulation study is then performed to determine the minimum sample sizes required to obtain possible estimates. The proposed efficient model is then extended to stratified random sampling and the extended model is compared with other stratified randomized response models. Finally, the method of double sampling for stratification is applied in case of unknown strata sizes.

Keywords: Simple random sampling, Stratified random sampling, Estimation of proportion, Randomized response technique, Two decks of card, Forced response model

1 INTRODUCTION

Data collected through direct questioning techniques can be reliable when measuring attitudes or opinions about nonsensitive topics. However, individuals are not so often willing to provide truthful answers about sensitive topics when such direct techniques are used.

Warner (1965) was the first to consider an indirect questioning technique known as randomized response technique. According to this technique, each

Handbook of Statistics, Vol. 34. http://dx.doi.org/10.1016/bs.host.2016.01.008

respondent in a simple random sample with replacement (SRSWR) of n respondents is provided with a random device having the two statements, (i) "I belong to the sensitive group A" with probability P and (ii) "I don't belong to the sensitive group A" with probability $(1-P)$. The respondent is required to provide a "yes" or "no" response depending on the outcome of the random device without reporting his actual classification, whether it is A or not A.

Richard Morton of the University of Sheffield, England, suggested a randomization design that is known as the "Forced Response Model" (Horvitz et al., 1976). Using this model, each respondent in an SRSWR of n respondents is forced by a random device either to answer the sensitive question with probability (P_0) or to give a forced response (yes or no) regardless of his actual situation with probability $(1-P_0)$. Under the case of providing a forced response, the probability of a "yes" response is (w) and the probability of a "No" response is $(1-w)$.

Odumade and Singh (OS) (2009) used two decks of cards. Deck (1) includes the two statements: (i) "I belong to the sensitive group A" with probability P^* and (ii) "I don't belong to the sensitive group A" with probability $(1-P^*)$. Deck (2) includes the two statements as in deck (1), but with probabilities T^* and $(1-T^*)$, respectively. Each respondent in an SRSWR of n respondents is requested to draw two cards simultaneously, one card from each deck, and read the statements in order. The respondent first matches his status with the statement written on the card taken from deck (1), and then he matches his status with the statement written on the card taken from deck (2).

Abdelfatah et al. (2011) modified the structure of one of the decks used by Odumade and Singh (2009). Deck (1) includes the two statements: (i) "I belong to the sensitive group A" with probability W_0 and (ii) "I don't belong to the sensitive group A" with probability $(1-W_0)$. The modified deck includes the statements: (i) "yes" with probability Q_0 and (ii) "no" with probability $(1-Q_0)$. The main advantage of Abdelfatah et al. (2011) model is that it can increase the respondents' cooperation as they will be required to answer the sensitive question only once and not twice as it is the case with the OS (2009) model. In addition, Abdelfatah et al. (2011) estimator can be adjusted to be more efficient than the OS (2009) estimator.

An unbiased estimator of the population proportion π belonging to the sensitive group A is:

$$\hat{\pi}_a = \frac{1}{2} + \frac{Q_0\left(\frac{n_{11}}{n} - \frac{n_{01}}{n}\right) + (1-Q_0)\left(\frac{n_{10}}{n} - \frac{n_{00}}{n}\right)}{2(2W_0-1)\left[Q_0{}^2 + (1-Q_0)^2\right]}, \quad W_0 \neq 0.5$$

where n_{11}/n, n_{10}/n, n_{01}/n, and n_{00}/n are the observed proportions of (yes, yes), (yes, no), (no, yes), and (no, no) responses, respectively.

The variance of the estimator $\hat{\pi}_a$ is given by:

$$V(\hat{\pi}_a) = \frac{1}{4n}\left\{\frac{Q_0{}^3 + (1-Q_0)^3}{(2W_0-1)^2\left[Q_0{}^2 + (1-Q_0)^2\right]^2} - (2\pi-1)^2\right\},\ W_0 \neq 0.5 \quad (1)$$

Abdelfatah and Mazloum (2015) extended Abdelfatah et al. (2011) and Odumade and Singh (2009) randomized response models to stratified random sampling. When comparing the stratified models, it was found that the stratified Abdelfatah et al. model, presented in Fig. 1, can be adjusted to be more efficient than the stratified Odumade and Singh model, and at the same time, it can improve the respondents' cooperation as they will be required to answer the sensitive question only once and not twice as it is the case with the stratified Odumade and Singh model.

According to the stratified Abdelfatah et al. model, an unbiased estimator of the population proportion π is:

$$\hat{\pi}_{sa} = \sum_{h=1}^{H} Z_h \hat{\pi}_{ha} \quad (2)$$

where

$$\hat{\pi}_{ha} = \frac{1}{2} + \frac{Q_{h0}\left(\frac{n_{h11}}{n_h} - \frac{n_{h01}}{n_h}\right) + (1-Q_{h0})\left(\frac{n_{h10}}{n_h} - \frac{n_{h00}}{n_h}\right)}{2(2W_{h0}-1)\left[Q_{h0}^2 + (1-Q_{h0})^2\right]},\quad W_{h0} \neq 0.5,$$

N_h, number of units in stratum h; $N = \sum_{h=1}^{H} N_h$, number of units in the whole population; and $Z_h = \frac{N_h}{N}$, weight of stratum h in the population, $\sum_{h=1}^{H} Z_h = 1$.

Under proportional and Neyman allocations of the sample size, the variances of the stratified estimator $\hat{\pi}_{sa}$ are given, respectively, by:

$$V_{prop}(\hat{\pi}_{sa}) = \frac{1}{4n}\sum_{h=1}^{H} Z_h\left[A_h - (2\pi_h - 1)^2\right], \quad (3)$$

Deck (1*h*)	Deck (2*h*)
$I \in A$ with probability (W_{h0})	Forced **Yes** with probability (Q_{h0})
$I \in A^c$ with probability ($1 - W_{h0}$)	Forced **No** with probability ($1 - Q_{h0}$)

FIG. 1 Statements used in the stratified Abdelfatah et al. model (Abdelfatah and Mazloum, 2015).

$$V_{ney}(\hat{\pi}_{sa}) = \frac{1}{4n}\left[\sum_{h=1}^{H} Z_h \sqrt{A_h - (2\pi_h - 1)^2}\right]^2, \tag{4}$$

where

$$A_h = \frac{Q_{h0}^3 + (1 - Q_{h0})^3}{(2W_{h0} - 1)^2\left[Q_{h0}^2 + (1 - Q_{h0})^2\right]^2}, \quad W_{h0} \neq 0.5.$$

In Section 2, a new randomized response model is proposed as a modification to Abdelfatah et al. (2011) model and the efficiency of the proposed model is examined. Section 3 presents a simulation study to determine the minimum sample sizes required to obtain possible estimates. The extension of the proposed model to stratified random sampling together with an efficiency comparison of the proposed stratified model with other stratified randomized response models is presented in Section 4. Finally, in Section 5, the method of double sampling for stratification is applied to the proposed stratified randomized response model.

2 EFFICIENT RANDOMIZED RESPONSE MODEL UNDER SIMPLE RANDOM SAMPLING

In the present section, a new randomized response model is proposed as a modification to Abdelfatah et al. (2011) model. The idea behind developing this model is to find a model that can improve the efficiency of Abdelfatah et al. (2011) model without losing the advantage of the latter regarding its ability to increase the respondents' cooperation. The proposed model is presented in Section 2.1 and its efficiency comparison with other randomized response models is carried out in Section 2.2.

2.1 The Proposed Model

Each respondent in an SRSWR of n respondents is requested to draw two cards simultaneously: one card from each of the two decks "deck (1) and deck (2)," as shown in Fig. 2, and read the statements in order.

The respondent first matches his actual status with the statement written on the card drawn from deck (1), and then he is supposed to either respond

Deck (1)
$I \in A$ with probability (W_0)
$I \in A^c$ with probability ($1-W_0$)

Deck (2)
$I \in A$ with probability (Q_1)
Forced **Yes** with probability [$0.5(1-Q_1)$]
Forced **No** with probability [$0.5(1-Q_1)$]

FIG. 2 Statements used in the proposed model.

to the sensitive question or say "yes" or "no" regardless of his actual status based on the card drawn from deck (2).

The proposed model can increase the respondents' cooperation because in some cases the respondents will be required to answer the question regarding the sensitive attribute only once and not twice as with the OS (2009) model.

Two situations in which the selected respondent belongs to group A and his response is (yes, yes) are as follows:

- The respondent draws the first card with the statement "$I \in A$" with probability (W_0) from deck (1) and the second card with the statement "$I \in A$" with probability (Q_1) from deck (2).
- The respondent draws the first card with the statement "$I \in A$" with probability (W_0) from deck (1) and the second card with the statement "yes" with probability $[0.5(1-Q_1)]$ from deck (2).

If the respondent belongs to group A^c, his response will be (yes, yes) if he draws the first card with the statement "$I \in A^c$" with probability $(1-W_0)$ from deck (1) and the second card with the statement "yes" with probability $[0.5(1-Q_1)]$ from deck (2).

As shown, the response (yes, yes) can be obtained from respondents belonging to either group A or A^c and hence the privacy of the person reporting (yes, yes) will not be violated.

The probability of getting a (yes, yes) response is given by:

$$
\begin{aligned}
P(\text{Yes, Yes}) = \theta_{11} &= \{W_0[Q_1 + 0.5(1-Q_1)]\}\pi + 0.5(1-W_0)(1-Q_1)(1-\pi) \\
&= [0.5(2W_0-1)(1-Q_1) + W_0Q_1]\pi + 0.5(1-W_0)(1-Q_1).
\end{aligned}
$$

In the same way, the following probabilities can be obtained:

$$
\begin{aligned}
P(\text{Yes, No}) = \theta_{10} &= [0.5(2W_0-1)(1-Q_1) - (1-W_0)Q_1]\pi \\
&\quad + (1-W_0)[Q_1 + 0.5(1-Q_1)] \\
P(\text{No, Yes}) = \theta_{01} &= [(1-W_0)Q_1 - 0.5(2W_0-1)(1-Q_1)]\pi + 0.5W_0(1-Q_1) \\
P(\text{No, No}) = \theta_{00} &= [0.5(1-2W_0)(1-Q_1) - W_0Q_1]\pi + W_0[Q_1 + 0.5(1-Q_1)].
\end{aligned}
$$

The observed proportions of responses (n_{ij}/n) are unbiased estimators for the true proportions of such responses (θ_{ij}) $\forall i,j=0,1$, where $\sum_{i=0}^{1}\sum_{j=0}^{1}\theta_{ij}=1$ and $\sum_{i=0}^{1}\sum_{j=0}^{1}n_{ij}/n=1$.

The estimator $(\hat{\pi}_b)$ of π is chosen so that the squared distance between the observed proportions and the true proportions is minimized or alternatively D is minimized where D is defined as:

$$
D = \frac{1}{2}\sum_{i=0}^{1}\sum_{j=0}^{1}\left(\theta_{ij} - \frac{n_{ij}}{n}\right)^2.
$$

Note that

$$\frac{\partial D}{\partial \pi}=2\left\{[0.5(2W_0-1)(1-Q_1)+W_0Q_1]^2+[0.5(2W_0-1)(1-Q_1)-(1-W_0)Q_1]^2\right\}\pi$$
$$+[0.5(2W_0-1)(1-Q_1)+W_0Q_1]\left(\frac{n_{00}}{n}-\frac{n_{11}}{n}\right)$$
$$+[0.5(2W_0-1)(1-Q_1)-(1-W_0)Q_1]\left(\frac{n_{01}}{n}-\frac{n_{10}}{n}\right)$$
$$-[0.5(2W_0-1)(1-Q_1)+W_0Q_1]^2-[0.5(2W_0-1)(1-Q_1)-(1-W_0)Q_1]^2.$$

Consequently, we have the following theorem.

Theorem 1. *An unbiased estimator of the population proportion* π *is:*

$$\hat{\pi}_b=\frac{1}{2}+\frac{(B-E)\left(\frac{n_{11}}{n}-\frac{n_{00}}{n}\right)+(F-G)\left(\frac{n_{10}}{n}-\frac{n_{01}}{n}\right)}{2\left((B-E)^2+(F-G)^2\right)}. \tag{5}$$

where

$$B=W_0[Q_1+0.5(1-Q_1)],\qquad F=0.5W_0(1-Q_1),$$
$$E=0.5(1-W_0)(1-Q_1),\qquad G=(1-W_0)[Q_1+0.5(1-Q_1)].$$

Theorem 2. *The variance of the estimator* $\hat{\pi}_b$ *is given by:*

$$V(\hat{\pi}_b)=\frac{(B-E)^2(B+E)+(F-G)^2(F+G)}{4n\left[(B-E)^2+(F-G)^2\right]^2}-\frac{(2\pi-1)^2}{4n}. \tag{6}$$

Proof.

$$V\left(\hat{\pi}_b\right)=\frac{(B-E)^2\,V\left(\frac{n_{11}}{n}-\frac{n_{00}}{n}\right)+(F-G)^2\,V\left(\frac{n_{10}}{n}-\frac{n_{01}}{n}\right)+2(B-E)(F-G)\,Cov\left(\frac{n_{11}}{n}-\frac{n_{00}}{n},\frac{n_{10}}{n}-\frac{n_{01}}{n}\right)}{4\left[(B-E)^2+(F-G)^2\right]^2}. \tag{7}$$

Using the following results from the standard multinomial distribution in Eq. (7), it is easy to prove Theorem 2.

$$V(n_{11}/n)=\theta_{11}(1-\theta_{11})/n \qquad Cov(n_{11}/n,n_{10}/n)=-\theta_{11}\theta_{10}/n$$
$$Cov(n_{10}/n,n_{01}/n)=-\theta_{10}\theta_{01}/n \qquad V(n_{10}/n)=\theta_{10}(1-\theta_{10})/n$$
$$Cov(n_{11}/n,n_{01}/n)=-\theta_{11}\theta_{01}/n \qquad Cov(n_{10}/n,n_{00}/n)=-\theta_{10}\theta_{00}/n$$
$$V(n_{01}/n)=\theta_{01}(1-\theta_{01})/n \qquad Cov(n_{11}/n,n_{00}/n)=-\theta_{11}\theta_{00}/n$$
$$Cov(n_{01}/n,n_{00}/n)=-\theta_{01}\theta_{00}/n \qquad \text{and } V(n_{00}/n)=\theta_{00}(1-\theta_{00})/n.$$

It is obvious from Eq. (6) that the variance of $\hat{\pi}_b$ is symmetric about 0.5 for each of the parameters (W_0 and π).□

Theorem 3. *An unbiased estimator of the variance of $\hat{\pi}_b$ is:*

$$\hat{V}(\hat{\pi}_b) = \frac{1}{4(n-1)}\left[\frac{(B-E)^2(B+E)+(F-G)^2(F+G)}{\left[(B-E)^2+(F-G)^2\right]^2} - (2\hat{\pi}_b - 1)^2\right]. \quad (8)$$

The proof is immediate by taking the expected values on both sides of Eq. (8).

2.2 Efficiency Comparison

The relative efficiency of the proposed model with respect to Abdelfatah et al. (2011) model is

$$\mathrm{RE}(a) = \frac{V(\hat{\pi}_a)}{V(\hat{\pi}_b)} \times 100\%$$

where $V(\hat{\pi}_a)$ and $V(\hat{\pi}_b)$ are as given in Eqs. (1) and (6), respectively.

Using the values of W_0, Q_0, and π for which Abdelfatah et al. (2011) model is more efficient than the Odumade and Singh (OS) (2009) model and the values of Q_1 ranging from 0.1 to 0.9 with a step of 0.1, a numerical study to examine the relative efficiency of the proposed model with respect to Abdelfatah et al. (2011) model is carried out.

For all these combinations of parameters, it is found that the proposed model is always more efficient than the Abdelfatah et al. (2011) model and hence more efficient than the OS (2009) model; however, for the sake of simplicity, Table 1 presents the results for only some chosen values of Q_1 (0.1, 0.4, 0.6, and 0.9).

The numerical study shows that:

(1) The value of the RE(a) is symmetric about 0.5 for each of the parameters Q_0, W_0, and π.

(2) The value of the RE(a) increases for values of W_0 close to 0.3, π close to 0, Q_0 close to 0.2, and Q_1 close to 1.

(3) In general, the RE(a) reaches its maximum for each value of π at $W_0=0.3$, $Q_0=0.2$, and $Q_1=0.9$ (as shown in bold in Table 1). Specifically, for $\pi=0.1$, the RE(a) reaches a maximum value of about 959%.

(4) For $Q_0=Q_1$, the RE(a) reaches its maximum for each value of π at $W_0=0.3$ and Q_0 (Q_1)$=0.9$ (as shown in bold in Table 1). In particular, at $\pi=0.1$, the RE(a) reaches a maximum value of about 922%.

(5) A high level of respondents' cooperation can be expected under the case where deck (2) includes a reasonable percentage of the forced statements (yes and no) and deck (1) as well includes reasonable number of both questions. This case can occur for values of W_0 and Q_1 close to 0.5. For instance, at $W_0=0.3$ and $Q_1=0.4$, the proposed model is still more efficient than Abdelfatah et al. (2011) model for all values of Q_0 and π.

TABLE 1 Percent Relative Efficiency of the Proposed Model over Abdelfatah et al. (2011) Model

			π				
W_0	Q_0	Q_1	*0.1*	*0.2*	*0.3*	*0.4*	*0.5*
0.1	0.1	0.1	113.74	110.56	109.06	108.35	108.13
		0.4	113.59	110.44	108.96	108.26	108.05
		0.6	128.70	121.40	118.11	116.58	116.12
		0.9	194.62	162.42	150.21	144.94	143.42
	0.2	0.1	120.28	115.58	113.37	112.32	112.01
		0.4	120.12	115.47	113.27	112.23	111.92
		0.6	136.10	126.92	122.78	120.86	120.29
		0.9	205.82	169.81	156.16	150.26	148.56
	0.3	0.1	116.13	112.39	110.63	109.80	109.55
		0.4	115.97	112.28	110.54	109.71	109.46
		0.6	131.40	123.42	119.82	118.14	117.65
		0.9	198.71	165.12	152.38	146.88	145.30
	0.4	0.1	105.30	104.07	103.49	103.22	103.14
		0.4	105.16	103.96	103.40	103.14	103.06
		0.6	119.15	114.28	112.08	111.06	110.76
		0.9	180.18	152.89	142.55	138.08	136.79
0.2	0.1	0.1	112.31	110.87	110.03	109.59	109.45
		0.4	139.00	133.51	130.45	128.87	128.38
		0.6	190.23	173.70	165.18	160.95	159.66
		0.9	370.78	288.44	254.81	239.84	235.47
	0.2	0.1	117.42	115.38	114.20	113.57	113.37
		0.4	145.33	138.95	135.39	133.55	132.98
		0.6	198.88	180.77	171.43	166.79	165.38
		0.9	387.65	300.17	264.45	248.55	243.91
	0.3	0.1	114.18	112.52	111.56	111.05	110.89
		0.4	141.31	135.50	132.26	130.58	130.06
		0.6	193.39	176.29	167.46	163.08	161.75
		0.9	376.95	292.72	258.33	243.02	238.55

Continued

TABLE 1 Percent Relative Efficiency of the Proposed Model over Abdelfatah et al. (2011) Model—Cont'd

			π				
W_0	Q_0	Q_1	*0.1*	*0.2*	*0.3*	*0.4*	*0.5*
0.2	0.4	0.1	105.73	105.06	104.67	104.46	104.40
		0.4	130.85	126.51	124.09	122.83	122.45
		0.6	179.07	164.59	157.12	153.41	152.28
		0.9	349.04	273.30	242.38	228.61	224.59
0.3	0.1	0.1	114.80	114.07	113.58	113.31	113.22
		0.4	205.88	196.80	191.21	188.16	187.18
		0.6	359.57	323.03	302.65	292.12	288.85
		0.9	**922.17**	**678.92**	**577.93**	**532.64**	**519.39**
	0.2	0.1	119.34	118.38	117.75	117.39	117.28
		0.4	214.02	204.24	198.23	194.94	193.89
		0.6	373.79	335.25	313.76	302.65	299.20
		0.9	**958.65**	**704.62**	**599.14**	**551.84**	**538.01**
	0.3	0.1	116.46	115.64	115.11	114.80	114.70
		0.4	208.85	199.52	193.77	190.63	189.63
		0.6	364.77	327.50	306.71	295.97	292.63
		0.9	935.50	688.31	585.68	539.66	526.19
	0.4	0.1	108.94	108.50	108.21	108.04	107.99
		0.4	195.37	187.19	182.16	179.41	178.53
		0.6	341.22	307.27	288.33	278.54	275.50
		0.9	875.12	645.80	550.58	507.88	495.40

3 SIMULATION STUDY

To reduce the possibility of obtaining impossible estimated values of π that lie outside the unit interval [0,1], Lee et al. (2013) determined the minimum sample size required to obtain possible estimated values of π based on both a good guess of π and the values of the randomization parameters used in the design through a simulation study.

The proposed estimator $\hat{\pi}_b$ in Eq. (5) also can take values outside the unit interval. For instance at $W_0=0.3$, $Q_1=0.4$, $n=30$, $n_{11}=2$, $n_{10}=22$, $n_{01}=2$,

TABLE 2 Minimum Sample Sizes Recommended for the Proposed Model

		π								
W_0	Q_1	*0.1*	*0.2*	*0.3*	*0.4*	*0.5*	*0.6*	*0.7*	*0.8*	*0.9*
0.3	0.4	1320	380	200	130	90	100	140	350	1420
0.3	0.9	280	100	50	40	40	40	60	100	280

and $n_{00}=4$, the proposed estimator $\hat{\pi}_b$ takes a value of -0.333. Following the lines of Lee et al. (2013), a simulation study is performed to determine the minimum sample sizes required to obtain possible values of the proposed estimator $\hat{\pi}_b$.

In the simulation study, the sample sizes were chosen to range from 10 to 10,000 with a step of 10 and the values of π ranged from 0.1 to 0.9 with a step of 0.1. For each combination of n, π, and a specific value of the randomization parameters (W_0 and Q_1), 100,000 samples were simulated, from the multinomial distribution, and each time, the estimate $\hat{\pi}_b$ was computed. The sample sizes beyond which the proportions of impossible estimated values become less than 0.00001 were eliminated and the others were retained; then the minimum sample size required to obtain possible estimated values of π was determined. The minimum sample sizes recommended for the proposed model under different combinations of the randomization parameters are given in Table 2. In addition, Fig. 3 displays a visual representation of the proportions of the impossible estimated values together with the sample sizes for all values of π at $W_0=0.3$ and $Q_1=0.4$.

As shown in Table 2 and Fig. 3, the minimum sample sizes required to obtain possible values of the proposed estimator $\hat{\pi}_b$ increase as $\pi \to 0$ or 1. For instance, the recommended minimum size of the sample at $W_0=0.3$ and $Q_1=0.4$ if the sensitive attribute is rare or predominant ranges approximately from 350 to 1420 depending on the available resources.

4 EFFICIENT RANDOMIZED RESPONSE MODEL UNDER STRATIFIED RANDOM SAMPLING

In Section 4.1, the proposed randomized response model is extended to stratified random sampling and its efficiency is examined relative to other stratified randomized response models in Section 4.2.

4.1 The Proposed Stratified Model

It is assumed that the population is divided into H strata, with known sizes to get the full benefit from stratification. An SRSWR of n_h respondents is independently selected from stratum h, $h=1, 2, \ldots, H$.

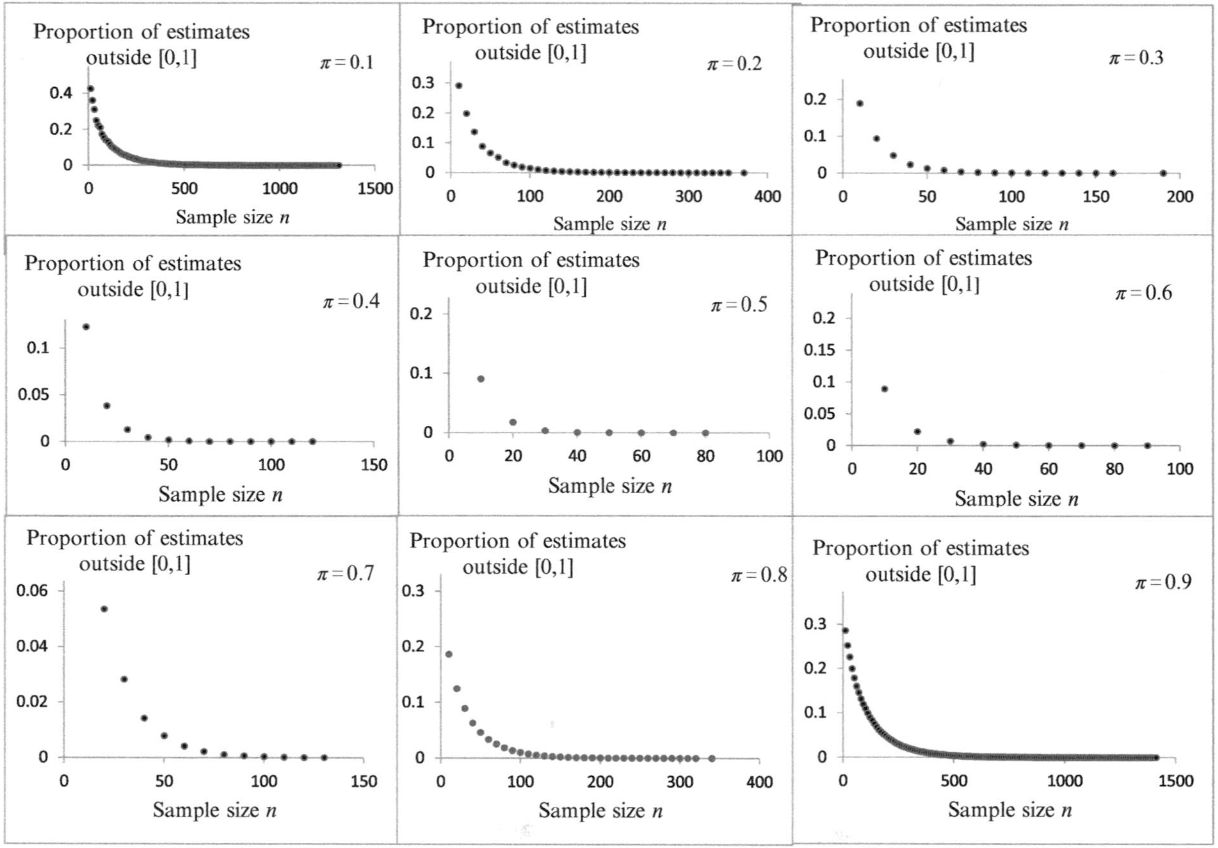

FIG. 3 Proportion of estimates outside [0,1] vs sample size (n) for all values of π at $W_0=0.3$ and $Q_1=0.4$.

Deck (1*h*)
$I \in A$ with probability (W_{h0})
$I \in A^c$ with probability($1-W_{h0}$)

Deck (2*h*)
$I \in A$ with probability (Q_{h1})
Forced **Yes** with probability [$0.5(1-Q_{h1})$]
Forced **No** with probability [$0.5(1-Q_{h1})$]

FIG. 4 Statements used in the proposed stratified model.

To estimate π_h, the proportion of the population having a sensitive attribute in stratum h, each respondent in the SRSWR from stratum h is provided with two decks as shown in Fig. 4. He is requested to draw two cards simultaneously; one card from each of the two decks "deck (1*h*) and deck (2*h*)" and read the statements in order.

The probability of getting a (yes, yes) response from stratum h is given by:

$$\begin{aligned} P(\text{yes, yes}) &= \theta_{h11} \\ &= [0.5(2W_{h0}-1)(1-Q_{h1})+W_{h0}Q_{h1}]\pi_h+0.5(1-W_{h0})(1-Q_{h1}). \end{aligned}$$

The rest of the probabilities are given by:

$$\begin{aligned} P(\text{yes, no}) = \theta_{h10} &= [0.5(2W_{h0}-1)(1-Q_{h1})-(1-W_{h0})Q_{h1}]\pi_h \\ &+ (1-W_{h0})[Q_{h1}+0.5(1-Q_{h1})]. \end{aligned}$$

$$P(\text{no, yes}) = \theta_{h01} = [(1-W_{h0})Q_{h1}-0.5(2W_{h0}-1)(1-Q_{h1})]\pi_h+0.5W_{h0}(1-Q_{h1}).$$

$$P(\text{no, no}) = \theta_{h00} = [0.5(1-2W_{h0})(1-Q_{h1})-W_{h0}Q_{h1}]\pi_h+W_{h0}[Q_{h1}+0.5(1-Q_{h1})].$$

Let n_{h11}/n_h, n_{h10}/n_h, n_{h01}/n_h, and n_{h00}/n_h be the observed proportions of (yes, yes), (yes, no), (no, yes), and (no, no) responses in stratum h, respectively. Such observed proportions, n_{hij}/n_h, are unbiased estimators for the corresponding true probabilities θ_{hij}, $i,j=0,1$, $h=1, 2, \ldots, H$ such that $\sum_{i=0}^{1}\sum_{j=0}^{1} n_{hij}=n_h$ and $\sum_{i=0}^{1}\sum_{j=0}^{1}\theta_{hij}=1\ \forall\ h=1, 2, \ldots, H$.

π_h, the proportion of the population having a sensitive attribute in stratum h, is estimated by minimizing the squared distance between the observed proportions and the true probabilities or alternatively by minimizing D_h where

$$D_h = \frac{1}{2}\sum_{i=0}^{1}\sum_{j=0}^{1}\left(\theta_{hij}-\frac{n_{hij}}{n_h}\right)^2.$$

The following estimator $\hat{\pi}_{hb}$ is obtained:

$$\hat{\pi}_{hb} = \frac{1}{2}+\frac{(B_h-E_h)\left(\dfrac{n_{h11}}{n_h}-\dfrac{n_{h00}}{n_h}\right)+(F_h-G_h)\left(\dfrac{n_{h10}}{n_h}-\dfrac{n_{h01}}{n_h}\right)}{2\left[(B_h-E_h)^2+(F_h-G_h)^2\right]},\quad W_{h0}\neq 0.5 \quad (9)$$

where

$$B_h = W_{h0}[Q_{h1} + 0.5(1 - Q_{h1})], \qquad F_h = 0.5W_{h0}(1 - Q_{h1}),$$
$$E_h = 0.5(1 - W_{h0})(1 - Q_{h1}), \qquad G_h = (1 - W_{h0})[Q_{h1} + 0.5(1 - Q_{h1})]. \tag{10}$$

The estimator $\hat{\pi}_{hb}$ is unbiased with variance given by:

$$V(\hat{\pi}_{hb}) = \frac{(B_h - E_h)^2(B_h + E_h) + (F_h - G_h)^2(F_h + G_h)}{4n_h\left[(B_h - E_h)^2 + (F_h - G_h)^2\right]^2} - \frac{(2\pi_h - 1)^2}{4n_h}. \tag{11}$$

Consequently, for estimating the population proportion π, we have the following theorems:

Theorem 4. *An unbiased estimator of the population proportion π is:*

$$\hat{\pi}_{sb} = \sum_{h=1}^{H} Z_h \hat{\pi}_{hb}, \tag{12}$$

where $\hat{\pi}_{hb}$ is defined in Eq. (9), N_h, number of units in stratum h; $N = \sum_{h=1}^{H} N_h$, number of units in the whole population; $Z_h = \frac{N_h}{N}$, weight of stratum h in the population, $\sum_{h=1}^{H} Z_h = 1$.

Theorem 5. *The variance of the estimator $\hat{\pi}_{sb}$ is given by:*

$$V(\hat{\pi}_{sb}) = \sum_{h=1}^{H} \frac{Z_h^2}{4n_h}\left[\eta_h - (2\pi_h - 1)^2\right], \tag{13}$$

$$\text{where } \eta_h = \frac{(B_h - E_h)^2(B_h + E_h) + (F_h - G_h)^2(F_h + G_h)}{\left[(B_h - E_h)^2 + (F_h - G_h)^2\right]^2}, \tag{14}$$

B_h, E_h, F_h, and G_h are defined in Eq. (10).

Proof.

$$V(\hat{\pi}_{sb}) = \sum_{h=1}^{H} Z_h^2\, V(\hat{\pi}_{hb}) + 2\sum_{h=1}^{H}\sum_{l>h}^{H} Z_h Z_l\, Cov(\hat{\pi}_{hb}, \hat{\pi}_{lb})$$

Since the selections in different strata are made independently, all covariance terms will vanish.

$$V(\hat{\pi}_{sb}) = \sum_{h=1}^{H} Z_h^2\, V(\hat{\pi}_{hb}). \tag{15}$$

Substituting $V(\hat{\pi}_{hb})$ of Eq. (11) into Eq. (15) yields Eq. (13).

The variance of $\hat{\pi}_{sb}$ in Eq. (13) is symmetric around $\pi_h = 0.5$ for $h = 1, 2, \ldots, H$ and in case π_h is replaced by $(1 - \pi_h)$ $\forall\, h = 1, 2, \ldots, H$, the variance of $\hat{\pi}_{sb}$ is found to be symmetric around $\pi = 0.5$.□

Theorem 6. *An unbiased estimator of the variance of $\hat{\pi}_{sb}$ is:*

$$\hat{V}(\hat{\pi}_{sb}) = \sum_{h=1}^{H} \frac{Z_h^2}{4(n_h - 1)} \left[\eta_h - (2\hat{\pi}_h - 1)^2\right].$$

Now, we will consider two methods of allocation to different strata, proportional allocation and optimal allocation.

a. *Proportional allocation:*

Under proportional allocation of the sample size, we have $n_h = n\frac{N_h}{N}$. In this case, the variance of the estimator $\hat{\pi}_{sb}$ is given by:

$$V_{prop}(\hat{\pi}_{sb}) = \frac{1}{4n}\sum_{h=1}^{H} Z_h\left[\eta_h - (2\pi_h - 1)^2\right], \tag{16}$$

where η_h is defined in Eq. (14).

Under proportional allocation, the estimator $\hat{\pi}_{sb}$ in (12) is always more efficient than its counterpart under simple random sampling $\hat{\pi}_b$ in (5) when $W_{h0} = W_0$ and $Q_{h1} = Q_1$ for $h = 1, 2, \ldots, H$.

b. *Optimal allocation:*

Under optimal allocation, we want to minimize the variance of $\hat{\pi}_{sb}$ for a specified cost. The total cost is assumed to be in the form:

$$C = c_0 + \sum_{h=1}^{H} c_h n_h, \tag{17}$$

where c_0 is the overhead cost and c_h is the cost per unit for stratum h.

Theorem 7. *The optimal allocation of the sample size n to obtain the minimum variance of $\hat{\pi}_{sb}$ for a specified cost C in Eq. (17) is given by:*

$$n_h = (C - c_0)\frac{N_h\sqrt{\frac{1}{c_h}\left[\eta_h - (2\pi_h - 1)^2\right]}}{\sum_{h=1}^{H} N_h\sqrt{c_h\left[\eta_h - (2\pi_h - 1)^2\right]}}. \tag{18}$$

where η_h is defined in Eq. (14).

Proof. By applying Cauchy–Schwarz inequality to the product $(C - c_0)\cdot V(\hat{\pi}_{sb})$, we get:

$$n_h = n\frac{N_h\sqrt{\frac{1}{c_h}\left[\eta_h - (2\pi_h - 1)^2\right]}}{\sum_{h=1}^{H} N_h\sqrt{\frac{1}{c_h}\left[\eta_h - (2\pi_h - 1)^2\right]}}. \tag{19}$$

Substituting n_h of Eq. (19) in the cost function of Eq. (17), we get:

$$n = \frac{(C - c_0)\sum_{h=1}^{H} N_h \sqrt{\frac{1}{c_h}\left[\eta_h - (2\pi_h - 1)^2\right]}}{\sum_{h=1}^{H} N_h \sqrt{c_h\left[\eta_h - (2\pi_h - 1)^2\right]}}. \quad (20)$$

Substituting n of Eq. (20) in Eq. (19) yields the proof.□

Theorem 8. *Under optimal allocation of the sample size given in Eq. (18), the minimal variance of the estimator $\hat{\pi}_{sb}$ in (12) is*

$$V_{opt}(\hat{\pi}_{sb}) = \frac{1}{4(C - c_0)}\left\{\sum_{h=1}^{H} Z_h \sqrt{c_h\left[\eta_h - (2\pi_h - 1)^2\right]}\right\}^2$$

where η_h is defined in Eq. (14).

The proof is immediate by substituting n_h of Eq. (18) into $V(\hat{\pi}_{sb})$ of Eq. (13).

Corollary 1. *A special case of optimal allocation, known as "Neyman Allocation," occurs when the costs per unit are the same for all strata. In this case, the optimal allocation of the sample size is given by:*

$$n_h = n\frac{N_h\sqrt{\eta_h - (2\pi_h - 1)^2}}{\sum_{h=1}^{H} N_h \sqrt{\eta_h - (2\pi_h - 1)^2}}$$

and the minimum variance of the estimator $\hat{\pi}_{sb}$ is:

$$V_{ney}(\hat{\pi}_{sb}) = \frac{1}{4n}\left[\sum_{h=1}^{H} Z_h \sqrt{\eta_h - (2\pi_h - 1)^2}\right]^2. \quad (21)$$

Under Neyman allocation, the estimator $\hat{\pi}_{sb}$ in (12) is always more efficient than its counterpart under simple random sampling $\hat{\pi}_b$ in (5) when $W_{h0} = W_0$ and $Q_{h1} = Q_1$ for $h = 1, 2, \ldots, H$.

4.2 Efficiency Comparison

In this section, an efficiency comparison of the proposed stratified model with Abdelfatah and Mazloum (2015) models is performed.

Assuming that there are two strata in the population with $W_{10} = W_{20} = W_0$, $Q_{10} = Q_{20} = Q_0$, $Q_{11} = Q_{21} = Q_1$; the relative efficiencies are calculated using the values of W_0, Q_0 for which the stratified Abdelfatah et al. estimator $\hat{\pi}_{sa}$ is more efficient than the stratified OS estimator as shown by Abdelfatah and Mazloum (2015) and the values of Q_1 ranging from 0.1 to 0.9 with a step of 0.1; $Z_1 = 0.1, 0.3, 0.5, 0.7, 0.9$; $Z_2 = 1 - Z_1$; and (π_1, π_2) take three combinations: (0.08, 0.13), (0.28, 0.33), and (0.48, 0.53). The relative efficiencies reported in Tables 3 and 4 will be obtained at other combinations of (π_1, π_2) because $V(\hat{\pi}_{sa})$ and $V(\hat{\pi}_{sb})$ are symmetric about $\pi_h = 0.5$ for $h = 1,2$. In this comparison, proportional and Neyman allocations of the sample size are considered.

a. Proportional allocation:

Under proportional allocation of the sample size, the relative efficiency of the proposed estimator $\hat{\pi}_{sb}$ in (12) with respect to the estimator $\hat{\pi}_{sa}$ in (2) is given by:

$$\text{RE}_1(sa) = \frac{V_{\text{prop}}(\hat{\pi}_{sa})}{V_{\text{prop}}(\hat{\pi}_{sb})} \times 100\%,$$

where $V_{prop}(\hat{\pi}_{sa})$ and $V_{prop}(\hat{\pi}_{sb})$ are given in Eqs. (3) and (16), respectively.

It is found that the estimator $\hat{\pi}_{sb}$ is always more efficient than the estimator $\hat{\pi}_{sa}$. Table 3 presents the summary statistics of the percent relative efficiency. It is found that the relative efficiency increases as π approaches 0 or 1. At $\pi = 0.085$, the relative efficiency ranges from 104% to 1028% with median relative efficiency of 132%, an average relative efficiency of 172%, and standard deviation of 113%. For each value of π, the maximum relative efficiency is attained at $W_0 = 0.3$, $Q_0 = 0.2$, and $Q_1 = 0.9$.

b. Neyman allocation:

Assuming Neyman allocation, the relative efficiency of the proposed estimator $\hat{\pi}_{sb}$ in (12) with respect to the estimator $\hat{\pi}_{sa}$ in (2) is given by:

$$\text{RE}_2(sa) = \frac{V_{ney}(\hat{\pi}_{sa})}{V_{ney}(\hat{\pi}_{sb})} \times 100\%,$$

where $V_{ney}(\hat{\pi}_{sa})$ and $V_{ney}(\hat{\pi}_{sb})$ are given in Eqs. (4) and (21), respectively.

It is found that the estimator $\hat{\pi}_{sb}$ is always more efficient than the estimator $\hat{\pi}_{sa}$. Table 4 presents the summary statistics of the percent relative efficiency. It is found that the relative efficiency increases as π approaches 0 or 1. At $\pi = 0.115$, the relative efficiency ranges from 103% to 907% with an average of 164% and standard deviation of 100%, and the median relative efficiency in such case is 130%. For each value of π, the maximum relative efficiency is attained at $W_0 = 0.3$, $Q_0 = 0.2$, and $Q_1 = 0.9$.

5 DOUBLE SAMPLING FOR THE PROPOSED STRATIFIED MODEL

In case of unknown strata sizes, the method of double sampling for stratification is applied to the proposed stratified model.

In double sampling, two samples are drawn. The first sample is an SRSWR of size n' and it is used to estimate the strata weights. Each of the n' respondents is asked to answer the direct question "Do you belong to stratum h?"

TABLE 3 Descriptive Statistics of the Percent Relative Efficiency of the Proposed Stratified Model with Respect to the Stratified Abdelfatah et al. Model (Abdelfatah and Mazloum (2015)) under Proportional Allocation

π	π_1	π_2	Z_1	Z_2	Frequency	Mean	Standard Deviation	Median	Minimum	Maximum
0.085	0.08	0.13	0.9	0.1	504	171.69	113.21	131.71	103.56	1028.00
0.095	0.08	0.13	0.7	0.3	504	168.87	108.36	131.18	103.44	982.76
0.105	0.08	0.13	0.5	0.5	504	166.28	103.95	130.61	103.33	941.73
0.115	0.08	0.13	0.3	0.7	504	163.90	99.92	129.77	103.22	904.34
0.125	0.08	0.13	0.1	0.9	504	161.71	96.21	128.72	103.12	870.13
0.285	0.28	0.33	0.9	0.1	504	144.19	67.40	120.32	102.28	610.63
0.295	0.28	0.33	0.7	0.3	504	143.69	66.59	120.05	102.26	603.62
0.305	0.28	0.33	0.5	0.5	504	143.20	65.81	119.80	102.23	596.80
0.315	0.28	0.33	0.3	0.7	504	142.73	65.05	119.55	102.21	590.16
0.325	0.28	0.33	0.1	0.9	504	142.26	64.31	119.30	102.18	583.69
0.485	0.48	0.53	0.9	0.1	504	138.98	59.08	117.67	102.02	538.61
0.495	0.48	0.53	0.7	0.3	504	138.99	59.10	117.67	102.02	538.74
0.505	0.48	0.53	0.5	0.5	504	139.00	59.11	117.68	102.02	538.88
0.515	0.48	0.53	0.3	0.7	504	139.01	59.13	117.68	102.02	539.01
0.525	0.48	0.53	0.1	0.9	504	139.02	59.14	117.69	102.02	539.15

TABLE 4 Descriptive Statistics of the Percent Relative Efficiency of the Proposed Stratified Model with Respect to the Stratified Abdelfatah et al. Model (Abdelfatah and Mazloum, 2015) under Neyman Allocation

π	π_1	π_2	Z_1	Z_2	Frequency	Mean	Standard Deviation	Median	Minimum	Maximum
0.085	0.08	0.13	0.9	0.1	504	171.79	113.35	131.72	103.57	1029.37
0.095	0.08	0.13	0.7	0.3	504	169.08	108.67	131.21	103.45	985.67
0.105	0.08	0.13	0.5	0.5	504	166.51	104.28	130.64	103.34	944.89
0.115	0.08	0.13	0.3	0.7	504	164.08	100.17	129.87	103.23	906.78
0.125	0.08	0.13	0.1	0.9	504	161.78	96.31	128.76	103.13	871.09
0.285	0.28	0.33	0.9	0.1	504	144.19	67.40	120.32	102.28	610.69
0.295	0.28	0.33	0.7	0.3	504	143.70	66.61	120.06	102.26	603.76
0.305	0.28	0.33	0.5	0.5	504	143.22	65.83	119.81	102.23	596.96
0.315	0.28	0.33	0.3	0.7	504	142.74	65.07	119.55	102.21	590.29
0.325	0.28	0.33	0.1	0.9	504	142.27	64.31	119.31	102.18	583.75
0.485	0.48	0.53	0.9	0.1	504	138.98	59.08	117.67	102.02	538.61
0.495	0.48	0.53	0.7	0.3	504	138.99	59.10	117.67	102.02	538.74
0.505	0.48	0.53	0.5	0.5	504	139.00	59.11	117.68	102.02	538.88
0.515	0.48	0.53	0.3	0.7	504	139.01	59.13	117.68	102.02	539.01
0.525	0.48	0.53	0.1	0.9	504	139.02	59.14	117.69	102.02	539.15

and then the respondents are classified into the different strata with n'_h respondents in stratum h. The weight of stratum h in the first sample, $z_h = n'_h/n'$, is an unbiased estimator of Z_h, the population weight of stratum h, $h = 1, 2, \ldots, H$.

The second sample is a stratified random sample of size n where n_h respondents are drawn from stratum h. The second sample in stratum h is usually an SRSWR from n'_h. Thus, $n_h = v_h n'_h$, $0 < v_h \leq 1$.

The second sample is used to obtain the estimator $\hat{\pi}_{hb}$ of π_h, the proportion of the population having a sensitive attribute in stratum h.

Theorem 9. *An unbiased estimator $\hat{\pi}_{db}$ of π is:*

$$\hat{\pi}_{db} = \sum_{h=1}^{H} z_h \hat{\pi}_{hb},$$

where

$\hat{\pi}_{hb}$: unbiased estimator of π_h from the second sample as defined in Eq. (9),
$z_h = n'_h/n'$: weight of stratum h from the first sample.

Proof. Let E_2 refer to an average over all random samples of n_h units that can be drawn from a given n'_h units in which z_h is fixed and E_1 refers to an average over repeated selections of the first sample:

$$E(\hat{\pi}_{db}) = E_1 E_2 \left(\sum_{h=1}^{H} z_h \hat{\pi}_{hb} \right) = E_1 \left(\sum_{h=1}^{H} z_h \pi_h \right) = \sum_{h=1}^{H} Z_h \pi_h = \pi \; \square$$

Theorem 10. *The variance of the estimator $\hat{\pi}_{db}$ is:*

$$V(\hat{\pi}_{db}) = \frac{1}{n'} \left\{ \sum_{h=1}^{H} \frac{Z_h}{4v_h} \left[\eta_h - (2\pi_h - 1)^2 \right] + \sum_{h=1}^{H} Z_h (\pi_h - \pi)^2 \right\}, \quad (22)$$

where η_h is defined in Eq. (14).

Proof.

$$V(\hat{\pi}_{db}) = E_1 V_2(\hat{\pi}_{db}) + V_1 E_2(\hat{\pi}_{db}). \quad (23)$$

$$\begin{aligned} E_1 V_2(\hat{\pi}_{db}) &= E_1 \left(\sum_{h=1}^{H} z_h^2 V_2(\hat{\pi}_{hb}) \right) \\ &= E_1 \left(\sum_{h=1}^{H} \frac{z_h^2}{4n_h} \left[\eta_h - (2\pi_h - 1)^2 \right] \right) \end{aligned}$$

Since $n_h = v_h n'_h = v_h z_h n'$, we have

$$\begin{aligned} E_1 V_2(\hat{\pi}_{db}) &= E_1 \left(\frac{1}{n'} \sum_{h=1}^{H} \frac{z_h}{4v_h} \left[\eta_h - (2\pi_h - 1)^2 \right] \right) \\ &= \frac{1}{n'} \sum_{h=1}^{H} \frac{Z_h}{4v_h} \left[\eta_h - (2\pi_h - 1)^2 \right] \end{aligned} \quad (24)$$

$$V_1E_2(\hat{\pi}_{db}) = V_1\left(\sum_{h=1}^{H} z_h\pi_h\right) = \sum_{h=1}^{H}\pi_h^2 V(z_h) + \sum_{h\neq l}\pi_h\pi_l Cov(z_h, z_l)$$
$$= \sum_{h=1}^{H}\pi_h^2\frac{Z_h(1-Z_h)}{n'} - \sum_{h\neq l}\pi_h\pi_l\frac{Z_hZ_l}{n'} = \frac{1}{n'}\sum_{h=1}^{H}Z_h(\pi_h-\pi)^2. \tag{25}$$

Substituting Eqs. (24) and (25) into Eq. (23) yields the proof.□

Theorem 11. *An unbiased estimator of the variance of $\hat{\pi}_{db}$ is:*

$$\hat{V}(\hat{\pi}_{db}) = \frac{1}{(n'-1)}\left\{\sum_{h=1}^{H}\frac{z_h}{4}\left(\frac{n_h'-1}{n_h-1}\right)\left[\eta_h - (2\hat{\pi}_{hb}-1)^2\right] + \sum_{h=1}^{H} z_h(\hat{\pi}_{hb}-\hat{\pi}_{db})^2\right\}. \tag{26}$$

The proof is obtained by taking the expected values on both sides of Eq. (26). Now, we will consider proportional and optimal allocations.

a. Proportional allocation:

With proportional allocation, $v_h = \frac{n}{n'}$ and the variance of the estimator $\hat{\pi}_{db}$ is given by:

$$V_{prop}(\hat{\pi}_{db}) = \frac{1}{4n}\sum_{h=1}^{H} Z_h\left[\eta_h - (2\pi_h-1)^2\right] + \frac{1}{n'}\sum_{h=1}^{H} Z_h(\pi_h-\pi)^2. \tag{27}$$

The only difference between $V_{prop}(\hat{\pi}_{sb})$ in Eq. (16) and $V_{prop}(\hat{\pi}_{db})$ in Eq. (27) is the term $\frac{1}{n'}\sum_{h=1}^{H} Z_h(\pi_h-\pi)^2$ which represents the variance added due to unknown strata sizes.

b. Optimal allocation:

Under optimal allocation, we will choose v_h and n' that minimize $V(\hat{\pi}_{da})$ for fixed cost C.

For a specified sample, consider the cost function:

$$C = c'n' + \sum_{h=1}^{H} c_h n_h, \tag{28}$$

where

c': classification cost per unit,
c_h: measurement cost per unit in stratum h.

We will use the expected cost C^* in Eq. (29) because the cost C in Eq. (28) is random:

$$C^* = E(C) = c'n' + \sum_{h=1}^{H} c_h v_h Z_h n', \tag{29}$$

where

$Z_h = \dfrac{N_h}{N}$: weight of stratum h in the population,
$0 < v_h = n_h/n'_h \leq 1$ is fixed.

Applying Cauchy–Schwarz inequality to the product $C^* \; V(\hat{\pi}_{db})$, the optimal v_h which minimizes $V(\hat{\pi}_{db})$ for fixed cost is:

$$v_h = \sqrt{\frac{c'\left[\eta_h - (2\pi_h - 1)^2\right]}{4c_h \sum_{h=1}^{H} Z_h(\pi_h - \pi)^2}}. \tag{30}$$

The optimal n' obtained from the expected cost function in Eq. (29) is given by:

$$n' = \frac{C^*}{c' + \sum_{h=1}^{H} Z_h \sqrt{\dfrac{c' c_h\left[\eta_h - (2\pi_h - 1)^2\right]}{4\sum_{h=1}^{H} Z_h(\pi_h - \pi)^2}}}. \tag{31}$$

By substituting the optimal v_h of Eq. (30) and n' of Eq. (31) in $V(\hat{\pi}_{db})$ of Eq. (22), the minimum variance for fixed cost is given by:

$$V_{\text{opt}}(\hat{\pi}_{db}) = \frac{1}{C^*}\left\{\sum_{h=1}^{H} Z_h \sqrt{\frac{c_h}{4}\left[\eta_h - (2\pi_h - 1)^2\right]} + \sqrt{c' \sum_{h=1}^{H} Z_h(\pi_h - \pi)^2}\right\}^2.$$

6 CONCLUSION

A new randomized response model is proposed as a modification to the Abdelfatah et al. (2011) model. The proposed model is always more efficient than the Abdelfatah et al. (2011) model under both simple and stratified random sampling.

REFERENCES

Abdelfatah, S., Mazloum, R., 2015. An improved stratified randomized response model using two decks of cards. Model Assist. Stat. Appl. 10, 309–320.

Abdelfatah, S., Mazloum, R., Singh, S., 2011. An alternative randomized response model using two decks of cards. Statistica 71, 381–390.

Horvitz, D.G., Greenberg, B.G., Abernathy, J.R., 1976. Randomized response: a data gathering device for sensitive questions. Int. Stat. Rev. 44, 181–196.

Lee, C.-S., Singh, S., Sedory, S.A., 2013. Simulated minimum sample size requirements in various randomized response models. Commun. Stat. Simul. Comput. 42, 771–789.

Odumade, O., Singh, S., 2009. Efficient use of two decks of cards in randomized response sampling. Commun. Stat.Theory Methods 38, 439–446.

Warner, S.L., 1965. Randomized response: a survey technique for eliminating evasive answer bias. J. Am. Stat. Assoc. 60, 63–69.

Chapter 10

Software for Randomized Response Techniques

M. Rueda*[,1], B. Cobo*, A. Arcos* and R. Arnab[†]
*University of Granada, Granada, Spain
[†]University of Botswana, Gaborone, Botswana
[1]Corresponding author: e-mail: mrueda@ugr.es

ABSTRACT

The rapid evolution of technology in last decades has allowed conducting studies for complex surveys. Over time, the software is used in all steps of the survey: survey design, data collection, statistical data edition, statistical analysis, and publication results. In this chapter we describe software that can be used in conducting surveys using randomized response techniques. In particular we will focus on how technology can help to implement randomization devices and estimation procedures.

Keywords: Software, Complex surveys, Randomizing devices, R packages

1 INTRODUCTION

Warner (1965) developed a data collection procedure, the randomized response (RR) technique that allows researchers to obtain sensitive information while guaranteeing privacy to respondents. This method encourages greater cooperation from respondents and reduces their motivation to falsely report their attitudes. The most important claim made for RR is that it yields more valid point estimates of sensitive behavior.

Warner's study generated a rapidly expanding body of research literature on alternative techniques for eliciting suitable RR schemes in order to estimate a population proportion (see Arnab, 2002, 2004; Arnab and Singh, 2010; Bouza, 2009; Chang et al., 2004; Christofides, 2003; Fox and Wyrick, 2008; Singh and Tarray, 2015; Tracy and Mangat, 1996; Van den Hout et al., 2010). A good revision of RR techniques can be seen in Bouza et al. (2010), in Chaudhuri (2011), or in Chaudhuri and Christofides (2013). Half a century after the original RRT model was introduced, these models continue to be used in a variety of disciplines.

Handbook of Statistics, Vol. 34. http://dx.doi.org/10.1016/bs.host.2016.01.009

There have been many reports that RR provides more accurate estimates of the prevalence of socially undesirable behavior than does asking the sensitive question directly. Numerous empirical studies have shown that RR obtains higher estimates of sensitive characteristics than are produced by direct questioning (Lara et al., 2006; van der Heijden et al., 2000). However, using RR incurs extra costs, and the advantage of using RR will only outweigh these extra costs if the estimates are substantially better than those derived from straightforward question-and-answer designs (Lensvelt-Mulders et al., 2006).

Other disadvantage of randomized response techniques is the cost for the need to use randomization device. Randomization devices must be provided to the responding party without producing mistrust in the interviewee. The device suggested by Warner was a spinner. Other devices are coins, cards, box with poker cards. For continuous variables the randomization device is even more complex. Eichhorn and Hayre (1983) used the multiplicative approach to produce a randomize response device. In this approach we need a random variable that is independent of the main variable "y," with known mean and variance.

In this chapter we will see some software that the interviewee can used to generate his random response easily and not mistrust of the scrambling system.

Currently, for the estimation procedures there are many programs and programming languages for working with complex surveys, but there are few that have implemented modules to work with randomized response. We will discuss both programs specifically designed to analyze data obtained from randomized response techniques, as codes available in the literature that implement certain specific estimation techniques with randomized data. We will emphasize language R for being the free software that is most commonly used in the scientific community nowadays.

2 SOFTWARE FOR HELPING TO CONDUCT A SURVEY WITH RR

Traditionally, surveys have been carried out using three main methods of data collection: face-to-face interviews, telephone interviews, and mail questionnaires. Among these methods, face-to-face interviews are preferred for surveys by randomized response, given the complexity of understanding the mechanism of randomization and also the sensitive character of the questions that diminishes the collaboration of the interviewee ones.

A difficulty with implementing the RRT in a face-to-face setting is the choice of an appropriate randomizing device. The randomizing device must be implemented in a way that makes the protection offered by the technique clear to the respondent. The randomizing devices most commonly employed (Coutts and Jann, 2011) are serial numbers on a bill, the flip of a coin, a spinner, cards, numbers selected from the phone book, and the respondent's month of birth. Today the new technologies offer more attractive alternatives for the users. Among them we emphasize the mobile phones.

The list of countries by smartphone penetration in 2014 is measured by Google's Our Mobile Planet and shows that in the United States the percentage of the population owning and using a smartphone is 56,4% being higher in other countries as United Arab emirates (73.8%) or South Korea (73%). The number of smartphone users worldwide will surpass 2 billion in 2016, according to new figures from eMarketer, after nearly getting there in 2015. Next year, there will be over 1.91 billion smartphone users across the globe, a figure that will increase another 12.6% to near 2.16 billion in 2016. For the first time, more than one-quarter of the global population will use smartphones in 2015, and by 2018, eMarketer estimates, over one-third of consumers worldwide, or more than 2.56 billion people, will do so.

Thanks to smartphones we have access to many interesting applications. In particular we are interested in applications that help us randomize telephone and personal surveys (we name this as "Smartphone calculator"). There are many, but we highlight:

- *RANDOM.ORG*: currency of any country, dices, poker cards, lottery numbers, uniform random generator. It notes that the *RANDOM.ORG* application is also available for PC and iPhone.
- *True Random Generator*: integer (with or without replacement), real numbers (by means uniform, Pareto, normal and exponential distributions), dices, and several lottery games.

These new technologies can be also useful for the generation of the scrambling variables used for some models. For example, distributions with different degrees of freedom are used by Eichhorn and Hayre (1983), Diana and Perri (2010, 2012), and Arcos et al. (2015). Bar-Lev et al. (2004) use negative exponential distributions, Saha (2007) uses continuous uniform distributions and Gjestvang and Singh (2007) use normal distributions with different means and variances. The app *RandomNumbers* generates random numbers from a wide range of distributions, both discrete and continuous.

Over the last decade a number of new methods of computer administration have supplanted the traditional modes of administration. Among the new modes, there are computer-assisted self-interviewing and web surveys.

The incorporation of computer-assisted self-interviewing and RR in computer-assisted randomized response interviewing has been proposed and successfully tested by Musch et al. (2001), Lensvelt-Mulders et al. (2006), and Lensvelt-Mulders and Boeije (2007). Moreover, Boeije and Lensvelt-Mulders (2002) and Lensvelt-Mulders and Boeije (2007) have shown that respondents prefer computer-assisted RR above paper-and-pencil and face-to-face RR.

The use of computers to collect sensitive data has multiple advantages: one of the advantages is the increase in the respondent's perception of privacy protection. Other advantage is that the use of computer-assisted self-interviewing makes possible the use of digital randomizers. Peeters et al.

(2010) use digital spinners and Lensvelt-Mulders and Boeije (2007) use virtual dices to implement the forced-response method.

In the last decade, there has been a rapid growth in the use of web surveys for a wide variety of data collection activities (Couper and Miller, 2008). Web surveys are appealing alternatives to other interview techniques because (a) they are a simple means to get access to a large group of people, (b) questionnaires can be distributed at very low costs, (c) no interviewers are needed, and there are no mailing and printing costs, (d) surveys can be launched very quickly, (e) offer respondents more anonymity and privacy than interviewer-administered surveys.

Web surveys offer new attractive possibilities, such as the use of multimedia that can be very helpful in the implementation of random devices.

Few methodological studies comparing direct questioning and online questioning in RR are made. Höglinger et al. (2014) perform an experimental evaluation of the RRT in online surveys. In this study the web survey did not yield higher estimates than direct questioning and the authors question the viability of the RRT for online surveys. In the authors' view, the reason for the low RRT estimates might lie in respondents' noncompliance with the RRT instruction.

Effectively, it is difficult to find a suitable randomizing device for online mode with two characteristics: (a) that is at respondents' immediate disposition and (b) is perceived as trustworthy.

In the first aspect, there are a lot of web pages to compute, online, random numbers from several probability distributions. We refer this as "web calculator." Several applications to randomize the response and to help the surveyed to compute the response in a web version are:

- Web's apps as http://www.bgfl.org/virtualdice and https://www.random.org/.
- Shiny package: This is a web application framework for R. It is possible to turn all R functionalities into interactive web applications.

But the interviewee sees increased its privacy if the software tool to hide the answer is different from that used to answer: to include in the same web questionnaire responses and a web calculator can do mistrust the interviewee. Effectively, electronic devices in the web can be manipulated by experimenters, and thus might not be judged trustworthy by the respondents. To conceal individual responses and to protect respondent privacy, the Smartphone's apps can be made an important role.

3 SOFTWARE FOR THE ESTIMATION WITH DATA OBTAINED USING RR TECHNIQUES

Usually, RR methods are developed assuming that sample is obtained using simple random sampling. Most of the surveys in practice are complex surveys involving stratification, clustering and unequal probability of selection of sample. Data from complex survey designs require special consideration with regard to estimation for finite population parameters and corresponding variance estimation procedures, as a consequence of significant departures from simple random sampling assumption. In such a complex survey design, unbiased variance estimation is not easy because of clustering and involvement of second-order inclusion probabilities which are generally complex. Several software packages have been developed to facilitate the analysis of complex survey data and implements some of these estimators as SAS, SPSS, Systa, Stata, SUDAAN, or PCCarp. CRAN contains several R packages that include these design-based methods typically used in survey methodology to treat samples obtained from a one sampling frame, eg, *survey*, *sampling*, *laeken*, or *TeachingSampling* among others (see Templ, 2015, for a detailed list of packages that includes methods to analyze complex surveys).

Standard software packages for complex surveys cannot be used directly when the sample is obtained from randomized response techniques. The analyses with standard statistical software, with certain modifications in the randomized variables, can yield correct point estimates of population parameters but still yield incorrect results for estimated standard errors. Recently some authors have developed R packages or Stata modules for estimation with randomized response surveys:

- **rr** (Blair et al., 2015): Statistical methods for the randomized response technique

  Conducts multivariate regression analyses for the sensitive item under four standard randomized response designs, including:
  - Warner model (Warner, 1965),
  - forced-question model (Boruch, 1972),
  - Kuk model (Kuk, 1990),
  - unrelated-question model (Greenberg et al., 1969; Horvitz et al., 1967).

The package performs regression with the randomized response as the outcome and logistic regression with the randomized response item as a predictor. In addition, tools for conducting power analysis for designing randomized response items are included.

The package generates predicted probabilities of answering affirmatively to the sensitive item for each respondent and also allows users to use the sensitive item as a predictor in an outcome regression under the forced-response design.

The package has four functions to perform various analyzes:

- rrreg: This function allows users to perform multivariate regression analysis on data from the randomized response technique. The method implemented by this function is the Maximum Likelihood (ML) estimation for the Expectation-Maximization (EM) algorithm.

- rrreg.predictor: This function allows users to perform multivariate regression analysis with the randomized response item as a predictor for a separate outcome of interest. It does so by jointly modeling the randomized response item as both outcome and predictor for an additional outcome given the same set of covariates.
- predict.rrreg: This function allows users to generate predicted probabilities for the randomized response item given an object of class "rrreg" from the previous rrreg function.
- predict.rrreg.predictor: This function allows users to generate predicted probabilities for the additional outcome variables with the randomized response item as a covariate given an object of class "rrreg.predictor" from the rrreg.predictor function.

In addition, the package includes the function power.rr.test. This function allows users to conduct power analysis for randomized response survey designs, both for the four designs previously described.

- **RRreg** (Heck and Moshagen, 2015): Conduct correlation and regression analysis for randomized response designs, simple univariate analysis, bivariate correlations including RR variables, logistic regression with an RR variable, and linear regression with RR variables as predictors. In **rr** package, estimation strategy is based on the EM algorithm whereas **RRreg** uses the standard optimization routine.

Unrelated question model (Greenberg et al., 1969; Horvitz et al., 1967), Mangat's model (Mangat, 1992), Kuk's model (Kuk, 1990), and forced-response model (Boruch, 1972) are included in the package. In addition, some RR and non-RR models are also included:

- Crosswise (Tian and Tang, 2014): The Crosswise model is technically identical to the Warner model. However, instead of answering to the sensitive question directly, respondents are instructed to respond to the sensitive and an irrelevant question simultaneously.
- Cheating detection model (Clark and Desharnais, 1998): This model is based on the forced-response design and prompts a proportion of participants to respond "yes" regardless of their true state. In contrast to the forced-response design, the cheating detection model divides the population into three distinct groups: compliant carriers of the sensitive attribute, compliant non carriers, and cheating respondents who always respond "no."
- Symmetric cheating detection model (Ostapczuk et al., 2009): In contrast to the previous cheating detection, in the symmetric some respondents are also prompted to respond "no" with a given probability, not only "yes."
- The stochastic lie detector (Moshagen et al., 2012): This model is based on Mangat's model and adds a second parameter which gives the probability of honest responding within the group carrying the sensitive attribute.

The package also includes functions to generate and to analyze RR data repeatedly for simulation and bootstrap purposes.

- **list** (Blair and Imai, 2011): **list** is a publicly available R package that allows researchers to conduct a multivariate statistical analysis for the item count technique (Chaudhuri and Christofides, 2007). This survey methodology is also known as the list experiment or the unmatched count technique. Under a standard design, it asks respondents questions about a set of actions or views at once rather than the sensitive one in isolation. To assess the prevalence of sensitive attitudes and behavior, the researcher randomizes whether the sensitive item of interest is added to the list of control items.

The package implements the methods described in Imai (2011) and Blair and Imai (2012), a Bayesian MCMC implementation of regression for the standard and multiple sensitive item list experiment designs including ceiling and floor effects and a random effects setup, and a Bayesian MCMC hierarchical regression model with up to three hierarchical groups. The package also includes two dataset contain applications of the item count technique in two national surveys.

- **rrlogit** (Jann, 2011): Stata module to estimate logistic regression for randomized response data.

This software fits a maximum-likelihood logistic regression for randomized response data (Maddala, 1983). It is suited for the analysis of data from Warner's model, the forced-response model, or the unrelated-question model with a known distribution for the nonsensitive question. Several postestimation tools are also included.

The methods implemented in these packages are used under the assumption on simple random sampling and do not explore various theoretical and practical issues that may arise when adopting different survey sampling methods. **RRTCS** tries to fill this gap by providing functions for point and interval estimation from randomized response surveys.

- **RRTCS** (Cobo et al., 2015) is a new R package to calculate point and interval estimation for linear parameters with data obtained from randomized response surveys. The package works with a wide range of sampling designs including simple random sampling with and without replacement (SRSWR and SRSWOR), stratified sampling, cluster sampling, unequal probabilities sampling, and combination of them.

It consists of 20 main functions each of them implementing one of these RR procedures for complex surveys:

- Randomized response procedures to estimate parameters of a qualitative stigmatizing characteristic:

Christofides model (Christofides, 2003), Devore model (Devore, 1977), forced-response model (Boruch, 1972), Horvitz model (Greenberg et al., 1969; Horvitz et al., 1967), Horvitz model with unknown B (Chaudhuri, 2011, p. 42), Kuk model (Kuk, 1990), Mangat model (Mangat, 1992), Mangat model with unknown B (Chaudhuri, 2011, p. 53), Mangat and Singh model (Mangat and Singh, 1990), Mangat, Singh, and Singh model (Mangat, Singh and Singh, 1992), Mangat, Singh, and Singh model with unknown B (Chaudhuri, 2011, p. 54), Singh and Joarder model (Singh and Joarder, 1997), Soberanis model (Soberanis Cruz et al., 2008), and Warner model (Warner, 1965).

- Randomized response procedures to estimate parameters of a quantitative stigmatizing characteristic:

  Pioneering work for quantitative randomized response techniques was due to Eriksson model (Eriksson, 1973). Others include BarLev model (Bar-Lev et al., 2004), Chaudhuri and Christofides model (Chaudhuri and Christofides, 2013, p. 97), Diana and Perri 1 model (Diana and Perri, 2010, p. 1877), Diana and Perri2 model (Diana and Perri, 2010, p. 1879), Eichhorn and Hayre model (Eichhorn and Hayre, 1983), and Saha model (Saha, 2007).

The package also includes an additional function called ResamplingVariance which provides estimates variance of the randomized response estimators using some resampling methods.

Finally, the package includes 20 dataset contain observations from different surveys conducted in real and simulated populations using different randomized response techniques.

Table 1 shows a comparative summary of the previous software according to their availabilities.

The above packages are specifically designed to work with data from randomized surveys. Others authors have generally used software to analyze this type of data adapting other methodologies. For example, Hox and Lensvelt-Mulders (2004) present a way to analyze the relations between randomized response estimates and explanatory variables using the standard structural equation modeling software *Mplus*. Sehra (2008) develops SAS code to analyze a two-stage additive optional model. The book of Tian and Tang (2014) also include numerous R code for replicate some analysis included in some chapters.

4 SUMMARY

In his ground breaking paper in 1965, Warner proposed a very interesting idea of how to deal with evasive answer bias, especially when it comes to controversial survey questions. While immense methodological progress has been made in RR, there are only a handful of published studies that use the RR method to answer substantive question. The lack of substantive applications

TABLE 1 Comparative Summary of the RR Software

Software	Sampling Designs for Which Analyses Are Made	Statistical Analyses That can be Performed	RR Models Included	Other Characteristics of Interest	Platforms and License
list	Simple random sampling	Multivariate regression analysis	Unmatched count technique	Includes two datasets of real surveys with RR, includes Bayesian MCMC techniques	Windows, Mac Os, Unix, Linux. Software license: GNU
RRreg	Simple random sampling	Univariate estimates of proportions, correlation and regression analysis	Unrelated question, Mangat, Kuk, forced response, Crosswise, cheating detection, stochastic lie detector	Generates data according to a specified RR model for simulation purposes	Windows, Mac Os, Unix, Linux. Software license: GNU
rr	Simple random sampling	Logistic regression and linear regression analysis	Warner, Kuk, forced response, and unrelated question	Includes a dataset of a survey with RR	Windows, Mac Os, Unix, Linux. Software license: GNU
rrlogit	Simple random sampling	Logistic regression analysis	Warner, forced response, unrelated question		Windows, Mac Os, Unix, Linux. Software license: Proprietary
RRTCS	Ranging from simple random samples to more complex designs with stratified, unequal probability, and cluster	Point and interval estimation for linear parameters	Warner, forced response, unrelated question, Kuk, Christofides, Devore, Mangat, Mangat and Singh, Mangat, Singh, and Singh, Singh and Joarder, BarLev, Chaudhuri, and Christofides, Diana and Perri, Eichhorn and Hayre, Eriksson, Saha	Includes resampling methods to variance estimation; includes datasets of surveys with RR, includes analysis for quantitative RR variables	Windows, Mac Os, Unix, Linux. Software license: GNU

suggests the need for new methods regarding the implementation of the RRT and also the analysis of data obtained with this methodology.

There may be many opportunities to use new technologies (computer assisted surveys, web surveys, tablet, mobile phones) as a mode of data collection in RR surveys. New technologies let us to get access to a large group of people at very low costs, makes the protection offered by the technique clear to the respondent, and they can incorporate multimedia (sound, pictures, animation, and movies) that facilitate the use of digital randomizers. Adding to the attractiveness of practicality, multimedia randomizers can make the interview more fun and to increase the collaboration of the interviewed ones.

On the other hand, very recently there have appeared free programs that allow performing complex statistical analyses with data obtained by different randomized response techniques.

In summary, new technologies offer both new challenges and new opportunities in all steps of the randomized response surveys.

In this work we have focused our exposition on the skills of RR, nevertheless there have been included also some references to other recent models proposed as alternative to treat with sensitive questions as Crosswise model (Yu, Tian, and Tang, 2008) and the stochastic lie detector (Moshagen et al., 2012), which may overcome some of the drawbacks of the RRT.

ACKNOWLEDGMENTS

This study was partially supported by Ministerio de Educación, Cultura y Deporte (grant MTM2012-35650 and FPU grant program, Spain), by Consejería de Economía, Innovación, Ciencia y Empleo (grant SEJ2954, Junta de Andalucía, Spain).

REFERENCES

Arcos, A., Rueda, M., Singh, S., 2015. A generalized approach to randomized response for quantitative variables. Qual. Quant. 49, 1239–1256.

Arnab, R., 2002. Optimum sampling strategies under randomized response surveys. Biom. J. 44, 1–16.

Arnab, R., 2004. Optional randomized response techniques for complex survey designs. Biom. J. 46, 114–124.

Arnab, R., Singh, S., 2010. Randomized response techniques: an application to the Botswana AIDS impact survey. J. Stat. Plann. Inference 140 (4), 941–953.

Bar-Lev, S.K., Bobovitch, E., Boukai, B., 2004. A note on randomized response models for quantitative data. Metrika 378 (60), 255–260.

Blair, G., Imai, K., 2011. list: Statistical Methods for the Item Count Technique and List Experiment. http://CRAN.Rproject.org/package=list.

Blair, G., Imai, K., 2012. Statistical analysis of list experiments. Polit. Anal. 20, 47–77.

Blair, G., Imai, K., Zhou, Y.Y., 2015. Package 'rr': Statistical Methods for the Randomized Response. http://CRAN.R-project.org/package=rr.

Boeije, H.R., Lensvelt-Mulders, G.J.L.M., 2002. Honest by chance: a qualitative interview study to clarify respondents' (non)-compliance with computer assisted randomized response. Bull. Am. Meteorol. Soc. 75, 24–39.

Boruch, R.F., 1972. Relations among statistical methods for assuring confidentiality of social research data. Soc. Sci. Res. 1, 403–414.

Bouza, C., 2009. Ranked set sampling and randomized response procedures for estimating the mean of a sensitive quantitative character. Metrika 70, 267–277.

Bouza, C., Herrera, C., Mitra, P., 2010. A review of randomized responses procedures: the qualitative variable case. Rev. Investig. Oper. 31, 240–247.

Chang, H.J., Wang, C.L., Huang, K.C., 2004. On estimating the proportion of a qualitative sensitive character using randomized response sampling. Qual. Quant. 38, 675–680.

Chaudhuri, A., 2011. Randomized Response and Indirect Questioning Techniques in Surveys. Chapman & Hall, London.

Chaudhuri, A., Christofides, T.C., 2007. Item count technique in estimating the proportion of people with a sensitive feature. J. Stat. Plann. Inference 137, 589–593.

Chaudhuri, A., Christofides, T.C., 2013. Indirect Questioning in Sample Surveys. Springer-Verlag, Berlin Heidelberg.

Christofides, T.C., 2003. A generalized randomized response technique. Metrika 57, 195–200.

Clark, S.J., Desharnais, R.A., 1998. Honest answers to embarrassing questions: detecting cheating in the randomized response model. Psychol. Methods 3, 160–168.

Cobo, B., Rueda, M., Arcos, A., 2015. Package 'RRTCS': Randomized Response Techniques for Complex Surveys. http://cran.r-project.org/web/packages/RRTCS/.

Couper, M.P., Miller, P.V., 2008. Web survey methods. Public Opin. Q. 72, 831–835.

Coutts, E., Jann, B., 2011. Sensitive questions in online surveys: experimental results for the randomized response technique (RRT) and the unmatched count technique (UCT). Sociol. Methods Res. 40, 169–193.

Devore, J.L., 1977. A note on the randomized response technique. Commun. Stat. Theory Methods 6, 1525–1529.

Diana, G., Perri, P.F., 2010. A new scrambled response models for estimating the mean of a sensitive quantitative character. J. Appl. Stat. 37, 1875–1890.

Diana, G., Perri, P.F., 2012. Calibration-based approach to sensitive data: a simulation study. J. Appl. Stat. 39, 53–65.

Eichhorn, B., Hayre, L.S., 1983. Scrambled randomized response methods for obtaining sensitive quantitative data. J. Stat. Plann. Inference 7, 307–316.

Eriksson, S.A., 1973. A new model for randomized response. Int. Stat. Rev. 41, 40–43.

Fox, J.P., Wyrick, C., 2008. A mixed effects randomized item response model. J. Educ. Behav. Stat. 33, 389–415.

Gjestvang, C.R., Singh, S., 2007. Forced quantitative randomized response model: a new device. Metrika 66, 243–257.

Greenberg, B.G., Abdel-Latif, A., Simmons, W.R., Horvitz, D.G., 1969. The unrelated question randomized response model for human surveys. J. Am. Stat. Assoc. 64, 520–539.

Heck, D.W., Moshagen, M., 2015. Package 'RReg' Correlation and Regression Analyses for Randomized Response Data. http://psycho3.uni-mannheim.de/Home/Research/Software/RReg/.

Höglinger, M., Jann, B., Diekmann, A., 2014. Sensitive Questions in Online Surveys: An Experimental Evaluation of the Randomized Response Technique and the Crosswise Model. University of Bern Social Sciences Working Paper, Bern, Switzerland. http://ideas.repec.org/p/bss/wpaper/9.html.

Horvitz, D.G., Shah, B.V., Simmons, W.R., 1967. The unrelated question RR model. In: Proceedings of the Social Statistics Section of the American Statistical Association. ASA, Alexandria, VA, pp. 65–72.

Hox, J., Lensvelt-Mulders, G.J.L.M., 2004. Randomized response analysis in Mplus. Struct. Equ. Model. 11 (4), 615–620.

Imai, K., 2011. Multivariate regression analysis for the item count technique. J. Am. Stat. Assoc. 106 (494), 407–416.

Jann, B., 2011. RRLOGIT: Stata Module to Estimate Logistic Regression for Randomized Response Data. Available at Research Papers in Economics (RePEc). https://ideas.repec.org/c/boc/bocode/s456203.html.

Kuk, A.Y.C., 1990. Asking sensitive questions indirectly. Biometrika 77, 436–438.

Lara, D., García, S.G., Ellertson, C., Camlin, C., Suárez, J., 2006. The measure of induced abortion in Mexico using random response technique. Sociol. Methods Res. 35, 279–301.

Lensvelt-Mulders, G.J.L.M., Boeije, H.R., 2007. Evaluating compliance with a computer assisted randomised response technique: a qualitative study into the origins of lying and cheating. Comput. Hum. Behav. 23, 591–608.

Lensvelt-Mulders, G.J.L.M., van der Heijden, P.G.M., Laudy, O., van Gils, G., 2006. A validation of a computer-assisted randomized-response survey to estimate the prevalence of fraud in social security. J. R. Stat. Soc. A169, 305–318.

Maddala, G.S., 1983. Limited Dependent and Qualitative Variables in Econometrics. Cambridge University Press, Cambridge.

Mangat, N.S., 1992. Two stage randomized response sampling procedure using unrelated question. J. Indian Soc. Agric. Stat. 44, 82–87.

Mangat, N.S., Singh, R., 1990. An alternative randomized response procedure. Biometrika 77, 439–442.

Mangat, N.S., Singh, R., Singh, S., 1992. An improved unrelated question randomized response strategy. Calcutta Stat. Assoc. Bull. 42, 277–281.

Moshagen, M., Musch, J., Erdfelder, E., 2012. A stochastic lie detector. Behav. Res. Methods 44, 222–231.

Musch, J., Broder, A., Klauer, K., 2001. Improving survey research on the world-wide web using the randomized response technique. In: Reips, U.D., Bosnjak, M. (Eds.), Dimensions of Internet Science. Pabst Science Publishers, Lengerich, Germany, pp. 179–192.

Ostapczuk, M., Moshagen, M., Zhao, Z., Musch, J., 2009. Assessing sensitive attributes using the randomized response technique: evidence for the importance of response symmetry. J. Educ. Behav. Stat. 34, 267–287.

Peeters, C.F.W., Lensvelt-Mulders, G.J.L.M., Lasthuizen, K., 2010. A note on a simple and practical randomized response framework for eliciting sensitive dichotomous and quantitative information. Sociol. Methods Res. 39 (2), 283–296.

Saha, A., 2007. A simple randomized response technique in complex surveys. Metron LXV, 59–66.

Sehra, S., 2008. Two-stage optional randomized response models. Master's Thesis, The University of North Carolina, Greensboro.

Singh, S., Joarder, A.H., 1997. Unknown repeated trials in randomized response sampling. J. Indian Stat. Assoc. 30, 109–122.

Singh, H., Tarray, T., 2015. An efficient alternative mixed randomized response procedure. Sociol. Methods Res. 44, 706–722.

Soberanis Cruz, V., Ramírez Valverde, G., Pérez Elizalde, S., González Cossio, F., 2008. Muestreo de respuestas aleatorizadas en poblaciones finitas: Un enfoque unificador. Agrociencia 42 (5), 537–549.

Templ, M., 2015. CRAN Task View: Official Statistics and Survey Methodology. https://cran.r-project.org/web/views/OfficialStatistics.html.

Tian, G.L., Tang, M.L., 2014. Incomplete Categorical Data Design: Non-Randomized Response Techniques for Sensitive Questions in Surveys. Chapman & Hall/CRC (Statistics in the Social and Behavioral Sciences), Boca Raton, USA.

Tracy, D., Mangat, N., 1996. Some development in randomized response sampling during the last decade—a follow up of review by Chaudhuri and Mukerjee. J. Appl. Stat. Sci. 4, 533–544.

Van den Hout, A., Bockenholt, U., Van der Heijden, P.G.M., 2010. Estimating the prevalence of sensitive behaviour and cheating with a dual design for direct questioning and randomized response. J. R. Stat. Soc.: Ser. C: Appl. Stat. 59, 723–736.

Van der Heijden, P.G.M., van Gils, G., Bouts, J., Hox, J.J., 2000. A comparison of randomized response, computer-assisted self-interview, and face-to-face direct questioning. Eliciting sensitive information in the context of welfare and unemployment benefit. Sociol. Methods Res. 28, 505–537.

Warner, S.L., 1965. RR: A survey technique for eliminating evasive answer bias. J. Am. Stat. Assoc. 60, 63–69.

Yu, J., Tian, G., Tang, M., 2008. Two new models for survey sampling with sensitive characteristic: Design and analysis. Metrika 67, 251–263.

Chapter 11

Poststratification Based on the Choice of Use of a Quantitative Randomization Device

O. Odumade*,[1], R. Arnab† and S. Singh‡
**Risk Analytics, Deloitte & Touche, Minneapolis, MN, United States*
†University of Botswana, Gaborone, Botswana
‡Texas A&M University-Kingsville, Kingsville, TX, United States
[1]Corresponding author: e-mail: sodumade@deloitte.com; oluseunodumade@yahoo.com

ABSTRACT
In this chapter, we use the idea of poststratification based on the respondents' choice of a particular randomization device in order to estimate the population mean of a sensitive quantitative variable. The proposed idea gives freedom to the respondents and is expected to result in greater cooperation from them as well as to provide some increase in the relative efficiency of the newly proposed estimator.

Keywords: Respondent's cooperation, Poststratification, Quantitative sensitive variables, Protection and efficiency

1 INTRODUCTION

Warner (1965) proposed a structured interviewing/surveying technique, called "randomized response technique" (RRT), to protect survey responder's privacy and to thus reduce a major source of bias (evasive answers or refusing to respond) in estimating the prevalence of sensitive characteristics in surveys of human populations.

An extensive amount of literature on the developments in the field of RRT can be found in Chaudhuri (2011) and Tracy and Mangat (1996). Leysieffer and Warner (1976) and Lanke (1975, 1976) studied different randomized response procedures at equal level of protection of the respondents, and later Nayak (1994), Bhargava (1996), Zou (1997), Bhargava and Singh (2001, 2002), and Moors (1997) found that Mangat and Singh (1990) and Warner (1965) models remain equally efficient at equal protection of respondents. While doing a face-to-face interview survey, experience shows that

Handbook of Statistics, Vol. 34. http://dx.doi.org/10.1016/bs.host.2016.02.001

respondent's cooperation is as important as privacy/confidentiality. For one respondent, a question is too sensitive while for other respondent it is not sensitive. A respondent could prefer to use of one randomization device over the other, thus freedom to choose a randomization device out of a couple of options of such devices, or the option of responding without the use of a randomization device, will certainly increase a respondent's cooperation.

In the next section, we suggest a new method to use randomization devices in real practice which may increase respondent cooperation in addition to some increase in the relative efficiency of the resultant estimator.

2 POSTSTRATIFICATION BASED ON THE CHOICE OF A QUANTITATIVE RANDOMIZATION DEVICE

Experience shows that in every face-to-face interview survey, there are respondents who are unwilling to treat the randomization device at hand but might be willing to trust a different device. Some respondents prefer to use the additive model due to Himmelfarb and Edgell (1980), some prefer to use multiplicative model due to Eichhorn and Hayre (1983), and some would be willing to respond directly without using any randomization device. Suppose we selected a simple random sample with replacement of n respondents. To every selected respondent in the sample, we ask for his/her choice of method of answering out of the above three options. Assume out of n respondents: n_1 respondents prefer the additive model, n_2 respondents prefer the multiplicative model, and n_3 respondents prefer to respond directly without using any randomization device. Suppose $Z_{1i}=X_i+S_1$, $i=1,2,3,\ldots,n_1$ be the scrambled responses obtained from n_1 respondents using the additive model, $Z_{2i}=X_iS_2$, $i=1,2,3,\ldots,n_2$ be the scrambled responses obtained from n_2 respondents using the multiplicative model, and $Z_{3i}=X_i$, $i=1,2,3,\ldots,n_3$ be the direct responses obtained from n_3 respondents by using no randomization device. Here, S_1 and S_2 are two independent scrambling variables and their distributions are assumed to be known. In other words, $E_R(S_1)=\theta_1$, $E_R(S_2)=\theta_2$, $V_R(S_1)=\gamma_1^2$, and $V_R(S_2)=\gamma_2^2$ are assumed to be known. Here, E_R and V_R denote, respectively, expectation and variance with respect to the randomization device. Note that here we know preference of a respondent, unlike the optional randomization devices studied as listed in Chaudhuri (2011) and among others. If more cooperation is required then a researcher can use more options of choosing a randomization device as listed in Perri (2008) and Diana and Perri (2009, 2010).

Let $\hat{w}_1=n_1/n$, $\hat{w}_2=n_2/n$, and $\hat{w}_3=n_3/n$ be the unbiased estimators of the true population proportions $W_1=N_1/N$, $W_2=N_2/N$, and $W_3=N_3/N$ who prefer to use the additive model, multiplicative model, and respond directly, respectively. Unfortunately the values of W_h, $h=1,2,3$, remain unknown in a survey, thus their estimators $\hat{w}_h$ are used in constructing the estimators. The suffix h refers to the h-th poststratum. Also let $\mu_{x_1}=\frac{1}{N_1}\sum_{i=1}^{N_1}X_{1i}$ and

$\sigma_{x_1}^2 = \frac{1}{N_1}\sum_{i=1}^{N_1}(X_{1i} - \mu_{x_1})^2$ be the population mean and variance of the sensitive variable of interest X in the first poststratum; $\mu_{x_2} = \frac{1}{N_2}\sum_{i=1}^{N_2} X_{2i}$ and $\sigma_{x_2}^2 = \frac{1}{N_2}\sum_{i=1}^{N_2}(X_{2i} - \mu_{x_2})^2$ be the population mean and variance, respectively, of the sensitive variable X in the second poststratum; and $\mu_{x_3} = \frac{1}{N_3}\sum_{i=1}^{N_3} X_{3i}$ and $\sigma_{x_3}^2 = \frac{1}{N_3}\sum_{i=1}^{N_3}(X_{3i} - \mu_{x_3})^2$ be the population mean and variance of the sensitive variable X in the third poststratum.

Now we have the following theorems:

Theorem 1. *An unbiased estimator of the population mean μ_x is given by:*

$$\hat{\mu}_{xp} = \sum_{h=1}^{3} \hat{w}_h \hat{\mu}_{xh} \tag{1}$$

where $\hat{\mu}_{x_1} = \frac{1}{n_1}\sum_{i=1}^{n_1} Z_{1i} - \theta_1$, $\hat{\mu}_{x_2} = \frac{1}{n_2\theta_2}\sum_{i=1}^{n_2} Z_{2i}$, *and* $\hat{\mu}_{x_3} = \frac{1}{n_3}\sum_{i=1}^{n_3} Z_{3i}$ *such that* $n_h > 0$ *are defined.*

Proof. Let E_2 denote the expected value for a given values of n_h, $h = 1,2,3$ over a given randomization device and E_1 is the expected value over all possible samples of size n, then we have:

$$\begin{aligned} E(\hat{\mu}_{xp}) &= E_1 E_2(\hat{\mu}_{xp} | n_h > 0) = E_1 E_2\left[\sum_{h=1}^{3} \hat{w}_h \hat{\mu}_{xh} | n_h > 0\right] \\ &= E_1\left[\sum_{h=1}^{3} \hat{w}_h E_2(\hat{\mu}_{xh} | n_h > 0)\right] = E_1\left[\sum_{h=1}^{3} \hat{w}_h \mu_{xh}\right] \\ &= \left[\sum_{h=1}^{3} \mu_{xh} E_1(\hat{w}_h)\right] = \left[\sum_{h=1}^{3} \mu_{xh} W_h\right] = \mu_x \end{aligned}$$

which proves the theorem. □

Theorem 2. *The variance of the unbiased estimator* $\hat{\mu}_{xp}$ is given by:

$$V(\hat{\mu}_{xp}) = \frac{1}{n}\left(\sum_{h=1}^{3} \sigma_h^2 W_h\right) + \frac{1}{n}\sum_{h=1}^{3} W_h(\mu_{xh} - \mu_x)^2 \tag{2}$$

where $\sigma_1^2 = \sigma_{x_1}^2 + \gamma_1^2$, $\sigma_2^2 = \sigma_{x_2}^2((1+\gamma_2^2)/\theta_2^2) + \mu_{x_2}^2$, and $\sigma_3^2 = \sigma_{x_3}^2$.

Proof. Here, we assume that the sample size is so large $P(n_h = 0) = 0$ (Stephan, 1945). Let V_2 denote the variance for a given values of n_h,

$h=1,2,3$, over a given randomization device and V_1 is the variance over all possible samples of size n, then we have

$$
\begin{aligned}
V(\hat{\mu}_{xp}) &= E_1V_2(\hat{\mu}_{xp}|n_h>0)+V_1E_2(\hat{\mu}_{xp}|n_h>0)\\
&= E_1V_2\left(\sum_{h=1}^{3}\hat{w}_h\hat{\mu}_{xh}|n_h>0\right)+V_1E_2\left(\sum_{h=1}^{3}\hat{w}_h\hat{\mu}_{xh}|n_h>0\right)\\
&= E_1\left(\sum_{h=1}^{3}\hat{w}_h^2V_2(\hat{\mu}_{xh}|n_h>0)\right)+V_1\left(\sum_{h=1}^{3}\hat{w}_hE_2(\hat{\mu}_{xh}|n_h>0)\right)\\
&= E_1\left(\sum_{h=1}^{3}\hat{w}_h^2\frac{\sigma_h^2}{n_h}\right)+V_1\left(\sum_{h=1}^{3}\hat{w}_h\mu_{xh}\right)\\
&= \frac{1}{n}E_1\left(\sum_{h=1}^{3}\hat{w}_h\sigma_h^2\right)+\left(\sum_{h=1}^{3}\mu_{xh}^2V_1(\hat{w}_h)+\sum_{h}^{3}\sum_{\neq h'=1}^{3}\mu_{xh}\mu_{xh'}\mathrm{Cov}_1(\hat{w}_h,\hat{w}_{h'})\right)\\
&= \frac{1}{n}\left(\sum_{h=1}^{3}\sigma_h^2E_1(\hat{w}_h)\right)+\frac{1}{n}\left[\sum_{h=1}^{3}\mu_{xh}^2W_h(1-W_h)-\sum_{h\neq}^{3}\sum_{h=1}^{3}W_hW_{h'}\mu_{xh}\mu_{xh'}\right]\\
&= \frac{1}{n}\left(\sum_{h=1}^{3}\sigma_h^2E_1(\hat{w}_h)\right)+\frac{1}{n}\left[\sum_{h=1}^{3}\mu_{xh}^2W_h\right]-\left(\sum_{h=1}^{3}\mu_{xh}W_h\right)^2\\
&= \frac{1}{n}\left(\sum_{h=1}^{3}\sigma_h^2E_1(\hat{w}_h)\right)+\frac{1}{n}\left[\sum_{h=1}^{3}\mu_{xh}^2W_h-\mu_x^2\right],\quad \text{where } \mu_x=\sum_{h=1}^{3}\mu_{xh}W_h\\
&= \frac{1}{n}\left(\sum_{h=1}^{3}\sigma_h^2W_h\right)+\frac{1}{n}\sum_{h=1}^{3}W_h(\mu_{xh}-\mu_x)^2
\end{aligned}
$$

which proves the theorem. □

Now we have the following corollary:

Corollary 1. *An estimator of $V(\hat{\mu}_{\mathrm{xp}})$ is suggested as:*

$$\hat{V}(\hat{\mu}_{xp})=\frac{1}{n}\sum_{h=1}^{3}\hat{w}_h\left[\hat{\sigma}_h^2+(\hat{\mu}_{xh}-\hat{\mu}_x)^2\right] \tag{3}$$

where $\hat{\sigma}_h^2$ is an unbiased estimator of σ_h^2 based on the information in the h-th poststratum.

Corollary 2. *If $W_1=1$, $W_2=W_3=0$ then the proposed estimator reduces to the additive model. The variance of the estimator based on the additive model is given by:*

$$V(\hat{\mu}_{\text{add}}) = \frac{1}{n}\left(\sigma_x^2 + \gamma_1^2\right) \tag{4}$$

Corollary 3. *If* $W_1 = 0$, $W_2 = 1$, *and* $W_3 = 0$ *then the proposed estimator reduces to the multiplicative model. The variance of the estimator based on the multiplicative model is given by:*

$$V(\hat{\mu}_{mult}) = \frac{1}{n}\left[\sigma_x^2\left(1 + C_{\gamma_2}^2\right) + \mu_x^2\right] \tag{5}$$

In the next section, we consider the comparison of the proposed estimator with the other competitors considered above.

3 RELATIVE EFFICIENCY

We define the percent relative efficiency of the proposed estimator with respect to the additive model as:

$$\text{RE}(1) = \frac{V(\hat{\mu}_{add})}{V(\hat{\mu}_{xp})} \times 100 \tag{6}$$

and the percent relative efficiency of the proposed estimator with respect to the multiplicative model as:

$$\text{RE}(2) = \frac{V(\hat{\mu}_{mult})}{V(\hat{\mu}_{xp})} \times 100 \tag{7}$$

We simulated the percent relative efficiency values for different choice of parameters of the study variable and the randomization device. We have listed many results in the Appendix and we also see there is no end of such results based on a choice of parameters. We discuss only a few of these results as follows. Let $W_1 = 0.3$, $W_2 = 0.69$, and $W_3 = 0.01$, which means that 30% people in like additive model, 69% people like multiplicative model, and only 1% people give direct response to a sensitive question in a population, then for $\gamma_1^2 = 169.3$, $C_{\gamma_2}^2 = 0.1$, $\mu_x = 22$, $\mu_{x_1} = 50$, $\mu_{x_2} = 10$, $\mu_{x_3} = 10$, $\sigma_{x_1}^2 = 2.5$, $\sigma_{x_2}^2 = 2.5$, $\sigma_{x3}^2 = 2.5$, and $\sigma_x^2 = 338.5$, the percent relative efficiency of the proposed estimator over the additive and multiplicative models is given by 110.8% and 186.8%, respectively. Thus, from this we can see that if only 1% of the respondents agree to provide a direct response then the proposed method can be benefitted over the use of only either additive or multiplicative model. In addition, due to many options for a respondent, more cooperation is expected.

APPENDIX

TABLE A.1 Relative Efficiency of the Proposed Estimator with Respect to the Additive Model and the Multiplicative Model for Different Choice of Parameters

W_1	W_2	W_3	γ_1^2	$C_{\gamma_2}^2$	μ_x	μ_{x_1}	μ_{x_2}	μ_{x_3}	$\sigma_{x_1}^2$	$\sigma_{x_2}^2$	$\sigma_{x_3}^2$	σ_x^2	RE(1)	RE(2)
0.3	0.69	0.01	169.3	0.1	22.0	50	10	10	2.5	2.5	2.5	338.5	110.8	186.8
0.3	0.69	0.01	169.3	0.3	22.0	50	10	10	2.5	2.5	2.5	338.5	110.7	201.4
0.3	0.69	0.01	169.3	0.5	22.0	50	10	10	2.5	2.5	2.5	338.5	110.6	216.0
0.3	0.69	0.01	169.3	0.1	22.0	50	10	10	2.5	2.5	4.0	338.5	110.8	186.8
0.3	0.69	0.01	169.3	0.3	22.0	50	10	10	2.5	2.5	4.0	338.5	110.7	201.4
0.3	0.69	0.01	169.3	0.5	22.0	50	10	10	2.5	2.5	4.0	338.5	110.6	216.0
0.3	0.69	0.01	169.8	0.1	22.0	50	10	10	2.5	4.0	2.5	339.5	110.8	186.5
0.3	0.69	0.01	169.8	0.3	22.0	50	10	10	2.5	4.0	2.5	339.5	110.7	201.0
0.3	0.69	0.01	169.8	0.5	22.0	50	10	10	2.5	4.0	2.5	339.5	110.5	215.5
0.3	0.69	0.01	169.8	0.1	22.0	50	10	10	2.5	4.0	4.0	339.6	110.8	186.5
0.3	0.69	0.01	169.8	0.3	22.0	50	10	10	2.5	4.0	4.0	339.6	110.7	201.0
0.3	0.69	0.01	169.8	0.5	22.0	50	10	10	2.5	4.0	4.0	339.6	110.5	215.5
0.3	0.69	0.01	169.5	0.1	22.0	50	10	10	4.0	2.5	2.5	339.0	110.8	186.7
0.3	0.69	0.01	169.5	0.3	22.0	50	10	10	4.0	2.5	2.5	339.0	110.7	201.3

0.3	0.69	0.01	169.5	0.5	22.0	50	10	10	4.0	2.5	2.5	339.0	110.6	215.9
0.3	0.69	0.01	169.5	0.1	22.0	50	10	10	4.0	2.5	4.0	339.0	110.8	186.7
0.3	0.69	0.01	169.5	0.3	22.0	50	10	10	4.0	2.5	4.0	339.0	110.7	201.3
0.3	0.69	0.01	169.5	0.5	22.0	50	10	10	4.0	2.5	4.0	339.0	110.6	215.9
0.3	0.69	0.01	170.0	0.1	22.0	50	10	10	4.0	4.0	2.5	340.0	110.8	186.4
0.3	0.69	0.01	170.0	0.3	22.0	50	10	10	4.0	4.0	2.5	340.0	110.7	201.0
0.3	0.69	0.01	170.0	0.5	22.0	50	10	10	4.0	4.0	2.5	340.0	110.5	215.4
0.3	0.69	0.01	170.0	0.1	22.0	50	10	10	4.0	4.0	4.0	340.0	110.8	186.4
0.3	0.69	0.01	170.0	0.3	22.0	50	10	10	4.0	4.0	4.0	340.0	110.7	200.9
0.3	0.69	0.01	170.0	0.5	22.0	50	10	10	4.0	4.0	4.0	340.0	110.5	215.4
0.3	0.69	0.01	168.5	0.1	22.1	50	10	20	2.5	2.5	2.5	337.1	110.7	188.1
0.3	0.69	0.01	168.5	0.3	22.1	50	10	20	2.5	2.5	2.5	337.1	110.6	202.7
0.3	0.69	0.01	168.5	0.5	22.1	50	10	20	2.5	2.5	2.5	337.1	110.5	217.3
0.3	0.69	0.01	168.6	0.1	22.1	50	10	20	2.5	2.5	4.0	337.1	110.7	188.1
0.3	0.69	0.01	168.6	0.3	22.1	50	10	20	2.5	2.5	4.0	337.1	110.6	202.7
0.3	0.69	0.01	168.6	0.5	22.1	50	10	20	2.5	2.5	4.0	337.1	110.5	217.3
0.3	0.69	0.01	169.1	0.1	22.1	50	10	20	2.5	4.0	2.5	338.1	110.7	187.8
0.3	0.69	0.01	169.1	0.3	22.1	50	10	20	2.5	4.0	2.5	338.1	110.6	202.3
0.3	0.69	0.01	169.1	0.5	22.1	50	10	20	2.5	4.0	2.5	338.1	110.4	216.8

Continued

TABLE A.1 Relative Efficiency of the Proposed Estimator with Respect to the Additive Model and the Multiplicative Model for Different Choice of Parameters—Cont'd

W_1	W_2	W_3	γ_1^2	$C_{\gamma_2}^2$	μ_x	μ_{x_1}	μ_{x_2}	μ_{x_3}	$\sigma_{x_1}^2$	$\sigma_{x_2}^2$	$\sigma_{x_3}^2$	σ_x^2	RE(1)	RE(2)
0.3	0.69	0.01	169.1	0.1	22.1	50	10	20	2.5	4.0	4.0	338.1	110.7	187.8
0.3	0.69	0.01	169.1	0.3	22.1	50	10	20	2.5	4.0	4.0	338.1	110.6	202.3
0.3	0.69	0.01	169.1	0.5	22.1	50	10	20	2.5	4.0	4.0	338.1	110.5	216.8
0.3	0.69	0.01	168.8	0.1	22.1	50	10	20	4.0	2.5	2.5	337.5	110.7	188.0
0.3	0.69	0.01	168.8	0.3	22.1	50	10	20	4.0	2.5	2.5	337.5	110.6	202.6
0.3	0.69	0.01	168.8	0.5	22.1	50	10	20	4.0	2.5	2.5	337.5	110.5	217.2
0.3	0.69	0.01	168.8	0.1	22.1	50	10	20	4.0	2.5	4.0	337.6	110.7	188.0
0.3	0.69	0.01	168.8	0.3	22.1	50	10	20	4.0	2.5	4.0	337.6	110.6	202.6
0.3	0.69	0.01	168.8	0.5	22.1	50	10	20	4.0	2.5	4.0	337.6	110.5	217.2
0.3	0.69	0.01	169.3	0.1	22.1	50	10	20	4.0	4.0	2.5	338.6	110.7	187.7
0.3	0.69	0.01	169.3	0.3	22.1	50	10	20	4.0	4.0	2.5	338.6	110.6	202.2
0.3	0.69	0.01	169.3	0.5	22.1	50	10	20	4.0	4.0	2.5	338.6	110.5	216.7
0.3	0.69	0.01	169.3	0.1	22.1	50	10	20	4.0	4.0	4.0	338.6	110.7	187.7
0.3	0.69	0.01	169.3	0.3	22.1	50	10	20	4.0	4.0	4.0	338.6	110.6	202.2
0.3	0.69	0.01	169.3	0.5	22.1	50	10	20	4.0	4.0	4.0	338.6	110.5	216.7
0.3	0.69	0.01	168.8	0.1	22.2	50	10	30	2.5	2.5	2.5	337.7	110.7	188.9

0.3	0.69	0.01	168.8	0.3	22.2	50	10	30	2.5	2.5	2.5	337.7	110.6	203.5
0.3	0.69	0.01	168.8	0.5	22.2	50	10	30	2.5	2.5	2.5	337.7	110.6	218.1
0.3	0.69	0.01	168.8	0.1	22.2	50	10	30	2.5	2.5	4.0	337.7	110.7	188.9
0.3	0.69	0.01	168.8	0.3	22.2	50	10	30	2.5	2.5	4.0	337.7	110.6	203.5
0.3	0.69	0.01	168.8	0.5	22.2	50	10	30	2.5	2.5	4.0	337.7	110.6	218.1
0.3	0.69	0.01	169.4	0.1	22.2	50	10	30	2.5	4.0	2.5	338.7	110.7	188.6
0.3	0.69	0.01	169.4	0.3	22.2	50	10	30	2.5	4.0	2.5	338.7	110.6	203.2
0.3	0.69	0.01	169.4	0.5	22.2	50	10	30	2.5	4.0	2.5	338.7	110.5	217.6
0.3	0.69	0.01	169.4	0.1	22.2	50	10	30	2.5	4.0	4.0	338.7	110.7	188.6
0.3	0.69	0.01	169.4	0.3	22.2	50	10	30	2.5	4.0	4.0	338.7	110.6	203.2
0.3	0.69	0.01	169.4	0.5	22.2	50	10	30	2.5	4.0	4.0	338.7	110.5	217.6
0.3	0.69	0.01	169.1	0.1	22.2	50	10	30	4.0	2.5	2.5	338.1	110.7	188.8
0.3	0.69	0.01	169.1	0.3	22.2	50	10	30	4.0	2.5	2.5	338.1	110.7	203.4
0.3	0.69	0.01	169.1	0.5	22.2	50	10	30	4.0	2.5	2.5	338.1	110.6	218.0
0.3	0.69	0.01	169.1	0.1	22.2	50	10	30	4.0	2.5	4.0	338.1	110.7	188.8
0.3	0.69	0.01	169.1	0.3	22.2	50	10	30	4.0	2.5	4.0	338.1	110.7	203.4
0.3	0.69	0.01	169.1	0.5	22.2	50	10	30	4.0	2.5	4.0	338.1	110.6	218.0
0.3	0.69	0.01	169.6	0.1	22.2	50	10	30	4.0	4.0	2.5	339.2	110.8	188.5
0.3	0.69	0.01	169.6	0.3	22.2	50	10	30	4.0	4.0	2.5	339.2	110.6	203.1

Continued

TABLE A.1 Relative Efficiency of the Proposed Estimator with Respect to the Additive Model and the Multiplicative Model for Different Choice of Parameters—Cont'd

W_1	W_2	W_3	γ_1^2	$C_{\gamma_2}^2$	μ_x	μ_{x_1}	μ_{x_2}	μ_{x_3}	$\sigma_{x_1}^2$	$\sigma_{x_2}^2$	$\sigma_{x_3}^2$	σ_x^2	RE(1)	RE(2)
0.3	0.69	0.01	169.6	0.5	22.2	50	10	30	4.0	4.0	2.5	339.2	110.5	217.5
0.3	0.69	0.01	169.6	0.1	22.2	50	10	30	4.0	4.0	4.0	339.2	110.8	188.5
0.3	0.69	0.01	169.6	0.3	22.2	50	10	30	4.0	4.0	4.0	339.2	110.6	203.1
0.3	0.69	0.01	169.6	0.5	22.2	50	10	30	4.0	4.0	4.0	339.2	110.5	217.5
0.3	0.69	0.01	170.1	0.1	22.3	50	10	40	2.5	2.5	2.5	340.2	110.8	189.3
0.3	0.69	0.01	170.1	0.3	22.3	50	10	40	2.5	2.5	2.5	340.2	110.8	203.9
0.3	0.69	0.01	170.1	0.5	22.3	50	10	40	2.5	2.5	2.5	340.2	110.7	218.5
0.3	0.69	0.01	170.1	0.1	22.3	50	10	40	2.5	2.5	4.0	340.2	110.8	189.3
0.3	0.69	0.01	170.1	0.3	22.3	50	10	40	2.5	2.5	4.0	340.2	110.8	203.9
0.3	0.69	0.01	170.1	0.5	22.3	50	10	40	2.5	2.5	4.0	340.2	110.7	218.5
0.3	0.69	0.01	170.6	0.1	22.3	50	10	40	2.5	4.0	2.5	341.2	110.9	189.0
0.3	0.69	0.01	170.6	0.3	22.3	50	10	40	2.5	4.0	2.5	341.2	110.7	203.6
0.3	0.69	0.01	170.6	0.5	22.3	50	10	40	2.5	4.0	2.5	341.2	110.6	218.1
0.3	0.69	0.01	170.6	0.1	22.3	50	10	40	2.5	4.0	4.0	341.3	110.9	189.0
0.3	0.69	0.01	170.6	0.3	22.3	50	10	40	2.5	4.0	4.0	341.3	110.7	203.5
0.3	0.69	0.01	170.6	0.5	22.3	50	10	40	2.5	4.0	4.0	341.3	110.6	218.1

0.3	0.69	0.01	170.3	0.1	22.3	50	10	40	4.0	2.5	2.5	340.7	110.9	189.2
0.3	0.69	0.01	170.3	0.3	22.3	50	10	40	4.0	2.5	2.5	340.7	110.8	203.8
0.3	0.69	0.01	170.3	0.5	22.3	50	10	40	4.0	2.5	2.5	340.7	110.7	218.4
0.3	0.69	0.01	170.3	0.1	22.3	50	10	40	4.0	2.5	4.0	340.7	110.9	189.2
0.3	0.69	0.01	170.3	0.3	22.3	50	10	40	4.0	2.5	4.0	340.7	110.8	203.8
0.3	0.69	0.01	170.3	0.5	22.3	50	10	40	4.0	2.5	4.0	340.7	110.7	218.4
0.3	0.69	0.01	170.9	0.1	22.3	50	10	40	4.0	4.0	2.5	341.7	110.9	188.9
0.3	0.69	0.01	170.9	0.3	22.3	50	10	40	4.0	4.0	2.5	341.7	110.8	203.4
0.3	0.69	0.01	170.9	0.5	22.3	50	10	40	4.0	4.0	2.5	341.7	110.6	218.0
0.3	0.69	0.01	170.9	0.1	22.3	50	10	40	4.0	4.0	4.0	341.7	110.9	188.9
0.3	0.69	0.01	170.9	0.3	22.3	50	10	40	4.0	4.0	4.0	341.7	110.8	203.4
0.3	0.69	0.01	170.9	0.5	22.3	50	10	40	4.0	4.0	4.0	341.7	110.6	218.0
0.3	0.69	0.01	172.4	0.1	22.4	50	10	50	2.5	2.5	2.5	344.7	111.1	189.2
0.3	0.69	0.01	172.4	0.3	22.4	50	10	50	2.5	2.5	2.5	344.7	111.0	203.9
0.3	0.69	0.01	172.4	0.5	22.4	50	10	50	2.5	2.5	2.5	344.7	110.9	218.5
0.3	0.69	0.01	172.4	0.1	22.4	50	10	50	2.5	2.5	4.0	344.8	111.1	189.2
0.3	0.69	0.01	172.4	0.3	22.4	50	10	50	2.5	2.5	4.0	344.8	111.0	203.9
0.3	0.69	0.01	172.4	0.5	22.4	50	10	50	2.5	2.5	4.0	344.8	110.9	218.5
0.3	0.69	0.01	172.9	0.1	22.4	50	10	50	2.5	4.0	2.5	345.8	111.1	188.9

Continued

TABLE A.1 Relative Efficiency of the Proposed Estimator with Respect to the Additive Model and the Multiplicative Model for Different Choice of Parameters—Cont'd

W_1	W_2	W_3	γ_1^2	$C_{\gamma_2}^2$	μ_x	μ_{x_1}	μ_{x_2}	μ_{x_3}	$\sigma_{x_1}^2$	$\sigma_{x_2}^2$	$\sigma_{x_3}^2$	σ_x^2	RE(1)	RE(2)
0.3	0.69	0.01	172.9	0.3	22.4	50	10	50	2.5	4.0	2.5	345.8	111.0	203.5
0.3	0.69	0.01	172.9	0.5	22.4	50	10	50	2.5	4.0	2.5	345.8	110.8	218.0
0.3	0.69	0.01	172.9	0.1	22.4	50	10	50	2.5	4.0	4.0	345.8	111.1	188.9
0.3	0.69	0.01	172.9	0.3	22.4	50	10	50	2.5	4.0	4.0	345.8	111.0	203.5
0.3	0.69	0.01	172.9	0.5	22.4	50	10	50	2.5	4.0	4.0	345.8	110.8	218.0
0.3	0.69	0.01	172.6	0.1	22.4	50	10	50	4.0	2.5	2.5	345.2	111.1	189.1
0.3	0.69	0.01	172.6	0.3	22.4	50	10	50	4.0	2.5	2.5	345.2	111.0	203.8
0.3	0.69	0.01	172.6	0.5	22.4	50	10	50	4.0	2.5	2.5	345.2	110.9	218.4
0.3	0.69	0.01	172.6	0.1	22.4	50	10	50	4.0	2.5	4.0	345.2	111.1	189.1
0.3	0.69	0.01	172.6	0.3	22.4	50	10	50	4.0	2.5	4.0	345.2	111.0	203.8
0.3	0.69	0.01	172.6	0.5	22.4	50	10	50	4.0	2.5	4.0	345.2	110.9	218.4
0.3	0.69	0.01	173.1	0.1	22.4	50	10	50	4.0	4.0	2.5	346.2	111.1	188.8
0.3	0.69	0.01	173.1	0.3	22.4	50	10	50	4.0	4.0	2.5	346.2	111.0	203.4
0.3	0.69	0.01	173.1	0.5	22.4	50	10	50	4.0	4.0	2.5	346.2	110.8	217.9
0.3	0.69	0.01	173.1	0.1	22.4	50	10	50	4.0	4.0	4.0	346.2	111.1	188.8
0.3	0.69	0.01	173.1	0.3	22.4	50	10	50	4.0	4.0	4.0	346.2	111.0	203.4

0.3	0.69	0.01	173.1	0.5	22.4	50	10	50	4.0	4.0	4.0	346.2	110.8	217.9
0.3	0.68	0.02	169.3	0.1	22.0	50	10	10	2.5	2.5	2.5	338.5	111.0	187.2
0.3	0.68	0.02	169.3	0.3	22.0	50	10	10	2.5	2.5	2.5	338.5	110.9	201.9
0.3	0.68	0.02	169.3	0.5	22.0	50	10	10	2.5	2.5	2.5	338.5	110.8	216.5
0.3	0.68	0.02	169.3	0.1	22.0	50	10	10	2.5	2.5	4.0	338.5	111.0	187.2
0.3	0.68	0.02	169.3	0.3	22.0	50	10	10	2.5	2.5	4.0	338.5	110.9	201.9
0.3	0.68	0.02	169.3	0.5	22.0	50	10	10	2.5	2.5	4.0	338.5	110.8	216.5
0.3	0.68	0.02	169.8	0.1	22.0	50	10	10	2.5	4.0	2.5	339.5	111.0	186.9
0.3	0.68	0.02	169.8	0.3	22.0	50	10	10	2.5	4.0	2.5	339.5	110.9	201.5
0.3	0.68	0.02	169.8	0.5	22.0	50	10	10	2.5	4.0	2.5	339.5	110.8	216.0
0.3	0.68	0.02	169.8	0.1	22.0	50	10	10	2.5	4.0	4.0	339.6	111.0	186.9
0.3	0.68	0.02	169.8	0.3	22.0	50	10	10	2.5	4.0	4.0	339.6	110.9	201.5
0.3	0.68	0.02	169.8	0.5	22.0	50	10	10	2.5	4.0	4.0	339.6	110.8	216.0
0.3	0.68	0.02	169.5	0.1	22.0	50	10	10	4.0	2.5	2.5	339.0	111.0	187.1
0.3	0.68	0.02	169.5	0.3	22.0	50	10	10	4.0	2.5	2.5	339.0	110.9	201.8
0.3	0.68	0.02	169.5	0.5	22.0	50	10	10	4.0	2.5	2.5	339.0	110.9	216.4
0.3	0.68	0.02	169.5	0.1	22.0	50	10	10	4.0	2.5	4.0	339.0	111.0	187.1
0.3	0.68	0.02	169.5	0.3	22.0	50	10	10	4.0	2.5	4.0	339.0	110.9	201.8
0.3	0.68	0.02	169.5	0.5	22.0	50	10	10	4.0	2.5	4.0	339.0	110.9	216.4

Continued

TABLE A.1 Relative Efficiency of the Proposed Estimator with Respect to the Additive Model and the Multiplicative Model for Different Choice of Parameters—Cont'd

W_1	W_2	W_3	γ_1^2	$C_{\gamma_2}^2$	μ_x	μ_{x_1}	μ_{x_2}	μ_{x_3}	$\sigma_{x_1}^2$	$\sigma_{x_2}^2$	$\sigma_{x_3}^2$	σ_x^2	RE(1)	RE(2)
0.3	0.68	0.02	170.0	0.1	22.0	50	10	10	4.0	4.0	2.5	340.0	111.0	186.8
0.3	0.68	0.02	170.0	0.3	22.0	50	10	10	4.0	4.0	2.5	340.0	110.9	201.4
0.3	0.68	0.02	170.0	0.5	22.0	50	10	10	4.0	4.0	2.5	340.0	110.8	215.9
0.3	0.68	0.02	170.0	0.1	22.0	50	10	10	4.0	4.0	4.0	340.0	111.1	186.8
0.3	0.68	0.02	170.0	0.3	22.0	50	10	10	4.0	4.0	4.0	340.0	110.9	201.4
0.3	0.68	0.02	170.0	0.5	22.0	50	10	10	4.0	4.0	4.0	340.0	110.8	215.9
0.3	0.68	0.02	167.8	0.1	22.2	50	10	20	2.5	2.5	2.5	335.7	110.9	189.8
0.3	0.68	0.02	167.8	0.3	22.2	50	10	20	2.5	2.5	2.5	335.7	110.8	204.4
0.3	0.68	0.02	167.8	0.5	22.2	50	10	20	2.5	2.5	2.5	335.7	110.7	219.0
0.3	0.68	0.02	167.9	0.1	22.2	50	10	20	2.5	2.5	4.0	335.7	110.9	189.8
0.3	0.68	0.02	167.9	0.3	22.2	50	10	20	2.5	2.5	4.0	335.7	110.8	204.4
0.3	0.68	0.02	167.9	0.5	22.2	50	10	20	2.5	2.5	4.0	335.7	110.7	219.0
0.3	0.68	0.02	168.3	0.1	22.2	50	10	20	2.5	4.0	2.5	336.7	110.9	189.5
0.3	0.68	0.02	168.3	0.3	22.2	50	10	20	2.5	4.0	2.5	336.7	110.8	204.1
0.3	0.68	0.02	168.3	0.5	22.2	50	10	20	2.5	4.0	2.5	336.7	110.6	218.6
0.3	0.68	0.02	168.4	0.1	22.2	50	10	20	2.5	4.0	4.0	336.7	110.9	189.5

0.3	0.68	0.02	168.4	0.3	22.2	50	10	20	2.5	4.0	4.0	336.7	110.8	204.1
0.3	0.68	0.02	168.4	0.5	22.2	50	10	20	2.5	4.0	4.0	336.7	110.6	218.6
0.3	0.68	0.02	168.1	0.1	22.2	50	10	20	4.0	2.5	2.5	336.1	110.9	189.7
0.3	0.68	0.02	168.1	0.3	22.2	50	10	20	4.0	2.5	2.5	336.1	110.8	204.3
0.3	0.68	0.02	168.1	0.5	22.2	50	10	20	4.0	2.5	2.5	336.1	110.7	218.9
0.3	0.68	0.02	168.1	0.1	22.2	50	10	20	4.0	2.5	4.0	336.1	110.9	189.7
0.3	0.68	0.02	168.1	0.3	22.2	50	10	20	4.0	2.5	4.0	336.1	110.8	204.3
0.3	0.68	0.02	168.1	0.5	22.2	50	10	20	4.0	2.5	4.0	336.1	110.7	218.9
0.3	0.68	0.02	168.6	0.1	22.2	50	10	20	4.0	4.0	2.5	337.1	110.9	189.4
0.3	0.68	0.02	168.6	0.3	22.2	50	10	20	4.0	4.0	2.5	337.1	110.8	204.0
0.3	0.68	0.02	168.6	0.5	22.2	50	10	20	4.0	4.0	2.5	337.1	110.6	218.5
0.3	0.68	0.02	168.6	0.1	22.2	50	10	20	4.0	4.0	4.0	337.2	110.9	189.4
0.3	0.68	0.02	168.6	0.3	22.2	50	10	20	4.0	4.0	4.0	337.2	110.8	204.0
0.3	0.68	0.02	168.6	0.5	22.2	50	10	20	4.0	4.0	4.0	337.2	110.6	218.5
0.3	0.68	0.02	168.4	0.1	22.4	50	10	30	2.5	2.5	2.5	336.7	110.9	191.5
0.3	0.68	0.02	168.4	0.3	22.4	50	10	30	2.5	2.5	2.5	336.7	110.8	206.1
0.3	0.68	0.02	168.4	0.5	22.4	50	10	30	2.5	2.5	2.5	336.7	110.8	220.8
0.3	0.68	0.02	168.4	0.1	22.4	50	10	30	2.5	2.5	4.0	336.8	110.9	191.5
0.3	0.68	0.02	168.4	0.3	22.4	50	10	30	2.5	2.5	4.0	336.8	110.8	206.1

Continued

TABLE A.1 Relative Efficiency of the Proposed Estimator with Respect to the Additive Model and the Multiplicative Model for Different Choice of Parameters—Cont'd

W_1	W_2	W_3	γ_1^2	$C_{\gamma_2}^2$	μ_x	μ_{x_1}	μ_{x_2}	μ_{x_3}	$\sigma_{x_1}^2$	$\sigma_{x_2}^2$	$\sigma_{x_3}^2$	σ_x^2	RE(1)	RE(2)
0.3	0.68	0.02	168.4	0.5	22.4	50	10	30	2.5	2.5	4.0	336.8	110.8	220.8
0.3	0.68	0.02	168.9	0.1	22.4	50	10	30	2.5	4.0	2.5	337.8	110.9	191.2
0.3	0.68	0.02	168.9	0.3	22.4	50	10	30	2.5	4.0	2.5	337.8	110.8	205.8
0.3	0.68	0.02	168.9	0.5	22.4	50	10	30	2.5	4.0	2.5	337.8	110.7	220.3
0.3	0.68	0.02	168.9	0.1	22.4	50	10	30	2.5	4.0	4.0	337.8	110.9	191.2
0.3	0.68	0.02	168.9	0.3	22.4	50	10	30	2.5	4.0	4.0	337.8	110.8	205.8
0.3	0.68	0.02	168.9	0.5	22.4	50	10	30	2.5	4.0	4.0	337.8	110.7	220.3
0.3	0.68	0.02	168.6	0.1	22.4	50	10	30	4.0	2.5	2.5	337.2	110.9	191.4
0.3	0.68	0.02	168.6	0.3	22.4	50	10	30	4.0	2.5	2.5	337.2	110.9	206.0
0.3	0.68	0.02	168.6	0.5	22.4	50	10	30	4.0	2.5	2.5	337.2	110.8	220.7
0.3	0.68	0.02	168.6	0.1	22.4	50	10	30	4.0	2.5	4.0	337.2	110.9	191.4
0.3	0.68	0.02	168.6	0.3	22.4	50	10	30	4.0	2.5	4.0	337.2	110.9	206.0
0.3	0.68	0.02	168.6	0.5	22.4	50	10	30	4.0	2.5	4.0	337.2	110.8	220.7
0.3	0.68	0.02	169.1	0.1	22.4	50	10	30	4.0	4.0	2.5	338.2	111.0	191.1
0.3	0.68	0.02	169.1	0.3	22.4	50	10	30	4.0	4.0	2.5	338.2	110.8	205.7
0.3	0.68	0.02	169.1	0.5	22.4	50	10	30	4.0	4.0	2.5	338.2	110.7	220.2

0.3	0.68	0.02	169.1	0.1	22.4	50	10	30	4.0	4.0	4.0	338.2	111.0	191.1
0.3	0.68	0.02	169.1	0.3	22.4	50	10	30	4.0	4.0	4.0	338.2	110.8	205.7
0.3	0.68	0.02	169.1	0.5	22.4	50	10	30	4.0	4.0	4.0	338.2	110.7	220.2
0.3	0.68	0.02	170.9	0.1	22.6	50	10	40	2.5	2.5	2.5	341.7	111.2	192.3
0.3	0.68	0.02	170.9	0.3	22.6	50	10	40	2.5	2.5	2.5	341.7	111.1	206.9
0.3	0.68	0.02	170.9	0.5	22.6	50	10	40	2.5	2.5	2.5	341.7	111.0	221.6
0.3	0.68	0.02	170.9	0.1	22.6	50	10	40	2.5	2.5	4.0	341.8	111.2	192.3
0.3	0.68	0.02	170.9	0.3	22.6	50	10	40	2.5	2.5	4.0	341.8	111.1	206.9
0.3	0.68	0.02	170.9	0.5	22.6	50	10	40	2.5	2.5	4.0	341.8	111.0	221.6
0.3	0.68	0.02	171.4	0.1	22.6	50	10	40	2.5	4.0	2.5	342.8	111.2	192.0
0.3	0.68	0.02	171.4	0.3	22.6	50	10	40	2.5	4.0	2.5	342.8	111.1	206.6
0.3	0.68	0.02	171.4	0.5	22.6	50	10	40	2.5	4.0	2.5	342.8	110.9	221.1
0.3	0.68	0.02	171.4	0.1	22.6	50	10	40	2.5	4.0	4.0	342.8	111.2	192.0
0.3	0.68	0.02	171.4	0.3	22.6	50	10	40	2.5	4.0	4.0	342.8	111.1	206.6
0.3	0.68	0.02	171.4	0.5	22.6	50	10	40	2.5	4.0	4.0	342.8	110.9	221.1
0.3	0.68	0.02	171.1	0.1	22.6	50	10	40	4.0	2.5	2.5	342.2	111.2	192.2
0.3	0.68	0.02	171.1	0.3	22.6	50	10	40	4.0	2.5	2.5	342.2	111.1	206.8
0.3	0.68	0.02	171.1	0.5	22.6	50	10	40	4.0	2.5	2.5	342.2	111.0	221.5
0.3	0.68	0.02	171.1	0.1	22.6	50	10	40	4.0	2.5	4.0	342.2	111.2	192.2

Continued

TABLE A.1 Relative Efficiency of the Proposed Estimator with Respect to the Additive Model and the Multiplicative Model for Different Choice of Parameters—Cont'd

W_1	W_2	W_3	γ_1^2	$C_{\gamma_2}^2$	μ_x	μ_{x_1}	μ_{x_2}	μ_{x_3}	$\sigma_{x_1}^2$	$\sigma_{x_2}^2$	$\sigma_{x_3}^2$	σ_x^2	RE(1)	RE(2)
0.3	0.68	0.02	171.1	0.3	22.6	50	10	40	4.0	2.5	4.0	342.2	111.1	206.8
0.3	0.68	0.02	171.1	0.5	22.6	50	10	40	4.0	2.5	4.0	342.2	111.0	221.5
0.3	0.68	0.02	171.6	0.1	22.6	50	10	40	4.0	4.0	2.5	343.2	111.2	191.9
0.3	0.68	0.02	171.6	0.3	22.6	50	10	40	4.0	4.0	2.5	343.2	111.1	206.5
0.3	0.68	0.02	171.6	0.5	22.6	50	10	40	4.0	4.0	2.5	343.2	110.9	221.0
0.3	0.68	0.02	171.6	0.1	22.6	50	10	40	4.0	4.0	4.0	343.2	111.2	191.9
0.3	0.68	0.02	171.6	0.3	22.6	50	10	40	4.0	4.0	4.0	343.2	111.1	206.5
0.3	0.68	0.02	171.6	0.5	22.6	50	10	40	4.0	4.0	4.0	343.2	110.9	221.0
0.3	0.68	0.02	175.3	0.1	22.8	50	10	50	2.5	2.5	2.5	350.7	111.6	192.1
0.3	0.68	0.02	175.3	0.3	22.8	50	10	50	2.5	2.5	2.5	350.7	111.5	206.8
0.3	0.68	0.02	175.3	0.5	22.8	50	10	50	2.5	2.5	2.5	350.7	111.4	221.5
0.3	0.68	0.02	175.4	0.1	22.8	50	10	50	2.5	2.5	4.0	350.7	111.6	192.1
0.3	0.68	0.02	175.4	0.3	22.8	50	10	50	2.5	2.5	4.0	350.7	111.5	206.8
0.3	0.68	0.02	175.4	0.5	22.8	50	10	50	2.5	2.5	4.0	350.7	111.4	221.5
0.3	0.68	0.02	175.8	0.1	22.8	50	10	50	2.5	4.0	2.5	351.7	111.6	191.8
0.3	0.68	0.02	175.8	0.3	22.8	50	10	50	2.5	4.0	2.5	351.7	111.5	206.5

0.3	0.68	0.02	175.8	0.5	22.8	50	10	50	2.5	4.0	2.5	351.7	111.3	221.1
0.3	0.68	0.02	175.9	0.1	22.8	50	10	50	2.5	4.0	4.0	351.7	111.6	191.8
0.3	0.68	0.02	175.9	0.3	22.8	50	10	50	2.5	4.0	4.0	351.7	111.5	206.4
0.3	0.68	0.02	175.9	0.5	22.8	50	10	50	2.5	4.0	4.0	351.7	111.3	221.1
0.3	0.68	0.02	175.6	0.1	22.8	50	10	50	4.0	2.5	2.5	351.1	111.6	192.0
0.3	0.68	0.02	175.6	0.3	22.8	50	10	50	4.0	2.5	2.5	351.1	111.5	206.7
0.3	0.68	0.02	175.6	0.5	22.8	50	10	50	4.0	2.5	2.5	351.1	111.4	221.4
0.3	0.68	0.02	175.6	0.1	22.8	50	10	50	4.0	2.5	4.0	351.1	111.6	192.0
0.3	0.68	0.02	175.6	0.3	22.8	50	10	50	4.0	2.5	4.0	351.1	111.5	206.7
0.3	0.68	0.02	175.6	0.5	22.8	50	10	50	4.0	2.5	4.0	351.1	111.4	221.4
0.3	0.68	0.02	176.1	0.1	22.8	50	10	50	4.0	4.0	2.5	352.1	111.6	191.7
0.3	0.68	0.02	176.1	0.3	22.8	50	10	50	4.0	4.0	2.5	352.1	111.5	206.4
0.3	0.68	0.02	176.1	0.5	22.8	50	10	50	4.0	4.0	2.5	352.1	111.4	221.0
0.3	0.68	0.02	176.1	0.1	22.8	50	10	50	4.0	4.0	4.0	352.2	111.6	191.7
0.3	0.68	0.02	176.1	0.3	22.8	50	10	50	4.0	4.0	4.0	352.2	111.5	206.3
0.3	0.68	0.02	176.1	0.5	22.8	50	10	50	4.0	4.0	4.0	352.2	111.4	221.0
0.3	0.67	0.03	169.3	0.1	22.0	50	10	10	2.5	2.5	2.5	338.5	111.2	187.6
0.3	0.67	0.03	169.3	0.3	22.0	50	10	10	2.5	2.5	2.5	338.5	111.2	202.3
0.3	0.67	0.03	169.3	0.5	22.0	50	10	10	2.5	2.5	2.5	338.5	111.1	217.0

Continued

TABLE A.1 Relative Efficiency of the Proposed Estimator with Respect to the Additive Model and the Multiplicative Model for Different Choice of Parameters—Cont'd

W_1	W_2	W_3	γ_1^2	$C_{\gamma_2}^2$	μ_X	μ_{X_1}	μ_{X_2}	μ_{X_3}	$\sigma_{X_1}^2$	$\sigma_{X_2}^2$	$\sigma_{X_3}^2$	σ_X^2	RE(1)	RE(2)
0.3	0.67	0.03	169.3	0.1	22.0	50	10	10	2.5	2.5	4.0	338.6	111.2	187.6
0.3	0.67	0.03	169.3	0.3	22.0	50	10	10	2.5	2.5	4.0	338.6	111.2	202.3
0.3	0.67	0.03	169.3	0.5	22.0	50	10	10	2.5	2.5	4.0	338.6	111.1	217.0
0.3	0.67	0.03	169.8	0.1	22.0	50	10	10	2.5	4.0	2.5	339.5	111.3	187.3
0.3	0.67	0.03	169.8	0.3	22.0	50	10	10	2.5	4.0	2.5	339.5	111.1	201.9
0.3	0.67	0.03	169.8	0.5	22.0	50	10	10	2.5	4.0	2.5	339.5	111.0	216.5
0.3	0.67	0.03	169.8	0.1	22.0	50	10	10	2.5	4.0	4.0	339.6	111.3	187.3
0.3	0.67	0.03	169.8	0.3	22.0	50	10	10	2.5	4.0	4.0	339.6	111.1	201.9
0.3	0.67	0.03	169.8	0.5	22.0	50	10	10	2.5	4.0	4.0	339.6	111.0	216.5

REFERENCES

Bhargava, M., 1996. An investigation into the efficiencies of certain randomized response strategies. Unpublished Ph.D. thesis submitted to Punjab Agricultural, University, Ludhiana, India.

Bhargava, M., Singh, R., 2001. Efficiency comparison of certain randomized response schemes with U-model. J. Indian Soc. Agric. Stat. 54 (1), 19–28.

Bhargava, M., Singh, R., 2002. On the efficiency comparison of certain randomized response strategies. Metrika 55 (3), 191–197.

Chaudhuri, A., 2011. Randomized Response and Indirect Questioning Techniques in Surveys. Chapman & Hall, Boca Raton, FL.

Diana, G., Perri, P.F., 2009. Estimating a sensitive proportion through randomized response procedures based on auxiliary information. Stat. Pap. 50, 661–672.

Diana, G., Perri, P.F., 2010. New scrambled response models for estimating the mean of a sensitive quantitative character. J. Appl. Stat. 37, 1875–1890.

Eichhorn, B.H., Hayre, L.S., 1983. Scrambled randomized response methods for obtaining sensitive quantitative data. J. Stat. Plann. Infer. 7, 307–316.

Himmelfarb, S., Edgell, S.E., 1980. Additive constants model: a randomized response technique for eliminating evasiveness to quantitative response questions. Psychol. Bull. 87, 525–530.

Lanke, J., 1975. On the choice of the unrelated question in Simons version of randomized response. J. Am. Stat. Assoc. 70, 80–83.

Lanke, J., 1976. On the degree of protection in randomized interviews. Int. Stat. Rev. 44, 197–203.

Leysieffer, F.W., Warner, S.L., 1976. Respondent jeopardy and optimal designs in randomized response models. J. Am. Stat. Assoc. 71, 649–656.

Mangat, N.S., Singh, R., 1990. An alternative randomized response procedure. Biometrika 77 (2), 439–442.

Moors, J.J.A., 1997. A critical evaluation of Mangat's two-step procedure in randomized response. Discussion Paper at Center for Economic Research, Tilburg University, The Netherlands.

Nayak, T.K., 1994. On randomized response surveys for estimating a proportion. Commun. Stat. Theory Methods 23 (1), 3303–3321.

Perri, P.F., 2008. Modified randomized devices for Simmons' model. Model. Assist. Stat. Appl. 3 (3), 233–239.

Stephan, F.F., 1945. The expected value and variance of the reciprocal and other negative powers of a positive Bernoulli variate. Ann. Math. Stat. 16, 50–61.

Tracy, D.S., Mangat, N.S., 1996. Some developments in randomized response sampling during the last decade—a follow up of review by Chaudhuri and Mukerjee. J. Appl. Stat. Sci. 4 (2/3), 147–158.

Warner, S.L., 1965. Randomized response: a survey technique for eliminating evasive answer bias. J. Am. Stat. Assoc. 60, 63–69.

Zou, G., 1997. Two-stage randomized response procedures as single stage procedures. Aust. J. Stat. 39, 235–236.

Chapter 12

Variance Estimation in Randomized Response Surveys

A.K. Adhikary[1]
Indian Statistical Institute, Kolkata, West Bengal, India
[1]Corresponding author: e-mail: arunadhikary@yahoo.co.in

ABSTRACT

Warner (1965) pioneered the randomized response technique to collect information on a sensitive qualitative character. Following Adhikary (2013) we propose a straightforward procedure to derive an exact expression for the variance of a homogeneous linear unbiased estimator of a population total of a sensitive quantitative character such as expenditure on drinking alcohol, additional income earned over usual salary to hide income tax in the income tax return, expenditure on gambling, number of induced abortions, income earned through prostitution, expenditure incurred toward payment of prostitutes in a red light area, and some such others which we do not want to disclose even to our wives. Following Adhikary (2013) an attempt has been made to derive an unbiased estimator of the variance of the estimator. This procedure is illustrated with Horvitz–Thompson (1952) estimator, Hansen–Hurwitz (1943) estimator based on probability proportional to size with replacement (PPSWR) sampling and also an improved version of it obtained by Rao-Blackwellization as considered by Adhikary (2009a,b), Raj's (1956) ordered estimator and Murthy's (1957) unordered estimator both based on PPSWOR sampling, ratio estimator based on Lahiri (1951), Midzuno (1952), and Sen's (1953) sampling scheme, and Hartley–Ross's (1954) unbiased ratio-type estimator based on SRSWOR sampling scheme considering in each case the form of the estimator suitable for the randomized response survey.

Keywords: Distinct units, Homogeneous linear unbiased estimator, Ordered estimator, Unordered estimator, PPSWR sampling, PPSWOR sampling, Ratio estimator, Rao-Blackwellization

1 INTRODUCTION

Let $U = (1, 2, \ldots, N)$ denote a finite population of size N and let y_i denote the value of a study variable y assumed on the ith unit of the population. Our problem is to estimate the population total $Y=\sum_{i=1}^{N} y_i$ based on a sample s drawn from a population of size N with a probability $p(s)$. If the study variable y is a sensitive quantitative character, then the value of y_i is not ascertainable in a direct response (DR) survey. Then we may adopt a randomized response (RR) survey in which a large number of real numbers X_j, $j=1,2,\ldots,m$ not necessarily distinct, be chosen and the ith sampled person will be asked to report the value Z_i where Z_i equals y_i with a preassigned probability C and equals X_j with a probability q_j, which is also preassigned, $j=1,2,\ldots,m$ such that

$$\mathrm{C}+\sum_{j=1}^{m} q_j=1.$$

Then denoting by $E_R(V_R, C_R)$ the operator for expectation (variance, covariance) with respect to the randomized procedure of generating RR, we have

$$E_R(Z_i)=Cy_i+\sum_{j=1}^{m} q_j X_j=Cy_i+(1-C)\mu, \text{ say,}$$

where $\mu=\sum_{j=1}^{m} q_j X_j / \sum_{j=1}^{m} q_j$.

Let r_i denote the transformed RR given by $r_i=[Z_i-(1-C)\mu]/C$ so that we have

$$\begin{aligned} E_R(r_i)=y_i,\ V_R(r_i)&=\frac{V_R(Z_i)}{C^2}\\ &=\frac{\left[C(1-C)y_i^2-2C(1-C)\mu y_i+\sum_{j=1}^{m} q_j X_j^2-(1-C)^2\mu^2\right]}{C^2}=V_i, \text{ say} \end{aligned}$$

and

$$C_R(r_i, r_j)=0\,\forall i\neq j.$$

Let a homogeneous linear unbiased estimator (HLUE) of the population total Y is given by

$$t(s, \mathbf{Y})=\sum_{i-1}^{N} d_{si} y_i,$$

where $\mathbf{Y}=(y_1, y_2, \ldots, y_N)$ and d_{si}'s are independent of y_i's but may depend on x_i's where x_i is the value of an auxiliary variable x highly correlated with the study variable y assumed on the ith unit and $d_{si}=0$ if i does not belong to s or s does not contain i.

In case of RR survey the corresponding estimator is

$$e(s, \mathbf{R})=\sum_{i=1}^{N} d_{si} r_i,$$

where $\mathbf{R}=(r_1, r_2, \ldots, r_N)$ is the corresponding vector of RRs so that we have

$$E_R e(s, \mathbf{R})=\sum_{i=1}^{N} d_{si} y_i=t(s, \mathbf{Y}).$$

Writing E_p as the operator for expectation with respect to the sampling design $p(s)$ we have the variance of the estimator $e = e(s, \mathbf{R})$ as

$$Var(e) = E_p E_R (e - Y)^2 = E_p E_R \left(e^2\right) - Y^2 = E_p E_R \left(\sum_{i=1}^{N} d_{si} r_i\right)^2 - Y^2$$

$$= E_p E_R \left[\sum_{i=1}^{N} d_{si}^2 r_i^2 + \sum_{i=1}^{N}\sum_{j\neq i=1}^{N} d_{si} d_{sj} r_i r_j\right] - Y^2$$

$$= E_p \left[\sum_{i=1}^{N} d_{si}^2 E_R\left(r_i^2\right) + \sum_{i=1}^{N}\sum_{j\neq i=1}^{N} d_{si} d_{sj} E_R\left(r_i r_j\right)\right] - Y^2$$

$$= E_p \left[\sum_{i=1}^{N} d_{si}^2 \left\{V_R(r_i) + E_R^2(r_i)\right\} + \sum_{i=1}^{N}\sum_{j\neq i=1}^{N} d_{si} d_{sj} y_i y_j\right] - Y^2$$

$$= E_p \left[\sum_{i=1}^{N} d_{si}^2 V_i + \sum_{i=1}^{N} d_{si}^2 y_i^2 + \sum_{i=1}^{N}\sum_{j\neq i=1}^{N} d_{si} d_{sj} y_i y_j\right] - Y^2$$

$$= \sum_{i=1}^{N} E_p\left(d_{si}^2\right) V_i + \sum_{i=1}^{N}\left[E_p\left(d_{si}^2\right) - 1\right] y_i^2 + \sum_{i=1}^{N}\sum_{j\neq i=1}^{N}\left[E_p\left(d_{si} d_{sj}\right) - 1\right] y_i y_j.$$

Adhikary (2013) demonstrated a straightforward procedure of deriving the variance of an HLUE of the population total Y and an unbiased estimator of the variance in case of a DR survey.

Following Adhikary (2013) an unbiased estimator of $Var(e)$ is given by

$$\widehat{Var(e)} = \sum_{i\in s} d_{si}^2 v_i + \sum_{i\in s}\left[d_{si}^2 - f_i(s)\right]\left(r_i^2 - v_i\right) + \sum\sum_{i\neq j\in s}\left[d_{si} d_{sj} - f_{ij}(s)\right] r_i r_j,$$

where v_i is an unbiased estimator of V_i with respect to the randomization device of generating RRs r_i and $f_i(s)$ and $f_{ij}(s)$ are such that $\sum_{s\ni i} f_i(s) p(s) = 1$ and $\sum_{s\ni i,j} f_{ij}(s) p(s) = 1$

$$= \sum_{i\in s} d_{si}^2 r_i^2 + \sum\sum_{i\neq j\in s} d_{si} d_{sj} r_i r_j - \sum_{i\in s} f_i(s)\left(r_i^2 - v_i\right) - \sum\sum_{i\neq j\in s} f_{ij}(s) r_i r_j$$

$$= \left(\sum_{i\in s} d_{si} r_i\right)^2 - \sum_{i\in s} f_i(s) r_i^2 - \sum\sum_{i\neq j\in s} f_{ij}(s) r_i r_j + \sum_{i\in s} f_i(s) v_i.$$

The possible choices of $f_i(s)$ and $f_{ij}(s)$ are

$$f_i(s) = \frac{1}{\pi_i},\quad f_i(s) = \frac{1}{M_1 p(s)}$$

$$f_{ij}(s) = \frac{1}{\pi_{ij}},\quad f_{ij}(s) = \frac{1}{M_2 p(s)}$$

where π_i and π_{ij} denote, respectively, the first-order and second-order inclusion probabilities and $M_r = \binom{N-r}{n-r}$, $r = 1, 2$.

This procedure is applied to find the variance of the Horvitz–Thompson (1952) estimator, Hansen–Hurwitz (1943) estimator and also an improved version of it as considered by Adhikary (2009a,b) obtained by Rao-Blackwellization of the Hansen–Hurwitz (1943) estimator, Raj's (1956) ordered estimator, Murthy's (1957) unordered estimator, ratio estimator based on Lahiri (1951), Midzuno (1952), and Sen's (1953) sampling scheme, and Hartley–Ross's (1954) unbiased ratio-type estimator based on simple random sampling without replacement (SRSWOR) sampling scheme and also to find an exact expression for an unbiased estimator of the variance considering in each case the RR version of the estimator.

2 VARIANCE OF HORVITZ–THOMPSON (1952) ESTIMATOR

In case of RR surveys, the Horvitz–Thompson estimator of the population total Y is given by

$$\widehat{Y_{HTE}} = \sum_{i \in s} \frac{r_i}{\pi_i}.$$

Thus $\widehat{Y_{HTE}}$ can be written as $\widehat{Y_{HTE}} = \sum_{i=1}^{N} d_{si} r_i$, where d_{si} is zero if i does not belong to s or s does not contain i and $d_{si} = (1/\pi_i)$ if $i \in s$ or $s \ni i$.

The variance of $\widehat{Y_{HTE}}$ is

$$\begin{aligned} Var\left(\widehat{Y_{HTE}}\right) &= \sum_{i=1}^{N} y_i^2 \left[\sum_{s \ni i} d_{si}^2 p(s) - 1\right] + \sum_{i=1}^{N} \sum_{j \neq i=1}^{N} y_i y_j \\ &\quad \left[\sum_{s \ni i, j} d_{si} d_{sj} p(s) - 1\right] + \sum_{i=1}^{N} \sum_{s \ni i} d_{si}^2 p(s) V_i \\ &= \sum_{i=1}^{N} y_i^2 \left(\frac{1}{\pi_i} - 1\right) + \sum_{i=1}^{N} \sum_{j \neq i=1}^{N} y_i y_j \left(\frac{\pi_{ij}}{\pi_i \pi_j} - 1\right) + \sum_{i=1}^{N} \frac{V_i}{\pi_i}. \end{aligned}$$

An unbiased estimator of $Var\left(\widehat{Y_{HTE}}\right)$ is

$$\begin{aligned} \widehat{Var}\left(\widehat{Y_{HTE}}\right) &= \left(\sum_{i \in s} \frac{r_i}{\pi_i}\right)^2 - \sum_{i \in s} \frac{r_i^2}{\pi_i} - \sum \sum_{i \neq j \in s} \frac{r_i r_j}{\pi_{ij}} + \sum_{i \in s} \frac{v_i}{\pi_i} \\ &= \sum_{i \in s} r_i^2 \left(\frac{1 - \pi_i}{\pi_i^2}\right) + \sum_{i \neq j \in s} r_i r_j \left(\frac{\pi_{ij} - \pi_i \pi_j}{\pi_{ij} \pi_i \pi_j}\right) + \sum_{i \in s} \frac{v_i}{\pi_i} \end{aligned}$$

This expression is due to Horvitz–Thompson (1952) modified in case of RR surveys.

Now to derive the other expressions for the variance of the Horvitz–Thompson (1952) estimator available in the literature, let us consider the following results.

Theorem 1.

$$\sum_{i=1}^{N} a_{ii}x_i^2 + \sum_{i=1}^{N}\sum_{j\neq i=1}^{N} a_{ij}x_ix_j = \sum_{i=1}^{N}\frac{x_i^2}{\alpha_i}A_i - \sum_{i<j=1}^{N}\sum_{j=1}^{N} a_{ij}\alpha_i\alpha_j\left(\frac{x_i}{\alpha_i}-\frac{x_j}{\alpha_j}\right)^2,$$

where $A_i = \sum_{j=1}^{N} a_{ij}\alpha_j$.

Proof.

$$\sum_{i<j=1}^{N}\sum_{j=1}^{N} a_{ij}\alpha_i\alpha_j\left(\frac{x_i}{\alpha_i}-\frac{x_j}{\alpha_j}\right)^2 = \frac{1}{2}\sum_{i\neq j=1}^{N}\sum_{j=1}^{N} a_{ij}\alpha_i\alpha_j\left[\frac{x_i^2}{\alpha_i^2}+\frac{x_j^2}{\alpha_j^2}-\frac{2x_ix_j}{\alpha_i\alpha_j}\right]$$

$$= \frac{1}{2}\left[2\sum_{i=1}^{N}\sum_{j\neq i=1}^{N} a_{ij}\frac{x_i^2}{\alpha_i}\alpha_j - 2\sum_{i\neq j=1}^{N}\sum_{j=1}^{N} a_{ij}x_ix_j\right]$$

$$= \sum_{i=1}^{N}\frac{x_i^2}{\alpha_i}\sum_{j\neq i=1}^{N} a_{ij}\alpha_j - \sum_{i\neq j=1}^{N}\sum_{j=1}^{N} a_{ij}x_ix_j$$

$$= \sum_{i=1}^{N}\frac{x_i^2}{\alpha_i}\sum_{j=1}^{N} a_{ij}\alpha_j - \sum_{i=1}^{N} a_{ii}x_i^2 - \sum_{i\neq j=1}^{N}\sum_{j=1}^{N} a_{ij}x_ix_j.$$

Then it follows that

$$\sum_{i=1}^{N} a_{ii}x_i^2 + \sum_{i=1}^{N}\sum_{j\neq i=1}^{N} a_{ij}x_ix_j = \sum_{i=1}^{N}\frac{x_i^2}{\alpha_i}A_i - \sum_{i<j=1}^{N}\sum_{j=1}^{N} a_{ij}\alpha_i\alpha_j\left(\frac{x_i}{\alpha_i}-\frac{x_j}{\alpha_j}\right)^2,$$

where $A_i = \sum_{j=1}^{N} a_{ij}\alpha_j$. That completes the proof. □

Theorem 2.

$$\text{If } \sum_{i=1}^{N}\alpha_i = 1, \quad \sum_{i<j=1}^{N}\sum_{j=1}^{N}\alpha_i\alpha_j\left(\frac{x_i}{\alpha_i}-\frac{x_j}{\alpha_j}\right)^2 = \sum_{i=1}^{N}\alpha_i\left(\frac{x_i}{\alpha_i}-X\right)^2,$$

where $X = \sum_{i=1}^{N} x_i$.

Proof. Putting $a_{ij} = 1$ for all i and j in Theorem 1, we get

$$\sum_{i<j=1}^{N}\sum_{j=1}^{N}\alpha_i\alpha_j\left(\frac{x_i}{\alpha_i}-\frac{x_j}{\alpha_j}\right)^2 = \sum_{i=1}^{N}\frac{x_i^2}{\alpha_i} - \sum_{i=1}^{N} x_i^2 - \sum_{i\neq j=1}^{N}\sum_{j=1}^{N} x_ix_j$$

$$= \sum_{i=1}^{N}\frac{x_i^2}{\alpha_i} - X^2 \text{ where } X = \sum_{i=1}^{N} x_i$$

$$= \sum_{i=1}^{N}\alpha_i\left(\frac{x_i}{\alpha_i}-X\right)^2$$

That completes the proof. □

Theorem 3. *If either* $A_i = 0$ *for every* $i = 1,2,\ldots,N$ *or* $\sum_{i=1}^{N} \frac{x_i^2}{\alpha_i} A_i = 0$, *then*

$$\sum_{i=1}^{N} a_{ii} x_i^2 + \sum_{i \neq j=1}^{N} \sum_{j=1}^{N} a_{ij} x_i x_j = -\sum_{i<j=1}^{N} \sum_{j=1}^{N} a_{ij} \alpha_i \alpha_j \left(\frac{x_i}{\alpha_i} - \frac{x_j}{\alpha_j} \right)^2.$$

Proof. Follows from Theorem 1 by putting $A_i = 0$ for every $i = 1,2,\ldots,N$.

Now by virtue of Theorem 1 the variance of the Horvitz–Thompson (1952) estimator can be written as

$$Var\left(\widehat{Y_{HTE}}\right) = \sum_{i=1}^{N} \frac{y_i^2}{\pi_i} \sum_{j=1}^{N} \frac{\pi_{ij} - \pi_i \pi_j}{\pi_i \pi_j} \pi_j - \sum_{i<j=1}^{N} \sum_{j=1}^{N} \frac{\pi_{ij} - \pi_i \pi_j}{\pi_i \pi_j} \pi_i \pi_j \left(\frac{y_i}{\pi_i} - \frac{y_j}{\pi_j} \right)^2 + \sum_{i=1}^{N} \frac{V_i}{\pi_i}$$

by putting $x_i = y_i$, $\alpha_i = \pi_i$ and $\alpha_{ij} = \frac{\pi_{ij} - \pi_i \pi_j}{\pi_i \pi_j}$ in Theorem 1

$$= \sum_{i=1}^{N} \frac{y_i^2}{\pi_i} \sum_{j=1}^{N} \frac{\pi_{ij} - \pi_i \pi_j}{\pi_i} + \sum_{i<j=1}^{N} \sum_{j=1}^{N} (\pi_i \pi_j - \pi_{ij}) \left(\frac{y_i}{\pi_i} - \frac{y_j}{\pi_j} \right)^2 + \sum_{i=1}^{N} \frac{V_i}{\pi_i}$$

$$= \sum_{i=1}^{N} \frac{y_i^2}{\pi_i} \beta_i + \sum_{i<j=1}^{N} \sum_{j=1}^{N} (\pi_i \pi_j - \pi_{ij}) \left(\frac{y_i}{\pi_i} - \frac{y_j}{\pi_j} \right)^2 + \sum_{i=1}^{N} \frac{V_i}{\pi_{ii}},$$

say, where $\beta_i = \sum_{j=1}^{N} \frac{\pi_{ij} - \pi_i \pi_j}{\pi_i} = 1 + \frac{1}{\pi_i} \sum_{j \neq i=1}^{N} \pi_{ij} - \sum_{j=1}^{N} \pi_j$. □

Now if we denote by $\vartheta(s)$ the number of distinct units in s and if $\vartheta(s)$ is constant for every sample s with $p(s) > 0$ then $\beta_i = 0$ for every $i = 1, 2, \ldots, N$ because $\beta_i = 1 + \frac{1}{\pi_i} \sum_{j \neq i=1}^{N} \pi_{ij} - \sum_{j=1}^{N} \pi_j = 1 + \frac{(n-1)\pi_i}{\pi_i} - n = 1 + (n-1) - n = 0$ as for fixed effective sample size (n) design we have $\sum_{j \neq i=1}^{N} \pi_{ij} = (n-1)\pi_i$ and $\sum_{j=1}^{N} \pi_j = n$.

Thus for a fixed effective sample size (n) design, we have

$$Var\left(\widehat{Y_{HTE}}\right) = \sum_{i<j=1}^{N} \sum_{j=1}^{N} (\pi_i \pi_j - \pi_{ij}) \left(\frac{y_i}{\pi_i} - \frac{y_j}{\pi_j} \right)^2 + \sum_{i=1}^{N} \frac{V_i}{\pi_i}$$

and an unbiased estimator of the variance of $\widehat{Y_{HTE}}$ is given by

$$\widehat{Var}\left(\widehat{Y_{HTE}}\right) = \sum_{i<j \in s} \sum \frac{(\pi_i \pi_j - \pi_{ij})}{\pi_{ij}} \left(\frac{r_i}{\pi_i} - \frac{r_j}{\pi_j} \right)^2 + \sum_{i \in s} \frac{v_i}{\pi_i}$$

This expression is due to Yates and Grundy (1953) modified in case of RR surveys.

Now if $\vartheta(s)$ is not a constant for every sample s with $p(s) > 0$, then a third expression for the variance of the Horvitz–Thompson (1952) estimator is given by

$$Var\left(\widehat{Y_{HTE}}\right)=\sum_{i=1}^{N}\frac{y_i^2}{\pi_i}\beta_i+\sum_{i<j=1}^{N}\sum_{j=1}^{N}\left(\pi_i\pi_j-\pi_{ij}\right)\left(\frac{y_i}{\pi_i}-\frac{y_j}{\pi_j}\right)^2+\sum_{i=1}^{N}\frac{V_i}{\pi_i}$$

and an unbiased estimator of the variance of $\widehat{Y_{HTE}}$ is given by

$$\widehat{Var}\left(\widehat{Y_{HTE}}\right)=\sum_{i\in s}\frac{r_i^2\beta_i}{\pi_i\pi_i}+\sum\sum_{i<j\in s}\frac{\pi_i\pi_j-\pi_{ij}}{\pi_{ij}}\left(\frac{r_i}{\pi_i}-\frac{r_j}{\pi_j}\right)^2+\sum_{i\in s}\frac{v_i}{\pi_i}$$

This expression is due to Chaudhuri (2000) and Chaudhuri and Pal (2002) modified in case of RR surveys.

3 VARIANCE OF HANSEN–HURWITZ (1943) ESTIMATOR

In case of RR surveys, the Hansen–Hurwitz (1943) estimator is

$$\widehat{Y_{HHE}}=\sum_{i=1}^{N}\frac{r_if_i}{np_i},$$

where p_i is the normed size measure of the ith unit $\left(0<p_i<1,\sum_{i=1}^{N}p_i=1\right)$ and f_i is the frequency with which the ith unit occurs in the sample, $i=1, 2, \ldots, N$.

Thus $\widehat{Y_{HHE}}$ can be written as

$$\widehat{Y_{HHE}}=\sum_{i=1}^{N}d_{si}r_i,$$

where d_{si} is zero if i does not belong to s or s does not contain i and $d_{si}=\frac{f_i}{np_i}$ if $i\in s$ or $s\ni i$.

The variance of the Hansen–Hurwitz (1943) estimator is

$$\begin{aligned}
Var\left(\widehat{Y_{HHE}}\right)&=\sum_{i=1}^{N}y_i^2\left[E_p\left(d_{si}^2\right)-1\right]+\sum_{i\neq j=1}^{N}\sum_{j=1}^{N}y_iy_j\left[E_p\left(d_{si}d_{sj}\right)-1\right]+\sum_{i=1}^{N}E_p\left(d_{si}^2\right)V_i\\
&=\sum_{i=1}^{N}y_i^2\left[E_p\left(\frac{f_i^2}{n^2p_i^2}\right)-1\right]+\sum_{i\neq j=1}^{N}\sum_{j=1}^{N}y_iy_j\left[E_p\left(\frac{f_if_j}{n^2p_ip_j}\right)-1\right]+\sum_{i=1}^{N}E_p\left(\frac{f_i^2}{n^2p_i^2}\right)V_i\\
&=\sum_{i=1}^{N}y_i^2\left[\frac{n(n-1)p_i^2}{n^2p_i^2}+\frac{np_i}{n^2p_i^2}-1\right]+\sum_{i\neq j=1}^{N}\sum_{j=1}^{N}y_iy_j\left[\frac{n(n-1)p_ip_j}{n^2p_ip_j}-1\right]\\
&\quad+\sum_{i=1}^{N}\left[\frac{n(n-1)p_i^2}{n^2p_i^2}+\frac{np_i}{n^2p_i^2}\right]V_i=\sum_{i=1}^{N}y_i^2\left[\frac{n-1}{n}+\frac{1}{np_i}-1\right]+\sum_{i\neq j=1}^{N}y_iy_j\left[\frac{n-1}{n}-1\right]\\
&\quad+\sum_{i=1}^{N}\left[\frac{n-1}{n}+\frac{1}{np_i}\right]V_i=\frac{1}{n}\sum_{i=1}^{N}\frac{y_i^2}{p_i}-\frac{1}{n}Y^2+\frac{1}{n}\sum_{i=1}^{N}\frac{V_i}{p_i}+\frac{n-1}{n}\sum_{i=1}^{N}V_i\\
&=\frac{1}{n}\sum_{i=1}^{N}p_i\left(\frac{y_i}{p_i}-Y\right)^2+\frac{1}{n}\sum_{i=1}^{N}\frac{V_i}{p_i}+\frac{n-1}{n}\sum_{i=1}^{N}V_i.
\end{aligned}$$

An unbiased estimator of the variance of $\widehat{Y_{HHE}}$ is given by

$$\widehat{Var}\left(\widehat{Y_{HHE}}\right)=\left(\sum_{i=1}^{N}\frac{f_ir_i}{np_i}\right)^2-\sum_{i=1}^{N}a_i(s)r_i^2-\sum_{i\neq j=1}^{N}\sum_{j=1}^{N}a_{ij}(s)r_ir_j+\sum_{i=1}^{N}\frac{f_iv_i}{np_i},$$

where $a_i(s)$ and $a_{ij}(s)$ are such that $E_p[a_i(s)]=1$ and $E_p\left[a_{ij}(s)\right]=1$ and v_i is an unbiased estimator of V_i with respect to the randomization device stated earlier

$$=\left(\sum_{i=1}^{N}\frac{f_ir_i}{np_i}\right)^2-\sum_{i=1}^{N}\frac{f_i(f_i-1)r_i^2}{n(n-1)p_i^2}-\sum_{i\neq j=1}^{N}\sum_{j=1}^{N}\frac{f_if_jr_ir_j}{n(n-1)p_ip_j}+\sum_{i=1}^{N}\frac{f_iv_i}{np_i}$$

on noting that $E_p[f_i(f_i-1)]=n(n-1)p_i^2$ and $E_p\left[f_if_j\right]=n(n-1)p_ip_j$

$$=\sum_{i=1}^{N}\frac{f_i^2r_i^2}{n^2p_i^2}+\sum_{i\neq j=1}^{N}\sum_{j=1}^{N}\frac{f_if_jr_ir_j}{n^2p_ip_j}-\sum_{i=1}^{N}\frac{f_i(f_i-1)r_i^2}{n(n-1)p_i^2}-\sum_{i\neq j=1}^{N}\sum_{j=1}^{N}\frac{f_if_jr_ir_j}{n(n-1)p_ip_j}+\sum_{i=1}^{N}\frac{f_iv_i}{np_i}$$

$$=\sum_{i=1}^{N}\frac{f_i^2r_i^2}{n^2p_i^2}-\sum_{i=1}^{N}\frac{f_i(f_i-1)r_i^2}{n(n-1)p_i^2}-\sum_{i\neq j=1}^{N}\sum_{j=1}^{N}\frac{f_if_jr_ir_j}{n^2(n-1)p_ip_j}+\sum_{i=1}^{N}\frac{f_iv_i}{np_i}$$

$$=\frac{1}{n(n-1)}\sum_{i=1}^{N}\frac{f_ir_i^2}{p_i^2}-\frac{1}{n^2(n-1)}\sum_{i=1}^{N}\frac{f_i^2r_i^2}{p_i^2}-\frac{1}{n^2(n-1)}\sum_{i\neq j=1}^{N}\frac{f_if_jr_ir_j}{p_ip_j}+\sum_{i=1}^{N}\frac{f_iv_i}{np_i}$$

$$=\frac{1}{n(n-1)}\sum_{i=1}^{N}\frac{f_ir_i^2}{p_i^2}-\frac{1}{n^2(n-1)}\left[\sum_{i=1}^{N}\frac{f_i^2r_i^2}{p_i^2}+\sum_{i\neq j=1}^{N}\sum_{j=1}^{N}\frac{f_if_jr_ir_j}{p_ip_j}\right]+\sum_{i=1}^{N}\frac{f_iv_i}{np_i}$$

$$=\frac{1}{n(n-1)}\left[\sum_{i=1}^{N}\frac{f_ir_i^2}{p_i^2}-n\left(\frac{1}{n}\sum_{i=1}^{N}\frac{r_if_i}{p_i}\right)^2\right]+\sum_{i=1}^{N}\frac{f_iv_i}{np_i}$$

$$=\frac{1}{n(n-1)}\sum_{i=1}^{N}f_i\left(\frac{r_i}{p_i}-\widehat{Y_{HHE}}\right)^2+\sum_{i=1}^{N}\frac{f_iv_i}{np_i}$$

which matches with the traditional unbiased estimator of $Var\left(\widehat{Y_{HHE}}\right)$.

In case of a DR survey Adhikary (2009a,b) attempted to improve upon the Hansen–Hurwitz (1943) estimator by Rao-Blackwellization considering the distinct units in a sample s drawn in n draws by PPSWR sampling and deriving the sampling distribution of the number of distinct units ϑ in a sample s drawn in n draws by PPSWR sampling.

In case of an RR survey, the corresponding modified version of the improved Hansen–Hurwitz (1943) estimator is

$$t^*(s)=\frac{1}{n}\sum_{j=1}^{\vartheta}b_{i_j}\frac{r_{i_j}}{P(s)},$$

where ϑ denotes the number of distinct units in the sample s,

$$P(s) = \sum_{s' \sim s} p(s), \quad p(s) = p_{i_1} p_{i_2} \ldots p_{i_n}$$

and $b_{i_j} = \frac{\delta p(s)}{\delta p_{i_j}}$, where $p(s)$ denotes the probability of selecting the sample s by PPSWR sampling scheme and $s' \sim s$ stands for all possible samples s' which are equivalent to s in the sense of sharing the same set of distinct units in s.

We may note that $t^*(s)$ can alternatively be written as

$$t^*(s) = \sum_{i=1}^{N} d_{si} r_i,$$

where $d_{si} = \frac{b_i}{nP(s)}$ if $i \in s$ and zero otherwise. Following Adhikary (2009a,b) the variance of $t^*(s)$ can be written as

$$\begin{aligned} Var[t^*(s)] &= E_p E_R [t^*(s) - Y]^2 \\ &= -\sum_{i<j=1}^{N} \sum_{j=1}^{N} d_{ij} p_i p_j \left(\frac{y_i}{p_i} - \frac{y_j}{p_j} \right)^2 + \sum_{i=1}^{N} d_{si} V_i, \end{aligned}$$

where $d_{ij} = E_p \left[d_{si} d_{sj} \right] - 1$.

An unbiased estimator of $Var[t^*(s)]$ is given by

$$\widehat{Var}[t^*(s)] = -\sum_{i<j=1}^{N} \sum_{j=1}^{N} d_{ij}(s) p_i p_j \left[\left(\frac{r_i}{p_i} - \frac{r_j}{p_j} \right)^2 - \left(\frac{v_i}{p_i^3} + \frac{v_j}{p_j^2} \right) \right] + \sum_{i=1}^{N} d_{si} v_i,$$

where $d_{ij}(s) = 0$ if i and j do not belong to s and $E_p \left[d_{ij}(s) \right] = d_{ij}$

$$\Rightarrow \sum_{s \ni i, j} d_{ij}(s) p(s) = d_{ij}.$$

Chaudhuri and Stenger (2005) considered an improved version of the Hansen–Hurwitz (1943) estimator, assuming that a PPSWR sample is drawn using normed size measures $p_i \left(0 < p_i < 1, \sum_{i=1}^{N} p_i = 1 \right)$ and each time a person appears in the sample, an independent RR r_k is obtained. Writing y_k, r_k, and p_k for the corresponding y_i, r_i, and p_i value for the individual i if chosen on the kth draw, then corresponding to the Hansen–Hurwitz (1943) estimator $\frac{1}{n} \sum_{i=1}^{N} \frac{f_i y_i}{p_i}$ for Y, the derived estimator is $e_{HH} = \frac{1}{n} \sum_{k=1}^{n} \frac{r_k}{p_k}$ having the variance

$$M = \frac{1}{n} \left[\sum_{i=1}^{N} \frac{y_i^2}{p_i} - Y^2 \right] + \frac{1}{n} \sum_{i=1}^{N} \frac{V_i}{p_i}$$

and an unbiased variance estimator is

$$m = \frac{1}{n(n-1)} \sum_{k=1}^{n} \left(\frac{r_k}{p_k} - \frac{1}{n} \sum_{k=1}^{n} \frac{r_k}{p_k} \right)^2.$$

4 VARIANCE OF RAJ'S (1956) ORDERED ESTIMATOR

Suppose a sample of size two is drawn from a finite population of size N by PPSWOR sampling scheme, ie, on the first draw we select the unit i with a probability $p_i\left(0<p_i<1, \sum_{i=1}^{N} p_i=1\right)$ and assuming that the unit i is selected on the first draw, we select the unit j on the second draw with a conditional probability $\frac{p_j}{1-p_i}$, $j \neq i$. Then in case of an RR survey Raj's (1956) ordered estimator is

$$t_D^{(1)}=\frac{1}{2}\left[\frac{1+p_i}{p_i} r_i+\frac{r_j}{p_j}(1-p_i)\right]$$

which is known to be unbiased for the population total Y based on the ordered sample $s_1=(i,j)$ in the sense that $E_p E_R\left(t_D^{(1)}\right)=Y$. On the other hand if we select the unit j on the first draw with a probability p_j and assuming the unit j is selected on the first draw, we select the unit i on the second draw with a conditional probability $\frac{p_i}{1-p_j}$, $i \neq j$. Then corresponding to the ordered sample $s_2=(j,i)$, the ordered estimator will be

$$t_D^{(2)}=\frac{1}{2}\left[\frac{1+p_j}{p_j} r_j+\frac{r_i}{p_i}(1-p_j)\right]$$

which is also unbiased for the population total Y based on the ordered sample $s_2=(j,i)$ in the sense that $E_p E_R\left(t_D^{(2)}\right)=Y$.

Thus in general Raj's (1956) ordered estimator can be written as

$$t_D=\sum_{i=1}^{N} d_{si} r_i$$

where d_{si} is zero if i does not belong to s or s does not contain i and

$$d_{si}=\begin{cases}\dfrac{1+p_i}{2p_i} & \text{if } i \in s_1 \\[2ex] \dfrac{1-p_j}{2p_i} & \text{if } i \in s_2\end{cases}$$

Thus it is easy to check that d_{si}'s satisfy the unbiasedness condition

$$E_p[d_{si}]=1 \forall i$$

because $E_p[d_{si}]=\sum_{j \neq i=1}^{N} \frac{1}{2} \frac{1+p_i}{p_i} \frac{p_i p_j}{1-p_i}+\sum_{j \neq i=1}^{N} \frac{1}{2} \frac{1-p_j}{p_i} \frac{p_j p_i}{1-p_j}=\frac{1+p_i}{2}+\frac{1-p_i}{2}=1.$

The variance of Raj's (1956) ordered estimator is

$$Var(t_D)=\sum_{i=1}^{N}\left[E_p\left(d_{si}^2\right)-1\right]y_i^2+\sum_{i\neq j=1}^{N}\sum_{j=1}^{N}\left[E_p\left(d_{si}d_{sj}\right)-1\right]y_iy_j+\sum_{i=1}^{N}E_p\left(d_{si}^2\right)V_i$$

$$=\sum_{i=1}^{N}\left[\sum_{j\neq i=1}^{N}\frac{(1+p_i)^2}{4p_i^2}\frac{p_ip_j}{1-p_i}+\sum_{j\neq i=1}^{N}\frac{(1-p_i)^2}{4p_i^2}\frac{p_jp_i}{1-p_j}-1\right]y_i^2$$

$$+\sum_{i\neq j=1}^{N}\sum_{j=1}^{N}\left[\frac{1+p_i}{2p_i}\frac{1-p_i}{2p_j}\frac{p_ip_j}{1-p_i}+\frac{1+p_j}{2p_j}\frac{1-p_j}{2p_i}\frac{p_jp_i}{1-p_j}-1\right]y_iy_j$$

$$+\sum_{i=1}^{N}\left[\sum_{j\neq i=1}^{N}\frac{(1+p_i)^2}{4p_i^2}\frac{p_ip_j}{1-p_i}+\sum_{j\neq i=1}^{N}\frac{(1-p_i)^2}{4p_i^2}\frac{p_jp_i}{1-p_j}\right]V_i$$

$$=\sum_{i=1}^{N}\left[\frac{2+p_i+p_i^2-\sum_{j\neq i=1}^{N}p_j^2}{4p_i}-1\right]y_i^2+\sum_{i\neq j=1}^{N}\sum_{j=1}^{N}\left[\frac{1+p_i}{4}+\frac{1+p_j}{4}-1\right]y_iy_j$$

$$+\sum_{i=1}^{N}\frac{2+p_i+p_i^2-\sum_{j\neq i=1}^{N}p_j^2}{4p_i}V_i$$

$$=\frac{1}{4}\sum_{i<j=1}^{N}\sum_{j=1}^{N}(2-p_i-p_j)p_ip_j\left(\frac{y_i}{p_i}-\frac{y_j}{p_j}\right)^2+\sum_{i=1}^{N}\frac{2+p_i+p_i^2-\sum_{j\neq i=1}^{N}p_j^2}{4p_i}V_i$$

An unbiased estimator of $Var(t_D)$ is given by

$$\widehat{Var}(t_D)=\frac{(1-p_i)(1-p_j)}{4}\left[\left(\frac{r_i}{p_i}-\frac{r_j}{p_j}\right)^2-\left(\frac{v_i}{p_i^2}+\frac{v_j}{p_j^2}\right)\right]+\sum_{i\in s}d_{si}v_i,$$

where v_i is an unbiased estimator of V_i with respect to the randomisation device stated earlier.

Now suppose a sample of size n is drawn according to PPSWOR sampling scheme in which a unit i_1 is selected on the first draw with a probability p_{i_1}, a unit i_2 is selected on the second draw with a probability $\frac{p_{i_2}}{1-p_{i_1}}$, and so on. Finally on the nth draw a unit i_n is selected with a probability $\frac{p_{i_n}}{1-p_{i_1}-p_{i_2}-\ldots-p_{i_{n-1}}}$. Then consider the following estimators

$$t_1=\frac{r_{i_1}}{p_{i_1}}$$

$$t_2=r_{i_1}+\frac{r_{i_2}}{p_{i_2}}(1-p_{i_1})$$

$$t_n=r_{i_1}+r_{i_2}+\ldots+r_{i_{n-1}}+\frac{r_{i_n}}{p_{i_n}}(1-p_{i_1}-p_{i_2}-\ldots-p_{i_{n-1}})$$

Then the Raj's (1956) ordered estimator for the ordered sample $s=(i_1, i_2, \ldots, i_n)$ is given by

$$\bar{t}=\frac{1}{n}\sum_{i=1}^{n} t_i$$

It is easy to check that $E_pE_R(\bar{t})=Y$ and $Var(\bar{t})=\frac{1}{n^2}\sum_{i=1}^{n} Var(t_i)$ on noting that $E_pE_R(t_i)=Y\forall i$ and t_i's are pairwise uncorrelated.

An unbiased estimator of $Var(\bar{t})$ is given by

$$\widehat{Var}(\bar{t})=\bar{t}^2-\frac{1}{n(n-1)}\sum_{i\neq j=1}^{n}\sum_{j=1}^{n} t_it_j$$

$$=\frac{1}{n(n-1)}\sum_{i=1}^{n}(t_i-\bar{t})^2$$

5 VARIANCE OF MURTHY'S (1957) UNORDERED ESTIMATOR

In case of an RR survey the Murthy's (1957) unordered estimator is

$$t_M=\sum_{i\in s}\frac{r_ip(s|\,i)}{p(s)},$$

where $p(s|\,i)$ is the conditional probability of selecting a sample s given that the i th unit is selected on the first draw and $p(s)$ is the probability of selecting the unordered sample s.

Thus the Murthy's (1957) unordered estimator can be written as

$$t_M=\sum_{i=1}^{N} d_{si}r_i$$

where $d_{si}=0$ if i does not belong to s or s does not contain i and $d_{si}=\frac{p((s|\,i))}{p(s)}$ if $i\in s$ or $s\ni i$.

The variance of the Murthy's (1957) unordered estimator is

$$Var(t_M)=\sum_{i=1}^{N} y_i^2\left[\sum_{s\ni i} d_{si}^2p(s)-1\right]+\sum_{i\neq j=1}^{N}\sum_{j=1}^{N} y_iy_j\left[\sum_{s\ni i,j} d_{si}d_{sj}p(s)-1\right]+\sum_{i=1}^{N}\sum_{s\ni i} d_{si}^2p(s)V_i$$

$$=\sum_{i=1}^{N} y_i^2\left[\sum_{s\ni i}\frac{p^2(s|\,i)}{p(s)}-1\right]+\sum_{i\neq j=1}^{N}\sum_{j=1}^{N} y_iy_j\left[\sum_{s\ni i,j}\frac{p(s|\,i)p(s|\,j)}{p(s)}-1\right]+\sum_{i=1}^{N}\sum_{s\ni i}\frac{p^2(s|\,i)}{p(s)}V_i$$

$$=\sum_{i=1}^{N}\frac{y_i^2}{p_i}A_i-\sum_{i<j=1}^{N}\sum_{j=1}^{N}\left[\sum_{s\ni i,j}\frac{p(s|\,i)p(s|\,j)}{p(s)}-1\right]p_ip_j\left(\frac{y_i}{p_i}-\frac{y_j}{p_j}\right)^2+\sum_{i=1}^{N}\sum_{s\ni i}\frac{p^2(s|\,i)}{p(s)}V_i$$

on putting $x_i=y_i$, $\alpha_i=p_i$, and

$$A_i = \sum_{j=1}^{N} a_{ij}\alpha_j = \sum_{j=1}^{N}\left[\sum_{s\ni i,j}\frac{p(s|\,i)p(s|\,j)}{p(s)} - 1\right]p_j$$

In Theorem 1 now

$$\sum_{i=1}^{N}\frac{y_i^2}{p_i}A_i = \sum_{i=1}^{N}\frac{y_i^2}{p_i}\sum_{j=1}^{N}\sum_{s\ni i,j}\frac{p(s|\,i)p(s|\,j)}{p(s)}p_j - \sum_{i=1}^{N}\frac{y_i^2}{p_i}$$

$$= \sum_{s\in S}\sum_{i,j\in s}\frac{y_i^2}{p_i}p_j\frac{p(s|\,i)p(s|\,j)}{p(s)} - \sum_{i=1}^{N}\frac{y_i^2}{p_i}$$

where S is the collection of all possible samples s

$$= \sum_{s\in S}\sum_{i\in s}\frac{y_i^2}{p_i}\sum_{j\in s}\frac{p_j p(s|j)}{p(s)}p(s|i) - \sum_{i=1}^{N}\frac{y_i^2}{p_i}$$

$$= \sum_{s\in S}\sum_{i\in s}\frac{y_i^2}{p_i}p(s|i) - \sum_{i=1}^{N}\frac{y_i^2}{p_i}$$

$$= \sum_{i=1}^{N}\frac{y_i^2}{p_i}\sum_{s\ni i}p(s|i) - \sum_{i=1}^{N}\frac{y_i^2}{p_i}$$

$$= \sum_{i=1}^{N}\frac{y_i^2}{p_i} - \sum_{i=1}^{N}\frac{y_i^2}{p_i} = 0.$$

Hence by virtue of Theorem 3, the variance of Murthy's (1957) unordered estimator can be written as

$$Var(t_M) = \sum_{i<j=1}^{N}\sum_{j=1}^{N}\left[1 - \sum_{s\ni i,j}\frac{p(s|\,i)p(s|\,j)}{p(s)}\right]p_ip_j\left(\frac{y_i}{p_i} - \frac{y_j}{p_j}\right)^2 + \sum_{i=1}^{N}\sum_{s\ni i}\frac{p^2(s|\,i)}{p(s)}V_i$$

An unbiased estimator of $Var(t_M)$ is given by

$$\widehat{Var}(t_M) = \sum_{i<j\in s}\sum_{j\in s}\left[\frac{p(s|\,i,j)}{p(s)} - \frac{p(s|\,i)p(s|\,j)}{p^2(s)}\right]p_ip_j\left[\left(\frac{r_i}{p_i} - \frac{r_j}{p_j}\right)^2 - \left(\frac{v_i}{p_i^2} + \frac{v_j}{p_j^2}\right)\right] + \sum_{i\in s}\frac{p(s|\,i)v_i}{p(s)}$$

on noting that

$$\sum_{s\ni i,j}p(s|\,i,j) = 1,$$

where $p(s|\,i,j)$ is the conditional probability of selecting the sample s given that the units i and j are selected on the first two draws. After some simplifications, we have

$$\widehat{Var}(t_M)=\frac{1}{[p(s)]^2}\sum_{i<j\in s}\sum_{j\in s}[p(s|\,i,j)p(s)-p(s|\,i)p(s|\,j)]p_ip\left[\left(\frac{r_i}{p_i}-\frac{r_j}{p_j}\right)^2-\left(\frac{v_i}{p_i^2}+\frac{v_j}{p_j^2}\right)\right]$$
$$+\sum_{i\in s}\frac{p(s|\,i)}{p(s)}v_i,$$

where v_i is an unbiased estimator of V_i with respect to the randomisation device stated earlier.

6 VARIANCE OF RATIO ESTIMATOR BASED ON LAHIRI (1951), MIDZUNO (1952), AND SEN'S (1953) SAMPLING SCHEME

In case of an RR survey the ratio estimator of the population total Y is given by

$$t_R=\frac{\bar{r}}{\bar{x}}X=\frac{\sum_{i\in s}r_i}{\sum_{i\in s}x_i}X=\frac{\sum_{i\in s}r_i}{\sum_{i\in s}p_i}=\frac{\sum_{i\in s}r_i}{p_s},$$

where $p_s=\sum_{i\in s}p_i$. The ratio estimator can be written as

$$t_R=\sum_{i=1}^{N}d_{si}r_i,$$

where d_{si} is 0 if i does not belong to s or s does not contain i and

$$d_{si}=\frac{1}{p_s}\quad\text{if } i\in s \text{ or } s\ni i.$$

The probability of selecting a sample s according to Lahiri (1951), Midzuno (1952), and Sen's (1953) sampling scheme is given by

$$p(s)=\frac{p_s}{M_1}$$

where $M_1=\binom{N-1}{n-1}$.

The variance of the ratio estimator based on Lahiri (1951), Midzuno (1952), and Sen's sampling scheme is given by

$$Var(t_R)=\sum_{i=1}^{N}y_i^2\left[\sum_{s\ni i}d_{si}^2p(s)-1\right]+\sum_{i\neq j=1}^{N}\sum_{j=1}^{N}y_iy_j\left[\sum_{s\ni i,j}d_{si}d_{sj}p(s)-1\right]+\sum_{i=1}^{N}\sum_{s\ni i}d_{si}^2p(s)V_i$$
$$=\sum_{i=1}^{N}y_i^2\left[\sum_{s\ni i}\frac{1}{p_s}\frac{1}{M_1}-1\right]+\sum_{i\neq j=1}^{N}\sum_{j=1}^{N}y_iy_j\left[\sum_{s\ni i,j}\frac{1}{p_s}\frac{1}{M_1}-1\right]+\sum_{i=1}^{N}\sum_{s\ni i}\frac{1}{p_s}\frac{1}{M_1}V_i$$
$$=\sum_{i=1}^{N}\frac{y_i^2}{p_i}A_i-\sum_{i<j=1}^{N}\sum_{j=1}^{N}\left[\sum_{s\ni i,j}\frac{1}{p_s}\frac{1}{M_1}-1\right]p_ip_j\left(\frac{y_i}{p_i}-\frac{y_j}{p_j}\right)^2+\sum_{i=1}^{N}\sum_{s\ni i}\frac{1}{p_s}\frac{1}{M_1}V_i$$

on putting $x_i = y_i$, $\alpha_i = p_i$, and

$$A_i = \sum_{j=1}^{N} a_{ij}\alpha_j = \sum_{j=1}^{N}\left[\sum_{s\ni i,j}\frac{1}{p_s}\frac{1}{M_1} - 1\right]p_j$$

In Theorem 1 now let us consider the quantity

$$\sum_{i=1}^{N}\frac{y_i^2}{p_i}A_i = \sum_{i=1}^{N}\frac{y_i^2}{p_i}\sum_{j=1}^{N}\left[\sum_{s\ni i,j}\frac{1}{p_s}\frac{1}{M_1} - 1\right]p_j$$

$$= \sum_{i=1}^{N}\sum_{j=1}^{N}\sum_{s\ni i,j}\frac{1}{p_s}\frac{1}{M_1}\frac{y_i^2}{p_i}p_j - \sum_{i=1}^{N}\frac{y_i^2}{p_i}$$

$$= E_p\left[\sum_{i\in s}\sum_{j\in s}\frac{1}{p_s^2}\frac{y_i^2}{p_i}p_j\right] - \sum_{i=1}^{N}\frac{y_i^2}{p_i}$$

$$= E_p\left[\sum_{i\in s}\frac{y_i^2}{p_i}\frac{1}{p_s}\right] - \sum_{i=1}^{N}\frac{y_i^2}{p_i}$$

$$= \left[\sum_{s\in S}\sum_{i\in s}\frac{y_i^2}{p_i}\frac{1}{p_s}p(s) - \sum_{i=1}^{N}\frac{y_i^2}{p_i}\right],$$

where S is the collection of all possible samples s

$$= \sum_{s\in S}\sum_{i\in s}\frac{y_i^2}{p_i}\frac{1}{p_s}\frac{p_s}{M_1} - \sum_{i=1}^{N}\frac{y_i^2}{p_i} = \sum_{i=1}^{N}\frac{y_i^2}{p_i}\sum_{s\ni i}\frac{1}{M_1} - \sum_{i=1}^{N}\frac{y_i^2}{p_i} = \sum_{i=1}^{N}\frac{y_i^2}{p_i} - \sum_{i=1}^{N}\frac{y_i^2}{p_i} = 0$$

Hence by Theorem 3 the variance of the ratio estimator can be written as

$$Var(t_R) = \sum_{i<j=1}^{N}\sum_{j=1}^{N}\left[1 - \sum_{s\ni i,j}\frac{1}{p_s}\frac{1}{M_1}\right]p_ip_j\left(\frac{y_i}{p_i} - \frac{y_j}{p_j}\right)^2 + \sum_{i=1}^{N}\sum_{s\ni i}\frac{1}{p_s}\frac{1}{M_1}V_i$$

An unbiased estimator of the variance of t_R is given by

$$\widehat{Var}(t_R) = \sum_{i<j\in s}\sum_{j\in s}\left[\frac{1}{M_2p(s)} - \frac{1}{p_s^2}\right]p_ip_j\left[\left(\frac{r_i}{p_i} - \frac{r_j}{p_j}\right)^2 - \left(\frac{v_i}{p_i^2} + \frac{v_j}{p_j^2}\right)\right] + \sum_{i\in s}\frac{1}{p_s^2}v_i$$

$$= \sum_{i<j\in s}\sum_{j\in s}\left[\frac{M_1}{M_2p_s} - \frac{1}{p_s^2}\right]p_ip_j\left[\left(\frac{r_i}{p_i} - \frac{r_j}{p_j}\right)^2 - \left(\frac{v_i}{p_i^2} + \frac{v_j}{p_j^2}\right)\right] + \sum_{i\in s}\frac{v_i}{p_s^2}$$

where v_i is an unbiased estimator of V_i with respect to the randomization device stated earlier and $M_2 = \binom{N-2}{n-2}$.

7 VARIANCE OF HARTLEY–ROSS (1954) UNBIASED RATIO-TYPE ESTIMATOR

In case of RR surveys the Hartley–Ross (1954) unbiased ratio-type estimator of the population total Y is given by

$$\widehat{Y_{HR}} = \bar{u}X + \frac{n(N-1)}{n-1}(\bar{r} - \bar{u}\bar{x}),$$

where $\bar{u} = \frac{1}{n}\sum_{i=1}^{n}\frac{r_i}{x_i}$ is the mean of the ratios and $\bar{r} = \frac{1}{n}\sum_{i=1}^{n} r_i$, $\bar{x} = \frac{1}{n}\sum_{i=1}^{n} x_i$ are the sample means of the RRs and the auxiliary variable x, respectively.

The Hartley–Ross (1954) unbiased ratio-type estimator can be written as

$$\widehat{Y_{HR}} = \sum_{i=1}^{N} d_{si} r_i,$$

where $d_{si} = 0$ if i does not belong to s or s does not contain i and

$$d_{si} = \frac{1}{n}\frac{X}{x_i} + \frac{n(N-1)}{n-1}\left(\frac{1}{n} - \frac{1}{n}\frac{\bar{x}}{x_i}\right) \quad \text{if}\, i \in s\, \text{or}\, s \ni i.$$

The variance of the Hartley–Ross (1954) unbiased ratio-type estimator is given by

$$Var\left(\widehat{Y_{HR}}\right) = \sum_{i=1}^{N} y_i^2 \left[\sum_{s \ni i} d_{si}^2 p(s) - 1\right] + \sum_{i \neq j=1}^{N}\sum_{j=1}^{N} y_i y_j \left[\sum_{s \ni i,j} d_{si} d_{sj} p(s) - 1\right]$$

$$+ \sum_{i=1}^{N}\sum_{s \ni i} d_{si}^2 p(s) V_i$$

$$= \sum_{i=1}^{N}\frac{y_i^2}{x_i}A_i - \sum_{i<j=1}^{N}\sum_{j=1}^{N} a_{ij} x_i x_j \left(\frac{y_i}{x_i} - \frac{y_j}{x_j}\right)^2 + \sum_{i=1}^{N}\sum_{s \ni i} d_{si}^2 p(s) V_i$$

on putting $x_i = y_i$, $\alpha_i = x_i$, and $A_i = \sum_{j=1}^{N} a_{ij} x_j$ in Theorem 1 where

$$a_{ij} = \sum_{s \ni i,j} d_{si} d_{sj} p(s) - 1$$

Now consider the quantity

$$\sum_{i=1}^{N}\frac{y_i^2}{x_i}A_i = \sum_{i=1}^{N}\frac{y_i^2}{x_i}\sum_{j=1}^{N} a_{ij} x_j$$

$$= \sum_{i=1}^{N}\frac{y_i^2}{x_i}\sum_{j=1}^{N}\left[\sum_{s \ni i,j} d_{si} d_{sj} p(s) - 1\right] x_j$$

$$= \sum_{s \in S}\sum_{i \in s}\sum_{j \in s} d_{si} d_{sj} \frac{y_i^2}{x_i} x_j p(s) - \sum_{i=1}^{N}\frac{y_i^2}{x_i} X$$

where S is the collection of all possible samples

$$=\sum_{s\in S}\sum_{i\in s}\frac{y_i^2}{x_i}d_{si}p(s)\sum_{j\in s}d_{sj}x_j-\sum_{i=1}^{N}\frac{y_i^2}{x_i}X$$

$$=\sum_{s\in S}\sum_{i\in s}\frac{y_i^2}{x_i}d_{si}p(s)X-\sum_{i=1}^{N}\frac{y_i^2}{x_i}X$$

because of the calibration equation

$$\sum_{j\in s}d_{sj}x_j=X$$

$$=\sum_{i=1}^{N}\frac{y_i^2}{x_i}X\sum_{s\ni i}d_{si}p(s)-\sum_{i=1}^{N}\frac{y_i^2}{x_i}X$$

$$=\sum_{i=1}^{N}\frac{y_i^2}{x_i}X-\sum_{i=1}^{N}\frac{y_i^2}{x_i}X=0$$

using the unbiasedness condition

$$\sum_{s\ni i}d_{si}p(s)=1\forall i=1,2,\ldots,N$$

as the Hartley–Ross (1954) estimator is known to be unbiased under SRSWOR sampling scheme. Hence by Theorem 3 the variance of the Hartley–Ross (1954) unbiased ratio-type estimator is given by

$$Var\left(\widehat{Y_{HR}}\right)=\sum_{i<j=1}^{N}\sum_{j=1}^{N}\left[1-\sum_{s\ni i,j}d_{si}d_{sj}p(s)\right]x_ix_j\left(\frac{y_i}{x_i}-\frac{y_j}{x_j}\right)^2+\sum_{i=1}^{N}\sum_{s\ni i}d_{si}^2p(s)V_i$$

An unbiased estimator of the variance of Hartley–Ross (1954) unbiased ratio-type estimator is given by

$$\widehat{Var}\left(\widehat{Y_{HR}}\right)=\sum_{i<j\in s}\sum_{j\in s}\left[\frac{1}{\pi_{ij}}-d_{si}d_{sj}\right]x_ix_j\left[\left(\frac{r_i}{x_i}-\frac{r_j}{x_j}\right)^2-\left(\frac{v_i}{x_i^2}+\frac{v_j}{x_j^2}\right)\right]+\sum_{i\in s}d_{si}^2v_i,$$

where $\pi_{ij}=\frac{n(n-1)}{N(N-1)}$ is the inclusion probability of a pair of units i and j under SRSWOR sampling scheme and v_i is an unbiased estimator of V_i with respect to the randomisation device stated earlier.

REFERENCES

Adhikary, A.K., 2009a. Improving the Hansen-Hurwitz estimator in PPSWR sampling. In: SenGupta, A. (Ed.), Advances in Multivariate Statistical Methods. Statistical Sciences and Interdisciplinary Research, vol. 4. World Scientific Press, USA, pp. 455–477.

Adhikary, A.K., 2009. Improving the Hansen-Hurwitz estimator in PPSWR sampling. Proceedings Volume of Sixth International Triennial Calcutta Symposium on Probability and Statistics, December 25–31. Cal. Stat. Assoc. Bull. vol. 61 (241–244), 267–287.

Adhikary, A.K., 2013. An exact expression for the mean square error of ratio estimator, regression estimator and generalized regression estimator in finite population sampling. Int. J. Appl. Math. Stat. Sci. 2 (5), 125–142.

Chaudhuri, A., 2000. Network and adaptive sampling with unequal probabilities. Calcutta Stat. Assoc. Bull. 50, 237–253.

Chaudhuri, A., Pal, S., 2002. On certain alternative mean square errors in complex surveys. J. Stat. Plan. Inference 104 (2), 363–375.

Chaudhuri, A., Stenger, H., 2005. Survey Sampling—Theory and Methods, second ed. Chapman & Hall/CRC, Taylor & Francis Group, New York.

Raj, D., 1956. Some estimators in sampling with varying probabilities without replacement. J. Am. Stat. Assoc. 51, 269–284.

Hansen, M.H., Hurwitz, W.N., 1943. On the theory of sampling from finite populations. Ann. Math. Stat. 14, 333–362.

Hartley, H.O., Ross, A., 1954. Unbiased ratio estimators. Nature 174, 270–271.

Lahiri, D.B., 1951. A method of sample selection providing unbiased ratio estimators. Bull. Int. Stat. Inst. 33 (2), 133–140.

Midzuno, H., 1952. On the sampling system with probabilities proportionate to sum of sizes. Ann. Inst. Stat. Math. 3, 99–107.

Murthy, M.N., 1957. Ordered and unordered estimators in sampling without replacement. Sankhyā 18, 379–390.

Sen, A.R., 1953. On the estimation of variance in sampling with varying probabilities. J. Indian Soc. Agric. Stat. 5 (2), 119–127.

Warner, S.L., 1965. Randomized response: a survey technique for eliminating evasive answer bias. J. Am. Stat. Assoc. 60, 63–69.

Yates, F., Grundy, P.M., 1953. Selection without replacement from within strata with probability proportional to size. J. R. Stat. Soc. B. 15, 253–263.

Chapter 13

Behavior of Some Scrambled Randomized Response Models Under Simple Random Sampling, Ranked Set Sampling and Rao–Hartley–Cochran Designs

C.N. Bouza-Herrera[1]
Universidad de La Habana, La Habana, Cuba
[1]Corresponding author: e-mail: bouza@matcom.uh.cu

ABSTRACT

The behavior of some important scrambled response models is considered under the use of simple random sampling, ranked set sampling (RSS) and Rao–Hartley–Cochran designs. Their extensions to RSS are derived. The behavior of the scrambled randomized models is compared in terms of their sampling errors.

Keywords: Sensitive questions, Randomized responses, Efficiency

MSC: 62D05

1 INTRODUCTION

Let Y be a positive sensitive variable evaluated in a finite population $U=\{u_1, \ldots, u_N\}$. The individual u_i, with a value of Y that carries a stigma, will tend to give incorrect information or to refuse to answer. It is well known that when dealing with sensitive questions we should face the need to reduce the refusals to respond and the response bias. A possibility is to replace the direct response to the sensitive question by using a random response query. The seminal work is due to Warner (1965). It dealt with qualitative variables. The aim of the surveyor in such cases is to estimate the probability of having

Handbook of Statistics, Vol. 34. http://dx.doi.org/10.1016/bs.host.2016.01.011

the stigma. The basic Warner's method consists in placing the question associated with the stigma together with some insensitive ones. The respondent chooses randomly a question and answers it without revealing which was selected. This problem is important and extensions have been derived in recent papers of Singh et al. (2000) and Christofides (2003).

The use of randomized response (RR) models is still placing challenges both for the theory of survey sampling as well as for its application. Different extensions of RR have been introduced, see Chaudhuri and Mukerjee (1987) for the earlier proposals. The interest in RR methods to estimate the mean of a sensitive quantitative variable is increasing, see Bouza (2010), Gjestvang and Singh (2007), and Hussain (2012). RR methods for scrambling the responses have been proposed by different authors, see for example Eichhorn and Hayre (1983), Gupta et al. (2002), Bar-Lev et al. (2004), Ryu et al. (2006), and Bouza and Singh (2009).

Rao–Hartley–Cochran (1962), RHC, proposed an unequal probability selection sampling design. The use of it and the application of Warner's RR procedure were considered by Soberanis Cruz and Cuevas Domínguez (2010).

Ranked set sampling (RSS) was proposed by McIntyre (1952). It is an alternative to simple random sampling with replacement (SRSWR) and outperforms the estimation of the mean with respect to SRSWR in many situations by reducing the variance of an estimator.

In this chapter we study the behavior of some scrambling procedures when RSS is used as an alternative to SRSWR and RHC.

Section 2 is devoted to the presentation of the scrambled variables procedures of Eichhorn and Hayre (1983), Bar-Lev et al. (2004), and Ryu et al. (2006). Their behavior in the case of using SRSWR is developed. Section 3 is concerned with the extension of these RR procedures when RHC sampling design is used. Section 4 is concerned with the study of the effect of using RSS instead of SRSWR or RHC. A study of the efficiency of our proposals is developed in Section 5.

2 SCRAMBLING PROCEDURES

Let us consider a sensitive quantitative variable Y. A known probability distribution P is determined for selecting a value

$$Z \in Z^* = \{Z^*_1, \ldots, Z^*_k\}$$

independent of Y. As Z^* is determined by the statistician both the mean and variance of Z, μ_Z, and σ_Z^2 are known a priori. Each selected person $i \in s,\ i = 1, \ldots, n$, responds with a coded value.

Eichhorn–Hayre (1983), EH, proposed an RR that establishes that the respondent must report the scrambled variable $R_i = Z_i Y_i$. Its expectation is

$$E(E(Z_i Y_i \mid EH)) = E(Y_i \mu_Z) = \mu_Y \mu_Z.$$

Hence R_i/μ_Z is unbiased for μ_Y and

$$V(E(Z_iY_i|\,EH))+E(V(Z_iY_i|\,EH))=\mu_Z^2V(Y_i)+\sigma_Z^2E\left(Y_i^2\right)=\mu_Z^2\sigma_Y^2+\sigma_Z^2\left(\mu_Y^2+\sigma_Y^2\right).$$

A naïve estimator of μ_Y is

$$\mu_{EH}=\frac{\sum_{i=1}^{n}R_i}{n\mu_Z}. \tag{1}$$

It is unbiased as $E(E(R_i|\,EH))=\mu_Y\mu_Z$ and its variance is given by:

$$V(\mu_{EH})=\frac{\sum_{i=1}^{n}V(R_i)}{n^2\mu_Z^2}=\frac{1}{n\mu_Z^2}\left(\mu_Z^2\sigma_Y^2+\sigma_Z^2\left(\mu_Y^2+\sigma_Y^2\right)\right)=\frac{\sigma_Y^2+C_Z^2\mu_Y^2\left(1+C_Y^2\right)}{n}$$
$$=\frac{V_{EH}}{n},C_H=\frac{\sigma_H}{\mu_H},H=Z,Y. \tag{2}$$

Bar-Lev et al. (2004), BLBB, proposed another RR technique for dealing with a sensitive quantitative variable. The sampler determines a probability P and a Bernoulli trial is performed, by the i respondent, for determining if the response will be the true value of Y_i or the scrambled variable $R_i=Z_iY_i$. The procedure must include the value 1 with a strictly positive probability.

Denote the corresponding Bernoulli random variable by α. The response of the respondent is represented by the equation

$$B_i=\alpha Y_i+(1-\alpha)R_i.$$

We have that

$$E(E(B_i|\,Y))=P\mu_Y+Q\mu_Y\mu_Z=\mu_Y(P+Q\mu_Z),Q=1-P. \tag{3}$$

From Bar-Lev et al. (2004) we also have that

$$V(B_i)=\mu_Y^2\left(1+C_Y^2\right)\left[P+\mu_Z^2Q\left(1+C_Z^2\right)\right]-\left[\mu_Z(P+\mu_ZQ)\right]^2.$$

Hence we may use the expression

$$V_B=V(B_i)=\mu_Y^2C^2+\sigma_Y^2C^2-\mu_Y^2U, \tag{4}$$

where

$$C^2=Q\mu_Z^2\left(1+C_Z^2\right)+P,U=(P+\mu_ZQ)^2.$$

An unbiased estimator of the population mean is

$$\mu_B=\frac{\sum_{i=1}^{n}B_i}{n(P+Q\mu_Z)}, \tag{5}$$

with variance

$$V(\mu_B)=\frac{\sum_{i=1}^{n}V_{0i}}{(n(P+Q\mu_Z))^2}=\frac{V_B}{n(P+Q\mu_Z)^2}. \tag{6}$$

The RR model of Ryu et al. (2006), RKHP, again assumes that, as Y is sensitive, the surveyor organizes a set of experiments allowing that the interviewees give a response where the true value is not reported. Each person i interviewed performs a Bernoulli experiment that generates a random variable α with success probability P_1. That is, α follows the Bernoulli distribution $B(P_1)$. The respondent is asked to consider that

$$\alpha = \begin{cases} 1 \rightarrow \text{report the value of Y} \\ 0 \rightarrow \text{perform a second experiment}. \end{cases}$$

The respondent does not inform if a second stage experiment was needed. In this stage again a Bernoulli experiment generates a random variable β distributed $B(P_2)$ and

$$\beta = \begin{cases} 1 \rightarrow \text{report the value of Y} \\ 0 \rightarrow ZY \end{cases},$$

Z is selected with equal probabilities from $Z^* = \{Z_1^*, \ldots, Z_k^*\}$ and the response is

$$R_i = aY_i + (1-\alpha)\beta Y_i + (1-\alpha)(1-\beta)Z_i Y_i.$$

As in the previous procedure, Z^* is fixed by the surveyor but granting that $\mu_Z = 1$. A naïve estimator of μ_Y, if $Q_t = 1 - P_t$, $t = 1,2$, is

$$\hat{\mu}_R = \frac{\sum_{i=1}^{n} R_i}{n} \tag{7}$$

and its variance is given by

$$V(\hat{\mu}_R) = \frac{\sigma_Y^2 + Q_1 Q_2 \sigma_Z^2 \left(\mu_Y^2 + \sigma_Y^2\right)}{n} = \frac{V_R}{n}.$$

3 BEHAVIOR UNDER RHC UNEQUAL PROBABILITY MODEL

Rao et al. (1962) proposed using a set of unequal probabilities for selecting a sample of size n. An auxiliary variable X is used for determining the set of selection probabilities. Their well known drawing procedure is the following:

Drawing Procedure of RHC

Divide the population $U = \{1, \ldots, N\}$ into n disjoint sets, U_i, $i = 1, 2, \ldots n$, of size N_i each.
Select a unit k_i from U_i with probability $\frac{x_{k_i}}{t_{xi}}$, $t_{xi} = \sum_{t=1}^{N_i} x_{it}$.
Repeat for $i < n$.□

Let us study the RHC's model when we use scrambled variables. We need to calculate the corresponding conditional expectation under each procedure and deal with the basic formulae to derive adequate estimators.

Take the RR of EH and develop the RHC estimator under this procedure. We have

$$\hat{t}_{EH,RHC} = \frac{1}{n\mu_Z}\sum_{i=1}^{n}\frac{R_{k_i}}{p_{k_i}}\pi_i.$$

The conditional expectation of this estimator is given by

$$E_{RC}\left(\hat{t}_{EH,RHC}\right) = E_{RC}\left(\sum\nolimits_{i=1}^{n}\frac{R_i}{\mu_Z p_i/_{\pi_i}}\right) = \sum\nolimits_{i=1}^{n}\frac{\mu_Y}{p_i/_{\pi i}}.$$

We have that an unbiased estimator of μ_Y is $\overline{y}_{EH,RHC} = \frac{1}{N\mu_Z}\sum_{i=1}^{n}\frac{R_i}{p_{k_i}}\pi_i$. The derivation of the variance is somewhat tricky. Take

$$\begin{aligned}(N\mu_Z)^2 V\left(\hat{t}_{EH,RHC}\right) &= E_1\left\{E_2\left[V\left(\hat{t}_{EH,RHC}\right)\right]\right\} \\ &\quad + E_1\left\{V_2\left[E_{RC}\left(\hat{t}_{EH,RHC}\right)\right]\right\} + V_1\left\{E_2\left[E_{RC}\left(\hat{t}_{EH,RHC}\right)\right]\right\} \\ &= E_1\left\{V_2\left[E_{RC}\left(\hat{t}_{EH,RHC}\right)\right]\right\} + V_1\left\{E_2\left[E_{RC}\left(\hat{t}_{EH,RHC}\right)\right]\right\} \\ &\quad + E_1\left\{E_2\left[V_{RC}\left(\hat{t}_{EH,RHC}\right)\right]\right\} \\ &= E_1 V_2\left(\sum\nolimits_{i=1}^{n}\frac{\mu_Z\mu_Y}{p_i/_{\pi_i}}\right) + E_1\left\{E_2\left[V_{RC}\left(\hat{t}_{EH,RHC}\right)\right]\right\}\end{aligned} \tag{8}$$

and introduce the Bernoulli variables

$$I_{G_{i;t}} = \begin{Bmatrix}1 & si\, G_i \ni t; \\ 0 & \text{otherwise}\end{Bmatrix}, I_{G_{i;t,t'}} = \begin{Bmatrix}1 & si\, G_i \ni t,t'; \\ 0 & \text{otherwise}\end{Bmatrix}.$$

Their conditional expectations are

$$E_1\left(I_{G_{i;t}}\right) = \Pr\left\{I_{G_{i;t}} = 1\right\} = \frac{N_i}{N}$$

and

$$E_1\left(I_{G_{i;t,t'}}\right) = \Pr\left\{I_{G_{i;t,t'}} = 1\right\} = \Pr\{G_i \ni t \text{ and } t'\} = \frac{N_i(N_i-1)}{N(N-1)}.$$

Hence

$$E_1 V_2\left(\sum\nolimits_{i=1}^{n}\frac{\mu_Y\mu_Z}{p_i/_{\pi_i}}\right) = (\mu_Y\mu_Z))^2\frac{\sum_{i=1}^{n}N_i^2 - N}{N(N-1)}$$

and

$$V\left(\hat{t}_{EH,RHC}\right) = \frac{\mu_Y^2\left(\sum_{i=1}^{n}N_i^2 - N\right)}{N(N-1)} + \frac{V_{EH}\left[N + \frac{\sum_{i=1}^{n}N_i^2 - N}{N(N-1)}\sum_{t=1}^{N}\sum_{\substack{t'=1\\t'\neq t}}^{N}\frac{p_t}{p_{t'}}\right]}{N^2\mu_Z^2}.$$

For Bar-Lev et al. (2004) procedure the expression of the estimator of the total is

$$\hat{t}_{A1,RHC}=\sum_{i=1}^{n}\frac{B_{k_i}}{p_{k_i}}\pi_i.$$

Taking the formula of V_B and developing similar calculations is derived that

$$E\left(\hat{t}_{A1,RHC}\right)=E_1\left\{E_2\left[E_{RC}\left(\hat{t}_{A1,RHC}\right)\right]\right\}=E_1E_2\left(\sum_{i=1}^{n}\frac{\mu_Y(P+Q\mu_Z)}{p_i/\pi_i}\right)$$
$$=N\mu_Y(P+Q\mu_Z).$$

Therefore an estimator of the population mean is

$$\overline{y}_{B,RHC}=\frac{1}{N(P+Q\mu_Z)}\sum_{i=1}^{n}\frac{B_{k_i}}{p_{k_i}}\pi_i, \tag{9}$$

which is unbiased for the estimation of the population mean. Using the same procedure for the derivation of the variance we obtain that

$$V\left(\overline{y}_{B,RHC}\right)=\frac{\mu_Y^2\left(\sum_{i=1}^{n}N_i^2-N\right)}{N^3(N-1)}+\frac{V_B\left[N+\frac{\sum_{i=1}^{n}N_i^2-N}{N(N-1)}\sum_{t=1}^{N}\sum_{\substack{t'=1\\t'\neq t}}^{N}\frac{p_t}{p_{t'}}\right]}{\left(N(P+Q\mu_Z)\right)^2}.$$

When RKHP-RR is used, the RHC estimator is

$$\hat{t}_{R,RHC}=\frac{1}{N}\sum_{i=1}^{n}\frac{R_{k_i}}{p_{k_i}}\pi_i. \tag{10}$$

Its unbiasedness is easily derived because the conditional expectation of the scrambled variable is equal to the variable of interest. Using again the Bernoulli variables $I_{G_{i;t}}$ and $I_{G_{i;t,t'}}$ as $V_R=\sigma_Y^2+Q_1Q_2\sigma_Z^2\left(\mu_Y^2+\sigma_Y^2\right)$ we proceed as in the case of BLBB-RR and, after some algebraic manipulations, we have that

$$V\left(\hat{t}_{R,RHC}\right)=\frac{\mu_Y^2\left[\frac{\sum_{i=1}^{n}N_i^2-N}{N(N-1)}\right]+V_R\left[N+\frac{\sum_{i=1}^{n}N_i^2-N}{N(N-1)}\sum_{t=1}^{N}\sum_{\substack{t'=1\\t'\neq t}}^{N}\frac{p_t}{p_{t'}}\right]}{N^2}.$$

4 BEHAVIOR UNDER RSS

RSS was proposed by McIntyre (1952) for estimating pasture yields and claimed that it was better than using simple random sampling with replacement (SRSWR). The basic ideas on the method can be obtained in

Chen et al. (2004). The theoretical frame that permits to use the RSS model is based on the hypothesis:

- We wish to enumerate the variable of interest Y.
- The units can be ordered linearly without ties.
- Any sample $s \subset U$ of size m can be enumerated.
- To identify a unit, order the units in s and enumerate them in less costly than to evaluate $\{Y_i, i \in s\}$ or to order U.

The basic selection procedure can be described as follows:

Procedure RSS for a Ranked Set Sample of Size n Generation

While $j<n$ do
    While $t<n$ do
        Select a new sample independently from U using SRSWR.
            Each unit in the sample $s_{(t)}$is ranked and the Order Statistics (OS) $Y_{(t:t)}$is the response variable.
        END
    END

We assume that the ranking is made using the auxiliary information provided by X and we have a sample of n order statistics. As the samples are independent, the obtained order statistics are mutually independent.

Let us consider the RSS model for Eichhorn and Hayre (1983). As we are working with the i-th OS

$$E\left(E\left(Z_iY_{(i)}|EH\right)\right)=E\left(Y_{(i)}\mu_Z\right)=\mu_{Y_{(i)}}\mu_Z.$$

Hence $\frac{R_{(i)}}{\mu_Z}=\frac{Z_iY_{(i)}}{\mu_Z}$ is unbiased for $\mu_{Y_{(i)}}$ and

$$\begin{aligned} V\left(E\left(Z_iY_{(i)}|EH\right)\right)+E\left(V\left(Z_iY_{(i)}|EH\right)\right) &= \mu_Z^2V\left(Y_{(i)}\right)+\sigma_Z^2E\left(Y_{(i)}^2\right) \\ &= \mu_Z^2\sigma_{Y_{(i)}}^2+\sigma_Z^2\left(\mu_{Y_{(i)}}^2+\sigma_{Y_{(i)}}^2\right). \end{aligned}$$

As $\sigma_{Y_{(i)}}^2=\sigma_Y^2-\left(\mu_{Y_{(i)}}-\mu_Y\right)^2$,

$$\begin{aligned} V\left(E\left(Z_iY_{(i)}|EH\right)\right)+E\left(V\left(Z_iY_{(i)}|EH\right)\right) &= \mu_Z^2\left[\sigma_Y^2-\left(\mu_{Y_{(i)}}-\mu_Y\right)^2\right] \\ &\quad +\sigma_Z^2\left[\sigma_Y^2-\left(\mu_{Y_{(i)}}-\mu_Y\right)^2+\mu_{Y_{(i)}}^2\right]. \end{aligned}$$

The RSS counterpart of the EH naïve estimator of μ_Y is

$$\mu_{(EH)}=\frac{\sum_{i=1}^{n}R_{(i)}}{n\mu_Z}.$$

It is unbiased. Note that we selected independent samples and we rank the units of each one of them. Therefore, the order statistics $Y_{(i)}$, $i=1,\ldots,n$, are mutually independent. Using this fact, the variance of the RSS estimator is given by:

$$V\left(\mu_{(EH)}\right)=\frac{\sum_{i=1}^{n}V\left(R_{(i)}\right)}{n^2\mu_Z^2}=\frac{\sigma_Y^2}{n}+\frac{C_Z^2\sum_{i=1}^{n}\mu_{Y_{(i)}}^2}{n}-\left[\left(\frac{C_Z^2+1}{n}\right)\sum_{i=1}^{n}\left(\mu_{Y_{(i)}}-\mu_Y\right)^2\right],$$

$$C_H=\frac{\sigma_H}{\mu_H},H=Z,Y.$$

It is clear that the RSS model is more accurate than the SRSWR one because

$$V\left(\mu_{(EH)}\right)-V(\mu_{EH})=-\frac{1}{n}\sum_{i=1}^{n}\left(\mu_{Y_{(i)}}-\mu_Y\right)^2,\mu_{Y_{(i)}}-\mu_Y=\Delta_{(i)}.$$

Let us consider the use of the scrambling procedure of Bar-Lev et al. (2004). The response of the i-th respondent is

$$B_{(i)}=\alpha Y_{(i)}+(1-\alpha)R_{(i)}.$$

The expectation of the obtained response is

$$E\left(E\left(B_{(i)}|Y\right)\right)=P\mu_{Y_{(i)}}+Q\mu_Z\mu_{Y_{(i)}}=\mu_{Y_{(i)}}(P+Q\mu_Z).$$

Take the estimator

$$\mu_{(Brss)}=\frac{\sum_{i=1}^{n}B_{(i)}}{n(P+Q\mu_Z)}.$$

It is unbiased because

$$E\left(\mu_{(Brss)}\right)=\frac{\sum_{i=1}^{n}E\left(B_{(i)}\right)}{n(P+Q\mu_Z)}=\frac{\sum_{i=1}^{n}\mu_{Y_{(i)}}}{n(P+Q\mu_Z)}=\mu_Y.$$

Under RSS, the variance of a single response is

$$V\left(B_{(i)}\right)=\mu_{Y_{(i)}}^2C^2+\sigma_{Y_{(i)}}^2C^2-\mu_{Y_{(i)}}^2U,$$

as $C_{Y_{(i)}}^2=\frac{\sigma_{Y_{(i)}}^2}{\mu_{Y_{(i)}}^2}$ is the variation coefficient of the statistic of order i. Using the fact that the order statistics are independent

$$V\left(\mu_{(Brss)}\right)=\frac{\sum_{i=1}^{n}\mu_{Y_{(i)}}^2C^2+\sigma_{Y_{(i)}}^2C^2-\mu_{Y_{(i)}}^2U}{n^2(P+Q\mu_Z)^2}.$$

Taking the relationship $\sigma_Y^2-\Delta_{Y_{(i)}}^2=\sigma_{Y_{(i)}}^2$, $\Delta_{Y_{(i)}}^2=\left(\mu_{Y_{(i)}}-\mu_Y\right)^2$ we may rewrite the above expression as

$$V\left(\mu_{(Brss)}\right)=C^2\frac{\sigma_Y^2-\sum_{i=1}^{n}\Delta_{Y_{(i)}}^2}{n^2(P+Q\mu_Z)^2}+(C^2-U)\frac{\sum_{i=1}^{n}\mu_{Y_{(i)}}^2}{n^2(P+Q\mu_Z)^2}.$$

Note that $C^2-U>0$, see Bar-Lev et al. (2004), is the squared coefficient of variation of the distribution of the variable Z. Hence, a sufficient condition for accepting that the RSS proposal has a smaller variance than the SRSWR model is that

$$\frac{\sum_{i=1}^{n}\mu_{Y_{(i)}}^2 - n\mu_Y^2}{n^2(P+Q\mu_Z)^2} \le 0.$$

Hence we cannot accept that RSS outperforms SRSWR always.

Using that for the RR model of Ryu et al. (2006), the value of the i-th OS $Y_{i(i)}$ will not be the reported response. The response of the i-th respondent is the scrambled variable

$$R_{(i)} = \alpha Y_{(i)} + (1-\alpha)\beta Y_{(i)} + (1-\alpha)(1-\beta)Z_i Y_{(i)}.$$

Its expectation is $E\left(R_{(i)}\right) = \mu_{Y_{(i)}}$, because we constructed Z^* in such a way that $E(Z)=1$. $\mu_{Y_{(i)}}$ is the expected value of the i-th order statistic. Consider the arithmetic mean of the responses:

$$\hat{\mu}_{(R)} = \frac{\sum_{i=1}^{n} R_{(i)}}{n}.$$

It is unbiased because

$$E\left(\hat{\mu}_{(R)}\right) = \frac{\sum_{i=1}^{n} E\left(R_{(i)}\right)}{n} = \frac{\sum_{i=1}^{n}\mu_{Y_{(i)}}}{n} = \mu_Y.$$

Under this scrambling procedure the variance derived for the i-th OS of R is

$$V\left(R_{(i)}\right) = V_{(i)} = \sigma_{Y_{(i)}}^2 + Q_1Q_2\sigma_Z^2\left(\mu_{Y_{(i)}}^2 + \sigma_{Y_{(i)}}^2\right).$$

Due to the independence, the variance of the RSS estimator is

$$V\left(\hat{\mu}_{(R)}\right) = \frac{\sum_{i=1}^{n} V_{(i)}}{n^2}.$$

Using the relationship between the variance of an OS and of the population variance it can be rewritten as

$$V\left(\hat{\mu}_{(R)}\right) = \frac{\sigma_Y^2}{n} + Q_1Q_2\frac{\sigma_Z^2\sigma_Y^2}{n} + \frac{Q_1Q_2\sigma_Z^2\sum_{i=1}^{n}\mu_{Y_{(i)}}^2 - \left(Q_1Q_2\sigma_Z^2+1\right)\sum_{i=1}^{n}\Delta_{Y_{(i)}}^2}{n^2}.$$

Adding and extracting $W = \frac{Q_1Q_2\sigma_Z^2\mu_Y^2}{n}$ we have that

$$\begin{aligned}\frac{Q_1Q_2\sigma_Z^2\sum_{i=1}^{n}\mu_{Y_{(i)}}^2}{n^2} \mp W &= \frac{Q_1Q_2\sigma_Z^2\mu_Y^2}{n} + \frac{Q_1Q_2\sigma_Z^2}{n}\left(\frac{\sum_{i=1}^{n}\mu_{Y_{(i)}}^2 - n\mu_Y^2}{n}\right)\\ &= \frac{Q_1Q_2\sigma_Z^2\mu_Y^2}{n} + \frac{Q_1Q_2\sigma_Z^2}{n}\left(\frac{\sum_{i=1}^{n}\Delta_{Y_{(i)}}^2}{n}\right).\end{aligned}$$

Therefore

$$V\left(\hat{\mu}_{(R)}\right) = V(\hat{\mu}_R) - \frac{\sum_{i=1}^{n}\Delta_{Y_{(i)}}^2}{n^2}.$$

As a result, the RSS alternative is more precise always.

$$V\left(\hat{t}_{EH,RHC}\right)-V\left(\mu_{(EH)}\right)$$

$$=\frac{\sigma_Y^2}{n}\left[1-\frac{\left[N+\frac{\sum_{i=1}^{n}N_i^2-N}{N(N-1)}\sum_{t=1}^{N}\sum_{\substack{t'=1\\ t'\neq t}}^{N}\frac{p_t}{p_{t'}}\right]}{N^2\mu_Z^2}\right]$$

$$+\frac{C_Z^2}{n}\left[\sum_{i=1}^{n}\mu_{Y(i)}^2-\mu_Y^2\frac{\left[N+\frac{\sum_{i=1}^{n}N_i^2-N}{N(N-1)}\sum_{t=1}^{N}\sum_{\substack{t'=1\\ t'\neq t}}^{N}\frac{p_t}{p_{t'}}\right]}{N^2}\right]$$

$$+\frac{C_Z^2\sigma_Y^2}{n}\left[1-\frac{\left[N+\frac{\sum_{i=1}^{n}N_i^2-N}{N(N-1)}\sum_{t=1}^{N}\sum_{\substack{t'=1\\ t'\neq t}}^{N}\frac{p_t}{p_{t'}}\right]}{N^2}\right]-\frac{\mu_Y^2\left(\sum_{i=1}^{n}N_i^2-N\right)}{N(N-1)}.$$

Let us compare the behavior of the scrambling procedures under the sampling designs of RHC and RSS.

Consider the case of Eichhorn and Hayre (1983). The difference of the variances is

$$V\left(\hat{t}_{EH,RHC}\right)-V\left(\mu_{(EH)}\right)=\frac{\sigma_Y^2}{n}\left[1-\frac{\left[N+\frac{\sum_{i=1}^{n}N_i^2-N}{N(N-1)}\sum_{t=1}^{N}\sum_{\substack{t'=1\\ t'\neq t}}^{N}\frac{p_t}{p_{t'}}\right]}{N^2\mu_Z^2}\right]$$

$$+\frac{C_Z^2}{n}\left[\sum_{i=1}^{n}\mu_{Y(i)}^2-\mu_Y^2\frac{\left[N+\frac{\sum_{i=1}^{n}N_i^2-N}{N(N-1)}\sum_{t=1}^{N}\sum_{\substack{t'=1\\ t'\neq t}}^{N}\frac{p_t}{p_{t'}}\right]}{N^2}\right]$$

$$+\frac{C_Z^2\sigma_Y^2}{n}\left[1-\frac{\left[N+\frac{\sum_{i=1}^{n}N_i^2-N}{N(N-1)}\sum_{t=1}^{N}\sum_{\substack{t'=1\\ t'\neq t}}^{N}\frac{p_t}{p_{t'}}\right]}{N^2}\right]-\frac{\mu_Y^2\left(\sum_{i=1}^{n}N_i^2-N\right)}{N(N-1)}.$$

For Bar-Lev et al.'s (2004) we have that

$$V\left(\mu_{(Brss)}\right) - V\left(\bar{y}_{B,RHC}\right)$$
$$= \frac{\sigma_Y^2}{(P+Q\mu_Z)^2}\left[\frac{N^2-n}{N^2 n} + C^2\left(\frac{N^2-n}{N^2 n}\right)\left[N + \frac{\sum_{i=1}^{n} N_i^2 - N}{N(N-1)}\sum_{t=1}^{N}\sum_{\substack{t'=1\\ t'\neq t}}^{N}\frac{p_t}{p_{t'}}\right]\right]$$
$$+ \frac{C^2}{(P+Q\mu_Z)^2}\left[\frac{1}{n^2}\sum_{i=1}^{n}\mu_{Y(i)}^2 - \frac{\mu_Y^2}{N^2}\left[N + \frac{\sum_{i=1}^{n} N_i^2 - N}{N(N-1)}\sum_{t=1}^{N}\sum_{\substack{t'=1\\ t'\neq t}}^{N}\frac{p_t}{p_{t'}}\right]\right]$$
$$- \sum_{i=1}^{n}\frac{\Delta_{Y(i)}^2}{n^2(P+Q\mu_Z)^2} - \frac{\mu_Y^2\left(\sum_{i=1}^{n} N_i^2 - N\right)}{N^3(N-1)}.$$

In the case of Ryu et al. (2006) scrambling procedure we have that

$$V\left(\hat{\mu}_{(R)}\right) - V\left(\hat{t}_{R,RHC}\right)$$
$$= \frac{\sigma_Y^2}{n}\left[1 + Q_1 Q_2 \sigma_Z^2\right]\left[1 - \frac{\left[N + \frac{\sum_{i=1}^{n} N_i^2 - N}{N(N-1)}\sum_{t=1}^{N}\sum_{\substack{t'=1\\ t'\neq t}}^{N}\frac{p_t}{p_{t'}}\right]}{N^2}\right]$$
$$+ \frac{Q_1 Q_2 \sigma_Z^2}{n}\left[\sum_{i=1}^{n}\mu_{(i)}^2 - \frac{\mu_Y^2\left[N + \frac{\sum_{i=1}^{n} N_i^2 - N}{N(N-1)}\sum_{t=1}^{N}\sum_{\substack{t'=1\\ t'\neq t}}^{N}\frac{p_t}{p_{t'}}\right]}{N^2}\right]$$
$$- \left[Q_1 Q_2 \sigma_Z^2\right]\sum_{i=1}^{n}\frac{\Delta_{Y(i)}^2}{n^2} - \frac{\mu_Y^2\left(\sum_{i=1}^{n} N_i^2 - N\right)}{N^2(N-1)}.$$

Note that the values of the differences depend of the probabilities p_t's and of the parameters of the distribution of the scrambling variable. Hence the surveyor, using adequate values sets of scrambling and auxiliary variables, is able to tune the justification of preferring RSS of RHC. These results suggest that the stratification generated by RHC design may overcome the behavior of RSS.

5 CONCLUSIONS

The results of the analysis developed in this chapter establish that the use of RHC procedure may be considered as the best alternative. This fact seems to be a consequence of the efficiency of the stratification generated by the RHC selection procedure. That is, this stratification is better than the one generated by RSS.

The preference by one of the scrambled procedures can be manipulated by fixing adequately the distributions

ACKNOWLEDGMENTS

This work was supported by the Program MÊS-CUBA EDITAL 46/2013 project 209/13 and a PNCB of CITMA. The author thanks the suggestions of the referees which allowed improving the initial version of this chapter.

REFERENCES

Bar-Lev, S.K., Bobovitch, E., Boukai, B., 2004. A note on randomized response models for quantitative data. Metrika 60, 255–260.

Bouza, C., 2010. Behavior of a randomized response procedure under unequal selection model of Chaudhuri-Stenger for insensitive variables under ranked set sampling. Adv. Appl. Stat. Sci. 4, 136–144.

Bouza, C., Singh, L., 2009. Optimal aspects of a model based randomized responses procedure under unequal selection of insensitive variables. Investig. Oper. 30, 234–243.

Chaudhuri, A., Mukerjee, R., 1987. Randomized response technique: a review. Statistica Neerlandica 41, 27–44.

Chen, Z., Bai, Z., Sinha, B.K., 2004. Ranked set sampling: theory and applications. Lectures Notes in Statistics, vol. 176. Springer, New York, NY.

Christofides, T.C., 2003. A generalized response technique. Metrika 57, 195–200.

Eichhorn, B.S., Hayre, L.S., 1983. Scrambled randomized response methods for obtaining sensitive quantitative data. J. Stat. Plan. Inference 100, 239–247.

Gjestvang, C.R., Singh, S., 2007. Forced quantitative randomized response model: a new device. Metrika 66, 243–257.

Gupta, S., Gupta, B., Singh, S., 2002. Estimation of sensitivity level of personal interview survey questions. J. Stat. Plan. Inference 100, 239–247.

Hussain, Z., 2012. Improvement of the Gupta and Thornton scrambling model through double use of randomization device. Int. J. Acad. Res. Bus. Soc. Sci. 2, 91–99.

McIntyre, G.A., 1952. A method for unbiased selective sampling, using ranked sets. Aust. J. Agr. Res. 2, 385–390.

Rao, J.N.K., Hartley, H.O., Cochran, W.G., 1962. On a simple procedure of unequal probability sampling without replacement. J. R. Stat. Soc. B 24, 482–491.

Ryu, J.B., Kim, J.-M., Heo, T.Y., Park, C.G., 2006. On stratified randomized response sampling. Model. Assist. Stat. Appl. 1, 31–36.

Singh, S., Singh, R., Mangat, N.S., 2000. Some alternative strategies to Moors' model in randomized response sampling. J. Stat. Plan. Inference 83, 243–255.

Soberanis Cruz, V.H., Cuevas Domínguez, J.D., 2010. Muestreo de respuesta aleatorizadas con probabilidades desiguales: el estimador de Rao-Hartley-Cochran. Investig. Oper. 31 (3), 248–257.

Warner, S.L., 1965. Randomized response: a survey technique for eliminating evasive answer bias. J. Am. Stat. Assoc. 60, 63–69.

Chapter 14

Estimation of a Finite Population Variance Under Linear Models for Randomized Response Designs

P. Mukhopadhyay[1]
Indian Statistical Institute, Kolkata, India
[1]*Corresponding author: e-mail: parimal00@hotmail.com*

ABSTRACT

Considering a linear model which combines random permutation model and models applicable to a wide class of randomized response (RR) designs, as developed by Bellhouse (1980), we have examined here the problem of finding optimal sampling strategies for estimating a finite population variance. An optimal estimator with a constant variance has been obtained within the class of all nonhomogeneous quadratic design-model-unbiased (model-unbiased) estimators of population variance, for any fixed-size noninformative sampling design. For the class of sampling designs including *srswor*, the same estimator remains optimal within the class of all quadratic design unbiased estimators for a particular case of the assumed random permutation model. In case the random permutation assumption is dropped, it is proved that there does not exist any uniformly minimum variance estimator in the entire class of design model unbiased estimators. Under certain conditions and under a slightly different RR model a uniformly minimum variance unbiased estimator has been obtained.

Keywords: Random permutation models, Response errors, Randomized response designs

1 INTRODUCTION

Let $\mathcal{P}$ denote a finite population of N identifiable units labeled $1,\ldots, i,\ldots, N$. Associated with each i is a pair of real numbers (y_i,w_i), values of a study variable y and a closely related auxiliary variable w, respectively. Our primary interest is estimating the finite population variance

Handbook of Statistics, Vol. 34. http://dx.doi.org/10.1016/bs.host.2016.01.012

$$\begin{aligned} V = V(y) &= \frac{1}{N}\sum\nolimits_1^N (y_i - \bar{y})^2, \\ &= a_1 \sum\nolimits_1^N y_i^2 - a_2 \sum\nolimits_{i \neq i'=1}^N y_i y_{i'}, \end{aligned} \tag{1}$$

where

$$a_1 = \frac{1}{N}\left(1 - \frac{1}{N}\right), \quad a_2 = \frac{1}{N^2}, \quad \bar{y} = \frac{\sum y}{N}$$

by survey sampling for which a sample s (a subset of size $n(s)$ from $\mathcal{P}$) is selected according to a sampling design $p(s)$ with inclusion probabilities $\pi_k (= \sum_{s \ni k} p(s)), \pi_{kk'} (= \sum_{s \ni k, k'} p(s))$, etc. The problem of estimating V under unequal probability sampling in the design-based setup has been considered earlier by Liu (1974), Chaudhuri (1978), Isaki (1983), and in the model-based approach by Mukhopadhyay (1978), Mukhopadhyay (1982), Mukhopadhyay (1984), Mukhopadhyay (1990), Rodrigues et al. (1985), and Mukhopadhyay and Bhattacharyya (1989, 1991), among others. We shall consider here estimation of V under the following linear models, based on random permutation models and probability models to include a wide class of one-sample randomized response designs as developed by Rao and Bellhouse (1978) and Bellhouse (1980), restricting ourselves to the class of fixed size n noninformative sampling designs $\rho_n = \{p : p(s) > o \Rightarrow n(s) = n\}$ throughout the study.

Assume that $r_j = y_j/w_j$ and w_j are unrelated. In this case, for a fixed $(w_1, \ldots, w_N)$, the sequence $R = (R_1, \ldots, R_N)$ where R_i denotes the realized value of the variate $r(= y/w)$ on i (for fixed w_i) may be looked upon as a random permutation of the set of unknown numbers $(r_1, \ldots, r_N)$. In this case, the following model (m) seems appropriate on y:

$$\frac{y_j}{w_j} = \bar{R} + f_j, \quad \text{where } \bar{R} = \frac{1}{N}\sum_j r_j;$$

$$\begin{aligned} &E_m(f_j) = 0, && E_m(f_j^2) = \delta_1 (>0), && E_m(f_j f_k) = \delta_2; \\ &E_m(f_j^3) = \delta_3, && E_m(f_j^2 f_k) = \delta_3', && E_m(f_j f_k f_l) = \delta_3''; \\ &E_m(f_j^4) = \delta_4, && E_m(f_j^3 f_k) = \delta_4', && E_m(f_j^2 f_k^2) = \delta_4''; \\ &E_m(f_j^2 f_k f_l) = \delta_4''', && E_m(f_j f_k f_l f_t) = \delta_4'''', && j \neq k \neq l \neq t = 1, \ldots, N. \end{aligned} \tag{2}$$

Assume that if $j \in s$ and if the respondent j answers truthfully about the value of y_j, the interviewer gets a measurement y_j^* (the value of random variable) with $E_r(y_j^* | y_j) = y_j, E_r$ denoting the expectation operator *wrt* the probability distribution of measurement error (ME). Assume further that y is a sensitive character (too sensitive to be measured directly) and the data are obtained from the sample through some randomized response technique (RRT). From $j \in s$, one therefore obtains a value z_j which is a random variable reflecting the randomization due to the use of RRT with $E_c(z_j | y_j^*)$, a function of y_j^*,E_c denoting the expectation operator *wrt* the probability distribution generated by the RRT. Combining the model (2) on y and the model due to the joint probability distribution of MEs and randomized responses, we assume, as in Bellhouse (1980), the following model on z:

$$\frac{z_j}{w_j} = \alpha + \beta \bar{R} + e_j,$$

$$\begin{array}{lll} \mathcal{E}(e_j) = 0, & \mathcal{E}(e_j^2) = \gamma_1 (>0), & \mathcal{E}(e_j e) = \gamma_2; \\ \mathcal{E}(e_j^3) = \gamma_3, & \mathcal{E}(e_j^2 e_k) = \gamma_3', & \mathcal{E}(e_j e_k e_l) = \gamma_3''; \\ \mathcal{E}(e_j^4) = \gamma_4, & \mathcal{E}(e_j^3 e_k) = \gamma_4', & \mathcal{E}(e_j^2 e_k^2) = \gamma_4''; \\ \mathcal{E}(e_j^2 e_k e_l) = \gamma_4''', & \mathcal{E}(e_j e_k e_l e_t) = \gamma_4^{iv}, & j \neq k \neq l \neq t = 1, \ldots, N, \end{array} \tag{3}$$

where α,β are some known constants and $\mathcal{E}(.)$ denotes the composite expectation $E_m E_r E_c$.

We shall consider the class of nonhomogeneous quadratic estimators of V, having the form

$$q_s = b_s + \sum_{j \in s} b(s,jj) z_j^2 + \sum_{j \neq j' \in s} b(s,jj') z_j z_{j'}, \tag{4}$$

where the coefficients b_s, $b(s,jj)$, and $b(s,jj')$ defined for all s and $j,j' \in s$ could be functions of concomitant variable w.

An estimator q_s is model unbiased (m-unbiased) for V if

$$\mathcal{E}(q_s) = E_m(V) = \sigma^2 (\text{say}) \ \forall \ \ s : p(s) > 0; \tag{5}$$

design-unbiased (p-unbiased) for V if

$$E_p E_r E_c(q_s) = V, \tag{6}$$

where E_p is the expectation *wrt* the sampling design $p(s)$; and pm-unbiased for V if

$$E_p \mathcal{E}(q_s) = \sigma^2. \tag{7}$$

Two measures of uncertainty of a strategy (p, q_s) are:

$$E_p \mathcal{E}(q_s - V)^2 = M_1 (\text{say})$$

and

$$E_p \mathcal{E}(q_s - \sigma^2)^2 = M_2 (\text{say}).$$

For a noninformative design, the order of the operators E_p and $\mathcal{E}$ [or $(E_r E_c)$] can be interchanged. An estimator $\tilde{q}$ in a Class is said to be optimal for V if for every q in this class

$$E_p \mathcal{E}(\tilde{q} - V)^2 \leq E_p \mathcal{E}(q - V)^2$$

if the measure of uncertainty considered is M_1 (or $E_p\mathcal{E}(\tilde{q}-\sigma^2)^2 \le E_p\mathcal{E}(q-\sigma^2)^2$ if the measure of uncertainty is M_2). If $\tilde{q}$ is optimal in the wider class (7) and also belongs to the subclass (5) or (6), it is also optimal in that subclass.

2 OPTIMAL ESTIMATION OF V

We shall first state a theorem on UMV estimation (Rao, 1973) in the present context.

Let C denote the class of pm-unbiased estimators of σ^2 and C_0 the corresponding class of pm-unbiased estimators t_s of zero, ie, $E_p\mathcal{E}(t_s)=0$.

Theorem 1. *An estimator* $\tilde{q}$ in C *is optimal for* σ^2 *iff for every estimator* $t \in C_0, E_p\mathcal{E}(\tilde{q}t=0)$.

Denoting the class of m-unbiased estimators (5), p-unbiased estimators (6), and pm-unbiased estimators (7) as Q_m, Q_p, and Q_{pm}, respectively, we have $q_s \in Q_m$ if

$$b_s=0, \{(\alpha+\beta\bar{R})^2+\gamma_1\}\sum_{j\in s} b(s,jj)w_j^2 = a_1(\bar{R}^2+\delta_1)\sum_{j=1}^{N} w_j^2$$

$$\{(\alpha+\beta\bar{R})^2+\gamma_2\}\sum\sum_{j\neq j'\in s} b(s,jj')w_jw_{j'} = -a_2(\bar{R}^2+\delta_2)\sum\sum_{j\neq j' 1}^{N} w_jw_{j'} \quad \forall s. \tag{8}$$

Similarly, $q_s \in Q_p$ if

$$E_p(b_s)=0, \quad \sum_1^N z_j^2 c_j = a_1\sum_1^N y_j^2, \quad \sum\sum_{j\neq j'=1}^{N} z_jz_{j'}c_{jj'} = -a_2\sum\sum_{j\neq j'=1}^{N} y_jy_{j'}, \tag{9}$$

where

$$c_j=\sum_{s\ni j} b(s,jj)p(s), \quad c_{jj'}=\sum_{s\ni j,j'} b(s,jj')p(s).$$

Again, $q_s \in Q_{pm}$ if

$$E_p(b_s)=0, \quad \{(\alpha_1+\beta\bar{R})^2+\gamma_1\}\sum_1^N w_j^2c_j = a_1(\bar{R}^2+\delta_1)\sum_1^N w_j^2$$

$$\{(\alpha_1+\beta\bar{R})^2+\gamma_2\}\sum\sum_{j\neq j'=1}^{N} w_jw_{j'}c_{jj'} = -a_2(\bar{R}+\delta_2)\sum\sum_{j\neq j'=1}^{N} w_jw_{j'}. \tag{10}$$

Let

$$t_s = d_s + \sum_{j\in s} d(s,jj)z_j^2 + \sum\sum_{j\neq j'\in s} d(s,jj')z_jz_{j'}$$

be a nonhomogeneous quadratic unbiased estimator of zero. Denoting by Q_m^0, Q_p^0, and Q_{pm}^0 the corresponding class of *m*-unbiased, *p*-unbiased, and *pm*-unbiased estimators t_s of zero, we have

$$t_s \in Q_m^0 \Leftrightarrow d_s = 0, \ \sum_{j \in s} d(s, jj) w_j^2 = 0,$$

$$\sum \sum_{j \neq j' \in s} d(s, jj') w_j w_{j'} = 0, \ \forall \ s; \tag{11}$$

$$t_s \in Q_p^0 \leftrightarrow E_p(d_s) = 0, \ e_j = 0, \ j = 1, \ldots, N$$

$$e_{jj'} = 0, \ j \neq j' = 1, \ldots, N, \tag{12}$$

where

$$e_j = \sum_{s \ni j} d(s, jj) p(s), \ e_{jj'} = \sum_{s \ni j, j'} d(s, jj') p(s);$$

$$t_s \in Q_{pm}^0 \Leftrightarrow E_p(d_s) = 0, \ \sum_{j=1}^N w_j^2 e_j = 0, \ \sum \sum_{j \neq j' = 1}^N w_j w_{j'} e_{jj'} = 0. \tag{13}$$

Note that

$$E_m(V) = \sigma^2 = a_1(\bar{R}^2 + \delta_1) \sum_1^N w_j^2 - a_2(\bar{R}^2 + \delta_2) \sum \sum_{j \neq j' = 1}^N w_j w_{j'}. \tag{14}$$

It is found that the choice $b_s = h_0$, $b(s, jj) = h_1 / w_j^2 \ \forall \ j \in s$ and all $s(\ni j)$, $b(s,jj') = h_2/w_j w_{j'} \forall j \neq j' \in s$ and all $s(\ni (j,j'))$ make $E_p \mathcal{E}(qt) = 0 \ \forall q \in Q_{pm}, t \in Q_{pm}^0$.

Substitution of these values in the *pm*-unbiasedness conditions (10) gives

$$h_0 = 0,$$

$$h_1 = \frac{a_1(\bar{R}^2 + \delta_1) \sum_{j=1}^N w_j^2}{n\{(\alpha_1 + \beta \bar{R})^2 + \gamma_1\}}, \tag{15}$$

$$h_2 = -\frac{a_2(\bar{R}^2 + \delta_2) \sum \sum_{j \neq j' = 1}^N w_j w_{j'}}{n(n-1)\{(\alpha + \beta \bar{R})^2 + \gamma_2\}}.$$

Hence

$$q_s^* = \frac{a_1(\bar{R}^2 + \delta_1) \sum_{j=1}^N w_j^2}{n\{(\alpha + \beta \bar{R})^2 + \gamma_1\}} \sum_{j \in s} \frac{z_j^2}{w_j^2} - \frac{a_2(\bar{R}^2 + \delta_2) \sum \sum_{j \neq j' = 1}^N w_j w_{j'}}{n(n-1)\{(\alpha + \beta \bar{R})^2 + \gamma_2\}} \sum \sum_{j \neq j' \in s} \frac{z_j z_{j'}}{w_j w_{j'}} \tag{16}$$

is optimal for σ^2 in the *pm*-unbiased class Q_{pm}.

We also note that

$$
\begin{aligned}
&E_p\mathcal{E}(q_s^* - \sigma^2)^2 \\
&= E_p\mathcal{V}\left[(h_1 + h_2)\sum_{j\in s} e_j^2 + 2(\alpha + \beta\bar{R})\{h_1 - (n-1)h_2\}\sum_{j\in s} e_j - h_2 \sum_{j\in s} e_j^2\right],
\end{aligned}
$$

where $\mathcal{V}(\theta) = \mathcal{E}(\theta - E\mathcal{E}(\theta))^2$ is a constant for all $p \in \rho_n$. Hence the following.

Theorem 2. *Under models (2) and (3), the estimator q_s^* (as given in 16) is optimal for σ^2 in the pm-unbiased class (10) for every fixed sampling design $p \in \rho_n$, ie,*

$$E_p\mathcal{E}(q_s^* - \sigma^2)^2 \le E_p\mathcal{E}(q_s - \sigma^2)^2 \quad \forall q_s \in Q_{pm}, p \in \rho_n.$$

Again, any $p \in \rho_n$ is optimal for using q_s^*.

Note that q_s^* is p-unbiased if

$$E_pE_rE_c(q_s^*) = V.$$

The joint sufficient condition for this to occur is

$$E_pE_c\left[\frac{a_1(\bar{R}^2 + \delta_1)}{\{(\alpha + \beta\bar{R})^2 + \gamma_1\}}\sum_1^N z_j^2 - \frac{a_2(\bar{R} + \delta_2)}{\{(\alpha + \beta\bar{R})^2 + \gamma_2\}}\sum\sum_{j\neq j'=1}^N z_j z_{j'}\right] = V \tag{17}$$

and $\pi_j = nw_j^2/\sum_1^N w_k^2 (j = 1,\ldots,N)$ and $\pi_{jj'} = n(n-1)w_jw_{j'}/\sum\sum_{k\neq k'=1}^N w_kw_{k'}$ $(j \neq j' = 1,\ldots,N)$. Because of the constancy conditions $\sum_{j'(\neq j)=1}^N \pi_{jj'} = (n-1)\pi_j$ and $\sum_1^N \pi_j = n$, the latter set of conditions is satisfied iff $w_1 = \cdots = w_N$. In this case

$$\pi_j = n/N, \quad \pi_{jj'} = n(n-1)/N(N-1), \quad j \neq j' = 1,\ldots,N \tag{18}$$

which are satisfied by srswor design. Thus, for $w_j = 1\ \forall j$, $q_s^* = q_s'^*$ is p-unbiased for V and is optimal for V (in the minimum M_2 sense) for any p with $\pi_j,\pi_{jj'}$ given in (18) provided the condition (17) is satisfied.
Noting that for $q_s \in Q_p$,

$$
\begin{aligned}
E_p\mathcal{E}(q_s - V)^2 &= E_p\mathcal{E}(q_s - \sigma^2)^2 - \mathcal{E}(V - \sigma^2)^2 \\
&\ge E_p\mathcal{E}(q_s^* - \sigma^2)^2 - \mathcal{E}(V - \sigma^2)^2 \\
&= E_p(q_s^* - V)^2
\end{aligned}
$$

when $q_s^* \in Q_p$, we have the following theorem.

Theorem 3. *Consider a randomized response design with $w_j = 1\forall j$ that yields (17) for some α, β, δ_1, δ_2, γ_1, γ_2. The average variance $E_p\mathcal{E}(q_s - V)^2$ of any* *p-unbiased quadratic estimator of V for any design $p \in \rho_n$ is never smaller than that of $q_s'^*$ for any $p_0 \in \rho_n$ with π_j, $\pi_{jj'}$ as given in (18), ie,*

$$E_{p_o}\mathcal{E}(q_s'^* - V)^2 \leq E_p\mathcal{E}(q_s - V)^2 \quad \forall q_s \in Q_p,\ p,\ p_0 \in \rho_n.$$

When $\tilde{w} \neq (\theta,\ldots,\theta)$, where θ is some arbitrary constant, there does not exist any p-unbiased general quadratic estimator of the form $\beta_0 + \sum_{j\in s}\beta_j z_j^2 + \sum_{j\neq j' \in s}\beta_{jj'} z_j z_{j'}$ which is optimal for V in the class Q_p, where $p \in \rho_n$, even if (17) holds.

Again, noting that q_s^* is m-unbiased for any sample of size n, we have the following theorem.

Theorem 4. *Under models (2) and (3), q_s^* is optimal m-unbiased for V in the class Q_m for any fixed $p \in \rho_n$, ie,*

$$E_p\mathcal{E}(q_s^* - \sigma^2)^2 \leq E_p\mathcal{E}(q_s - \sigma^2)^2, \quad q_s \in Q_m, p \in \rho_n.$$

We now drop the random permutation model m on y and consider only randomization due to ME and RR, the composite expectation $E_r E_c$ being denoted as $\mathcal{E}'$. Assume that

$$\mathcal{E}'(z_i) = \phi(y_i),$$

a known function of y_i and also that there exists a (known) one-to-one function $\psi(z_i) = t_i$ such that

$$\mathcal{E}'[\psi(z_i)] = y_i, \quad i = 1,\ldots,N.$$

Let $\tilde{t} = (t_1,\ldots,t_N)$. An estimator $e = e(s,\tilde{t})$ is a function which depends on $\tilde{t}$ only through $i \in s$. Consider the class of estimators of V,

$$C: \{e(s,\tilde{t}) : E_p\mathcal{E}'(e(s,t)) = V(\tilde{y}) \ \ \forall \tilde{y} \in R_N\}.$$

Let V_{prc} denote the variance operator wrt combined randomization due to p, ME, and RR, and $\mathcal{V}'$ wrt ME and RR.

Theorem 5. *Among estimators in C based on any sampling design p, there does not exist any uniformly minimum variance estimator (UMVU) of V.*

Proof. Let $\tilde{y}_0 = (Y_{10},\ldots,Y_{N0})$ be a point in R_N with $\tilde{t}_0 = (t_{10},\ldots,t_{N0})$ as the consequent value of $\tilde{t}$ emerging from a survey. If possible, let $\bar{e}(s,\tilde{t})$ be a UMV estimator in C. Let

$$e'(s,\tilde{t}) = \bar{e}(s,\tilde{t}) - \mathcal{E}'[\bar{e}(s,\tilde{t}_0)] + V(\tilde{y}_0).$$

Clearly, $e'(s,\tilde{t}) \in C$. Again,

$$V_{prc}(\bar{e}(s,\tilde{t}_0)) = E_p\mathcal{V}'(\bar{e}(s,\tilde{t}_0)) + V_p\mathcal{E}'(\bar{e}(s,\tilde{t}_0))$$
$$> E_p\mathcal{V}'(\bar{e}(s,\tilde{t}_0)) = V_{prc}(e'(s,\tilde{t}_0)),$$

the two sides being equal *iff* $\mathcal{E}'(\bar{e}(s,\tilde{t}_0)) = E_p\mathcal{E}'(\bar{e}(s,\tilde{t}_0)) = V(\tilde{y}_0) \ \ \forall s : p(s) > 0$, which holds *iff* p is a census design. Hence the theorem. □

3 UMVU ESTIMATION

We have seen in Theorem 5 that if the random permutation model is dropped, there does not exist any UMVU for V in the class of all design model unbiased estimators, models being randomized response model and ME model as summarized in (2) and (3) and the sampling design being any fixed-size non-informative sampling design. However, under slightly different set-up a UMVU estimator exists, as we shall currently show.

We assume that the sampling design used in selecting s is an ordered design such that

(i) each s is a sequence of n distinct labels and has nominal size n;
(ii) all $n!$ sequences that are permutations of the same set of labels have equal probability of being selected;
(iii) $\pi_{ij} > 0 \ \forall \ i \neq j = 1,\ldots, N$.

Let $\mathcal{S} = \{K_1,\ldots,K_n\}$ be a random vector taking values $s = (k_1,\ldots,k_n)$ where K_i is a random variable denoting the label of the unit selected at the ith draw in $\mathcal{S}$. Data collected from the survey are $\mathcal{D} = ((K_1, z_{K_1}),\ldots,(K_n, z_{K_n}))$ available through two types of randomization, one due to sampling design and the other due to RRT. An estimator $e = e(s,\tilde{z})$ is a function defined on $\mathcal{S} \times R_N$ such that for a given s its value depends on $\tilde{z}(=(z_1,\ldots,z_N))$ only through $i \in s$. An estimator of t is unbiased for V if

$$E_pE_c(t(s,\tilde{z})) = V(\tilde{y}) \ \ \forall \tilde{y} \in R_N. \tag{19}$$

Following Eriksson (1973), we shall consider here a slightly different RR model. When the ith individual (with true value y_i) is selected in the sample, he chooses a quantity at random from the set $(a_1,\ldots,a_{M_i i})$ and a quantity b_i at random from the set $(b_{1i},\ldots,b_{Q_i i})$ and responds

$$z_i = a_i y_i + b_i. \tag{20}$$

It can be seen that

$$E_c(r_i) = y_i, \ \ E_c(R_i^2) = y_i^2, \ \ E_c(R_{ij}) = y_i y_j, \tag{21}$$

where

$$r_i = \frac{z_i - \bar{b}_i}{\bar{a}_i}$$

$$R_i^2 = \frac{z_i^2 - \tilde{b}_i - 2\bar{a}_i\bar{b}_i r_i}{\tilde{a}_i}$$

$$R_{ij} = \frac{z_i z_j - \bar{a}_i\bar{b}_j r_i - \bar{a}_j\bar{b}_i r_j - \bar{b}_i\bar{b}_j}{\tilde{a}_i\tilde{a}_j}$$

$$\bar{a}_i = \frac{1}{M_i}\sum_{j=1}^{M_i} a_{ji}, \quad \bar{b}_i = \frac{1}{Q_i}\sum_{j=1}^{Q_i} b_{ji},$$

$$\tilde{a}_i = \frac{1}{M_i}\sum_{j=1}^{M_i} a_{ji}^2, \quad \tilde{b}_i = \frac{1}{Q_i}\sum_{j=1}^{Q_i} b_{ji}^2.$$

Let $\tilde{e} = (e_1, \ldots, e_N) \in R_N$ be a vector of known constants. For a given sample $s = (i_i, \ldots, i_n)$ define the quantities

$$\frac{(R_{ik}^2 - e_{ik}^2)}{n\pi_{ik}} = T_{ik}^2 \quad (k = 1, \ldots, n)$$

$$\frac{(R_{i_k i_l} - e_{i_k} e_{i_l})}{n(n-1)\pi_{i_k i_l}} = T_{i_k i_l} \quad (k \neq l = 1, \ldots, n)$$

and the set of unlabeled data

$$T_s = (T_{i1}^2, \ldots, T_{in}^2, T_{i_1 i_2}, \ldots, T_{i_{n-1} i_n} | \text{labels dropped}).$$

Clearly the members of T_s may not be distinct. We shall consider the class of estimators that depend on $\mathcal{D}$ only through T's and call this class of estimator as Δ.

Lemma 1. *Let p be any fixed effective size sampling design. If* $\bar{a}_i (\neq 0), \bar{b}_i, \tilde{b}_i$ *are known constants and* $g(T_s)$ *is a real valued statistic such that*

$$E_p E_c\{g(T_s)\} = 0$$

for every RR model (19) and every $\tilde{y} \in R_N$, *then* $g(T_s) = 0$ *identically.*

Proof. The lemma follows from 3.7 of Cassel et al. (1977) whose proof is similar to a proof in Das (1962) and theorem 2.1 of Adhikari et al. (1984) (which can be easily extended to Eriksson's model 19). □

We now have the following theorem.

Theorem 6. *Under the RR models satisfying (19), for any ordered design satisfying conditions (i)–(iii) and for any known constants* $\tilde{e} \in R_N$, *the generalized difference estimator of population variance*

$$\begin{aligned} t_{GD}(\tilde{e}) &= a_1 \sum_{i \in s} T_i^2 - a_2 \sum\sum_{i \neq j \in s} T_{ij} + V(\tilde{e}) \\ &= w(T_s) + V(\tilde{e}) \end{aligned} \tag{22}$$

is UMV in the class of all unbiased estimators of $V(\tilde{y})$ *in* Δ.

Proof. It follows by virtue of assumptions (i)–(iii) on the class of sampling designs that all the random variables in $(T_{i1}^2, \ldots, T_{i_n}^2)$ have an exchangeable joint distribution with common expectation $\sum_{i=1}^N (y_i^2 - e_i^2)$. Similarly, for all permutations of $(i_1, \ldots, i_n)$ the random variables in $(T_{i_1 i_2}, \ldots, T_{i_{n-1} i_n})$ have the same joint distribution with common expectation $\sum\sum_{i \neq j=1}^N (y_i y_j - e_i e_j)$. Hence $E_p E_c(t_{GD}(\tilde{e})) = V(\tilde{y}) - V(\tilde{e})$. Now, it follows from lemma 3.7 of Cassel et al. (1977) that MVUE in the case of all unbiased estimators of $V(\tilde{y}) - V(\tilde{e})$ that depend on $\mathcal{D}$ only through T_s must be symmetric in its components. This combined with Lemma 1 states that $w(T_s)$ is the unique unbiased estimate of $V(\tilde{y}) - V(\tilde{e})$. Hence the result. □

REFERENCES

Adhikari, A.K., Chaudhuri, A., Vijayan, K., 1984. Optimum sampling strategies for randomized response trials. Int. Stat. Rev. 52, 115–129.

Bellhouse, D.R., 1980. Linear models for randomized response designs. J. Am. Stat. Assoc. 75, 1001–1004.

Cassel, C.M., Sarndal, C.E., Wretman, J.H., 1977. Foundations of Inference in Survey Sampling. John Wiley, New York.

Chaudhuri, A., 1978. On estimating the variance of a finite population. Metrika 25, 65–76.

Das, A.C., 1962. On MVU estimates of parameters of a nite population based on varying probability sampling. Cal. Statist. Assoc. Bull. 11, 39–48.

Eriksson, S.A., 1973. A new model for randomized response. Int. Stat. Rev. 41, 101–103.

Isaki, C.T., 1983. Variance estimation using auxiliary information. J. Am. Stat. Assoc. 78, 117–123.

Liu, T.P., 1974. A general unbiased estimator for the variance of a finite population. Sankhya 36, 23–52.

Mukhopadhyay, P., 1978. Estimating the variance of a finite population under a superpopulation model. Metrika 25, 115–122.

Mukhopadhyay, P., 1982. Optimum strategies for estimating the variance of a finite population under a superpopulation model. Metrika 29, 143–158.

Mukhopadhyay, P., 1984. Estimating the variance of a finite population under generalized random permutation models. Calcutta Stat. Assoc. Bull. 33, 93–104.

Mukhopadhyay, P., 1990. Asymptotic properties of a generalized predictor of a finite population variance under probability sampling. Sankhya B 52, 343–346.

Mukhopadhyay, P., Bhattacharyya, S., 1989. On estimating the variance of a finite population under a superpopulation model. J. Ind. Stat. Assoc. 27, 37–46.

Mukhopadhyay, P., Bhattacharyya, S., 1991. Estimating a finite population variance under general linear models with exchangeable errors. Calcutta Stat. Assoc. Bull. 40, 219–228.

Rao, C.R., 1973. Linear Statistical Inference and Its Applications. John Wiley, New York.

Rao, J.N.K., Bellhouse, D.R., 1978. Optimal estimation of a finite population mean under generalized random permutation models. J. Stat. Plan. Inference 2, 125–141.

Rodrigues, J., Bolfarine, H., Rogatko, A., 1985. A general theory of prediction in finite populations. Int. Stat. Rev. 3, 239–254.

Chapter 15

Randomized Response and New Thoughts on Politz–Simmons Technique

T.J. Rao[*,1], **J. Sarkar**[†] **and B.K. Sinha**[‡]
[*]*CRRao Advanced Institute of Mathematics, Statistics and Computer Science, Hyderabad, Telangana, India*
[†]*Indiana University–Purdue University Indianapolis, IN, United States*
[‡]*Retired Professor, Indian Statistical Institute, Kolkata, India*
[1]*Correponding author: e-mail: tjrao@hotmail.com*

ABSTRACT

An important source of nonsampling errors in sample surveys is "Not-at-Home's" for which Politz–Simmons technique [PST] is used to estimate parameters using data on first call itself, thus avoiding "call-backs," by asking respondents during the interview a question about their availability at home (or, otherwise) at the same time during the preceding five "week nights." In the present day context of ever-changing socioeconomic scenario of the society, it is to be noted that this question itself has become highly sensitive for the respondents who thereby may evade to answer this question truthfully. We have applied Warner's randomized response technique (RRT) in this situation and developed a nontrivial randomized response PST.

Keywords: Warner's randomized response technique, Not-at-Home's, Call-backs, Politz–Simmons technique, Randomized Response Hartley–Politz–Simmons technique

1 INTRODUCTION

Nonresponse in sample surveys is a major problem. Two of the main sources of nonresponse are evasion of answers to questions of a sensitive nature, and "Not-at-Home's." The technique of "call-backs" was proposed to survey the units [persons] corresponding to "not-at-home" category. Early attempts to tackle this type of nonresponse may be attributed to Yates (1946) and Hansen and Hurwitz (1946) who introduced the methodology for call-back procedures. In the latter work, the population is divided into *response stratum* where the units [persons] do respond and the *nonresponse stratum* in which the units do not respond in the

Handbook of Statistics, Vol. 34. http://dx.doi.org/10.1016/bs.host.2016.01.013

first attempt. Thereafter, a subset of nonrespondents is used for estimating the required population parameters.Deming (1953) considered a mathematical model for call-backs dealing with cost considerations as well.

In the discussion of the paper read by Yates (1946) at the Royal Statistical Society, Hartley (1946) suggested an "ingenious" alternative which is "decidedly cheaper" than call-backs. Later in the *Journal of American Statistical Association*, Politz and Simmons (1949) published a technique which is very similar to Hartley's suggestion to estimate parameters based on data at first call itself, thus avoiding call-backs. As discussed later in this paper, the technique employed by Politz and Simmons involves a question which is itself sensitive in the present day context, thus making the respondent evade an answer—thereby contributing to nonresponse or to wilful wrong response. In this paper we shall apply RRT (Warner, 1965) for this scenario and develop a nontrivial randomized response Hartley–Politz–Simmons (RR-HPS) technique.

We will take it for granted that the readers are familiar with Warner's RRT and its generalizations as documented in standard text books, viz., Chaudhuri and Mukerjee (1988) and Hedayat and Sinha (1991) (see also Chaudhuri, 2011; Chaudhuri and Christofides, 2013). We will mainly focus on Politz–Simmons Scheme and proceed from there.

2 NOT-AT-HOME'S

A source of nonresponse as we mentioned in Section 1 is "Not-at-Home's." This group of individuals are temporarily absent at home, either at work or on a short holiday or visit. During the 1950s much attention was given to the number of call-backs and the relative costs that are involved. Some of the early attempts of call-back technique include the works of Kish and Hess (1959) and Stephan and McCarthy (1958), among others. A detailed analysis of relative costs for first, second, and third call-backs of a UK Survey by Durbin and Stuart (1954) and Deming's model for call-backs dealing with costs is given in Cochran's (1977) textbook.

2.1 Randomized Response Hartley–Politz–Simmons [RR-HPS] Technique

As mentioned earlier, an "ingenious" alternative to call-backs which is "decidedly cheaper" than call-backs was first put forward by Hartley (1946). Hartley mentions: "Details of this scheme were given to the War-time Social Survey, but I understand that, owing to pressure of work, an opportunity of trying has, as yet, not arisen." Soon after Politz and Simmons (1949) published their work popularly known as Politz–Simmons technique in the *Journal of American Statistical Association* which is on similar lines to the suggestions of Hartley, Politz and Simmons (1949), while acknowledging the work of Hartley, mention thus: "During the past three years we have developed a plan for eliminating the need for call-backs and several experiments have been made applying this plan to market surveys."

Their technique consists of grouping the results [what Hartley calls as strata] of first call as follows: Assuming all calls are made at a particular time in the evening of six week nights, they proceed to question the respondent if he was at home at the same time on each of the five preceding week nights. If i is the number of nights the person was present at home, out of tonight, and the previous five nights, then $i/6$ gives an estimate of the probability f of the individual being at home at this time on a randomly chosen night. In an immediately following paper, Politz and Simmons (1950) discuss the difference in the effects of clustering in using the "nights-at-home" plan. Let n_i be the number of contacts who report being at home exactly i nights, and $\bar{y}_i$ be the mean of the study variable of interest for strata $i = 1, 2, \ldots, 6$. Noticing that the first call answers are mostly from those persons who are at home most of the time, the response is given a weight inversely proportional to f. Thus Politz–Simmons estimate for the population mean is given by

$$\bar{y} = \frac{\sum_{i=1}^{6} n_i \bar{y}_i/(i/6)}{\sum_{i=1}^{6} n_i/(i/6)} = \frac{\sum_{i=1}^{6} n_i \bar{y}_i/i}{\sum_{i=1}^{6} n_i/i}. \quad (1)$$

Here the question placed before the respondent is about his/her presence at home at the same time of the interview during the preceding "five week nights." The respondents may worry about such an enquiry thinking that a burglar posing as an investigator is trying to get information on their frequency of presence at home, or a detective is hired in a divorce case, or a relation is keeping tabs on their personal activities, and so on. For security reasons, personal privacy, family interactions, and similar situations, respondents may not be willing to answer such a question regarding their availability "at home," and try to evade the answer or give a wrong answer. Thus we note that this question is sensitive in the present day context of ever changing society. For an operational definition of sensitivity and its relationship to context with particular reference to identity, see Barnett (1998). When dealing with a sensitive question, the Response HPS technique may fail to give a correct result. Therefore, we will superimpose a RR approach on this, and discuss a Randomized Response HPS technique as suggested in Rao (2014).

Let i be the number of nights the respondent was at home [out of 6 nights including the night of the survey]. Clearly, $i = 0, 1, \ldots, 6$. Suppose that $P(i)$ denotes the true population proportion of households (HHs) who are at home on i nights. Ask the respondent to throw a regular 6-faced die. Instruct the respondent to say "Yes" if the number appearing on the face thrown is the same as i; otherwise, say "No." Let X be the number of successes ("Yes" responses) among the n respondents. Then X follows a binomial distribution with success probability $\phi = \sum_{i=1}^{6} P(i)/6 < 1/6$. Let $P(X = m) = \pi_m$, according to the binomial distribution (n, ϕ). For a study variable y, let $\bar{Y}$ denote the population mean to be estimated from a sample of size n. One possible estimate is:

$$t_1 = [m\, t_m + (n - m)\, t_{n-m}]/n; \tag{2}$$

where t_m is the mean of y for the m "yes" respondents and t_{n-m} is the mean of y for the $(n - m)$ "no" respondents. That is, t_1 is the overall sample mean, irrespective of the yes/no responses. Other possible estimates are:

$$t_2 = [w_1\, m\, t_m + (1 - w_1)(n - m)\, t_{n-m}]/n, \tag{3}$$

where $w_1 = \pi_m/[\pi_m + \pi_{n-m}]$;

$$t_3 = \frac{[m\, t_m/\pi_m + (n - m)t_{n-m}/\pi_{n-m}]}{n[1/\pi_m + 1/\pi_{n-m}]}. \tag{4}$$

Naturally, in the above, $\hat{\phi} = X/n$ and it serves as a "plug-in estimator" in computing the π_m's in the above formulas. However, the above technique does not give different relative weights to different number of "week nights at homes" because of the uniform distribution of the six faces of the die. In other words, we are not in a position to estimate the individual $P(i)$'s from the knowledge of X.

In the following section we propose to develop a new methodology to circumvent this problem and illustrate the same with an experimental data.

2.2 A New RR-HPS Technique

The point to note is that a simple-minded RRT will not provide desired results towards reasonable estimation of different "not-at-home" class probabilities. We defer the theoretical result to the Appendix and, in what follows, we take up the working methodology for a new technique in detail.

A random listing of n households (HHs) out of N HHs in a certain population is available according to SRSWOR method of sampling. Upon investigation, it is found that of these n HHs, n_a are not at home. Among the rest who are present on this day of the survey, there are all kinds of HHs in terms of the exact number of week nights they are at home at this hour during the immediately preceding five week nights.

Direst Question (DQ): How many week nights did you stay at home? [Please count today, plus the nights you stayed at home during the last 5 week nights.] A [truthful] answer would have been an integer between 1 and 6 (both inclusive). Thereby, we would have a frequency distribution of the responses from the respondents who are available at this particular time during week nights—varying between 1 and 6. However, this mode of data collection by asking a direct question is not within the purview of the study, as per our contention that the direct question is sensitive as explained earlier.

Indirect Response: Give each respondent a worksheet (see Appendix) that has a randomized sequence of capital letters A,B,C,D and another sequence of lower case letters w,x,y,z. These sequences are generated according to simple random sampling with replacement. The respondent need not know the percentage distribution of these letters in the worksheet. Each respondent is instructed to report two counting numbers ($T =$ ___, $t =$ ___) as follows:

You identify your truthful answer to DQ as posed above, but do not disclose it to anybody. If it is one of 1,2,3, your target upper case letter is A; otherwise, your target upper case letter is B. From a random starting point (where the row and the column are determined by rolling a fair die twice) in the array of printed Capital Letters, mark off 10 occurrences of your target capital letter. Report $T =$ # letters (in the sequence containing letters A,B,C,D) covered thus far to get your target capital letter exactly 10 times. [Note: You stop counting as soon as you find the 10th appearance of your target letter].

If your truthful answer to DQ as posed above is 1 *OR* 4, your target lower case letter is w; if your answer is 2 *OR* 5 your target lower case letter is x; if your answer is 3 *OR* 6, your target lower case letter is y. From a random starting point (again, the row and the column are determined by rolling a fair die twice more) on the printed array of lower case letters, mark off 10 occurrences of your target lower case letter. Report $t =$ # letters (in the sequence containing letters w,x,y,z) covered thus far to get your target lower case letter exactly 10 times.

The respondents must make sure that their TRUE responses [in terms of number of week nights they are at home] correspond exactly to the [Capital Letter, lower case letter] pairs as shown on the worksheet. For example, in Version 1, the correspondence is as follows: $1 = [A,w]$;$2 = [A,x]$; $3 = [A,y]$; $4 = [B,w]$; $5 = [B,x]$; $6 = [B,y]$.

Every respondent is supposed to adopt the RRT as explained above and provide a [truthful] response in terms of the response pair (T, t), without disclosing his/her true classification in terms of the actual number of week nights staying at home OR his/her identification in terms of [Capital Letter, lower case letter] pair so determined by him/her before adopting the RRT.

Conversion Rule: How to convert the reported randomized responses into a numerical answer to fit into a response under DQ? We summarize the conversion rule here. See details in the Appendix.

1. First, we determine the respondent's Capital and lower case classes using these rules:

   Rule 1: If $T \leq 27$, Capital Class is A; and if $T \geq 28$, Capital Class is B;

   Rule 2: If $t \leq 27$, lower class $= w$; if $27 < t \leq 72$, lower class $= x$; and if $72 < t$, lower class $= y$.

2. Then, we convert the classes into numerical value using the particular version used by the respondent. For example, in Version 1 the conversion is as follows: $[A,w]=1;[A,x]=2$; $[A,y]=3$; $[B,w]=4$; $[B,x]=5$; $[B,y]=6$.

Note that no respondent chooses as target class levels C, D, or z, nor is any respondent placed into such classes. These levels are used in the worksheet to maintain the probabilities of the other levels. Not to obscure the essential steps of reasoning, we proceed through the following steps:

Step 0: We give a general instruction to all the available respondents on the purpose and use of RRT involving this specific sensitive issue. This can be done collectively or separately, preferably in small groups. The RRT comprises of a working sheet with two sequences of letters as explained above. Furthermore, everybody can see at the top of the worksheet, a CHART to determine the target letter codes as mentioned above. In fact, there are six different versions of the correspondence CHART so prepared. See details in the Appendix. This knowledge is not essential for the respondents.

We are now ready to go ahead and adopt the RRT so explained.

Step I: Randomly divide the available HHs into approximately equal numbers in each of the six versions of the worksheet. This is not essential for our theory to work. But it is a good safeguard in case respondents might discuss matters with one another afterwards.

Step II: To each respondent in the j-th group of HHs, give "Version j" of the CHART prepared beforehand ($1 \leq j \leq 6$) and ask him/her to follow the instructions of the RRT and report the individual (T,t) pair of numbers, *without* disclosing his/her identity in terms of classification involving [Capital Letter, lower case letter] combination. The version number is already printed on the worksheet.

Step III: Collect the data and analyze according to the classification rule laid down above.

Step IV: Estimate the category-specific proportions. Obtain point estimates and interval estimates of individual category-specific proportions and also simultaneous inference in terms of confidence ellipsoid. Furthermore, some related meaningful inference questions will be addressed. All these inferences are demonstrated with reference to an illustrative example.

Example 1. We take a random sample of $N=300$ subjects. Among them, it turns out that 150 are not at home on the night of the survey. The other 150, if they had responded directly, would have fallen into types 1–6 with an assumed breakdown of $(4,8,32,40,36,30)$. We must reiterate that here we are dealing with a length-biased sampling. That is, units having a large response value to the direct question are proportionally more likely to have been included in the sample than units having a small response value. Therefore, the realized sample proportions of different numbers of week night "stay-at-home's" are not unbiased estimates of the corresponding true population stay-at-home's. In fact, under SRSWOR, the realized sample proportions $[p(0);p(1),p(2),\ldots,p(6)]$ are related to the unbiased estimates $[\hat{P}(0);\hat{P}(1),\hat{P}(2),\ldots,\hat{P}(6)]$, of the corresponding true population probabilities in these categories $[P(0);P(1),P(2),\ldots,P(6)]$, through linear equations $p(i)=\hat{P}(i)\,i/6$ for $1\le i\le 6$, and $p(0)=\hat{P}(0)+\sum_{i=1}^{6}\hat{P}(i)\,(1-i/6)$. Inverting these equations, we get

$$\hat{P}(i)=6\,\frac{p(i)}{i},1\le i\le 6;\ \text{ and }\ \hat{P}(0)=p(0)-\sum_{i=1}^{6}(6-i)\,\frac{p(i)}{i}. \tag{5}$$

It easily follows that $\sum_{i=0}^{6}\hat{P}(i)=1$ if and only if $\sum_{i=0}^{6}p(i)=1$. In Example 1, the realized sample proportions are (0.500; 0.013, 0.027, 0.107, 0.133, 0.120, 0.100). Consequently, a point estimate of the true vector of population proportions is (0.183; 0.080, 0.080, 0.213, 0.200, 0.144, 0.100), using (5). In particular, the first entry 0.183 corresponds to an unbiased estimate $\hat{P}(0)$ of $P(0)$, the proportion of people "never at home on week nights," given in (5).

On the other hand, instead of giving direct response if the respondents had provided randomized responses (T,t) following the instructions above, we would have classified them into types 1–6 according to our Conversion Rule. The breakdown p-vector would have been a random outcome, from which we could calculate the point estimate of P-vector using (5).

We carried out an application of the above-described RR-HPS scheme and derived a classification of the 150 HHs based on the above decision rules. Here are the detailed results:

1. First we randomly assign serial numbers to all the 150 respondents who are present on the night of the survey.
2. The breakdown (4, 8, 32, 40, 36, 30) gives the number of respondents possessing letter combinations $[A,w],[A,x],[A,y],[B,w],[B,x],[B,y]$, respectively. That is, all respondents use the same Version 1 of the Randomized Response Worksheet. The respondents are supposed to adhere to these specifications. Then we randomize the order of the respondents and assign them new serial numbers. The interviewer is totally unaware of the breakdown or the randomization.
3. The responses provided by the respondents and the corresponding classification by applying the Conversion Rule are given in Table 1. A summary of misclassifications is given in Table 2.

TABLE 1 Responses from 150 Respondents and Their Classification into Types 1–6 Using Our Conversion Rule, Compared Against Their True Types (Misclassified Respondents Are Marked with an "X")

No.	T	t	in	from	mis	No.	T	t	in	from	mis	No.	T	t	in	from	mis
1	40	24	4	4	.	51	86	238	6	6	.	101	33	16	4	4	.
2	50	178	6	6	.	52	42	37	5	4	X	102	64	22	4	4	.
3	13	146	3	3	.	53	24	53	2	2	.	103	47	149	6	6	.
4	65	59	5	5	.	54	47	49	5	5	.	104	57	20	4	4	.
5	27	160	3	3	.	55	50	17	4	4	.	105	55	26	4	4	.
6	13	135	3	3	.	56	49	37	5	5	.	106	22	136	3	3	.
7	24	140	3	3	.	57	59	46	5	5	.	107	56	133	6	6	.
8	70	23	4	4	.	58	57	22	4	4	.	108	78	22	4	4	.
9	55	107	6	6	.	59	37	20	4	4	.	109	55	43	5	5	.
10	31	37	5	5	.	60	16	91	3	3	.	110	35	62	5	5	.
11	21	206	3	3	.	61	26	162	3	3	.	111	42	27	4	4	.
12	20	23	1	1	.	62	57	223	6	6	.	112	20	48	2	2	.
13	13	184	3	3	.	63	14	90	3	3	.	113	60	103	6	6	.
14	46	60	5	5	.	64	60	153	6	6	.	114	15	134	3	3	.
15	38	94	6	6	.	65	40	199	6	6	.	115	22	121	3	3	.
16	13	110	3	3	.	66	31	25	4	4	.	116	48	31	5	5	.
17	45	36	5	6	X	67	22	85	3	3	.	117	28	16	4	4	.
18	18	112	3	3	.	68	22	63	2	2	.	118	37	38	5	5	.
19	51	23	4	4	.	69	52	46	5	5	.	119	68	106	6	6	.
20	22	64	2	2	.	70	47	12	4	4	.	120	56	137	6	6	.
21	46	55	5	5	.	71	65	73	6	5	X	121	66	14	4	4	.
22	57	16	4	4	.	72	50	172	6	6	.	122	36	21	4	4	.
23	65	22	4	4	.	73	53	56	5	5	.	123	67	165	6	6	.

24	35	120	6	6	.	74	17	122	3	3	.	124	66	52	5	5	.
25	23	21	1	4	X	75	46	14	4	4	.	125	28	198	6	3	X
26	40	99	6	6	.	76	70	54	5	5	.	126	46	64	5	5	.
27	25	148	3	3	.	77	29	34	5	5	.	127	22	80	3	3	.
28	17	23	1	1	.	78	22	171	3	3	.	128	27	60	2	5	X
29	18	56	2	2	.	79	39	36	5	5	.	129	22	122	3	3	.
30	50	102	6	6	.	80	20	125	3	3	.	130	34	17	4	4	.
31	55	33	5	5	.	81	58	252	6	6	.	131	61	69	5	5	.
32	19	90	3	3	.	82	31	177	6	6	.	132	47	199	6	6	.
33	16	95	3	3	.	83	52	43	5	5	.	133	38	20	4	4	.
34	63	23	4	4	.	84	22	18	1	1	.	134	65	93	6	6	.
35	39	22	4	4	.	85	42	21	4	4	.	135	51	67	5	5	.
36	19	99	3	3	.	86	61	20	4	5	X	136	15	115	3	3	.
37	18	143	3	3	.	87	41	21	4	4	.	137	43	23	4	4	.
38	81	16	4	4	.	88	44	47	5	5	.	138	51	36	5	5	.
39	62	26	4	4	.	89	64	25	4	4	.	139	20	140	3	3	.
40	34	148	6	6	.	90	33	26	4	4	.	140	16	133	3	3	.
41	59	58	5	5	.	91	30	31	5	5	.	141	14	98	3	3	.
42	58	114	6	6	.	92	38	108	6	6	.	142	27	56	2	5	X
43	60	46	5	5	.	93	62	16	4	4	.	143	62	28	5	4	X
44	39	103	6	6	.	94	22	62	2	2	.	144	54	23	4	4	.
45	35	176	6	6	.	95	15	142	3	3	.	145	23	126	3	3	.
46	59	15	4	4	.	96	53	18	4	4	.	146	37	97	6	6	.
47	39	99	6	5	X	97	16	41	2	2	.	147	85	92	6	6	.
48	59	51	5	5	.	98	28	20	4	4	.	148	60	40	5	5	.
49	42	19	4	4	.	99	19	70	2	2	.	149	48	106	6	6	.
50	29	22	4	1	X	100	44	52	5	5	.	150	61	59	5	5	.

TABLE 2 Our Conversion Rule Gave Good Estimates of the True Proportions, Though It Misclassified 11/150=7.3% of the Respondents

	True category assumed							Classified	
Using (T, t) classify respondents into	*0*	*1*	*2*	*3*	*4*	*5*	*6*	*Sum*	p-*vec*
0	150							150	0.500
1		3	.	.	1	.	.	4	0.013
2		.	8	.	.	2	.	10	0.033
3		.	.	31	.	.	.	31	0.103
4		1	.	.	37	1	.	39	0.130
5		.	.	.	2	31	1	34	0.113
6		.	.	1	.	2	29	32	0.107
Under direct response									
Realized freq.	150	4	8	32	40	36	30	300	1.000
Realized *p*-vector	0.500	0.013	0.027	0.107	0.133	0.120	0.100	1.000	
Est. true P-vector									
by direct response	0.183	0.080	0.080	0.213	0.200	0.144	0.100	1.000	
by RR-HPS	0.176	0.080	0.100	0.207	0.195	0.136	0.107	1.000	

Do the observed frequencies of classified respondents across types 1–6 match the assumed true frequencies? A chi-square test of goodness-of-fit yields: $\chi^2 = 0.801$ with $df = 5$ and p-value $= 0.977$. The fit is deemed satisfactory.

The Classification Rule laid down above is based on the supposition that each respondent will provide a response in the form $[T, t]$ based on 10 appearances of the target letters [both Capital and lower case]. Consequently, the actual values of $[T, t]$ may turn out to be quite large. We have endeavored to make the task of counting as simple as possible by printing on Randomized Response Worksheet the random sequences of [both Capital and lower case] letters in rows containing 50 letters segmented into 10 blocks of five letters each. The worksheet is given in the Appendix.

Note that we started with the decomposition $n = n_a + n_p = 150 + 150$ and an assumed split of 150 available and responding HHs into 6 response categories. We applied the prescribed RR-HPS scheme once and found an observed categorization of the 150 responding HHs—again into 6 categories. To quantify the amount of variation involved in this categorization, we took this observed categorization as the "stipulated distribution" of the 150 HHs, carried out 100 repetitions of the RR-HPS scheme on this data set, and arrived at 100 estimates of the stipulated distribution. Using these repeated estimates we calculated means, variances, and covariances of the resulting bootstrap sampling distribution shown in Tables 3–6. Table 3 further shows that the stipulated frequency in each class is reasonably close to the true frequency in terms of t-statistic.

From Table 4, we observe that the true population P-proportions differ from the estimated proportions (which were used for simulation) by less than twice the corresponding standard deviation of the bootstrap samples. Of course, the estimates are themselves correlated because their sum is a constant. The covariance matrices of estimated frequencies and estimated P-proportions are displayed in Tables 5 and 6, respectively.

It is a routine task to construct confidence intervals and confidence ellipsoid for the stipulated proportions in different categories. This is based on variances and covariances as computed from the bootstrap sample and displayed earlier. We skip this part. Instead, we ask and resolve two related issues:

(i) At least what percentage of HHs are at home three or more of the six week nights? Computations suggest a mean of 62.6%, with a standard deviation of 1.21%, and hence a one-sided lower confidence limit of 60.6% with a 95% confidence level.

(ii) At most what percentage of people are at home never or on only one of the six week nights? Again, we find an estimated mean of 26.1%, with a standard deviation of 1.41%, and hence a one-sided upper confidence limit of 28.3% with a 95% confidence level.

TABLE 3 Summary of Estimates of Class Frequencies Based on 100 Bootstrap Samples

Class	Med	Mean	StDe	LB_95%	UB_95%	Simu.	True	*t*-stat
1	5	5.09	1.29	2.56	7.62	4	4	0.00
2	11	11.40	2.06	7.36	15.44	10	8	– 0.97
3	29	28.92	1.99	25.02	32.82	31	32	0.50
4	37	36.62	2.13	32.45	40.79	39	40	0.47
5	35	34.67	3.37	28.06	41.28	34	36	0.59
6	33	33.30	2.38	28.64	37.96	32	30	– 0.84

TABLE 4 Summary of Estimates of Class Proportions Based on 100 Bootstrap Samples

Class	Median	Mean	StDe	LB_95%	UB_95%	Simu.	True	*t*-stat
P0	0.158	0.159	0.022	0.116	0.202	0.176	0.200	1.09
P1	0.100	0.102	0.026	0.051	0.153	0.080	0.050	– 1.15
P2	0.110	0.114	0.021	0.073	0.155	0.100	0.100	0.00
P3	0.193	0.193	0.013	0.168	0.218	0.207	0.200	– 0.54
P4	0.185	0.183	0.011	0.161	0.205	0.195	0.200	0.45
P5	0.140	0.139	0.013	0.114	0.164	0.136	0.150	1.08
P6	0.110	0.111	0.008	0.095	0.127	0.107	0.100	– 0.87

TABLE 5 (Singular) Covariance Matrix of Estimated Frequencies Based on 100 Bootstrap Samples

Class	1	2	3	4	5	6	Sum
1	1.6585	− 0.6020	0.5224	− 1.2382	0.2522	− 0.5929	0.0000
2	− 0.6020	4.2626	− 1.4121	− 0.2101	− 2.0283	− 0.0101	0.0000
3	0.5224	− 1.4121	3.9733	0.2824	− 1.5721	− 1.7939	0.0000
4	− 1.2382	− 0.2101	0.2824	4.5208	− 4.0358	0.6808	− 0.0001
5	0.2522	− 2.0283	− 1.5721	− 4.0358	11.3546	− 3.9707	− 0.0001
6	− 0.5929	− 0.0101	− 1.7939	0.6808	− 3.9707	5.6869	0.0000

TABLE 6 (Singular) Covariance Matrix ($\times 10^4$) of Estimated Population *P*-Proportions Based on 100 Bootstrap Samples

	P0	P1	P2	P3	P4	P5	P6	Sum
P0	4.7477	− 4.6955	− 1.1972	− 0.7973	0.8125	0.5485	0.5820	0.0008
P1	− 4.6955	6.6339	− 1.2040	0.6957	− 1.2382	0.2018	− 0.3955	− 0.0018
P2	− 1.1972	− 1.2040	4.2626	− 0.9403	− 0.1051	− 0.8113	− 0.0043	0.0005
P3	− 0.7973	0.6957	− 0.9403	1.7663	0.0941	− 0.4192	− 0.3985	0.0008
P4	0.8125	− 1.2382	− 0.1051	0.0941	1.1302	− 0.8072	0.1140	0.0003
P5	0.5485	0.2018	− 0.8113	− 0.4192	− 0.8072	1.8167	− 0.5297	− 0.0003
P6	0.5820	− 0.3955	− 0.0043	− 0.3985	0.1140	− 0.5297	0.6319	− 0.0001

ACKNOWLEDGMENTS

This research was initiated while the first and the third authors were visiting CRRao AIMSCS, Hyderabad, and it was completed while the third author was visiting the second author at IUPUI, Indiana. The authors acknowledge the excellent academic atmosphere and the facilities extended to them at both the places during the course of this investigation. The authors are grateful to the referee for pointing out several places where further clarification and justification were necessary. The authors also thank the associate editor for his guidance during revision.

APPENDIX

Here we give details of the Randomized Response Hartley-Politz-Simmons [RR-HPS] technique, and explain why it works. We also display the Randomized Response Worksheet given to the respondents, together with an illustration on how to fill out the worksheet.

In constructing the Randomized Response Worksheet, letters A, B, C, D (and also w, x, y, z) are drawn as random sample with replacement from discrete distributions with probabilities $\underline{\theta} = (0.50, 0.20, 0.08, 0.22)$ respectively. [Indeed, such worksheet can be generated individually for each respondent, so that if two respondents collude they will find their sheets are different!] Let the probability of occurrence of the target letter be denoted by θ_*. Then the distribution of T_*, the total number of letters until 10 occurrences of a target letter, is negative binomial $(r = 10, \theta_*)$. In particular, $T_A \sim NB(r = 10, \theta_A = 0.50)$ and $T_B \sim NB(r = 10, \theta_B = 0.20)$. We use Conversion Rule 1: Classify into A if $T <= 27$; otherwise, classify into B. The cut-off value 27 was so chosen in order to ensure that there is sufficiently high probability of correctly classifying both type A and type B subjects, but it is not so extremely high that the subjects' confidentiality is obliterated. Indeed, the probabilities of classification (and misclassification) for Capital Letter groups are as follow:

$\underline{\theta}=(0.50, 0.20, 0.08, 0.22)$ *Conditional Probabilities*	$r = 10$ *if subject belongs to*	
	A	*B*
Classified into A	0.9390	0.0304
Classified into B	0.0610	0.9696

Likewise, $t_w \sim NB(r = 10, \theta_w = 0.50)$, $t_x \sim NB(r = 10, \theta_x = 0.20)$ and $t_y \sim NB(r = 10, \theta_y = 0.08)$. We use Conversion Rule 2: Classify into w if $t <= 27$; into x if $27 < t <= 72$; and into y if $72 < t$. The probabilities of classification (and misclassification) for lower case letter groups are as follow:

$\underline{\theta}$=(0.50,0.20,0.08,0.22) Conditional Probabilities	r = 10		
	if subject belongs to		
	w	x	y
Classified into w	0.9390	0.0304	0.0000
Classified into x	0.0610	0.9007	0.0602
Classified into y	0.0000	0.0689	0.9397

In the example given in the Randomized Response Worksheet, $(T,t) = (50,62)$. Using the above classification rules, we classify the respondent into class $[B,x] = 5$, which is a correct classification.

One has a lot of flexibility in adjusting the RR-HPS technique: (1) Choice of the $\underline{\theta}$ probability vector for generating the random letter sequence, and (2) Choice of r, the number of occurrences of target letters. Every such choice of $(\underline{\theta},r)$ will determine the appropriate cut-off points for the conversion rules in order to control the probabilities of correct classification and misclassification. For example, keeping the selection probabilities of the four letters at $\underline{\theta} = (0.50,0.20,0.08,0.22)$, one can consider $r = 15$, together with Conversion Rule 2: Classify into w if $t <= 40$; into x if $40 < t <= 110$; and into y if $110 < t$. [To be complete, we must mention that we also adopt a parallel Conversion Rule 1 here and in later variations. For example, here Conversion Rule 1 is: Classify into A if $T <= 40$ and into B if $40 < T$.] Or, one can consider $r = 20$, together with Conversion Rule 2: Classify into w if $t <= 55$; into x if $55 < t <= 150$; and into y if $150 < t$ (and its parallel Conversion Rule 1). In these cases, the probabilities of (mis)classification are as follows.

$\underline{\theta}$=(0.50,0.20,0.08,0.22) Conditional Probabilities	r = 15			r = 20		
	if subject belongs to			if subject belongs to		
	w	x	y	w	x	y
Classified into w	0.9597	0.0079	0.0000	0.9850	0.0036	0.0000
Classified into x	0.0403	0.9601	0.0293	0.0150	0.9837	0.0168
Classified into y	0.0000	0.0320	0.9707	0.0000	0.0128	0.9832

Alternatively, one can modify the selection probabilities of the four letters to be $\underline{\theta} = (0.40,0.24,0.14,0.22)$, and consider $r = 40$, together with Conversion Rule 2: Classify into w if $t <= 122$; into x if $122 < t <= 215$; and into y if $215 < t$. Or, one can consider $r = 50$, together with Conversion Rule 2: Classify into w if $t <= 152$; into x if $152 < t <= 270$; and into y if $270 < t$. In these cases, the probabilities of (mis)classification are as follows.

$\underline{\theta}$=(0.40,0.24,0.14,0.22) *Conditional Probabilities*	$r = 40$			$r = 50$		
	if subject belongs to			*if subject belongs to*		
	w	*x*	*y*	*w*	*x*	*y*
Classified into w	0.9585	0.0174	0.0000	0.9705	0.0081	0.0000
Classified into x	0.0415	0.9585	0.0360	0.0295	0.9791	0.0231
Classified into y	0.0000	0.0241	0.9640	0.0000	0.0128	0.9769

Note that the number of letters covered until one gets r appearances of the target letter becomes (almost) proportionately larger as r increases, in view of the law of large numbers. Note also that, for every choice of selection probabilities, we can reduce the probabilities of misclassification by choosing r sufficiently large. However, one must keep in mind that while high probabilities of correct classification are desirable from the analysts' point of view, one must exercise proper restraint so as not to jeopardize the confidentiality of the respondents. For instance, in the extreme case when the analyst can tell with almost certainty what the direct answer would have been based only on the randomized responses, the respondents' confidentiality is not ensured. Furthermore, a very large value of r means that the respondent has to spend a lot of time and pay extremely careful attention to figure out the reported values of T and t correctly. This is not desirable. Therefore, as a guideline to the practitioner, we recommend that they settle on the maximum tolerable probabilities of misclassification and then choose $\underline{\theta}$ and r in order to keep the expected values of both T and t reasonably low. We leave to the interested readers the problem of optimal choice of these parameters.

As already mentioned above, after choosing the selection probabilities $\underline{\theta}$, one can, in principle, generate the Randomized Response Worksheet individually for each respondent. Another safeguard can be built into the scheme: One can generate several different versions by changing the CHART to determine target Capital Letter and lower case letter simply by interchanging A and B, and/or taking a permutation of (w,x,y), provided a Version Number is printed and saved together with responses (T,t). From the known version number, the interviewer can recover the correct classification of the respondent. If the frequencies of the different versions are carefully chosen all types of respondents will have roughly similar distributions for T and t. Hence, large groups of respondents colluding together will find no reason to suspect that our RR-HPS scheme violates their confidentiality. Also, should the responses (together with the Version Number) fall in the wrong hands, they will be unable to decipher the actual classification rule unless they happen to know how the different versions were formed.

Randomized Response Worksheet

See an example on the next page.

Direct Question: How many nights were you home around this time (out of tonight and the previous 5 week nights)?

Do not write the answer to this DQ. Just use it to determine your target Capital letter and lower case letter codes.

Version 1		If your answer to the Direct Question is …					
		1	2	3	4	5	6
then your target Capital and lower case letters are …	Capital Letter	A	A	A	B	B	B
	lower case letter	w	x	y	w	x	y

Your target letters are __ , __

You roll the die twice and get the first and the second roll as __ , __ (You can guess these)

1. **Roll a die twice to determine the starting point. Put a "[" mark at the starting point. Proceed row-wise until your target letter appears 10 times. Stop and count how many letters you have covered from start to stop.**

```
First  Second Roll
Roll   1     2     3     4     5     6
 [1]   CDAAC CDDAB ACCAB DABBA ABAAA BAABB ADAAA DDAAD DAABA AADBA
 [2]   DAABD DAACC DDADA DADCB DAAAA BACAB AADAA DABBD AAAAB BDDBB
 [3]   AADAB DBBAA ADBBA AAABD AAAAD BACAA DDBBA DBADA BADAD ACADA
 [4]   BDBAA BDABA DCDDD DAACA DABAD AACAD ACDAB BDABA DDADD DBDAA
 [5]   BADBA CDADD ADACB CABDD ACAAA AAAAB DCADB BDACB BDDAA AADDA
 [6]   ABBDB ABDAB DABAA DABAD BAAAB DADDC ADDAA CABAA BABBB CBCDB
       AADAC AABAA ACBDA BBAAD AAAAA AAACB AABAA AAADB ABDBD BCBAB
       BABAD ABAAA AAADA AADDA ADACA ADABA BADAB BAACA DDAAA AAADD
       AADAB ADBAA CADBA ADDAA ABCBD AAAAD BADAA AAAAB AACAA BAADD
       DDADA AACCA AABBD ADBAD AABBA BADAB DAADC ACACA BDBBA BDAAA
        Go to the top row, if needed. Note each line has 50 letters.
```

Cap = number of letters covered to get your target Capital Letter exactly 10 times = T = ______

You roll the die twice more and get the first and the second roll as __ , __ (You can guess these)

2. **Roll a die twice to determine the starting point. Put a "[" mark at the starting point. Proceed row-wise until your target letter appears 10 times. Stop and count how many letters you have covered from start to stop.**

```
First  Second Roll
Roll   1     2     3     4     5     6
 [1]   wxwwx zwwzw zzxwx wwzxx xwzxx wyzww yywww wxyxz xwxyx wwwzz
 [2]   xxwzy zzxwx zxywz zxwww wzxxw xzzyz wwywz wwwxw wzyzx xwyxx
 [3]   wwwww zzwwz wywzy zzwzw zzwwz zwwxx xwwwx wwwzw wxwzw wwwzw
 [4]   wzwwx wyzwx wwzzz wzyxx zwyyw wywww wwzww wxzzx wwwwz wzxwz
 [5]   zxwwx wxwxx wxwww wzwyw zwxwz zwwzw zwxxy wzwzw wyxxw zzwxy
 [6]   wwzwx wyxwz wxwxw ywzyz yxxww zzwyx xzwzw zxxyx xxzxw wxwwx
       wxzxz yxzxx wyyxw wyxww zzzww wzyxw wwzwx zwwzz ywwwz wwwwy
       zzwzx wxwxz xwzxw wxwww wwwww wxzww wzwxz xwwww xzywz wzwxw
       wxwww zzwzy zzzzz xwwzx wxzyx wzxxw xxzxy xxxxz zywyw wxyxy
       wzxww yxwzw yzwwx xxzww xxwwx wxwww zyzzw zzwwx wwwzz xzwwx
        Go to the top row, if needed. Note each line has 50 letters.
```

low = number of letters covered to get your target lower case letter exactly 10 times = t = ____

Cut along the dotted line below. Keep the upper part to yourself (and destroy).

Submit only this bottom part after filling in Cap and low values.

Do not identify yourself, or your answer to the direct question, or your target letters.

Number of letters covered to get your target letters exactly 10 times

Version 1 Cap = _____ low = _____

Randomized Response Worksheet: Example

Example: Suppose that your answer to the Direct Question is 5.

Direct Question: How many nights were you home around this time (out of tonight and the previous 5 week nights)?
Do not write the answer to this DQ. Just use it to determine your target Capital letter and lower case letter codes.

Version 1		If your answer to the Direct Question is …					
		1	2	3	4	5	6
then your target Capital and lower case letters are …	Capital Letter	A	A	A	B	B	B
	lower case letter	w	x	y	w	x	y

Then your target letters are B, x.

Suppose that you roll the die twice and get the first and the second roll as [5], 2

1. Roll a die twice to determine the starting point. Put a "[" mark at the starting point. Proceed row-wise until your target letter appears 10 times. Stop and count how many letters you have covered from start to stop.

```
First  Second Roll
Roll   1      2      3      4      5      6
 [1]   CDAAC  CDDAB  ACCAB  DABBA  ABAAA  BAABB  ADAAA  DDAAD  DAABA  AADBA
 [2]   DAABD  DAACC  DDADA  DADCB  DAAAA  BACAB  AADAA  DABBD  AAAAB  BDDBB
 [3]   AADAB  DBBAA  ADBBA  AAABD  AAAAD  BACAA  DDBBA  DBADA  BADAD  ACADA
 [4]   BDBAA  BDABA  DCDDD  DAACA  DABAD  AACAD  ACDAB  BDABA  DDADD  DBDAA
 [5]   BADBA [CDADD  ADACB  CABDD  ACAAA  AAAAB  DCADB  BDACB  BDDAA  AADDA
 [6]   ABBDB  ABDAB  DABAA  DABAD  BAAAB  DADDC  ADDAA  CABAA  BABBB  CBCDB
       AADAC  AABAA  ACBDA  BBAAD  AAAAA  AAACB  AABAA  AAADB  ABDBD  BCBAB
       BABAD  ABAAA  AAADA  AADDA  ADACA  ADABA  BADAB  BAACA  DDAAA  AAADD
       AADAB  ADBAA  CADBA  ADDAA  ABCBD  AAAAD  BADAA  AAAAB  AACAA  BAADD
       DDADA  AACCA  AABBD  ADBAD  AABBA  BADAB  DAADC  ACACA  BDBBA  BDAAA
       Go to the top row, if needed. Note each line has 50 letters.
```

Cap = number of letters covered to get your target Capital Letter exactly 10 times = T = 50

Suppose that you roll the die twice more and get the first and the second roll as [3], 6

2. Roll a die twice to determine the starting point. Put a "[" mark at the starting point. Proceed row-wise until your target letter appears 10 times. Stop and count how many letters you have covered from start to stop.

```
First  Second Roll
Roll   1      2      3      4      5      6
 [1]   wxwwx  zwwzw  zzxwx  wwzxx  xwzxx  wyzww  yywww  wxyxz  xwxyx  wwwzz
 [2]   xxwzy  zzxwx  zxywz  zxwww  wzxxw  xzzyz  wwywz  wwwxw  wzyzx  xwyxx
 [3]   wwwww  zzwwz  wywzy  zzwzw  zzwwz [zwwxx  xwwwx  wwwzw  wxwzw  wwwzw
 [4]   wzwwx  wyzwx  wwzzz  wzyxx  zwyyw  wywww  wwzww  wxzzx  wwwwz  wzxwz
 [5]   zxwwx  wxwxx  wxwww  wzwyw  zwxwz  zwwzw  zwxxy  wzwzw  wyxxw  zzwxy
 [6]   wwzwx  wyxwz  wxwxw  ywzyz  yxxww  zzwyx  xzwzw  zxxyx  xxzxw  wxwwx
       wxzxz  yxzxx  wyyxw  wyxww  zzzww  wzyxw  wwzwx  zwwzz  ywwwz  wwwwy
       zzwzx  wxwxz  xwzxw  wxwww  wwwww  wxzww  wzwxz  xwwww  xzywz  wzwxw
       wxwww  zzwzy  zzzzz  xwwzx  wxzyx  wzxxw  xxzxy  xxxxz  zywyw  wxyxy
       wzxww  yxwzw  yzwwx  xxzww  xxwwx  wxwww  zyzzw  zzwwx  wwwzz  xzwwx
       Go to the top row, if needed. Note each line has 50 letters.
```

low = number of letters covered to get your target lower case letter exactly 10 times = t = 62

Cut along the dotted line below. Keep the upper part to yourself (and destroy).

Submit only this bottom part after filling in Cap and low values.
Do not identify yourself, or your answer to the direct question, or your target letters.

Number of letters covered to get your target letters exactly 10 times

Version 1 Cap = 50 low = 62

REFERENCES

Barnett, J., 1998. Sensitive questions and response effects: An evaluation. J. Manag. Psychol. 13 (1/2), 63–76.

Chaudhuri, A., 2011. Randomized Response and Indirect Questioning Techniques in Surveys. CRC Press, Chapman and Hall, Taylor & Francis Group, Florida.

Chaudhuri, A., Christofides, T., 2013. Indirect Questioning in Sample Surveys. Springer, Germany.

Chaudhuri, A., Mukerjee, R., 1988. Randomized Response: Theory and Techniques. Marcel Dekker, New York.

Cochran, W., 1977. Sampling Techniques, Third edition, Wiley, New York.

Deming, W., 1953. On a probability mechanism to attain an economic balance between the resultant error of nonresponse and the bias of nonresponse. J. Am. Stat. Assoc. 48, 743–772.

Durbin, J., Stuart, A., 1954. Callbacks and clustering in sample surveys: an experimental study. J. R. Stat. Soc. A. 117, 387–428.

Hansen, M., Hurwitz, W., 1946. The problem of nonresponse in sample surveys. J. Am. Stat. Assoc. 41, 517–529.

Hartley, H., 1946. Discussion on paper by F Yates. J. R. Stat. Soc. 109, 37–38.

Hedayat, A., Sinha, B.K., 1991. Design and Inference in Finite Population Sampling. Wiley, New York.

Kish, L., Hess, I., 1959. A "replacement" procedure for reducing the bias of nonresponse. Am. Stat. 13, 17–19.

Politz, A., Simmons, W., 1949. An attempt to get "not at homes" into the sample without callbacks. J. Am. Stat. Assoc. 44, 9–16.

Politz, A., Simmons, W., 1950. An attempt to get "not at homes" into the sample without callbacks. J. Am. Stat. Assoc. 45, 136–137.

Rao, T., 2014. A revisit to Godambe's theorem and Warner's randomized response technique. In: Proceedings of International Conference on Statistics and Information Technology for a Growing Nation, 2. Communicated.

Stephan, F., McCarthy, P., 1958. Sampling Opinions. Wiley, New York.

Warner, S., 1965. Randomized response: a survey technique for eliminating evasive answer bias. J. Am. Stat. Assoc. 60, 63–69.

Yates, F., 1946. A review of recent statistical developments in sampling and sampling surveys. J. R. Stat. Soc. 109, 12–43.

Chapter 16

Optional Randomized Response: A Critical Review

R. Arnab*,†,1 and M. Rueda‡

*University of Botswana, Gaborone, Botswana

†University of Kwazulu-Natal, Durban, South Africa

‡University of Granada, Granada, Spain

1Corresponding author: e-mail: arnabr@mopipi.ub.bw

ABSTRACT

For the collection of data related to sensitive issues, randomized response (RR) techniques have been proposed. RR techniques provide more reliable data, protect respondents' confidentiality, and avoid an unacceptable rate of nonresponses. In optional RR, a section of people, those who feel that the subject of enquiry is not confidential enough and wish to provide direct response spontaneously, are given an option to provide direct responses. In this chapter, we have reviewed some important optional RR techniques available in the literature.

Keywords: Complex surveys, inclusion probabilities, randomized response, sampling strategies, sensitivity level, unbiased estimator, varying probability sampling

1 INTRODUCTION

In surveys we often gather information relating to highly sensitive issues such as induced abortion, drug addiction, HIV infection status, duration of suffering from AIDS, sexual behavior, incidence of domestic violence, and tax evasion, for example. In these situations the respondents often provide untrue responses or even refuse to respond because of social stigma and or fear. Under such circumstances randomized response (RR) techniques may be used to collect more reliable data, protect respondents' confidentiality, and avoid unacceptable rates of nonresponse. The RR technique was introduced by Warner (1965). Warner's technique was extended by a large number of researchers, viz. Horvitz et al. (1967), Greenberg et al. (1969), Franklin (1989), Mangat and Singh (1990), Kuk (1990), Arnab (1990, 1995, 1996, 1998), and Singh et al. (2000), among others, to increase the cooperation of respondents and efficiency. Applications of the RR technique in real life

Handbook of Statistics, Vol. 34. http://dx.doi.org/10.1016/bs.host.2016.01.014

situations were reported by Kerkvliet (1994), Goodstadt and Gruson (1975), and Van der Heijden et al. (1998) among others. Details are given by Chaudhuri and Mukherjee (1988), Singh (2003), and Chaudhuri (2011). In some situations a section of respondents feels that the subject of enquiry is not confidential enough and would like to respond directly. To accommodate such respondents, optional randomized response technique (ORT) has been proposed by Chaudhuri and Mukherjee (1988), Gupta et al. (2002), and Arnab (2004) among others. Our focus in this chapter is on various ORT methods only. Before describing ORT, we will describe a few RR techniques relevant to the present discussion.

1.1 Warner's Technique: The Pioneering Method

The RR technique was introduced by Warner (1965) for estimating π_A, the proportion of units (respondents) in a population possessing a certain stigmatized character A. In Warner method a sample s of size n is selected by simple random sampling with replacement (SRSWR) procedure. Since the character A is sensitive, the respondent (unit) need not supply information directly about whether he/she is a member of the group A, instead he/she is asked to provide an RR based on an RR device as follows: The respondent has to select a card at random from a pack of well-scaffolded cards. The pack consists of two types of cards with known proportions and is identical in appearance. Card type 1 with proportion $P_1(\neq 1/2)$ contains the question "Do you belong to the group A?" while card type 2 with proportion $1-P_1$ contains the question "Do you belong to the group $\bar{A}$?" where $\bar{A}$ is the complement of group A. The respondent will supply a truthful answer "Yes" or "No" for the question mentioned in the selected card. The experiment is performed in the absence of the interviewer and hence the privacy of the respondent is maintained because the interviewer will not know which of the two questions the respondent has answered. We will call this RR model with parameter P_1 as $R_w(P_1)$.

Let λ_{wA} be the proportion of "Yes" answers obtained from the respondents selected in the sample. Then we have the following theorem.

Theorem 1.

i. $\hat{\pi}_{Aw} = \frac{\lambda_{wA}-(1-P_1)}{2P_1-1}$ *is an unbiased estimator of* π_A

ii. $\mathrm{Var}(\hat{\pi}_{Aw}) = \dfrac{\pi_A(1-\pi_A)}{n} + \dfrac{P_1(1-P_1)}{n(2P_1-1)^2}$

iii. *An unbiased estimator of* $\mathrm{Var}(\hat{\pi}_{Aw})$ *is*

$$\hat{\mathrm{V}}\mathrm{ar}(\hat{\pi}_{Aw}) = \frac{\lambda_{wA}(1-\lambda_{wA})}{(n-1)(2P_1-1)^2}$$

In case the selected individual responded directly whether he/she belongs to the sensitive group or not and $\hat{\pi}_A$ *is the proportion of "Yes" answers obtained from the sample, then we would get* $E(\hat{\pi}_A) = \pi_A$ *and*

$\mathrm{Var}(\hat{\pi}_A) = \pi_A(1-\pi_A)/n$. *Clearly* $\mathrm{Var}(\hat{\pi}_{Aw})$ *exceeds* $\mathrm{Var}(\hat{\pi}_A)$ *and the loss of efficiency, using Warner RR technique used for the protection of confidentiality of the respondents, is*

$$1 - \frac{\mathrm{Var}(\hat{\pi}_A)}{\mathrm{Var}(\hat{\pi}_{Aw})} = \left[1 + \frac{\pi_A(1-\pi_A)}{P_1(1-P_1)}(2P_1-1)^2\right]^{-1}$$

1.2 Ericksson's Technique

Here we assume that y is a quantitative variable and can take any value in the known interval (a,b). In Eriksson's RR technique (1973), M values $Q_1(=a), Q_2, \ldots, Q_M(=b)$ are chosen in the interval (a,b). The vector $\mathbf{Q} = (Q_1, \ldots, Q_M)$ covers the range (a,b) and the value of M depends on the length of the interval. The respondent is supposed to report either the true value y_i with probability c or the Q_j value with probability $q_j\left(q_j > 0, \sum_j q_j = 1 - c\right)$. For example, in collecting information regarding expenditures on illegal drugs, a randomized device can be framed as follows: Take a jar containing capsules which appear identical. Each of the capsules contains exactly one ticket. Capsules of c proportions contain tickets written "Disclose your true expenditure" and the remaining capsules contain tickets marked, respectively, \$0 $(=Q_1)$, \$5 $(=Q_2)$, \$10 $(=Q_3), \ldots$, \$2000 $(=Q_{401})$ with proportions $q_1, q_2, \ldots, q_{401}$, respectively. Here it is assumed that the minimum expenditure on drugs is zero, and the maximum \$2000, and $M = 401$. The respondent is to shake the jar well and select a capsule at random. After opening the capsule, if a respondent finds the ticket "Disclose your true expenditure," then he/she has to disclose the true expenditure; otherwise, he/she will report the amount written in the ticket. The confidentiality of the respondent is preserved since the experiment is performed in the absence of the interviewer and the interviewer will receive some amount as a response and he/she will not be able to identify whether this figure is really his/her real expenditure or the value written on the ticket.

Let z_i be the response obtained from the ith respondent and let E_R, V_R, and C_R, respectively, denote the expectation, variance, and covariance operators with respect to the RR technique. Now note that RRs are independent, we obtain

$$E_R(z_i) = cy_i + \sum_j q_j Q_j,\quad E_R\left(z_i^2\right) = cy_i^2 + \sum q_j Q_j^2,\quad \text{and}\quad C_R\left(z_i, z_j\right) = 0\quad \text{for } i \neq j$$

Denoting

$$r_i = \frac{\left(z_i - \sum_j q_j Q_j\right)}{c}$$

as the revised RR for the ith unit, we get

$$E_R(r_i)=y_i,\ \ V_R(r_i)=V_R(z_i)/p^2=\alpha y_i^2+\beta y_i+\gamma=\sigma_i^2,\ \text{ and }\ C_R(r_i,r_j)=0 \text{ for } i\neq j$$

where $\alpha=\frac{1-c}{c}$, $\beta=-2\sum_j q_jQ_j/c$, and $\gamma=\left[\sum_j q_jQ_j^2-\left(\sum_j q_jQ_j\right)^2\right]/c^2$.

An unbiased estimator of σ_i^2 is

$$\hat{\sigma}_i^2=\frac{\alpha r_i^2+\beta r_i+\gamma}{1+\alpha}$$

1.3 A More General Model

Now we will consider a more general RR model proposed by Arnab (1990) as follows:

$$\text{Model } R:\ E_R(r_i)=y_i,\ \ V_R(r_i)=\phi_i=\phi(y_i),\ \text{ and }\ C_R(r_i,r_j)=0\ \text{ for } i\neq j \tag{1}$$

where α, β, and γ are constants free from y_is, and ϕ_i is a function of y_i only. Here we assume that a nonnegative unbiased estimator ϕ_i is available and it will be denoted by $\hat{\phi}_i$.

Most of the RR models for qualitative or quantitative characteristics satisfy the Model R and hence they may be considered as a special case of the Model R given in (1).

For a qualitative variable y, $y_i=1$ if the ith unit belongs to the sensitive group A and that $y_i=0$ otherwise. In this case proportion of individuals belong to the sensitive group is $\pi_A=\sum_{i\in U}y_i/N$. Now noting $E_R\left(\hat{\phi}_i\right)=E_R\left(r_i^2\right)-E_R(r_i)=\phi_i+y_i^2-y_i=\phi_i$ as $y_i^2=y_i$, one can find an unbiased estimator of $V_R(r_i)=\phi_i$ as $\hat{\phi}_i=r_i(r_i-1)$. For example, consider Warner $R_{w(P)}$ with z_i as an RR obtained from the ith respondent. Let us denote $z_i=1$ if the ith respondent answers "Yes" and $z_i=0$ if answers "No."

In this case we have $E_R(z_i)=y_iP+(1-y_i)(1-P)=E_R\left(z_i^2\right)$ and $V_R(z_i)=P(1-P)$.

Now writing $r_i=\frac{z_i-(1-P)}{(2P-1)}$ and noting P is a known constant, we find

$$E_R(r_i)=y_i,\ \ V_R(r_i)=\phi_i=\frac{P(1-P)}{(2P-1)^2}=\hat{\phi}_i=\phi,\ \text{a known constant.}$$

2 GENERAL METHOD OF ESTIMATION

Most of the surveys in practice are complex surveys involving stratification, varying probability sampling, and information of more than one character collected at a time. Some of them are of a confidential nature and others are not. Further, the sensitive characters need not be of a qualitative nature only, but they may also be quantitative. To cope with this situation, a unified method of estimation has been proposed by various authors including Chaudhuri (2011) and Arnab (1994). In this chapter, we will describe a general method of estimation of the population characteristic proposed by Arnab (1994). Let a sample s of size n be selected from a finite population U by some arbitrary sampling design p with probability $p(s)$. Let $\pi_i(>0)$ and $\pi_{ij}(>0)$ denote the inclusion probabilities of the ith, and ith and jth unit $(i \neq j)$ of the population U and y_i be the value of the sensitive character y for the ith unit which cannot be directly obtained from the respondent. Hence, an RR r_i is obtained from the ith respondent by some suitable randomized device. The RRs r_i are assumed to follow the model (1), viz.

$$E_R(r_i) = y_i, \quad V_R(r_i) = \phi_i, \quad \text{and} \quad C_R(r_i, r_j) = 0 \quad \text{for } i \neq j$$

Consider a linear homogeneous unbiased estimator for the total Y based on a direct survey as

$$t(s, y) = \sum_{i \in s} b_{si} y_i \tag{2}$$

where b_{si}s are constants free from y_is and r_is and satisfy the unbiasedness condition

$$\sum_{s \supset i} b_{si} p(s) = 1 \quad \forall\, i = 1, \ldots, N$$

Since y_is are not directly obtained from the respondent, we replace y_i by its estimate r_i in $t(s,y)$ and obtain the following estimator for RR survey:

$$t(s, r) = \sum_{i \in s} b_{si} r_i \tag{3}$$

The following theorem (Arnab, 1994) shows how one can use the variance of the estimator $t(s,y)$ and its unbiased estimator based on the direct surveys to obtain variance and unbiased estimators of the RR surveys.

Theorem 2.

i. $E[t(s,r)] = Y$

ii. $V[t(s,r)] = V_p[t(s,y)] + \sum_{i \in U} \alpha_{ii} \phi_i$

iii. Let $\hat{V}[t(s,y)]$ *be a homogeneous quadratic unbiased estimator of* $V_p[t(s,y)]$. *Then an unbiased estimator of* $V[t(s,r)]$ *is*

$$\hat{V}^*[t(s,r)] = \hat{V}[t(s,r)] + \sum_{i \in s} \hat{\phi}_i$$

where V_p *denotes variance with respect to the sampling design* p, $\alpha_{ii} = \sum_{s \supset i} b_{si}^2 p(s)$ *and* $\alpha_{ij} = \sum_{s \supset i,j} b_{si} b_{sj} p(s)$, $V_p[t(s,y)] = \sum_{i \in U} (\alpha_{ii} - 1) y_i^2 + \sum_{i \neq} \sum_{j \in U} (\alpha_{ij} i - 1) y_i y_j$, *and* $\hat{V}[t(s,r)]$ *is obtained by writing* r_i *in place of* y_i *in* $\hat{V}[t(s,y)]$.

3 OPTIONAL RANDOMIZED RESPONSE TECHNIQUES

In ORT it is assumed that the aspects of inquiry are felt to be sensitive by most of the respondents but some are more willing to answer directly. In ORT, respondents are given an option either to supply RR using a specified randomized device or to response directly according to the extent to which the respondent feels that the question is sensitive or not. Most of the methods developed for ORT are limited to SRSWR sampling only. A few of the ORT techniques are available for complex surveys. ORT can be classified into two categories: full ORT (FORT) and partial ORT (PORT). The ORT is more efficient than compulsory RRT (CRT) because the probability of obtaining true responses in ORT is much higher than the CRT (Vide Arnab, 2004).

Full ORT (FORT): Here respondents are given an option either to supply RR using a specified randomized device or to respond directly according to whether the respondent feels that the question is sensitive or not. In this method it is assumed that respondents who feel that the character under investigation is confidential, belong to certain group G and produce RR with probability one, while the respondents who feel the character is not confidential, belong to the complementary group $\bar{G}$ and supply direct response with probability one. FORT was proposed by Chaudhuri and Mukherjee (1988), Arnab (2004), Chaudhuri and Saha (2005), and Huang (2008) among others.

Partial ORT (PORT): Here it is assumed that the respondents may supply direct or RR with certain probability depending on their judgment of sensitivity (mood) at the particular time of answering the question. In other words, the respondent may sometimes supply RR and at other time direct response during the period of survey. Most of the researchers developed various methods of PORT, eg, Mangat and Singh (1994), Gupta (2001), Gupta et al. (2002, 2013), and Pal (2008) among others.

3.1 Full ORT

Arnab (2004) proposed a theory of estimation of population characteristics for ORT which is applicable for both qualitative and quantitative characteristics. Let $s_G = s \cap G$ be the set of respondents selected in the sample s that belong to group G and $s_{\bar{G}} = s \cap \bar{G} = s - s_G$.

3.1.1 Arnab (2004)

In this method, respondents belonging to group s_G provide RRs using some suitable randomized device, while respondents belonging to group $s_{\bar{G}}$ provide direct responses. Here we assume that the respondents do not disclose their membership to group G or $\bar{G}$.

Let r_i be the revised RR obtained from the ith respondent if he/she is included in the sample s_G and let y_i be the direct response if the ith unit belongs to $s_{\bar{G}}$. Here we suppose that the r_is follow model (1). Let us define

$$\widetilde{r}_i = \delta_i y_i + (1-\delta_i) r_i \tag{4}$$

where $\delta_i = 1$ if $i \in \bar{G}$ and $\delta_i = 0$ if $i \in G$.

Clearly, $\widetilde{r}_i$ follows the following model:

$$\text{Model } R^*:\ E_R(\widetilde{r}_i) = y_i,\ \ V_R(\widetilde{r}_i) = (1-\delta_i) V_R(r_i) = (1-\delta_i)\phi_i,\ \text{ and }\ C_R(\widetilde{r}_i, \widetilde{r}_j) = 0 \ \text{ for } i \neq j \tag{5}$$

Replacing r_i by $\widetilde{r}_i$ in Eq. (3), we find a linear unbiased estimator for the population total Y under ORR technique as

$$\widetilde{t}(s, \widetilde{r}) = \sum_{i \in s} b_{si} \widetilde{r}_i \tag{6}$$

Arnab (2004) derived the following results straight from Theorem 2.

Theorem 3.

i. *$\widetilde{t}(s,\widetilde{r})$ is an unbiased estimator for Y*

ii. $V\left[\widetilde{t}(s,\widetilde{r})\right] = \sum_{i \in U} (\alpha_i - 1) y_i^2 + \sum\sum_{i \neq j \in U} (\alpha_{ij} - 1) y_i y_j + \sum_{i \in G} \alpha_i \phi_i$

iii. *An unbiased estimator of $V\left[\widetilde{t}(s,\widetilde{r})\right]$ is*

$$\hat{V}\left[\widetilde{t}(s,\widetilde{r})\right] = \sum_{i \in s} c_{ii}(s) \widetilde{r}_i^2 + \sum\sum_{i \neq j \in s} c_{ij}(s) \widetilde{r}_i \widetilde{r}_j + \sum_{i \in s} (1-\delta_i) b_{si} \hat{\phi}_i$$

where $\alpha_i = \sum_{s \supset i} b_{si}^2 p(s)$, $\alpha_{ij} = \sum_{s \supset i} b_{si} b_{sj} p(s)$, $\sum_{s \supset i} c_{ii}(s) p(s) = \alpha_i - 1$, *and* $\sum_{s \supset i} c_{ij}(s) p(s) = \alpha_{ij} - 1$.

Remark 1. The variance of $\widetilde{t}(s,\widetilde{r})$ is smaller than that of $t(s,r)$ by an amount of $\sum_{i \in \bar{G}} \alpha_i \phi_i$. Hence the ORR technique is more efficient than the compulsory RR technique.

3.1.1.1 A Few Important Sampling Strategies

3.1.1.1.1 Horvitz–Thompson Estimator Based on a Fixed Sample Size Design Substituting $b_{si} = 1/\pi_i$ in (2), the Horvitz–Thompson estimator for the population total Y is obtained as

$$t(s, \widetilde{r}) = t_{ht} = \sum_{i \in s} \frac{\widetilde{r}_i}{\pi_i}$$

The expression of variance and its unbiased estimators obtained from Theorem 3 are obtained as follows:

$$V(t_{ht}) = \frac{1}{2}\sum_{i\neq}\sum_{j\in U}(\pi_i\pi_j - \pi_{ij})\left(\frac{y_i}{\pi_i} - \frac{y_j}{\pi_j}\right)^2 + \sum_{i\in G}\frac{\phi_i}{\pi_i}$$

and

$$\hat{V}(t_{ht}) = \frac{1}{2}\sum_{i\neq}\sum_{j\in s}\frac{(\pi_i\pi_j - \pi_{ij})}{\pi_{ij}}\left(\frac{\widetilde{r}_i}{\pi_i} - \frac{\widetilde{r}_j}{\pi_j}\right)^2 + \sum_{i\in s_G}\frac{\hat{\phi}_i}{\pi_i}$$

3.1.1.1.2 Simple Random Sampling Without Replacement For SRSWOR $\pi_i = n/N$ and $\pi_{ij} = \frac{n(n-1)}{N(N-1)}$ and we get

$$t(s,\widetilde{r}) = t_{\text{wor}} = N\bar{\widetilde{r}}$$

$$V(t_{\text{wor}}) = \frac{N(N-n)}{n}S_y^2 + \frac{N}{n}\sum_{i\in s_G}\phi_i$$

and

$$\hat{V}(t_{\text{wor}}) = \frac{N(N-n)}{n}s_r^2 + \frac{N}{n}\sum_{i\in s_G}\hat{\phi}_i$$

where $(n-1)s_r^2 = \sum_{i\in s}\left(\widetilde{r}_i - \bar{\widetilde{r}}\right)^2$ and $\bar{\widetilde{r}} = \sum_{i\in s}\widetilde{r}_i/n$.

3.1.1.1.3 Rao–Hartley–Cochran Sampling Design In the Rao et al. (1962) sampling design, at first, the population is divided at random into n groups, each of size N/n (assuming integer) units. Then, from each group, one unit is selected with probability proportional to the normed size measure $p_i\left(>0, \sum_{i\in U}p_i = 1\right)$. Let P_i be the sum of p_j values for the group containing the ith unit.

The expression for the unbiased estimator for the population total Y is obtained as

$$t(s,\widetilde{r}) = t_{rhc} = \sum_{i\in s}\frac{\widetilde{r}_i}{p_i}P_i$$

The expressions for the variance and its unbiased estimators are given by Arnab (2004) as:

$$V(t_{rhc}) = \frac{N-n}{n(N-1)}\left(\sum_{i\in U}\frac{y_i^2}{p_i} - Y^2\right) + \sum_{j\in G}a_j\phi_j$$

and

$$\hat{V}(t_{rhc}) = \frac{N-n}{N(n-1)} \sum_{i \in s} \left(\frac{\widetilde{r}_i}{p_i} - t_{rhc} \right)^2 P_i + \sum_{j \in s_G} a_j \frac{\hat{\phi}_j}{p_j}$$

where $a_j = \frac{1}{n(N-1)} \left(N(n-1) + \frac{N-n}{p_j} \right)$.

3.1.1.1.4 Probability Proportional to Size with Replacement Sampling Design Here we suppose that the sample s is selected using PPSWR method with normed size measure p_i attached to the ith unit. The proposed estimator for the total Y is

$$t(s, \widetilde{r}) = t_{pps} = \frac{1}{n} \sum_{k=1}^{n} \frac{\widetilde{r}(k)}{p(k)} \tag{7}$$

where $\widetilde{r}(k) = \widetilde{r}_i$, $p(k) = p_i$ if the kth draw produces ith unit, $i = 1, \ldots, N$.

In this case

$$V(t_{pps}) = \frac{1}{n} \left[\sum_{i \in U} p_i \left(\frac{y_i}{p_i} - Y \right)^2 + \sum_{i \in G} \frac{\phi_i}{p_i} \right] \tag{8}$$

and

$$\hat{V}(t_{pps}) = \frac{1}{n(n-1)} \sum_{k=1}^{n} \left(\frac{\widetilde{r}(k)}{p(k)} - t_{pps} \right)^2 \tag{9}$$

3.1.1.1.5 Simple Random Sampling with Replacement In case sample s is selected by SRSWR, an unbiased estimator of Y, its variance, and an unbiased estimator of variance are obtained by substituting $p_i = 1/N$ in Eqs. (7)–(9) as follows:

$$t(s, \widetilde{r}) = t_{wr} = \frac{N}{n} \sum_{k=1}^{n} \widetilde{r}(k)$$

$$V(t_{wr}) = \frac{N^2}{n} \left[\sigma_y^2 + \frac{1}{N} \sum_{i \in G} \phi_i \right]$$

and

$$\hat{V}(t_{wr}) = \frac{N^2}{n(n-1)} \sum_{k=1}^{n} (r(k) - t_{wr})^2$$

where $\sigma_y^2 = \frac{1}{N}(y_i - \bar{Y})^2$.

3.1.2 Chaudhuri and Saha (2005)

In this FORT, respondents belong to the sampled sensitive group s_G supplied RRs and the remaining sampled respondents belong to the nonsensitive group $s_{\bar{G}}$ revealed their identity and supplied direct responses. Since the respondents belonging to $s_{\bar{G}}$ are identified and their y-values are known, the sampler himself/herself produced RRs r_i, $i \in s_{\bar{G}}$ using the known y_is $i \in s_{\bar{G}}$. The proposed estimator for the total and its unbiased variance estimators are given as follows:

Theorem 4.

i. $t(s,\widetilde{r}) = \sum_{i \in s_G} b_{si} r_i + \sum_{i \in s_{\bar{G}}} b_{si} y_i = \sum_{i \in s} b_{si} \widetilde{r}_i$ *is an unbiased estimator of* Y.

ii. $\hat{V}(t(s,\widetilde{r})) = \hat{V}(t(s,y)) - [t(s,\widetilde{r}) - t(s,y)]^2$ *is an unbiased estimator of* $V(t(s,\widetilde{r}))$.

3.1.3 Chang and Huang (2001)

In RR techniques primary focus is the estimation of π_A, the proportion of individuals belonging to a sensitive group A. In Chang and Huang (2001) method one can estimate π_A as well as sensitivity level τ of the character y. The sensitivity level is defined as the proportion of the respondents in the population bearing the sensitive attribute A given a truthful answer under direct question method. In the proposed method, two independent subsamples of sizes $n_j (j=1,2)$ are selected by using SRSWR method. The person in the jth subsample is instructed to reply a direct question whether he/she belongs to the group A or not. If the answer is "No," the respondent is required to supply RR using Warner model with parameter P_j. Here it is assumed that the respondents in the nonsensitive group have no reason to tell a lie and thus are completely truthful in their answers for both the direct and RR responses. It is further supposed that the respondents belong to the sensitive group A supply honest responses for the RR model but with probability T for usual direct response. The proposed estimators for π_A and T are as follows:

$$\hat{\pi}_A = \frac{(1-P_2)\bar{z}_1 - (1-P_1)\bar{z}_2}{P_1 - P_2}, \quad P_1 \neq P_2$$

$$\hat{T} = \frac{(1-2P_2)\bar{z}_1 - (1-2P_1)\bar{z}_2 - (P_1 - P_2)}{P_1 - P_2}$$

where $\bar{z}_j$ is the proportion of "Yes" answers reported by the respondents in the jth subsample. Chang and Huang proved that $\hat{\pi}_A$ is unbiased for π_A with variance

$$V(\hat{\pi}_A) = \frac{1}{(P_1 - P_2)^2}\left[\frac{(1-P_2)^2\theta_1(1-\theta_1)}{n_1} + \frac{(1-P_1)^2\theta_2(1-\theta_2)}{n_2}\right]$$

where

$$\theta_j = \pi_A T + \pi_A (1-T) P_j + (1-\pi_A)\left(1-P_j\right), \quad j=1,2$$

Chang and Huang (2001) also derived approximate expression of the bias and mean square error of $\hat{T}$. Huang (2004) also proposed an alternative FORT similar to the above the difference in the proposed model is to take a single sample of size n instead of taking two independent samples of sizes n_1 and $n_2 = (n - n_1)$. Marcheslli and Barabesi (2006) extended Huang (2004) FORT by using Franklin (1989) RR model instead of Warner model. Barabesi (2008) generalized Marcheslli and Barabesi (2006) by incorporating varying probability sampling design.

3.1.4 Huang (2008)

Huang (2008) proposed two FORT which are described as follows.

3.1.4.1 Method I

Two independent samples s_1 and s_2 of sizes n_1 and $n_2 (= n - n_1)$ are selected by SRSWR method. In the jth sample $s_j (= 1,2)$ each of the respondents is given two options: (i) disclose the correct response y or (ii) the scrambled response $z_j = x_j y$, where x_j is a random sample from a population with mean $\mu_{xj} = 1$ and variance γ_j. Here the population is in nature portioned into two groups G and $\bar{G}$. The members of group G always provide scrambled responses, while the members of group $\bar{G}$ provide direct responses. Let W_G be the unknown proportion of individuals belong to group G; μ_G and σ^2_{yG}, respectively, denote the population mean and population variance of y. Let z_{ji} be the RR obtained from the ith member of the sample s_j and $\bar{z}_j = \frac{1}{n_j}\sum_{i=1}^{n_j} z_{ji}$.

Theorem 5.

i. *$\hat{\mu}_{yI} = \alpha \bar{z}_1 + (1-\alpha)\bar{z}_2$ with $0 < \alpha < 1$ is an unbiased estimator of μ_y*

ii. *The variance of $\hat{\mu}_{y1}$ is*

$$V\left(\hat{\mu}_{yI}\right) = \alpha^2 \frac{\sigma^2_{z1}}{n_1} + (1-\alpha)^2 \frac{\sigma^2_{z2}}{n_2}$$

iii. *An unbiased estimator of $V\left(\hat{\mu}_{y1}\right)$ is*

$$\hat{V}\left(\hat{\mu}_{yI}\right) = \alpha^2 \frac{s^2_{z1}}{n_1} + (1-\alpha)^2 \frac{s^2_{z2}}{n_2}$$

3.1.4.2 Method II

The method II is similar to the method I but with a little difference. The difference is that the μ_{xj} population mean of x_j is chosen such that $\mu_{xj} \neq 1$. In this case we have

Theorem 6.

i. $\hat{\mu}_{yII} = \frac{(1-\mu_{x2})\bar{z}_1 - (1-\mu_{x1})\bar{z}_2}{\mu_{x1}-\mu_{x2}}$ *with* $\mu_{x1} \neq \mu_{x2}$ *is an unbiased estimator of* μ_y

ii. *The variance of* $\hat{\mu}_{yII}$ *is*

$$V(\hat{\mu}_{yII}) = \frac{1}{(\mu_{x1}-\mu_{x2})^2}\left[\frac{(1-\mu_{x2})^2\sigma_{z1}^2}{n_1} + \frac{(1-\mu_{x1})^2\sigma_{z2}^2}{n_2}\right]$$

iii. *An unbiased estimator of* $V(\hat{\mu}_{yII})$ *is*

$$\hat{V}(\hat{\mu}_{yII}) = \frac{1}{(\mu_{x1}-\mu_{x2})^2}\left[\frac{(1-\mu_{x2})^2 s_{z1}^2}{n_1} + \frac{(1-\mu_{x1})^2 s_{z2}^2}{n_2}\right]$$

3.2 Partial ORT

PORT was originally proposed by Mangat and Singh (1994) for qualitative variables and it was extended by several authors including Singh and Joarder (1997), Gupta (2001), Gupta and Shabbir (2004), Gupta et al. (2002, 2006, 2010, 2013), Huang (2004, 2008), Pal (2008), and Chaudhuri and Dihidar (2009) among others. Most of the researchers proposed PORT for SWRWR sampling, while a few of them viz. Pal (2008) and Chaudhuri and Dihidar (2009) for complex survey designs. Some of the proposed techniques are described below starting with complex survey designs.

3.2.1 Pal (2008)

In this proposed technique, each respondent is to supply two independent RRs. In the first response z_{i1}, the ith person is to report the true value y_i with unknown probability $C_i (0 \leq C_i \leq 1)$ or an RR $(y_i x_{i1}/\theta_1) + u_i$ with probability $1 - C_i$. The second response z_{i2} from the ith respondent is either the true response y_i with unknown probability C_i or an RR (with probability $1 - C_i$) where he/she can either repeat z_{i1} with known probability $p(0 < p < 1)$ or scrambled response $(y_i x_{i2}/\theta_2) + u_{i2}$ with probability $1 - p$. Here x_{ij} and u_{ij} are random samples from populations with chosen means θ_j and α_j and variances σ_j^2 and v_j^2, respectively, $j = 1,2$.

From the ith respondent an unbiased estimator of y_i comes out as

$$\widetilde{r}_i = \frac{\widetilde{\alpha}_2 z_{i1} - \alpha_1 z_{i2}}{\widetilde{\alpha}_2 - \alpha_1}$$

where $\widetilde{\alpha}_2 = p(\alpha_1 - \alpha_2) + \alpha_2$ and $\widetilde{\alpha}_2 \neq \alpha_1$.

The unbiased estimator for the total, variance of the estimator, and its unbiased estimator can be derived from *Theorem 1*.

Chaudhuri and Dihidar (2009) extended Pal (2008) PORT further for qualitative and quantitative variables when the sample size n is not necessarily fixed.

We describe a few PORT based on the SRSWR sampling design as follows:

3.2.2 Mangat and Singh (1994)

In this PORT, a sample s of size n is selected by SRSWR method. Each of the selected respondents in the sample s is instructed to say "Yes" if he/she possesses the sensitive characteristic A. If the respondent does not possess the characteristic A was asked to supply RR using Warner (1965) model with parameter P ($R_w(P)$). The whole procedure is completed by the respondent, unobserved by the interviewer and hence confidentiality of the respondents was maintained. Let π_A be the proportion of the respondents in the population possess the attribute A and $\hat{\lambda}_A$ proportion of "Yes" answers in the sample, then

Theorem 7.

i. $\hat{\pi}_A = \frac{\hat{\lambda}_A-(1-P)}{P}$ *is unbiased for* π_A

ii. $V(\hat{\pi}_A) = \dfrac{\pi_A(1-\pi_A)}{n} + \dfrac{(1-\pi_A)(1-P)}{nP}$

Singh and Tarray (2014) extended Mangat (1994) PORT for stratified sampling.

3.2.3 Gupta et al. (2002)

Gupta et al. (2002) extended Mangat (1994) PORT for quantitative variables as follows. Each respondent selected in the SRSWR sample chooses one of the following two options:

(a) Report the true value of y or (b) report randomized response yx, where x is a random sample from a population with known mean $\mu_x = 1$ and known variance γ^2. Here also the interviewer will not know whether the respondents supplied the true response or randomized response. Let us denote the response obtained from the ith respondent as

$$z_i = x_i^k y_i$$

where $k = 1$, if the response is scrambled and $k = 0$, otherwise.

Gupta et al. (2002) derived expressions of the unbiased estimator of the μ_y population mean of y and its variance as follows:

Theorem 8.

i. $\hat{\mu}_y = \frac{1}{n}\sum_{i=1}^{n} z_i$ *is an unbiased estimator of the* μ_y

ii. $V(\hat{\mu}_y) = \dfrac{1}{n}\left[\sigma_y^2 + W_G^2\left(\sigma_y^2 + \mu_x^2\right)\right]$

where W_G is the probability that a person will report the scramble response, which is generally unknown.

W_G is called the degree of sensitivity of the attribute A.

It can be verified that

$$W_G = \frac{E(\ln(z)) - E(\ln(y))}{E(\ln(x))}$$

Taylor's approximation yields

$$\text{In}(y) \cong \text{In}(\mu_y) + (y-\mu_y)\frac{1}{\mu_y} - \frac{1}{2}(y-\mu_y)^2\frac{1}{\mu_y^2} \tag{10}$$

Using the first-order approximation, an estimator of W_G was obtained as

$$\hat{W}_G = \frac{\frac{1}{n}\sum_{i=1}^{n}\ln z_i - \ln\left(\frac{1}{n}\sum_{i=1}^{n} z_i\right)}{\delta}$$

where $\delta = E(\ln x)$.

The estimator is clearly biased and its variance is not easy to calculate.

3.2.4 Gupta et al. (2010)

In this technique W_G, the proportion of people who never provide direct response can be estimated without any approximation. The method is described as follows:

Two independent samples s_1 and s_2 of sizes n_1 and $n_2(=n-n_1)$ are selected by SRSWR method as in Huang (2008). In the $j(=1,2)$th sample s_j, a randomly selected proportion (T) of the respondents are asked (forced) to provide truthful responses, while the rest are to use optional scrambling method which provides two options: (i) disclose the correct response y or (ii) the scrambled response $z_j = x_j + y$, where x_j is a random sample from a population with known mean $\mu_{xj}(\neq 1)$ and known variance γ_j. Under the proposed model, partial optional RRs obtained from the sample s_j are

$$z_j = \begin{cases} y \text{ with probability } T + (1-T)(1-W_G) \\ (y + x_j) \text{ with probability } (1-T)W_G \end{cases} \quad \text{for } j = 1,2$$

Let $\bar{z}_j$ be the sample mean of the responses from the sample s_j. Then we have

Theorem 9.

i. $\hat{\mu}_y = \dfrac{\mu_{x1}\bar{z}_2 - \mu_{x2}\bar{z}_1}{\mu_{x1} - \mu_{x2}}$; $\mu_{x1} \neq \mu_{x2}$ *is an unbiased estimator of* μ_y

ii. *The variance of* $\hat{\mu}_y$ *is*

$$V(\hat{\mu}_y) = \frac{1}{(\mu_{x2} - \mu_{x1})^2}\left(\mu_{x2}^2 \frac{\sigma_{z1}^2}{n_1} + \mu_{x1}^2 \frac{\sigma_{z2}^2}{n_2}\right)$$

where $\sigma_{zj}^2 =$ variance *of* z_j, $j = 1,2$.

iii. *An unbiased estimator of W and its variance are*

$$\hat{W}_G = \frac{\bar{z}_1 - \bar{z}_2}{(\mu_{x1} - \mu_{x2})(1-T)} \text{ and } V(\hat{W}_G) = \frac{1}{(\mu_{x2} - \mu_{x1})^2}\left(\frac{\sigma_{z1}^2}{n_1} + \frac{\sigma_{z2}^2}{n_2}\right)$$

Gupta and Shabbir (2004) considered similar two-stage PORT using multiplicative scrambling response, while Sihm and Gupta (2015) proposed three-stage PORT for qualitative variables.

3.2.5 Huang (2010)

Here the ith respondent in the sample asked to report the true response y_i if the respondent feels that the character is not sensitive to him/her otherwise report the scramble response $\widetilde{x}_i = y_i x_i + d_i$, where x_i and d_i are random samples from distributions with known means and variances. Let $\bar{z}_j$ be the sample mean of the responses from the sample s_j. Then $\hat{\mu}_y = \frac{\mu_{d1}\bar{z}_2 - \mu_{d2}\bar{z}_1}{\mu_{d1} - \mu_{d2}}$; $\mu_{d1} \neq \mu_{d2}$ and $\hat{W}_G = \frac{\bar{z}_1 - \bar{z}_2}{(\mu_{d1} - \mu_{d2})}$ being μ_{dj} the mean of d_i. Sihm and Gupta (2015) extended Huang (2010) model to two-stage PORT, while Mehta et al. (2012) proposed three-stage PORT.

4 EFFICIENCY OF THE ORT

Consider two RR models R_A and R_B which yield the ϕ_i values as ϕ_{iA} and ϕ_{iB}, respectively. In order to compare R_A with R_B, one needs to compare the magnitude of the values of the variances ϕ_{iA} and ϕ_{iB} along with the degree of privacy protection provided by the models R_A and R_B. Different measures of privacy protections are available in the literature, eg, Lanke (1975, 1976), Leysieffer and Warne (1976), and Anderson (1975a,b). Hence one should compare one RR model to another keeping the confidentiality to certain level fixed. If we ignore the confidentiality factor, we find from Theorem 2 that for a given sampling strategy (sampling design and estimator), R_A will be better than R_B if $\phi_{iA} \leq \phi_{iB}$ for every $i \in U$. The expression ϕ_i depends on the parameter of the RR model. Generally, we cannot find an RR model which is uniformly better than the other keeping level of confidentiality to a certain level. However, some researchers pointed out that Kuk's (1990) model performs uniformly better than the Warner's (1965) model. This is not totally correct because Kuk's (1990) model is a combination of k-independent RR models. The right platform of comparison should be to compare Warner model repeated k times with Kuk's model.

The comparison between two ORTs ORR_A and ORR_B can be performed similarly by using Theorem 3: ORR_A will be better than ORR_B if $\phi_{iA} \leq \phi_{iB}$ for every $i \in U_G$.

The comparison between a compulsory model R_A and an optional model based on this randomized model R_A is simply obtained by using Theorems 2 and 3: the optional model will be more precise than the nonoptional model and the gain in precision will be higher as the number of individuals who disclose their true responses in the optional model is higher.

A complex problem is the comparison of a compulsory RR technique with an ORR response model based on a different RR technique. There are some studies that compare specific models. Eg, Barabesi (2008) claims that his RR technique based on Franklin's (1989) RR model is uniformly better than ORR techniques proposed by Arnab (1996) and Saha (2007) which is not true. Barabesi did compare variance of r_i with $\tilde{r}_i$ (Arnab, 2004, notation) which is not comparable. So, in the literature different researchers compared their proposed RR strategies numerically, indicating the circumstances under which one strategy performs better than the others as unconditional results are rarely available.

5 CONCLUSION

ORT is used where aspects of the enquiry are felt sensitive by most of the people but some of them feel that the subject of enquiry is not confidential enough and hence willing to provide truthful direct answer. ORT is classified in two categories: full optional response (FORT) and partial optional response (PORT). In a FORT, a section of respondents reveal truthful answer with probability one, while in a PORT, a section of respondents provide truthful answer with certain unknown probability W_G which is assumed to be constant for all the respondents. This assumption is obviously quite restrictive. Most of the researches on ORT are limited to SWSWR sampling only and cannot be used for complex survey designs. However, a PORT based on SRSWR sampling, one can obtain a suitable estimator of W_G. As efficiency is concerned, ORT is generally more efficient than compulsory RRT (CRT) because in ORT, a section of respondents provide direct responses. All considered estimators are based only on the information of the main sensitive variable y. The efficiency of these estimators can be improved by using auxiliary information. Eg, Gupta et al. (2014) propose ratio and regression estimators for the finite population mean of a sensitive study variable utilizing nonsensitive auxiliary information under an additive ORT model. Other important techniques as calibration (Diana and Perri, 2012) or logistic methods (Van den Hout et al., 2007) can also be adapted to the optional randomized scheme. No particular ORT is found unconditionally better than the other. Hence researchers compared performances of different ORT numerically and identified situations where one is preferable to the other. As per the authors' knowledge, no comparison of ORT is available in the literature which takes account of the confidentiality aspects of the respondent as considered by Lanke (1975, 1976), Leysieffer and Warne (1976), and Anderson (1975a,b) and hence some researchers in the area will add value to the literature.

REFERENCES

Anderson, H., 1975a. Efficiency versus protection in RR designs for estimating proportions. Tech. Report 9. University of Lund, Sweden.

Anderson, H., 1975b. Efficiency versus protection in a general RR model. Tech. Report 10. University of Lund, Sweden.

Arnab, R., 1990. On commutativity of design and model expectations in randomized response surveys. Commun. Stat. Theory Methods 19, 3751–3757.

Arnab, R., 1994. Non-negative variance estimation in randomized response surveys. Commun. Stat. Theory Methods 23, 1743–1752.

Arnab, R., 1995. Optimal estimation of a finite population total under randomized response surveys. Statistics 27, 175–180.

Arnab, R., 1996. Randomized response trials: a unified approach for qualitative data. Commun. Stat. Theory Methods 25 (6), 1173.

Arnab, R., 1998. Randomized response surveys: optimum estimation of a finite population total. Stat. Papers 39, 405–408.

Arnab, R., 2004. Optional randomized response techniques for complex survey designs. Biom. J. 46, 114–124.

Barabesi, L., 2008. A design-based randomized response procedure for estimation of population proportion and sensitivity level. J. Stat. Plan. Inference 138, 2308–2408.

Chang, H.J., Huang, K.C., 2001. Estimation of proportion and sensitivity of a qualitative character. Metrika 53, 269–280.

Chaudhuri, A., 2011. Randomized Response and Indirect Questioning Techniques in Surveys. Chapman & Hall/CRC, New York.

Chaudhuri, A., Dihidar, K., 2009. Estimating means of stigmatizing qualitative and quantitative variables from discretionary responses randomized or direct. Sankhya 71-B, 123–136.

Chaudhuri, A., Mukherjee, R., 1988. Randomized Response: Theory and Techniques. Marcel Dekker, New York.

Chaudhuri, A., Saha, A., 2005. Optional versus compulsory randomized response techniques in complex surveys. J. Stat. Plan. Inference 135, 516–527.

Diana, G., Perri, P.F., 2012. A calibration-based approach to sensitive data: a simulation. J. Appl. Stat. 0266-4763. 39:1, 53–65.

Eriksson, S.A., 1973. A new model for randomized response. Int. Stat. Rev. 41, 40–43.

Franklin, L.A., 1989. A comparison of estimators for randomized response sampling with continuous distribution from dichotomous populations. Commun. Stat. Theory Methods 18, 489–505.

Goodstadt, M.S., Gruson, V., 1975. The randomized response technique: a test on drug use. J. Am. Stat. Assoc. 70, 814–818.

Greenberg, B.D., Abul-Ela, A.L.A., Simmons, W.R., Horvitz, D.G., 1969. The unrelated question randomized response model: theoretical framework. J. Am. Stat. Assoc. 64, 520–539.

Gupta, S., 2001. Qualifying the sensitivity level of binary response personal interview survey questions. J. Comb. Inform. Syst. Sci. 26 (1–4), 101–109.

Gupta, S., Gupta, B., Singh, S., 2002. Estimation of sensitivity level of personal interview survey question. J. Stat. Plan. Inference 100, 239–247.

Gupta, S., Kalucha, G., Shabbir, J., Dass, B.K., 2014. Estimation of finite population mean using optional RRT models in the presence of nonsensitive auxiliary information. Am. J. Math. Manage. Sci. 33 (2), 147.

Gupta, S., Mehta, S., Shabbir, J., Dass, B.K., 2013. Generalized scrambling in quantitative optional randomized response models. Commun. Stat. Theory Methods 42, 4034–4042.

Gupta, S., Shabbir, J., 2004. Sensitivity estimation for personal interview survey questions. Statistica 64 (3), 643–653.

Gupta, S., Shabbir, J., Sehra, S., 2010. A comparison of multiplicative and additive optional RRT models. Commun. Stat. Theory Methods 140 (10), 2870–2874.

Gupta, S., Thornton, B., Shabbir, J., Singhal, S., 2006. A comparison of multiplicative and additive optional RRT models. J. Stat. Theory Appl. 64, 226–239.

Horvitz, D.G., Shah, B.V., Simmons, W.R., 1967. The unrelated question randomized response model. In: Proceedings of Social Statistical Section of the American Statistical Association, pp. 65–72.

Huang, K.C., 2004. A survey technique for estimating the proportion and sensitivity in a dichotomous finite population. Stat. Neeralandica 58, 75–82.

Huang, K.C., 2008. Estimation of sensitive characteristics using optional randomized technique. Qual. Quant. 42, 679–686.

Huang, K.C., 2010. Unbiased estimators of mean, variance and sensitivity level for quantitative characteristics in finite population sampling. Metrika 71, 341–352.

Kerkvliet, J.M., 1994. Estimating a logit model with randomized data: the case of cocaine use. Aust. J. Stat. 36, 9–20.

Kuk, A.Y., 1990. Asking sensitive question indirectly. Biometrika 77, 436–438.

Lanke, J., 1975. On the choice of unrelated questions in Simmons' version of RR. J. Am. Stat. Assoc. 68, 525–530.

Lanke, J., 1976. On the degree of protection in randomized interviews. Int. Stat. Rev. 44, 197–203.

Leysieffer, R.W., Warne, S.L., 1976. Respondent jeopardy and optimal designs in RR models. J. Am. Stat. Assoc. 71, 649–656.

Mangat, N.S., 1994. An improved ramdomized response strategy. J. Roy. Stat. Sosc., B 56, 93–95.

Mangat, N.S., Singh, R., 1990. An alternative randomized response procedure. Biometrika 77, 349–442.

Mangat, N.S., Singh, S., 1994. An optional randomized response sampling technique. J. Ind. Stat. Assoc. 32, 71–75.

Marcheslli, M., Barabesi, L., 2006. A generalization of Huang's randomized response procedure for estimation of population proportion and sensitivity level. Metron LXIV, 145–159.

Mehta, S., Dass, B.K., Shabbir, J., Gupta, S.N., 2012. A three-stage optional randomized response model. J. Stat. Theory Pract. 6 (3), 417–427.

Pal, S., 2008. Unbiasedly estimating the total of a stigmatizing variable from a complex survey on permitting options for direct or randomized responses. Stat. Papers 49, 157–164.

Rao, J.N.K., Hartley, H.O., Cochran, W.G., 1962. A simple procedure of unequal probability sampling without replacement. J. R. Stat. Soc. B 24, 482–491.

Saha, A., 2007. Optional randomized response in stratified unequal probability sampling. A simulation based numerical study with Kuk's method. Test 16, 346–354.

Sihm, J.S., Gupta, S., 2015. A two-stage binary optional response model. Commun. Stat. Simul. Comput. 44 (9), 2278–2296.

Singh, H.P., Tarray, T.A., 2014. A stratified mangat and singh's optional randomized response model using proportional and optimal allocation. Statistica, LXXIII, pp. 65–83.

Singh, S., 2003. Advanced Sampling Theory with Applications. Kluwer Academic Publishers, Netherlands.

Singh, S., Joarder, A.H., 1997. Optional randomized response technique for sensitive quantitative variable. Metron 55, 151.

Singh, S., Singh, R., Mangat, N.S., 2000. Some alternative strategies to Moor's model in randomized response sampling. J. Stat. Plan. Inference 83, 243–255.

Van den Hout, A., van der Heijden, P.G., Gilchrist, R., 2007. The logistic regression model with response variables subject to randomized response. Comput. Stat. Data Anal. 51 (12), 6060.

Van der Heijden, P.G.M., Van Gils, G., Bouts, J., Hox, J.J., 1998. A comparison of randomized response, CASAQ, and direct questioning: eliciting sensitive information in the context of social security fraud. Kwantitatieve Methoden 59, 15–34.

Warner, S.L., 1965. Randomize response: a survey technique for eliminating evasive answer bias. J. Am. Stat. Assoc. 60, 63–69.

Chapter 17

A Concise Theory of Randomized Response Techniques for Privacy and Confidentiality Protection[a]

T.K. Nayak[*,†,1], S.A. Adeshiyan[‡] and C. Zhang[†]
*Center for Disclosure Avoidance Research, US Census Bureau, Washington, DC, United States
†George Washington University, Washington, DC, United States
‡US Energy Information Administration, Washington, DC, United States
1Corresponding author: e-mail: tapan@gwu.edu

ABSTRACT

A variety of randomized response (RR) procedures for privacy and confidentiality protection have been proposed, studied, and compared in the literature. We describe statistically relevant attributes of RR mechanisms and use those to duly organize and unify existing estimation theories for diverse RR methods. Any RR procedure can be characterized by its transition probability matrix P, as it determines all statistical properties of the procedure. In RR surveys, P is fixed, but in post-randomization for confidentiality protection, P may depend on the original data. This affects statistical inferences significantly. We also review some optimality results in the comparison of RR surveys of a binary variable, based on both privacy protection and statistical efficiency.

Keywords: Categorical variable, Post-randomization, Random transformation, Sampling design, Transition probability matrix, Unbiased estimation, Variance inflation

1 INTRODUCTION

Concerns about privacy and confidentiality often yield substantial nonresponses and false responses, especially in surveys that ask direct questions

[a]The views expressed in this chapter are those of the authors and not necessarily those of the US Census Bureau. The analysis and conclusions contained in this chapter are those of the authors and do not represent the official position of the US Energy Information Administration (EIA) or the US Department of Energy (DOE).

Handbook of Statistics, Vol. 34. http://dx.doi.org/10.1016/bs.host.2016.01.015

about sensitive attributes such as criminal history, tax evasion, drug abuse, gambling, and abortion. To protect respondents' privacy and increase truthful respondent participation, Warner (1965) proposed the first randomized response (RR) procedure for a binary characteristic. Consider a population where each person belongs either to a sensitive group A or to its complement A^c, where the true proportion (π) of the population that belongs to A is unknown and needs to be estimated from survey data. In Warner's (1965) method, each respondent is instructed to select one of the two questions Q_1: Do you belong to A? and Q_2: Do you belong to A^c? in a prescribed random manner (eg, by drawing a card from a shuffled deck or using a spinner) and truthfully answer the question by "Yes" or "No" without disclosing the question, to protect his/her privacy. The probability (p) of selecting Q_1 is the method's design parameter and the survey organization needs to choose its value and devise an experiment to implement it. Basically, Warner's procedure converts each true "Yes" (or No) to a false "No" (or Yes) with probability $1 - p$.

Subsequent to Warner's (1965) pioneering paper, many other RR methods for both categorical and quantitative variables have been proposed and investigated (see Chaudhuri and Mukerjee, 1988; Chaudhuri, 2010; Chaudhuri and Christofides, 2013; and other articles in this book for excellent exposition and references). All RR methods modify the true responses in some stochastic fashion, which necessitates developing theory for (i) making inferences from randomized data and (ii) assessing how much privacy and confidentiality protection an RR method provides. The main goal of this chapter is to discuss these two aspects of RR methods for categorical data in a general and unified manner. We shall concentrate on important logical elements of RR theory, leaving aside related mathematical derivations that are readily available in the literature. Also, we shall not discuss RR procedures for quantitative variables, which mostly employ noise addition or multiplication (see Brand, 2002; Evans et al., 1998; Fuller, 1993; Nayak et al., 2011), as those are patently different from methods for categorical variables.

Privacy and confidentiality are often used synonymously, but incorrectly. Privacy is an individual's right to control access to his information and privacy protection means hiding a respondent's true values from everyone, including the interviewer. Thus, Warner's method was designed to give privacy protection. Protecting confidentiality means keeping information about each respondent confidential. Often respondents give their information to survey organizations trusting that their data will be used by researchers and policy makers only to learn about the population as a whole and not about any individual or survey unit. Privacy arises at data collection stage, whereas confidentiality emerges after the data have been collected. Confidentiality protection often prevents data agencies from releasing original data as that might enable others to gain much information about specific units in the survey. So, agencies typically release perturbed or masked versions of original data.

Randomization of the original values is one technique for protecting confidentiality. However, we should note that in RR surveys the randomization mechanism is chosen before data collection and hence it cannot depend on the data, but for confidentiality protection, the randomization process may be chosen based on the data. This is a significant difference and it requires us to treat privacy and confidentiality differently. Nayak et al. (2015) discuss additional differences between privacy and confidentiality protection.

Both data analysis and the protection of privacy or confidentiality depend on the randomization process and the sampling mechanism. Thus, we need to organize the theory by different types of sampling and RR mechanism. We shall consider two types of sampling: multinomial sampling and general probability sampling from a finite population. While RR methods may differ in many ways, only certain features are statistically relevant and affect data analysis or privacy and confidentiality protection. It is important to recognize those features for developing statistical theory and proper comparison of competing methods. The rest of this chapter is organized as follows. In Section 2, we examine RR experiments from a statistical perspective to discern their essential features and design parameters. This helps us to put different RR procedures in a common framework and grasp their substantive differences, if any. In Section 3, we discuss statistical estimation when the randomization experiment is specified prior to data collection. In Section 4, we review some recent results on estimation for invariant post-randomization, which is a particular type of data-dependent randomization technique that is useful for confidentiality protection. In Section 5, we discuss privacy protection and certain optimality results. Section 6 contains some closing remarks.

2 VITAL ATTRIBUTES OF RANDOMIZATION EXPERIMENTS

In this section, we examine randomization experiments for categorical variables to bring out their statistically essential features. Let X be a categorical variable with categories labeled $c_1,\ldots,c_k$, and let $\pi_i = P(X = c_i), i = 1,\ldots,k$, and $\Pi = (\pi_1,\ldots,\pi_k)'$. Also, when we randomize several categorical variables, X will represent their cross-classification and Π will represent all joint probabilities. Typically, Π is unknown and we wish to make inferences about Π and functions of it. The main purpose of RR surveys is to collect useful information while protecting respondent's privacy. Fundamentally, all RR methods ask each respondent to perform a given random experiment and use its outcome, the respondent's true category, and prespecified rules to compute an output and report it. Let Z denote the output variable. Evidently, the input space (or domain) of the random transformation from X to Z is $\{c_1,\ldots,c_k\}$. Consider an RR process with a finite output space (or range), containing m elements, labeled $d_1,\ldots,d_m$. Then, the randomization process produces different outputs, for given input, with specific probabilities. Thus, let $p_{ij} = P(Z = d_i | X = c_j), i = 1,\ldots,m, j = 1,\ldots,k$. To define m uniquely, we should require each $d_i (i=1,...,m)$ to occur with positive probability, ie, $p_{ij} > 0$ for some j. Then, the transition probability matrix of order $(m \times k)$ of the RR procedure is $P = ((p_{ij}))$ and $\sum_i p_{ij} = 1$ for $j = 1,...,k$.

Any randomization process is characterized by its output space and the transition probability matrix P (just as a random variable is fully described by its probability distribution). So, all effects of an RR procedure on the distribution of Z and privacy protection are determined only through its P. One important (and underappreciated) implication of this is that for statistical investigations we should consider only P, leaving aside ancillary features of the experiment, eg, whether it uses a deck of cards or a spinner. However, as Leysieffer and Warner (1976), Fligner et al. (1977), Nayak (1994), and others have noted, some papers have improperly compared different methods using other features that are not really comparable across those experiments. For example, using a common probability for asking Q_1 in Warner's and Simmons' (see below) methods do not make the two methods comparable in terms of privacy protection.

Next, all RR procedures can be divided into two groups depending on whether P is fully known or not. For example, we consider a widely known variation of Warner's procedure, called Simmons' unrelated question method (Greenberg et al., 1969), which uses an unrelated nonsensitive question Q_3: Do you belong to B? in place of Q_2. For the transformation in Warner's and Simmons' methods, the domain is $\{A, A^c\}$ and the range is $\{Yes, No\}$. However, in Warner's method, the transition probabilities are known, but in Simmons' method they are known only if $P(B)$, the probability of answering "yes" to Q_3, is known. If P is unknown, generally multiple samples from different settings (and correspondingly specialized theory) are needed for estimating π. In contrast, for known P, one can estimate Π from one sample fairly easily. For protecting privacy and confidentiality, we do not see a need for using an RR method with unknown P. Thus, we shall only consider RR procedures with known P.

Let $\lambda_i = P(Z = d_i), i = 1,...,m$, and $\lambda = (\lambda_1,...,\lambda_m)'$. Then, under multinomial sampling, ie, random sampling from an infinite population or simple random sampling with replacement if the population is finite, $\lambda_i = \sum_{j=1}^{k} p_{ij}\pi_j$. In matrix notation, we have

$$\lambda = P\Pi. \tag{1}$$

We can use RR data to make inferences about λ, which in turn can be used to make inferences about Π, using (1). In particular, if $m = k$ and P is nonsingular, an estimator $\hat{\lambda}$ of λ yields the estimator

$$\hat{\Pi} = P^{-1}\hat{\lambda}. \tag{2}$$

of Π. However, if $m \neq k$, generally (1) will yield no solution or multiple solutions for Π, causing difficulties in estimating Π from RR data. For $k = 2$,

Nayak and Adeshiyan (2009) showed that for any RR procedure with $m > 2$, there exists one RR procedure with $m = 2$ that is better in terms of privacy protection and statistical efficiency. Thus, in the following we shall only consider RR procedures for which $k = m$ and P is a known nonsingular matrix.

Next, note that in general, the labels $d_1,\ldots,d_m$ for possible outputs of an RR procedure are arbitrary. In particular, any permutation of the labels does not alter the procedure or its properties, but changes the transition probability matrix. This implies that two RR procedures with transition probability matrices P_1 and P_2 are statistically equivalent if P_1 can be obtained by permuting the rows of P_2. Obviously, P_1 and P_2 can be equivalent only if they have common dimensions, ie, the ranges of the two procedures contain the same number of elements. As an example, two Warner's methods, where Q_1 is asked with probabilities p and $(1 - p)$, respectively, are equivalent, as one can be obtained from the other by interchanging the two responses "Yes" and "No." For $k = m = 2$, ie, when the original and perturbed variables are binary, Nayak (1994) imposed $p_{11} > p_{12}$ and $p_{22} > p_{21}$ to characterize each RR procedure uniquely by its transition probabilities. It will be interesting to develop similar constraints to specify P uniquely for general k and m. However, in the following, we shall only consider RR procedures with output space $\{c_1,\ldots,c_k\}$. This is helpful in two ways. First, fixing the range to $\{c_1,\ldots,c_k\}$ makes all transition probability matrices directly comparable. In particular, P_1 and P_2 would be equivalent if and only if $P_1 = P_2$. Second, for data users, released data are easier (and less confusing) to understand when they are reported using the categories of original variables.

As we noted in Section 1, in the context of confidentiality protection, P can be chosen based on the original data. Whether P depends on the data or not is an important consideration in RR theory, because if P depends on the data, then it is a random matrix and this should be properly accounted for in statistical inferences. We shall elaborate this further in Section 4.

Logically, to design an RR survey, we should first choose a transition probability matrix P, considering both privacy protection and statistical efficiency, and then devise an experiment to implement it. First, any P can be implemented using k experiments, where the jth experiment is used only by respondents with $X = c_j$ and it outputs $c_1,\ldots,c_k$ with probabilities $p_{1j},\ldots,p_{kj}$. This can be accomplished, for example, by using k decks of cards, one for each category of X. This is similar to Kuk's (1990) RR procedure. In this approach, each respondent needs to select and use a randomization mechanism (eg, one of k decks of cards), unobserved by the interviewer, based on his true category of X. Nonetheless, respondents may be suspicious of such methods as a respondent's experiment selection reveals his true category of X.

Next, we discuss another view of RR procedures, following Nayak (1994), and show that it is also possible to devise a common experiment, independent of true X category, for implementing any given P. RR is often considered as a random transformation of the true response. Mathematically, a transformation is a function and a random transformation can be viewed as probabilistically selecting one function from a set of functions and then applying it to the true response to generate an RR. For $k = 2$, there are four possible functions from $\{c_1, c_2\}$ to $\{c_1, c_2\}$, given by $g_1 : c_1 \to c_1, c_2 \to c_2$ (ie, $g(c_1) = c_1$ and $g(c_2) = c_2$); $g_2 : c_1 \to c_2, c_2 \to c_1$; $g_3 : c_1 \to c_1, c_2 \to c_1$; and $g_4 : c_1 \to c_2, c_2 \to c_2$, and any RR procedure essentially uses $g_1,\ldots,g_4$ with some probabilities $\alpha_1,\ldots,\alpha_4$. Two procedures can be differentiated by their values of $\alpha_1,\ldots,\alpha_4$. Note that g_1 and g_2 correspond to using Q_1 and Q_2, and g_3 and g_4 represent *forced* A and *forced* A^c responses (labeling A and A^c by c_1 and c_2). Thus, for example, Warner's method uses only g_1 and g_2 with probabilities p and $(1 - p)$, and the triangular method of Tan et al. (2009) used only g_1 and g_3 (with given probabilities). As Quatember (2009) noted, many existing methods use different combinations of Q_1, Q_2, Q_3 and *forced* A and *forced* A^c responses. However, using Q_3 is equivalent to using g_3 and g_4 with probabilities $P(B)$ and $1 - P(B)$, respectively.

More generally, for X with k categories, there are k^k functions from $\{c_1,\ldots,c_k\}$ to $\{c_1,\ldots,c_k\}$, which we shall denote by $g_1,\ldots,g_{k^k}$. Let $\alpha = (\alpha_1,\ldots,\alpha_{k^k})$ be a probability distribution on the set of these functions. An RR procedure may be viewed as selecting one function according to α and applying it to the true response. Here, the randomization process is the same for all respondents. For given α, the transition probabilities are:

$$p_{ij} = P(Z = c_i|X = c_j) = \sum_{l=1}^{k^k} \alpha_l P(Z = c_i|X = c_j, g_l), \quad i,j = 1,\ldots,k, \tag{3}$$

where $P(Z = c_i|X = c_j,g_l)$ is the conditional probability of $Z = c_i$ given that $X = c_j$ and g_l is selected by the randomization process. Since each g_l is a (deterministic) function, this conditional probability is either 0 or 1. Specifically, $P(Z = c_i|X = c_j,g_l) = 1$ if $g_l(c_j) = c_i$; otherwise $P(Z = c_i|X = c_j,g_l) = 0$. Obviously, any α induces a unique transition probability matrix P. Conversely, given any transition probability matrix P, we can find α to satisfy (3). Note that there are $k(k - 1)$ equations in (3), in view of the fact that $\sum_i p_{ij} = 1$ for $j = 1,\ldots,k$, and $(k^k - 1)$ unknowns (in α). Thus, the number of unknowns is much larger than the number of equations, and for given P, (3) will have multiple solutions. We can use any of those solutions to implement P by an experiment that is common to all respondents.

3 STATISTICAL ESTIMATION FOR FIXED *P*

Most authors have considered estimating Π under multinomial sampling when P is known and fixed. Let n denote the sample size, T_i and S_i denote the sample frequencies of $X = c_i$ and $Z = c_i$, respectively, $T = (T_1,\ldots,T_k)'$, and $S = (S_1,\ldots,S_k)'$. Then, $T \sim Mult(n,\pi)$, $S \sim Mult(n,\lambda)$, and $\hat{\lambda} = S/n$ is the MLE (and UMVUE) of λ, which in turn yields

$$\hat{\Pi} = P^{-1}\hat{\lambda} = P^{-1}(S/n), \tag{4}$$

as an unbiased estimator of Π with

$$Var(\hat{\Pi}) = \frac{1}{n}(D_\Pi - \Pi\Pi') + \frac{1}{n}[P^{-1}D_\lambda(P^{-1})' - D_\Pi], \tag{5}$$

where D_Π is a diagonal matrix with diagonal elements being $\pi_1,\ldots,\pi_k$ and D_λ is defined similarly (see Chaudhuri and Mukerjee, 1988, p. 43). The first term on the right side of (5) is the sampling variance and the last term is the additional variance due to randomization. Most authors estimate (5) by replacing π by $\hat{\Pi}$ and give interval estimates only for large n when $\hat{\Pi}$ is approximately normally distributed. However for small and moderate n, both $\hat{\Pi}$ and interval estimates can take values outside the interval [0,1]. Frey and Prez (2012) nicely discuss this aspect for binary X and present a novel method for constructing exact confidence intervals.

While research on RR methods started long ago and in survey context, inferences under general sampling designs have been discussed only recently, initiated by Padmawar and Vijayan (2000). Their basic ideas have been used and further developed by Chaudhuri (2001, 2004), Nayak and Adeshiyan (2009), and others. In the following, we summarize some results from Adeshiyan (2011). Consider a finite population of N units, labeled $i = 1,\ldots,N$, and suppose a sample s, which is a subset of $\{1,\ldots,N\}$, is selected using a noninformative sampling design $p(s)$. As before, the survey variable X is categorical with k categories and we observe only a randomized version (Z) of it, where the transition probability matrix P is known, fixed, and nonsingular.

For unit $i(i = 1,\ldots,N)$, we introduce two k-dimensional indicator vectors I_i and J_i. Specifically, if the true category of unit i is c_j, then the jth component of I_i is 1 and all other components are 0. We define J_i similarly, to record the (randomized) response category of unit i, if it is sampled. Thus, the data can be represented as $\{(i, J_i);\ i \in s\}$. In this setup, consider estimating $\tau = \sum_{i=1}^{N} I_i$, which is equivalent to estimating $\Pi = (\sum_{i=1}^{N} I_i)/N$. If the true categories, ie, I_i, $i \in s$ were observed, Π can be estimated unbiasedly using the Horvitz–Thompson estimator. More generally, suppose $\hat{\tau} = \sum_{i\in s} w_{si} I_i$ is a homogeneous linear (design) unbiased estimator of τ based on original data, ie,

$$\sum_s \left(\sum_{i\in s} w_{si} I_i\right) p(s) = \sum_{i=1}^{N} I_i \quad \text{for all } I_1,\ldots,I_N.$$

Then, from the fact $E(J_i|I_i) = I_i$, where the expectation is with respect to the RR process, it follows that $\hat{\tau}^* = \sum_{i\in s} w_{si}\left(P^{-1}J_i\right)$ is a design unbiased estimator of τ based on RR survey data. Thus, any unbiased estimator for an open survey can be easily modified to obtain an unbiased estimator when RR is used.

Adeshiyan (2011) showed that the variance of $\hat{\tau}^*$ can be expressed as

$$V(\hat{\tau}^*) = E_p\left[\sum_{i\in s} w_{si}^2\left\{P^{-1}D_{PI_i}(P')^{-1}\right\} - D_{I_i}\right] + V_p(\hat{\tau}),$$

where E_p denotes expectation with respect to the sampling design,

$$V_p(\hat{\tau}) = \sum_{i=1}^{N} b_i D_{I_i} + \sum_{i,j=1,i\neq j}^{N} b_{ij} I_i I_j',$$

$b_i = \sum_{s\ni i} w_{si}^2 p(s) - 1$ and $b_{ij} = \sum_{s\ni i,j} w_{si} w_{sj} p(s) - 1$. An unbiased estimator of $V(\hat{\tau}^*)$ is also presented in Adeshiyan (2011).

4 ESTIMATION UNDER INVARIANT POST-RANDOMIZATION

Statistical agencies often release only a perturbed or masked version of original data to protect the confidentiality of respondent-level information. Doyle et al. (2001) and Willenborg and De Waal (2001) described a variety of methods, such as grouping, top coding, swapping, multiple imputation, and noise addition, for creating perturbed data.Warner (1971) indicated that RR techniques may be used for confidentiality protection. Gouweleeuw et al. (1998) developed the idea as the Post-randomization Method (PRAM) for perturbing categorical data. As before, consider a categorical variable X, which may also represent the cross-classification of several variables. Just as in RR surveys, PRAM converts the true categories to RRs (categories) using a known transition probability matrix P, also called the PRAM matrix. The randomization process is applied to each observation in the data set, independently of all other records. Also, this process is carried out by the survey organization rather than the respondents. For further discussion of PRAM and additional references, we refer the reader to Gouweleeuw et al. (1998), Van den Hout and Van der Heijden (2002), Van den Hout and Elamir (2006), Cruyff et al. (2008), Shlomo and Skinner (2010), Nayak and Adeshiyan (2016), and Nayak et al. (2015).

For fixed P, PRAM is mathematically equivalent to RR surveys and hence all inferential methods developed for RR surveys are applicable to data perturbed by PRAM. However, data agencies rarely release the values of the parameters used in their perturbation procedures, such as P for PRAM. Thus, data users usually would not know the PRAM matrix P and hence would not be able to use the results for known P, as in the preceding section. In response to this obstacle, statistical agencies try to use data perturbation procedures for which standard inferential methods for original data remain valid, at least approximately, for masked data. This motivated Gouweleeuw et al. (1998) to define a PRAM to be an invariant PRAM if P satisfies

$$PT = T \quad \text{or equivalently} \quad P\hat{\Pi}_0 = \hat{\Pi}_0, \tag{6}$$

where $\hat{\Pi}_0 = T/n$. Since $E[S|T] = PT$, under multinomial sampling, (6) implies that $\hat{\Pi}_* = S/n$, the relative frequency vector based on perturbed data, is an unbiased estimator of Π. Also, $\hat{\Pi}_*$ is always a probability vector and it can be calculated without using P or its inverse.

To apply invariant PRAM, we need to select and use one P satisfying (6). We should note that generally (6) has many solutions for P and the solution space of (6) is a nonempty convex set, which also contains the identity matrix. Gouweleeuw et al. (1998) and Nayak and Adeshiyan (2016) give some methods for solving (6). One important point to note is that solutions of (6) depend on original data through T. Thus, in invariant PRAM, P is a random matrix. Consequently, mathematical results for fixed P may not hold for invariant PRAM. In the following we summarize some properties of $\hat{\Pi}_*$, from Nayak and Adeshiyan (2016).

Let $P = [P_1 : \cdots : P_k]$, ie, denote the ith column of P by P_i, and rewrite (6) as

$$\sum_{i=1}^{k} T_i P_i = T. \tag{7}$$

Let F_{ij} denote the number of units whose category changed from c_i to c_j due to PRAM, and let $F_i = (F_{i1}, \ldots, F_{ik})'$. Then, $S = \sum_i^k F_i$, and given T and P, $F_1, \ldots, F_k$ are independently distributed with $F_i \sim Mult(T_i, P_i), i = 1, \ldots, k$. Now,

$$E(\hat{\Pi}_*|T,P) = \frac{1}{n}\sum_{i=1}^{k} E[F_i|T,P] = \frac{1}{n}\sum_{i=1}^{k} T_i P_i = \frac{T}{n}, \tag{8}$$

by (7). From (8), it follows easily that under multinomial sampling $E(\hat{\Pi}_*) = \Pi$, ie, $\hat{\Pi}_*$ is an unbiased estimator of Π. It can also be seen that

$$V(\hat{\Pi}_*|T,P) = \frac{1}{n^2}\sum_{i=1}^{k} T_i[D_{P_i} - P_i P_i'] = \frac{1}{n}\left[D_{\hat{\Pi}_0} - \sum_{i=1}^{k}\left(\frac{T_i}{n}\right)P_i P_i'\right], \tag{9}$$

and

$$\begin{aligned} V(\hat{\Pi}_*) &= V[E(\hat{\Pi}_*|T,P)] + E[V(\hat{\Pi}_*|T,P)] \\ &= V(\hat{\Pi}_0) + \frac{1}{n}\left[D_{\Pi} - E\left\{\sum_{i=1}^{k}(\frac{T_i}{n})P_i P_i'\right\}\right]. \end{aligned} \tag{10}$$

Notice the difference between the last terms of (5) and (10), the variance inflations due to PRAM with fixed P and invariant PRAM. The last expectation in (10) also involves the conditional distribution of P given T, which is determined by the data agency's algorithm (or method) for choosing a solution of (6) for P. Typically, data users would not know P or the agency's full process for selecting P and hence would not be able to evaluate or estimate

$V(\hat{\Pi}_*)$. These observations show that under invariant PRAM, data users will be able to calculate an unbiased estimate of Π (using $\hat{\Pi}_*$), but not its sampling variance (and hence reliability). However, as Nayak and Adeshiyan (2016) describe, the data agency can calculate (9) and also estimates of (10). They also proved that

$$V_{max}(\hat{\Pi}_*) = \left(2 - \frac{1}{n}\right)\left[\frac{D_\Pi - \Pi\Pi'}{n}\right] \tag{11}$$

is a tight upper bound of $V(\hat{\Pi}_*)$, in the sense that $[V_{max}(\hat{\Pi}_*) - V(\hat{\Pi}_*)]$ is nonnegative definite for all invariant PRAM and there exists an invariant PRAM for which $V(\hat{\Pi}_*) = V_{max}(\hat{\Pi}_*)$.

An obvious lower bound for $V(\hat{\Pi}_*)$ is $V(\hat{\Pi}_0) = [D_\Pi - \Pi\Pi]/n$. These upper and lower bounds can be estimated by replacing Π with $\hat{\Pi}_*$. Estimating linear combinations of Π is a common problem. A natural (and unbiased) estimator of $a'\Pi$ is $a'\hat{\Pi}_*$. The upper and lower bounds of $V(\hat{\Pi}_*)$ immediately yield bounds for $V(a'\hat{\Pi}_*) = a'[V(\hat{\Pi}_*)]a$. We conclude this section by mentioning that Nayak and Adeshiyan (2016) also present some results on estimation under invariant PRAM and general probability sampling.

5 ASSESSING PRIVACY AND CONFIDENTIALITY PROTECTION

The main reason for randomizing true responses, although it reduces data quality, is privacy and confidentiality protection. Thus, we must assess how well a method protects privacy or confidentiality. However, this is a very difficult task, because privacy and confidentiality are complex, multifaceted concepts. Disclosure of private or confidential information can occur in many different ways and forms (see Willenborg and De Waal, 2001), and it is difficult to develop criteria for measuring privacy and confidentiality protection (or lack of it) that are both credible and widely applicable. Lambert (1993) discusses several intrinsic challenges in measuring risk and consequences of disclosure. While many researchers have investigated this topic, precise and substantive results have been obtained only for some specific situations.

In RR survey context, most work on measuring and comparing privacy protection has focused on binary survey variables with one category being stigmatizing, as Warner (1965) considered. For reviewing the results, it will be convenient to use Warner's (1965) terminology (but not limited to his experiment). Suppose X is a binary variable with two categories A and A^c, with $\pi = P(A)$, and only A is sensitive or stigmatizing. Suppose the RR also has two categories, denoted Y and $\bar{Y}$ (for "Yes" and "No," but they can also mean other responses). Here, a transition probability matrix P is of order 2×2 and has two free elements, as each column adds to 1. Thus, each procedure can be characterized by the two probabilities embedded in it, viz.,

$\gamma = P(Y \mid A)$ (ie, the probability that a respondent answers "Yes" given that his/her true category is A) and $\delta = P(Y \mid A^c)$. The pair (γ, δ) defines the *design* of an RR procedure. However, considering the effects of interchanging Y and $\bar{Y}$ (as discussed in Section 2), it is seen that two procedures with designs (γ, δ) and (δ, γ) are equivalent. Following Nayak (1994), the *design space* can be specified appropriately as $\mathcal{D} = \{(\gamma, \delta) : 0 \leq \delta < \gamma \leq 1\}$, where we associate Y with A and $\bar{Y}$ with A^c (which seems natural).

Many authors, including Leysieffer and Warner (1976), Lanke (1976), Fligner et al. (1977), Nayak (1994), and Guerriero and Sandri (2007), have discussed measuring privacy protection in the preceding context. Generally, all measures involve the two posterior probabilities

$$P(A|Y) = \frac{\gamma\pi}{\gamma\pi + \delta(1-\pi)} \quad \text{and} \quad P(A|\bar{Y}) = \frac{(1-\gamma)\pi}{(1-\gamma)\pi + (1-\delta)(1-\pi)} \tag{12}$$

and suggest that protection increases as these two probabilities decrease. It can be seen that the estimator $\hat{\pi}$ of π based on (4) has variance $V(\hat{\pi}) = [\theta(1-\theta)]/[n(\gamma-\delta)^2]$, where $\theta = \delta + (\gamma - \delta)\pi$. Noting that we desire $P(A|Y), P(A|\bar{Y})$, and $V(\hat{\pi})$ to be small and all three quantities depend on unknown π, Nayak (1994) presented the following.

Definition 1. An RR design $D_1 = (\gamma_1, \delta_1)$ is said to be better than another design $D_2 = (\gamma_2, \delta_2)$, or D_1 dominates D_2, if $P_{D_1}(A|Y) \leq P_{D_2}(A|Y), P_{D_1}(A|\bar{Y}) \leq P_{D_2}(A|\bar{Y})$, and $V_{D_1}(\hat{\pi}) \leq V_{D_1}(\hat{\pi})$ for all $0 \leq \pi \leq 1$, with at least one strict inequality holding for some π.

An RR design D is said to be inadmissible if it is dominated by some other design D_*. Otherwise, D is said to be admissible.

Theorem 1. *An RR procedure with design (γ,δ) is admissible if and only if $\gamma = 1$.*

Nayak (1994) discussed some important implications of this result. It shows that any admissible procedure must require all respondents in A to respond Y. In particular, any procedure that uses Q_2 or Q_3 or "forced no" cannot be admissible. In terms of the functions g_1, g_2, g_3, g_4 in Section 2, an admissible procedure can use only g_1 and g_3. We can also see easily that of the 15 RR procedures in Table 1 of Quatember (2009), only 1 is admissible.

Now suppose X is binary and only A is stigmatizing, but the RR variable Z has m categories, $d_1, \ldots, d_m$ (eg, Christofides, 2003; Kuk, 1990; Leysieffer and Warner, 1976). Consider $\psi = \max\{P(A|Z=d_1), \ldots, P(A|Z=d_m)\}$ as a (summary) measure for respondent's risk (of being classified in A). With this, Nayak and Adeshiyan (2009) proved that for any RR procedure $\mathcal{R}$ with $m \geq 3$ response categories, there exists an RR procedure $\mathcal{R}_*$ with two response categories such that $\mathcal{R}_*$ is at least as good as $\mathcal{R}$ in terms of both privacy protection and statistical efficiency.

If both A and A^c are sensitive or stigmatizing, measuring and controlling privacy protection requires a different approach. Intuitively, for any response, say Y, we would like both $P(A|Y)$ and $P(A^c|Y)$ to be small, but that is not sensible as $P(A|Y) + P(A^c|Y) = 1$. However, some papers, eg, Quatember (2009) and Chaudhuri (2010), have discussed the topic. Similar issues arise in post-randomization for confidentiality protection, where true variables commonly have multiple categories of which many may be sensitive. Also, confidentiality protection goals can be quite different. Some methods for evaluating the efficacy of PRAM in disclosure control and ideas on how to choose the PRAM matrix have appeared in Gouweleeuw et al. (1998), Van den Hout and Elamir (2006), Shlomo and De Waal (2008), and Shlomo and Skinner (2010). We believe that additional theoretical and empirical research will help to develop PRAM as a valuable disclosure control method.

6 DISCUSSION

Many RR techniques for privacy and confidentiality protection have appeared in the literature. We discussed some basic elements of RR mechanisms and stressed that for proper understanding, unification, and fair comparison, RR procedures should be characterized and examined through their transition probability matrices. The main factors that affect inferential methods are the dimensions and rank of the transition probability matrix P, and whether P is (i) fully known or not and (ii) fixed or selected based on the data. However, RR procedures with known, square, and nonsingular P are most convenient for statistical estimation and they should also be adequate for protecting respondents' privacy and confidentiality.

A common assumption in mathematical theory of RR methods is that survey participants respond truthfully, which is unrealistic. Also, respondents' perception of privacy protection may depend on secondary features, such as familiarity and complexity, of the RR experiment. Thus, respondents' participation and truthfulness may be different in two procedures with a common P. In practice, one should assess respondents' willingness and concerns by conducting pilot studies, and utilize the findings in selecting an appropriate RR procedure. For devising an RR experiment, after choosing P, it may be helpful to think in terms of selecting a mathematical function probabilistically and then applying it to the respondent's true category, as discussed in Section 2. This approach offers much flexibility, as any P can be executed by many probability distributions on the set of all functions and one may use secondary criteria to choose one of those (see Nayak, 1994 for some specific proposals and related results for a binary variable).

ACKNOWLEDGMENTS

We thank Mr. Joseph Conklin and Prof. Lucio Barabesi for carefully reading an earlier draft and giving helpful suggestions for improving the presentation.

REFERENCES

Adeshiyan, S.A., 2011. Unification of randomized response designs and certain aspects of post-randomization for statistical disclosure control. PhD dissertation. The George Washington University.

Brand, R., 2002. Microdata protection through noise addition. In: Domingo-Ferrer, J. (Ed.), Inference Control in Statistical Databases. Springer, Berlin, pp. 97–116.

Chaudhuri, A., 2001. Using randomized response from a complex survey to estimate a sensitive proportion in a dichotomous finite population. J. Stat. Plan. Inference 94 (1), 37–42.

Chaudhuri, A., 2004. Christofides randomized response technique in complex sample surveys. Metrika 60 (3), 223–228.

Chaudhuri, A., 2010. Randomized Response and Indirect Questioning Techniques in Surveys. CRC Press, Boca Raton.

Chaudhuri, A., Christofides, T.C., 2013. Indirect Questioning in Sample Surveys. Springer, New York.

Chaudhuri, A., Mukerjee, R., 1988. Randomized Response: Theory and Techniques. Marcel Dekker, New York.

Christofides, T.C., 2003. A generalized randomized response technique. Metrika 57 (2), 195–200.

Cruyff, M.J.L.F., Van Den Hout, A., Van Der Heijden, P.G.M., 2008. The analysis of randomized response sum score variables. J. R. Stat. Soc. Ser. B (Stat. Methodol.) 70 (1), 21–30.

Doyle, P., Lane, J., Theeuwes, J., Zayatz, L. (Eds.), 2001. Confidentiality, Disclosure and Data Access: Theory and Practical Applications for Statistical Agencies. Elsevier, Amsterdam.

Evans, T., Zayatz, L., Slanta, J., 1998. Using noise for disclosure limitation of establishment tabular data. J. Off. Stat. 14 (4), 537–551.

Fligner, M.A., Policello, G.E., Singh, J., 1977. A comparison of two randomized response survey methods with consideration for the level of respondent protection. Commun. Stat. Theory Methods 6 (15), 1511–1524.

Frey, J., Prez, A., 2012. Exact binomial confidence intervals for randomized response. Am. Stat. 66 (1), 8–15.

Fuller, W.A., 1993. Masking procedures for microdata disclosure limitation. J. Off. Stat. 9 (2), 383–406.

Gouweleeuw, J.M., Kooiman, P., Willenborg, L.C.R.J., De Wolf, P.P., 1998. Post randomisation for statistical disclosure control: theory and implementation. J. Off. Stat. 14 (4), 463–478.

Greenberg, B.G., Abul-Ela, A.L.A., Simmons, W.R., Horvitz, D.G., 1969. The unrelated question randomized response model: theoretical framework. J. Am. Stat. Assoc. 64 (326), 520–539.

Guerriero, M., Sandri, M.F., 2007. A note on the comparison of some randomized response procedures. J. Stat. Plan. Inference 137 (7), 2184–2190.

Kuk, A.Y.C., 1990. Asking sensitive questions indirectly. Biometrika 77 (2), 436–438.

Lambert, D., 1993. Measure of disclosure risk and harm. J. Off. Stat. 9 (2), 313–331.

Lanke, J., 1976. On the degree of protection in randomized interviews. Int. Stat. Rev. 44 (2), 197–203.

Leysieffer, R.W., Warner, S.L., 1976. Respondent jeopardy and optimal designs in randomized response models. J. Am. Stat. Assoc. 71 (355), 649–656.

Nayak, T.K., 1994. On randomized response surveys for estimating a proportion. Commun. Stat. Theory Methods 23 (11), 3303–3321.

Nayak, T.K., Adeshiyan, S.A., 2009. A unified framework for analysis and comparison of randomized response surveys of binary characteristics. J. Stat. Plan. Inference 139 (8), 2757–2766.

Nayak, T.K., Adeshiyan, S.A., 2016. On invariant post-randomization for statistical disclosure control. Int. Stat. Rev. (in press). http://dx.doi.org/10.1111/insr.12092.

Nayak, T.K., Sinha, B., Zayatz, L., 2011. Statistical properties of multiplicative noise masking for confidentiality protection. J. Off. Stat. 27 (3), 527–544.

Nayak, T.K., Zhang, C., Adeshiyan, S.A., 2015. Emerging applications of randomized response concepts and some related issues. Model Assist. Stat. Appl. 10 (4), 335–344.

Padmawar, V.R., Vijayan, K., 2000. Randomized response revisited. J. Stat. Plan. Inference 90 (2), 293–304.

Quatember, A., 2009. A standardization of randomized response strategies. Surv. Methodol. 35 (2), 143–152.

Shlomo, N., De Waal, T., 2008. Protection of micro-data subject to edit constraints against statistical disclosure. J. Off. Stat. 24 (2), 229–253.

Shlomo, N., Skinner, C., 2010. Assessing the protection provided by misclassification-based disclosure limitation methods for survey microdata. Ann. Appl. Stat. 4 (3), 1291–1310.

Tan, M.T., Tian, G.L., Tang, M.L., 2009. Sample surveys with sensitive questions: a nonrandomized response approach. Am. Stat. 63 (1), 9–16.

Van den Hout, A., Elamir, E.A.H., 2006. Statistical disclosure control using post randomisation: variants and measures for disclosure risk. J. Off. Stat. 22 (4), 711–731.

Van den Hout, A., Van der Heijden, P.G.M., 2002. Randomized response, statistical disclosure control and misclassification: a review. Int. Stat. Rev. 70 (2), 269–288.

Warner, S.L., 1965. Randomized response: a survey technique for eliminating evasive answer bias. J. Am. Stat. Assoc. 60 (309), 63–69.

Warner, S.L., 1971. The linear randomized response model. J. Am. Stat. Assoc. 66 (366), 884–888.

Willenborg, L.C.R.J., De Waal, T., 2001. Elements of Statistical Disclosure Control. Springer, New York.

Chapter 18

A Review of Regression Procedures for Randomized Response Data, Including Univariate and Multivariate Logistic Regression, the Proportional Odds Model and Item Response Model, and Self-Protective Responses

M.J.L.F. Cruyff*, U. Böckenholt†, P.G.M. van der Heijden*,‡,1 and L.E. Frank*

*Utrecht University, Utrecht, The Netherlands

†Kellogg School of Management, Northwestern University, Evanston, IL, United States

‡S3RI, University of Southampton, Southampton, United Kingdom

1Corresponding author: e-mail: P.G.M.vanderHeijden@uu.nl

ABSTRACT

In survey research, it is often problematic to ask people sensitive questions because they may refuse to answer or they may provide a socially desirable answer that does not reveal their true status on the sensitive question. To solve this problem Warner (1965) proposed randomized response (RR). Here, a chance mechanism hides why respondents say *yes* or *no* to the question being asked. Thus far RR has been mainly used in research to estimate the prevalence of sensitive characteristics. It is not uncommon that researchers wrongly believe that the RR procedure has the drawback that it is not possible to relate the sensitive characteristics to explanatory variables. Here, we provide a review of the literature of regression procedures for dichotomous RR data. Univariate RR data can be analyzed with a version of logistic regression that is adapted so that it can handle data collected by RR. Subsequently the manuscript presents extensions towards repeated cross-sectional data that allowed for a change in the design with which the RR data are collected. We also review regression procedures for multivariate

Handbook of Statistics, Vol. 34. http://dx.doi.org/10.1016/bs.host.2016.01.016

dichotomous RR data, such as the model by Glonek and McCullagh (1995), a model for the sum of a set of dichotomous RR data, and a model from item response theory that assumes a latent variable that explains the answers on the RR variables. We end with a discussion of a recent development in the analysis of multivariate RR data, namely models that take into account that there may be respondents that do not follow the instructions of the RR design by answering *no* whatever the sensitive question asked. These are coined self-protective responses.

Keywords: Randomized response, Logistic regression, Multivariate, Randomized response data, Proportional odds model, Item response model, Self-protective responses

1 INTRODUCTION

Is it possible to measure sensitive behavior such as noncompliance with rules and regulations that govern public life using surveys? Because it is well known that questions about compliance behavior with rules and regulations may not yield truthful responses, the randomized response (RR) method has been proposed as a survey tool to get more honest answers to sensitive questions (Warner, 1965). In the original RR approach, respondents were provided with two statements, A and B, with statement A being the complement of statement B. For example, statement A is "I used hard drugs last year" and statement B is "I did not use hard drugs last year." A randomizing device, for instance, in the form of a pair of dice determines whether statement A or B is to be answered. The interviewer records the answer *yes* or *no* without knowing the outcome of the randomizing response device. Thus the interviewee's privacy is protected but it is still possible to calculate the probability that the sensitive question (A and not B) is answered positively.

Recent meta-analyses have shown that RR methods can outperform more direct ways of asking sensitive questions (Lensvelt-Mulders et al., 2005). Importantly, the relative improvements in the validity increased with the sensitivity of the topic under investigation.

This chapter will show recent developments in the analysis of RR data, focusing on regression models relating the sensitive question(s) measured with RR to explanatory variables and modeling multivariate relationships. Thus we show the developments in this field since a pioneering paper of Scheers and Dayton (1988), who showed how this relation with sensitive questions and explanatory variables can be studied by adjusting the logistic regression model.

In this manuscript, the examples stem from surveys that we conducted for the Dutch government on noncompliance in social welfare. In the Netherlands, Dutch employees must be insured under the Sickness Benefit Act, the Unemployment Insurance Act, the Health Insurance Act, and the Invalidity Insurance Act. Under each of these acts, a (previously) employed person is eligible for financial benefits provided certain conditions are met. For details on the design of the 2002 study, we refer to Lensvelt-Mulders et al. (2006).

Most of the examples focus on six RR questions, four of which are health related and the remaining two are work related. The health questions are:

1. Have you been told by your physician about a reduction in your disability symptoms without reporting this improvement to your social welfare agency?
2. On your last spot check by the social welfare agency, did you pretend to be in poorer health than you actually were?
3. Have you noticed personally any recovery from your disability complaints without reporting it to the social welfare agency?
4. Have you felt for some time now to be substantially stronger and healthier and able to work more hours, without reporting any improvement to the social welfare agency?

The work-related questions are:

1. In the last 12 months have you moonlighted while receiving your IIA benefits?
2. In the last 12 months have you taken on a small job alone or together with your friends that you got paid for without informing the social welfare agency?

The remainder of the manuscript is structured as follows. In Section 2, we discuss the analysis of RR data without explanatory variables. In Section 3, we discuss work on logistic regression adjusted for RR data. In Section 4, we describe extensions to the situation of multivariate RR data. Section 5 discusses how regression approaches are to be adjusted in the light of new models that adjust for self-protective responses, ie, the fact that part of the sample is not following the RR design laid out by the researcher and answers "no" to every randomized response question asked.

2 UNIVARIATE AND MULTIVARIATE RR DATA, NO EXPLANATORY VARIABLES

2.1 Theory

This section discusses the analysis of univariate and multivariate RR data, where *no* explanatory variables are involved. We start off using the forced response design (Boruch, 1971) as an example of an RR design.

Assume that the sensitive question asks for a *yes* or *no* answer. The forced response design is as follows. After the sensitive question is asked, the respondent throws two dice and keeps the outcome hidden from the interviewer. If the outcome is 2, 3, or 4, the respondent answers *yes*. If the outcome is 5, 6, 7, 8, 9, or 10, the respondent answers according to the truth. If the outcome is 11 or 12, the respondent answers *no*.

Let Y be the latent binary RR variable that denotes the true status on the sensitive item, and let Y^* be the observed RR variable that denotes the observed answer on the randomized sensitive question, with *yes* = 1 and *no* = 2. Then

$$\begin{aligned} P(Y^*=1) &= P(Y^*=1|Y=2)P(Y=2)+P(Y^*=1|Y=1)P(Y=1) \\ &= 1/6+3/4P(Y=1). \end{aligned} \quad (1)$$

If we write $P(Y^* = 1) = c + dP(Y = 1)$, other designs can be described in the same way (Böckenholt and van der Heijden, 2007). As a first example, in the original Warner design an individual is allocated to either statement A, with probability p, or to statement B, with probability $1 - p$, with $p \neq 0.5$. The sensitive question in group A is "do you belong to the sensitive population?" and in group B "don't you belong to group to the sensitive population?." Now $c = (1 - p)$ and $d = (2p - 1)$. As a second example that we have used in our research (see Van der Heijden et al., 2000), we have the Kuk (1990) design. Here, an (indirect) *yes*-or-*no* question is asked using stacks of cards. As an example, let two stacks of cards contain both black and red cards. In the right stack, the proportion of red cards is 8/10 and in the left stack 2/10. The respondent is asked to draw one card from each stack and keep the color of the cards hidden from the interviewer. Next, the question is asked. Instead of answering the question directly, the respondent names the color of the card he took from the related stack, ie, when the answer is *yes*, the respondent names the color of the card he took from the right stack (thus red is associated with *yes*), and when the answer is *no*, he names the color of the card from the left stack (thus black is associated with *no*). In this case we have $c = 1/5$ and $d = 3/5$.

Following Chaudhuri and Mukerjee (1988), we simplify notation by using matrix notation. We do this for the general situation that Y^* as well as Y have K categories, indexed by j,k; this will turn out to be useful in the situation of multivariate RR data, ie, the situation that more than one sensitive question is asked using an RR design. We collect the probabilities $P(Y^*=j)=\pi_j^*$ referring to the observed variable Y^* in a vector $\boldsymbol{\pi}^* = (\pi_1^*,\ldots,\pi_K^*)^t$, and we collect the probabilities $P(Y = k) = \pi_k$ referring to the latent variable Y in a vector $\boldsymbol{\pi} = (\pi_1,\ldots,\pi_K)^t$. We collect the randomizing probabilities given by the RR design in a matrix $\boldsymbol{P}$ with elements $p_{jk} = P(Y^* = j|Y = k)$. Then the general form of RR designs can be written as (Chaudhuri and Mukerjee, 1988; Van den Hout and van der Heijden, 2002):

$$\boldsymbol{\pi}^* = \boldsymbol{P}\boldsymbol{\pi}, \quad (2)$$

As a first example, consider the forced response design. Here, when the true status is "yes," the misclassification probability is 3/36 = 1/12, and, when the true status is "no," the misclassification probability is 6/36 = 2/12, and this yields

$$\boldsymbol{P}_{FR} = \begin{pmatrix} p_{11} & p_{12} \\ p_{21} & p_{22} \end{pmatrix} = \begin{pmatrix} 11/12 & 2/12 \\ 1/12 & 10/12 \end{pmatrix}. \quad (3)$$

As a second example, consider the Kuk design. Here, misclassification probabilities are 2/10, both when the true status is "yes" or "no":

$$\boldsymbol{P}_{Kuk} = \begin{pmatrix} p_{11} & p_{12} \\ p_{21} & p_{22} \end{pmatrix} = \begin{pmatrix} 8/10 & 2/10 \\ 2/10 & 8/10 \end{pmatrix}. \tag{4}$$

As a third example, let there be two binary RR variables, leading to a first couple (Y_1^*, Y_1) with matrix $\boldsymbol{P_1}$, and to a second couple (Y_2^*, Y_2) with matrix $\boldsymbol{P_2}$. Then the vector $\boldsymbol{\pi}^* = (\pi_{11}^*, \pi_{12}^*, \pi_{21}^*, \pi_{22}^*)^t$, the vector $\boldsymbol{\pi} = (\pi_{11}, \pi_{12}, \pi_{21}, \pi_{22})^t$ and $\boldsymbol{P} = \boldsymbol{P_1} \otimes \boldsymbol{P_2}$, where $\otimes$ is the Kronecker product.

2.2 Estimation

The estimation of Eq. (2) is straightforward when the parameters are in the interior of the parameter space. If we use sample proportions of *yes* and *no* answers as estimates of $\hat{\boldsymbol{\pi}}^*$, then

$$\hat{\boldsymbol{\pi}} = \boldsymbol{P}^{-1} \hat{\boldsymbol{\pi}}^* \tag{5}$$

yields the unbiased moment estimates for $\boldsymbol{\pi}$. If the parameters are in the interior of the parameter space, then the unbiased moment estimates are equal to the maximum likelihood estimates (compare Van den Hout and van der Heijden, 2002).

Parameters are not necessarily in the interior of the parameter space. An example where they are not in the interior of the parameter space can be derived from Eq. (1), when $P(Y^* = 1) < 1/6$. In other words, the observed proportions are below chance level. If this happens, the moment estimate for $P(Y = 1)$ will be negative and the moment estimate for $P(Y = 2)$ will be larger than one. As maximum likelihood estimates cannot be negative, they will be found at the boundary of the parameter space (ie, it is 0).

Maximum likelihood estimates can always be found by maximizing the kernel of the multinomial log-likelihood. Let $\boldsymbol{n} = (n_1, \ldots, n_K)^t$ be the vector of observed frequencies related to the probabilities for the observed response Y^* and let u be a unit vector of length K, then the kernel of the log-likelihood is

$$\ell = \boldsymbol{u}^t(\boldsymbol{n} \log \pi^*) = \boldsymbol{u}^t(\boldsymbol{n} \log \boldsymbol{P}\boldsymbol{\pi}). \tag{6}$$

Maximizing ℓ over the parameters $\boldsymbol{\pi}$ can be done using an EM algorithm, or by maximizing the likelihood directly (compare Van den Hout and van der Heijden, 2002). We note that, when only one sensitive question is involved and the two moment estimates are outside the parameter space, the two maximum likelihood estimates simply end up on the boundary as 0 and 1. When more than one sensitive question is involved and one or more moment estimates are outside the parameter estimates, then, in general, the maximum likelihood estimates cannot be derived analytically and ℓ has to be maximized using iterative methods.

Example. In the 2002 survey on regulatory noncompliance w.r.t. disability benefit the six RR questions introduced in Section 1 were asked. The sample size was 1,760. Point estimates with 95% bootstrap confidence intervals are, for the first health item 0.03 (0.004–0.052), for the second health item 0.04 (0.012–0.061), for the third health item 0.07 (0.046–0.098) and for the fourth health item 0.13 (0.102–0.156). Note that the estimates are larger for the later, less severe items. For work item 1, the point estimate was 0.16 (0.128–0.184) and for work item 2 0.08 (0.050–0.103).

3 LOGISTIC REGRESSION OF UNIVARIATE RR DATA

3.1 Theory

In logistic regression the dependent variable is predicted from one or more covariates. Let the z covariates of individual i be collected in covariate vector $\boldsymbol{x}_i$ of length $z \times 1$, and let the regression parameters be collected in a parameter vector $\boldsymbol{\beta}$ of length $z \times 1$. Then one way to formulate logistic regression is as

$$\pi_{1i} = \frac{\exp(\beta_0 + \boldsymbol{x}_i{}^t\boldsymbol{\beta})}{1 + \exp(\beta_0 + \boldsymbol{x}_i{}^t\boldsymbol{\beta})} \tag{7a}$$

$$\pi_{2i} = \frac{1}{1 + \exp(\beta_0 + \boldsymbol{x}_i{}^t\boldsymbol{\beta})}. \tag{7b}$$

Now assume that we deal with RR data, and let elements π_{1i} and π_{2i} refer to the probability of the true status 1 and 2 for individual i, and let π^*_{1i} and π^*_{2i} refer to the probabilities of the responses 1 and 2 for individual i. If we collect the elements π_{1i} and π_{2i} into a vector $\boldsymbol{\pi}_i$ and elements π^*_{1i} and π^*_{2i} into a vector $\boldsymbol{\pi}^*_i$, then

$$\pi^*_i = \boldsymbol{P}\boldsymbol{\pi}_i, \tag{8}$$

with $\boldsymbol{\pi}_i$ defined as in (7a) and (7b). Thus the probability of a sensitive true status of individual i is a function of covariates $\boldsymbol{x}_i$.

The outcome of the RR logistic regression procedure can be interpret in the usual way (see, for example, Agresti, 2002). The β parameters reveal the relation between the covariates and the probability of a sensitive true status, and significance of each parameter can be assessed using Wald tests. Chi-square tests can be used to assess the fit of the model if the observed RR data are counts that allow comparison with fitted values under the RR logistic regression model, and differences between models can be tested using likelihood ratio chi-square tests.

The earliest reference to logistic regression for RR data that we came across was Maddala (1983, pages 54-56), and an elaborate treatment can be found in Scheers and Dayton (1988), Van der Heijden and van Gils (1996), and Ahangar et al. (2012). Chaudhuri and Saha (2004) and Lensvelt-Mulders et al. (2006) extended the logistic regression procedure so that it can incorporate person weights that make it possible to weight a sample toward population characteristics, if known. Jann (2005) provided software for RR logistic regression in Stata. We refer to Ronning (2005) for a discussion of the randomized response probit model. Hsieh et al. (2009) discuss logistic regression for randomized response data when values on the covariates are missing. Applications can be found in the above papers and in Landsheer et al. (1999), St John et al. (2011), and Krumpal (2012).

3.2 Estimation

The logistic regression model for RR data is estimated by setting up the likelihood and maximizing over the parameters. Let the observed RR data for individual i be (n^*_{1i},n^*_{2i}), with (n^*_{1i},n^*_{2i}) = (1,0) if individual i has answer 1, and (n^*_{1i},n^*_{2i}) = (0,1) if individual i has answer 2. Then the log-likelihood

$$\ell(\beta_0,\boldsymbol{\beta}) = \sum_i \left(n^*_{1i}\log\pi^*_{1i} + n^*_{2i}\log\pi^*_{2i}\right). \tag{9}$$

For further model development it is useful to write this in matrix terms, similar as in (6). Let $\boldsymbol{u}$ be a unit vector of length 2 × 1, and collect (n^*_{1i},n^*_{2i}) in vector $\boldsymbol{n}^*_i$ and of length 2 × 1 respectively. Then Eq. (9) can be written as

$$\ell(\beta_0,\boldsymbol{\beta}) = \sum_i \boldsymbol{u}^t\left(\boldsymbol{n}^*_i\log\pi^*_i\right) = \sum_i \boldsymbol{u}^t\left(\boldsymbol{n}^*_i\log \boldsymbol{P}\pi_i\right), \tag{10}$$

where the elements of $\boldsymbol{\pi}_i$ are defined in (7). Thus maximizing $\ell(\beta_0,\boldsymbol{\beta})$ over the parameters $(\beta_0,\boldsymbol{\beta})$ yields the maximum likelihood estimates. The incorporation of person weights w_i, alluded to in Section 3.1, is simply accomplished by reformulating the log-likelihood as

$$\ell(\beta_0,\boldsymbol{\beta}) = \sum_i \boldsymbol{u}^t\left(w_i\boldsymbol{n}^*_i\log \boldsymbol{P}\pi_i\right). \tag{11}$$

Maddala (1983) provides first and second order derivatives of the log-likelihood and suggests to use the Newton–Raphson method to maximize the log-likelihood (see also Scheers and Dayton, 1988 and Van der Heijden and van Gils, 1996). Van den Hout et al. (2007) show that the model is a member of the family of generalized linear models and propose to fit the model with the iterative reweighted least-squares algorithm, which is a very stable fitting procedure.

Example. As an illustration, we report an example taken from Lensvelt-Mulders et al. (2006). The dependent variable is the work item “In the last 12 months have you taken on a small job alone or together with your friends that you got paid for without informing the social welfare agency?”. As an explanatory variable we choose “I think it is more beneficial to me not to follow the rules connected to my disability insurance benefit", abbreviated as "benefit", that is measured on a five-point scale and has mean 3.67 and standard deviation 0.777. We note that many more explanatory variables have been measured in this survey, motivated from a rational choice framework, and refer to Lensvelt-Mulders et al. (2006) and Elffers et al. (2003) for details. The logistic regression model logit(noncompliance) = constant + b * benefit has estimates 0.765 for the constant and 0.751 for b. In order to study the impact of these estimates, we compare the estimated probability of noncompliance for the mean value of benefit (ie, 3.67), and the mean plus or minus one standard deviation (ie, 3.67 + 0.78 = 4.45 and 3.67 − 0.78 = 2.89). For the mean value of benefit, the estimated probability of noncompliance is 12%, for 4.45 the estimated probability is 20% and for 2.89 the estimated probability is 7%. This shows that benefit has a strong relation with the decision not to comply with the above disability insurance benefit regulation: the more people perceive their benefit if they do not comply, the more they do not comply with this work regulation.

3.3 Extensions

3.3.1 Repeated Cross Sectional RR Data

We now discuss how to handle repeated cross sectional RR data, ie, data where at different points in time samples are taken using an RR design. In this situation, one of the points of interest will be whether the prevalence has changed over different time points.

Imagine the situation of two time points, and interest only goes out to the comparison of the prevalence at time point 1 and the prevalence at time point 2. An answer can be obtained in the context of the logistic regression model. For two time points, let x_i indicate the time at which individual i is sampled, with $x_i = 1$ if individual i is in the sample for time point 1 and $x_i = 0$ if individual i is in the sample for time point 2. Then the logistic regression model for the latent probabilities π_{1i} and π_{2i} is

$$\pi_{1i} = \frac{\exp(\beta_0 + \beta x_i)}{1 + \exp(\beta_0 + \beta x_i)} \tag{12a}$$

$$\pi_{2i} = \frac{1}{1 + \exp(\beta_0 + \beta x_i)}, \tag{12b}$$

where the latent probabilities π_{1i} and π_{2i} are plugged into the log-likelihood (10) in order to take the RR design into account. Thus if individual i is in the first sample, then prevalence π_{1i} is $\pi_{1i} = \exp(\beta_0 + \beta)/(1 + \exp(\beta_0 + \beta))$, and if individual i is in the second sample, then prevalence π_{1i} is $\pi_{1i} = \exp(\beta_0)/(1 + \exp(\beta_0))$. If additional covariates have to be taken into account, this format can be extended in a straightforward way by including these covariates into a covariate vector $\boldsymbol{x}_i$ (compare Eq. (7a) and (7b)).

Similarly, if there are more than two time points, the number of dummy variables can be enlarged to contrast specific time points, or linear contrasts can be used. For more details and a slightly different approach, we refer to Frank et al. (2009), who approach this problem in a contingency table format. We postpone an example to a slightly more complicated situation that is discussed in the next section.

3.3.2 Repeated Cross Sectional RR Data with a Changing RR Design

We now need to complicate the notation (compare Frank et al., 2009). For simplicity we assume that there are two RR designs, namely RR design 1 and RR design 2. There are $n^{(1)}$ individuals in RR design 1 and $n^{(2)}$ individuals in RR design 2. Individuals are still indexed by i, where $i = 1,\ldots,n^{(1)} + n^{(2)}$. The covariate values for individual i are still collected into covariate vector $\boldsymbol{x}_i$. Note that such covariates may include

1. covariates that will receive identical regression coefficients over the two designs
2. a dummy variable for RR design
3. interactions between 1 and 2 if the regression coefficients are assumed to be different over the two RR designs.

Thus we can still use the logistic regression model for the latent probabilities π_{1i} and π_{2i} defined in (7a) and (7b). The RR design plays a role in the definition of the log-likelihood. Let $\boldsymbol{P}^{(1)}$ define RR design 1 and $\boldsymbol{P}^{(2)}$ RR design 2. Similar to (10) we can now define the log-likelihood as

$$\ell(\beta_0,\boldsymbol{\beta}) = \sum_i \boldsymbol{u}^t\left(\boldsymbol{n}_i^* \log \pi_i^*\right) \tag{13}$$

$$= \sum_{i=1,\ldots,n^{(1)}} \boldsymbol{u}^t\left(\boldsymbol{n}_i^* \log \boldsymbol{P}^{(1)}\pi_i\right) + \sum_{i=n^{(1)}+1,\ldots,n^{(1)}+n^{(2)}} \boldsymbol{u}^t\left(\boldsymbol{n}_i^* \log \boldsymbol{P}^{(2)}\pi_i\right) \tag{14}$$

with the elements in $\boldsymbol{\pi}_i$ defined as in (7a) and (7b).

Example. We provide an illustration from the paper by Frank et al. (2009). For the survey discussed in the introduction, three waves were carried out in the years 2000, 2002, and 2004. In the year 2000, the Kuk design was used, and in 2002 and 2004, the forced choice design was used (see Section 1). The example here focuses on the health item "For periods of any length at all, do you ever feel stronger and healthier and able to work more hours without informing the Department of Social Services of this change?" For this item, Table 1 displays the observed, misclassified, frequencies of *yes* (n_1^*) and *no* (n_2^*) answers on the three time points, as well as the estimated probabilities of regulatory noncompliance ($\hat{\pi}_1$) corrected for the RR design.

TABLE 1 Observed Weighted Frequencies of *Yes* (n_1^*) and *No* (n_2^*) Answers, and Estimated Probabilities ($\hat{\pi}_1$) with 95% Confidence Intervals of Noncompliance Corrected for the RR Design for the Cross-sectional Data on Regulatory Noncompliance, Measured on Three Time Points

	2000 (n=*1308*)	2002 (n=*1760*)	2004 (n=*830*)
n_1^*	388	466	197
n_2^*	920	1294	633
$\hat{\pi}_1$	0.16	0.10	0.07
95% CI	[0.13, 0.19]	[0.07, 0.13]	[0.03, 0.11]

Person weights were used to weight the sample towards population characteristics (cf. Lensvelt-Mulders et al., 2006)

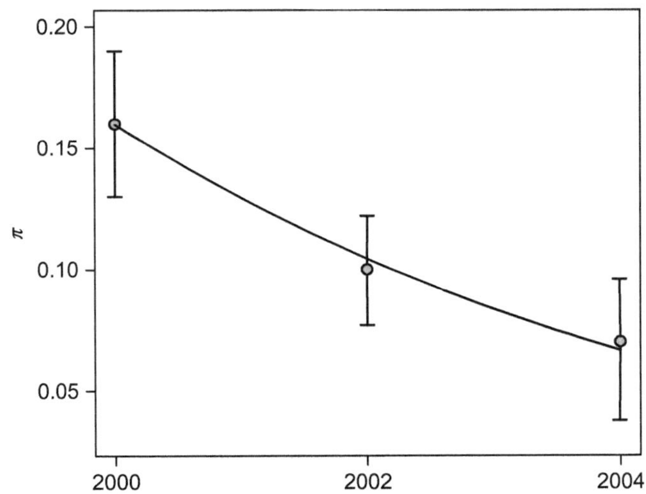

FIG. 1 Univariately estimated regulatory noncompliance probabilities (represented by the *dots*) with 95% confidence intervals, and fitted trend line on the RR question *For periods of any length at all, do you ever feel stronger and healthier and able to work more hours without informing the Department of Social Services of this change?*

Two models were fitted to the RR data: the independence model and the linear trend model. The goodness-of-fit of the models was evaluated with the likelihood ratio statistic. Fig. 1 shows that the estimated regulatory noncompliance probabilities decrease monotonically over the time points, and, accordingly, the estimated logistic regression parameter for the linear trend has a negative value (see Table 2). The value of the likelihood ratio statistic L^2 in Table 2 indicates that the independence model does not fit

TABLE 2 Results Trend Analyses (the Model of Choice is in Bold Face)

Model	L^2	df	p	$\hat{\beta}_0\ (\hat{\sigma}_{\hat{\beta}_0})$	$\hat{\beta}_1\ (\hat{\sigma}_{\hat{\beta}_1})$
[1] Independence	10.415	2	.001	− 2.09 (0.10)	
[2] **Linear**	**0.004**	**1**	**0.95**	− 1.17 **(0.28)**	− 0.49 **(0.16)**
	ΔL^2	Δdf	p		
[1]-[2]	10.411	1	0.001		

($L^2 = 10.415$, $df = 2$, $p = 0.001$), whereas the linear model produces good fit values: $L^2 = 0.004$, $df = 1$, $p = 0.95$. Testing the linear model against the independence model (see the ΔL^2 values in the lower part of Table 2) leads to the conclusion that the linear trend model forms a significant improvement. Also, both parameters of the linear trend model depart significantly from zero. Summarizing, this means that among the persons entitled to receive social benefit, the proportion of respondents who do not comply to the rule of informing the Department of Social Services about their health improvement, has significantly decreased in the period from 2000 to 2004.

3.3.3 RR for Explanatory Variable

Thus far we have assumed that the RR variable is the dependent variable. Therefore the model of analysis is logistic regression and the log-likelihood was adjusted to take the misclassification due to RR into account. However, it may also be possible that the RR variable is an exploratory variable in a regression context. An important set of applications where this may be the case is the context of statistical disclosure control (SDC). An aim of SDC is to prevent identification of the identity of individual units and this is a concern of, for example, national statistical institutes, see Van der Heijden and Böckenholt (2008) for a more elaborate discussion of this issue.

Van den Hout and Kooiman (2007) studied the linear regression model with categorical covariates subject to RR. They set up the likelihood for this model and use an EM algorithm to obtain the maximum likelihood estimates. For more details and an example, we refer to their paper.

4 EXTENSIONS OF REGRESSION APPROACHES TO MULTIVARIATE RR DATA

4.1 The Multivariate Logistic Regression Model Proposed by Glonek and McCullagh

The multivariate logistic regression model proposed by Glonek and McCullagh (1995) is worked out by Van den Hout et al. (2007). We closely follow the presentation of Van den Hout et al. (2007), who elaborate the model for two RR variables.

Van den Hout et al. (2007) discuss the model in the context of the well-known family of generalized linear models. In this family of models, we have to specify the random and the systematic component of a model. For two RR variables we showed at the end of Section 2.1 that there is a first couple (Y_1^*, Y_1) with matrix $\boldsymbol{P_1}$, and a second couple (Y_2^*, Y_2) with matrix $\boldsymbol{P_2}$. Then let the vector $\boldsymbol{\pi}^* = \left(\pi_{11}^*, \pi_{12}^*, \pi_{21}^*, \pi_{22}^*\right)^t$, let the vector $\boldsymbol{\pi} = (\pi_{11}, \pi_{12}, \pi_{21}, \pi_{22})^t$ and let $\boldsymbol{P} = \boldsymbol{P_1} \otimes \boldsymbol{P_2}$, where $\otimes$ is the Kronecker product. The two vectors $\boldsymbol{\pi}^*$ and $\boldsymbol{\pi}$ and the matrix $\boldsymbol{P}$ are related by $\boldsymbol{\pi}^* = \boldsymbol{P}\,\boldsymbol{\pi}$. We assume for the random component of the model of Glonek and McCullagh that the bivariate variables $\boldsymbol{Y}$ and $\boldsymbol{Y}^*$ are multinomially distributed with parameter vectors $\boldsymbol{\pi}$ and $\boldsymbol{\pi}^*$ respectively.

The link in the multivariate model between the linear predictor and the expected response vector is

$$\boldsymbol{\eta} = \boldsymbol{C}^t \log(\boldsymbol{L}\boldsymbol{\pi}), \tag{15}$$

where $\boldsymbol{C}$ is the contrast matrix and $\boldsymbol{L}$ is the marginal indicator. For the bivariate case that we are considering, with $\boldsymbol{\eta} = (0, \eta_{Y_1}, \eta_{Y_2}, \eta_{Y_1Y_2})^t$, the link functions are given by

$$\eta_{Y_1} = \log\frac{\pi_{1+}}{1-\pi_{1+}}, \quad \eta_{Y_2} = \log\frac{\pi_{+1}}{1-\pi_{+1}}, \quad \eta_{Y_1Y_2} = \log\frac{\pi_{00}\pi_{11}}{\pi_{01}\pi_{10}}, \tag{16}$$

where the plus subscript denotes summation, for example $\pi_{1+} = \pi_{10} + \pi_{11}$. The regression equations are given by

$$\eta_{Y_1} = \boldsymbol{\beta}_{Y_1}^t \boldsymbol{x}_{Y_1}, \quad \eta_{Y_2} = \boldsymbol{\beta}_{Y_2}^t \boldsymbol{x}_{Y_2}, \quad \eta_{Y_1Y_2} = \boldsymbol{\beta}_{Y_1Y_2}^t \boldsymbol{x}_{Y_1Y_2}. \tag{17}$$

This model is a marginal model—it implies univariate logistic models for both Y_1 and Y_2 marginally (Glonek and McCullagh, 1995). The odds ratio is used to model the dependence between Y_1 and Y_2. The contrast matrix and the marginal indicator for this model are given by

$$\boldsymbol{C}^t = \begin{pmatrix} 1 & 0 & 0 & 0 & 0 & 0 & 0 & 0 & 0 \\ 0 & -1 & 1 & 0 & 0 & 0 & 0 & 0 & 0 \\ 0 & 0 & 0 & -1 & 1 & 0 & 0 & 0 & 0 \\ 0 & 0 & 0 & 0 & 0 & 1 & -1 & -1 & 1 \end{pmatrix}$$

$$\text{and } \boldsymbol{L}^t = \begin{pmatrix} 1 & 1 & 0 & 1 & 0 & 1 & 0 & 0 & 0 \\ 1 & 1 & 0 & 0 & 1 & 0 & 1 & 0 & 0 \\ 1 & 0 & 1 & 1 & 0 & 0 & 0 & 1 & 0 \\ 1 & 0 & 1 & 0 & 1 & 0 & 0 & 0 & 1 \end{pmatrix}.$$

The first element of $\boldsymbol{\eta}$ represents the null contrast $\log(\pi_{++}) = 0$. For an observation i, the regression is given by $\boldsymbol{\eta}_i = \boldsymbol{X}_i\boldsymbol{\beta}$, where $\boldsymbol{\beta} = (\boldsymbol{\beta}^t_{Y_1}, \boldsymbol{\beta}^t_{Y_2}, \boldsymbol{\beta}^t_{Y_1Y_2})^t$ and

$$\boldsymbol{X}_i = \begin{pmatrix} 0 & 0 & 0 \\ \boldsymbol{x}^t_{Y_1} & 0 & 0 \\ 0 & \boldsymbol{x}^t_{Y_2} & 0 \\ 0 & 0 & \boldsymbol{x}^t_{Y_1Y_2} \end{pmatrix}.$$

For example, if there is one covariate x_1 which is the same for all three regressions and an intercept is included, then $\boldsymbol{x}^t_{Y_1} = \boldsymbol{x}^t_{Y_2} = \boldsymbol{x}^t_{Y_1Y_2} = (1, x_1)$, $\boldsymbol{X}_i$ is a 4×6 matrix, and $\boldsymbol{\beta} = (\beta_{Y_10}, \beta_{Y_11}, \beta_{Y_20}, \beta_{Y_21}, \beta_{Y_1Y_20}, \beta_{Y_1Y_20})^t$.

Given an RR design, the link between the linear predictor and the expected response vector is

$$\boldsymbol{\eta} = \boldsymbol{C}^t \log(\boldsymbol{L\pi}) = \boldsymbol{C}^t \log\left(\boldsymbol{LP}^{-1}\boldsymbol{\pi}^*\right). \tag{18}$$

To estimate the RR multivariate logistic regression model, a log-likelihood is set up (compare Van den Hout et al., 2007). Assume that data are given by independent observations $(\boldsymbol{y}_i^*, \boldsymbol{X}_i)$, $i = 1, \ldots, n$. Then the log-likelihood is

$$l(\boldsymbol{\beta} \ |(\boldsymbol{y}_i^*, \boldsymbol{X}_i), i \in \{1, \ldots, n\}, \boldsymbol{P} = \sum_{i=1}^{n} (\boldsymbol{y}_i^*)^t \log \boldsymbol{\pi}^* = \sum_{i=1}^{n} (\boldsymbol{y}_i^*)^t \log \boldsymbol{P\pi}. \tag{19}$$

The maximization of this log-likelihood is discussed in Van den Hout et al. (2007) and we refer to their paper for details.

Example. Van den Hout et al. (2007) report the following analysis. Data are taken from an RR survey concerning unemployment benefit in the Netherlands in 2004. The sample size is $n = 753$. The RR design is a slightly adapted form of the forced response design. There are two RR questions. First, Y_1^* denotes the observed answer to the RR question whether the number of job applications of respondent is below the required minimum (*yes* = 1, *no* = 0). Second, Y_2^* denotes the observed answer to the question whether respondent i did any voluntary work without reporting this. In addition, X_1 is gender (*male* = 1, *female* = 0), X_2 is age in years, and X_3 denotes whether the income of a respondent constitutes the larger part of the income of the household (*yes* = 1, *no* = 0).

Table 3 presents the results of the analysis. As an illustration, Model 8 is defined by (15), (16) and the regression equations $\boldsymbol{\eta}_i = \mathbf{X}_i\boldsymbol{\beta}$ which are given by $\boldsymbol{\beta} = (\beta_{Y_10}, \beta_{Y_11}, \beta_{Y_12}, \beta_{Y_13}, \beta_{Y_20}, \beta_{Y_21}, \beta_{Y_22}, \beta_{Y_23}, \beta_{Y_1Y_20})^t$ and

$$\mathbf{X}_i = \begin{pmatrix} 0 & 0 & 0 & 0 & 0 & 0 & 0 & 0 & 0 \\ 1 & x_{Y_11,i} & X_{Y_12,i} & X_{Y_13,i} & 0 & 0 & 0 & 0 & 0 \\ 0 & 0 & 0 & 0 & 1 & X_{Y_21,i} & X_{Y_22,i} & X_{Y_23,i} & 0 \\ 0 & 0 & 0 & 0 & 0 & 0 & 0 & 0 & 1 \end{pmatrix}.$$

TABLE 3 Deviances of Bivariate Logistic Regression Models Taken the RR Design into Account

Model		# Parameters	Deviance	AIC
1	Intercept(1,2,3)	3	1928.033	1934.033
2	Intercept(1,2,3), $X_2(1,2)$	5	1905.071	1915.071
3	Intercept(1,2,3), $X_1(2)$, $X_2(1,2)$	6	1903.411	1915.411
4	Intercept(1,2,3), $X_2(1,2)$, $X_3(2)$	6	1903.098	1915.098
5	Intercept(1,2,3), $X_2(1,2)$, $X_3(1,2)$, $\beta_{Y_13}=\beta_{Y_23}$	6	1901.638	1913.638
6	Intercept(1,2,3), $X_2(1,2)$, $X_3(1)$	6	1903.799	1915.799
7	Intercept(1,2,3), $X_2(1,2)$, $X_3(1,2)$	7	1901.611	1915.611
8	Intercept(1,2,3), $X_1(1,2)$, $X_2(1,2)$, $X_3(1,2)$	9	1900.889	1918.889

Sample size is 753. Notation: $X_j(k,..)$ means the covariate, X_j is included in the kth regression equation, where the order of the equations is given by (17).

So the odds ratio is estimated by using an intercept only. Models 1 up to 7 are defined by restrictions on the parameters, see Table 3. Model 3 vs Model 2 shows that adding X_1 does not seem necessary. With respect to adding X_3, the best model according to the AIC is Model 5 in which the regression coefficient of X_3 is restricted to be the same in the marginal models for Y_1 and Y_2, and $\beta_{Y_11}=\beta_{Y_21}=0$. Coefficient estimates for this model are given by

$$\begin{aligned}&\hat{\beta}_{Y_10}=-0.03\ \ (0.54), \quad \hat{\beta}_{Y_20}=-3.31\ \ (0.68),\\ &\hat{\beta}_{Y_12}=-0.03\ \ (0.01), \quad \hat{\beta}_{Y_22}=0.04\ \ (0.013),\\ &\hat{\beta}_{Y_13}=\hat{\beta}_{Y_23}=0.44\ \ (0.33), \text{and}\, \hat{\beta}_{Y_1Y_20}=1.30\ \ (0.54),\end{aligned}$$

where the estimated standard errors are within the brackets. The Wald 95% confidence interval of $\exp(\beta_{Y_13})=\exp(\beta_{Y_23})$ is (0.81; 2.99) which suggest that including x_3 in the model is not necessary. If we use the deviance to select a model, we would opt for model 2.

The interpretation of the parameters of the marginal models for Y_1 and Y_2 is the same as in a univariate logistic regression model. In the multivariate model, we can easily investigate whether an effect of a covariate is the same in the marginal models. This is illustrated by Model 5 which states that the effect of whether or not the income of the respondent constitutes the larger part of the income of the household is the same for the prevalence of regulatory noncompliance regarding job applications and the prevalence of noncompliance regarding voluntary work. More specific, Model 5 states that a change from $X_3 = 0$ to $X_3 = 1$, means that the odds on noncompliance regarding both Y_1 and Y_2 changes multiplicatively by $\exp(\hat{\beta}_{Y_13}) = \exp(\hat{\beta}_{Y_23}) = \exp(0.44) = 1.56$. So the odds on noncompliance is higher when a person's income constitutes the larger part of the income of the household.

In addition, the multivariate model describes the association between the two response variables Y_1 and Y_2 by estimating the odds ratio. In Model 5, the ratio of the odds in the two rows of the cross-classification of Y_1 and Y_2 is estimated to be $\exp(1.30) = 3.67$ with Wald 95% confidence interval (1.28; 10.49). Thus, interpreting this model, respondents appeared to be nearly $4 \times$ as likely to ignore rules concerning job applications as to do voluntary work without reporting them.

4.2 Proportional Odds Model

In Section 4.1, the multivariate RR data are analyzed directly using an appropriate regression model. Here, we discuss the analysis of a summary of the multivariate data, namely of a sum score (see Cruyff et al., 2008). For example, if there is a set of M sensitive questions indexed by m ($m = 1,\ldots,M$) it may be interesting to know how many sensitive questions are answered affirmatively. This problem can be considered as the problem of estimating a sum score variable Z from M RR variables Y_m, where the sum ranges from $0,\ldots,M$. A second question is then in what way this sum score can be related to explanatory variables.

We now approach this problem in a formal way (compare Cruyff et al., 2008). Let the sum score variable denoting the number of true *yes* responses be defined by

$$Z = \sum_{m=1}^{M} Y_m. \tag{20}$$

Analogously, let the sum score variable $Z^* = \sum_{m=1}^{M} Y_m^*$ denote the number of observed *yes* responses. The probability of observing sum score s on variable Z^*, for $s \in \{0,\ldots,M\}$, is given by the RR sum score model

$$\pi_s^* = \sum_{t=0}^{M} q_{st}\pi_t, \tag{21}$$

where $\pi_s^* = IP(Z^* = s)$, $\pi_t = IP(Z = t)$ and $q_{st} = IP(Z^* = s|Z = t)$. A general definition of the elements q_{st} can be found in Cruyff et al. (2008).

The model is estimated as follows. Note that, similar to Eq. (2) we can write Eq. (20) in matrix terms by collecting the elements q_{st} in a matrix $\boldsymbol{Q}$ and it follows that

$$\boldsymbol{\pi}^* = \boldsymbol{Q}\boldsymbol{\pi}. \tag{22}$$

Similar to Eq. (5), we can find a moment estimate of $\boldsymbol{\pi}$ by

$$\hat{\boldsymbol{\pi}} = \boldsymbol{Q}^{-1}\hat{\boldsymbol{\pi}}^*, \tag{23}$$

where $\boldsymbol{\pi} = (\pi_0, \ldots, \pi_M)', \boldsymbol{\pi}^* = (\pi_0^*, \ldots, \pi_M^*)'$ and π_s^* estimated by n_s^*/n, with n_s^* denoting the frequency of the observed sum score s on variable Z^*. The matrix $\boldsymbol{Q}$ is an $(M + 1) \times (M + 1)$ transition matrix with entries $(s + 1, t + 1)$ given by the conditional misclassification probabilities q_{st}, for $s, t \in \{0, \ldots, M\}$. The method of moment solution always fits the data, but can result in probability estimates outside the boundaries of parameter space defined by (0,1).

The maximum likelihood estimates of the RR sum score model are obtained by maximizing the kernel of the observed-data log-likelihood

$$\ln \ell\left(\boldsymbol{\pi} | n_0^*, \ldots, n_M^*\right), = \sum_{s=0}^{M} n_s^* \ln\left(\sum_{t=0}^{M} q_{s|t}\pi_t\right), \tag{24}$$

for $\pi_t \in (0,1)$. Kuha and Skinner (1997) have provided EM algorithms. Van den Hout and van der Heijden (2002) show that if the method of moments estimates are in the interior of the parameter space, the maximum likelihood solution is identical to the method of moments solution. Otherwise, one or more maximum likelihood estimates will be on the boundary.

We now present the model for the regression of an RR sum score variable on a set of covariates. Assume that the sum scores are on an ordinal scale and let $IP(Z = t|\mathrm{x})$ denote the probability that the sum score variable Z takes on the value t given the covariate vector $\mathbf{x}$. Define $\gamma_t = IP(Z \leq t|\mathrm{x})$. Then the proportional odds model (McCullagh, 1980) states that

$$\gamma_t = \frac{\exp\left(\alpha_t - \mathrm{x}'\boldsymbol{\beta}\right)}{1 + \exp\left(\alpha_t - \mathrm{x}'\boldsymbol{\beta}\right)}, \tag{25}$$

where the threshold parameters α_t can be thought of as the values on a latent trait variable that mark the transition from $Z = t - 1$ to $Z = t$. The threshold parameters satisfy the condition

$$-\infty < \alpha_0 \leq \alpha_1 \leq \ldots \leq \alpha_M \equiv \infty. \tag{26}$$

Note that for $M = 1$, the order of the threshold parameters is $-\infty < \alpha_0 \leq \alpha_1 \equiv \infty$, and expression (25) reduces to the binary logistic regression model (with a negative sign for $\boldsymbol{\beta}$).

In the RR design, Z is not directly observed. Therefore, the cumulative probabilities $IP(Z \leq t|\mathrm{x})$ are modeled through the observed variable Z^*, with the relation between Z^* and Z given by the RR sum score model. The RR proportional odds model is given by

$$\gamma_s^* = \sum_{j=0}^{s}\sum_{t=0}^{M} q_{j|t}\left(\gamma_t - \gamma_{t-1}\right), \tag{27}$$

where $\gamma_s^* = IP(Z^* \leq s|\mathrm{x})$.

The maximum likelihood estimator (MLE) of model (27) is obtained by maximization of the kernel of the observed data log-likelihood, given by

$$\ln \ell\left(\boldsymbol{\beta},\boldsymbol{\alpha}|z_i^*,\ldots,z_n^*,\mathrm{x}_i,\ldots,\mathrm{x}_n\right)=\sum_{i=1}^{n}\ln\left(\sum_{t=0}^{M}q_{z_i^*|t}(\gamma_t-\gamma_{t-1})\right), \tag{28}$$

where $\gamma_{-1} = 0$ and $\gamma_M = 1$. To identify the model, we use the convention $\alpha_0 = 0$. For the maximization of Eq. (28), Cruyff et al. (2008) use standard optimization routines and we refer to their paper for details.

Example. Cruyff et al. (2008) report the following example concerning the following three sensitive questions discussed in the introduction: "At a Social Services checkup, have you ever acted as if you were sicker or less able to work than you actually were?", "For periods of any length at all, do you ever feel stronger and healthier and able to work more hours without informing the Department of Social Services?", and "Have you done any small jobs for or via friends or acquaintances in the past year, or paid jobs of any size without reporting it to the Department of Social Services? (This only pertains to monetary payments.)". They analyzed the sum score variable $Z=\sum_{m=1}^{3}Y_m$, denoting the number of *yes* responses to these three questions of the Social Security Survey, with the RR sum score model and the RR proportional odds model. The frequencies of the sum scores 0,1,2,3 observed in the sample are given by the vector $\boldsymbol{n}^* = (811,649,245,55)$.

The respective method of moments (MM) sum score probability estimates of the RR sum score model are $\hat{\pi}=(0.850,0.075,0.058,0.017)$. Since the MM estimates are all in the interior of the parameter space, the ML solution is identical. The log-likelihood of ML solution is -1949.54. The same probability estimates and log-likelihood can also be obtained with the RR proportional odds null model, ie, the model without any covariates except the intercept. The parameter estimates of the null model are $\hat{\beta}_0=-1.74$, $\hat{\alpha}_1=0.77$ and $\hat{\alpha}_2=2.32$, and the sum score probabilities are found by plugging these estimates into $\hat{\gamma}_t$ defined in (25), and using expression $\hat{\pi}_t=\hat{\gamma}_t-\hat{\gamma}_{t-1}$.

Table 4 presents the parameter estimates of the RR proportional odds model with all six covariates. The log-likelihood of this model is -1937.84, yielding a likelihood ratio test statistic of 23.4 with 6 degrees of freedom in relation to the null model. The parameter estimates of the covariates gender, age, last job contract and education are significant. To interpret these results, the property of the proportional odds model is used that, for all t, the odds of noncompliance with more than t rules change with a factor $\exp(-\beta_j)$ for each unit increase in covariate j, holding all other covariates constant. The parameter estimate for gender indicates that for men the odds of noncompliance are about 2.3 × those of women. Similarly, the odds of noncompliance for people above the age of 45 and for people who had a regular job contract are about 1.8 × that of younger people and people who had a different kind of job contract, respectively. Finally, the odds of noncompliance decrease with a factor 0.73 for each increase in the level of education.

TABLE 4 Parameter Estimates of the RR Proportional Odds Model

Parameters	Estimates (se)	*t* value
α_1	0.99 (0.31)	3.10
α_2	2.46 (0.38)	6.46
Intercept	− 0.85 (0.46)	− 1.84
Gender	− 0.81 (0.26)	− 3.14
Education	0.32 (0.16)	2.05
Age	− 0.57 (0.28)	− 2.23
Time unemployed	0.13 (0.16)	0.80
Last job contract	− 0.57 (0.29)	− 1.99
Degree of disability	− 0.26 (0.25)	− 1.05

5 MODELS INCLUDING SELF-PROTECTIVE RESPONSES

This section of the chapter differs of the sections before. Where the earlier sections provided lists of models, in this section we discuss the item response model adjusted for RR items. Then we show how this model can be used to tackle the problem that there may be respondents who do not follow the RR design but say "no" to every RR item they are presented with. This latter phenomenon is referred to as "self-protective" responses, abbreviated by SP. Then we discuss an example, where we also include a regression component to the item response model. We end with some remaining issues concerning models with parameters to model SP.

5.1 Item Response Models

Where in Section 4.2 a sum score is used to summarize the answers on a set of sensitive items, here we discuss the so-called Rasch model (Rasch, 1980). This model assumes, among others, a latent variable for the persons, and given the latent variable the answers to the items are independent. In psychometrics the Rasch model is a popular tool for measurement. For RR data, this model was independently adapted by Böckenholt and van der Heijden (2004, 2007) and Fox (2005). Under the Rasch model, the probability that item j is answered affirmatively by person i is written as

$$\Pr(x_{ij}=1)=\Pr_j(\theta_i)=\frac{1}{1+\exp\left(-\alpha(\theta_i-\gamma_j)\right)}, \tag{29}$$

where α and γ_j are parameters characterizing the so-called item response function.

Under the previously explained Forced Choice response format, respondents answer "yes" or "no" by chance with probabilities $\frac{1}{6}$ and $\frac{1}{12}$, respectively. As a result, we obtain

$$\Pr^{(FC)}(x_{ij}=1)=\frac{1}{6}+\frac{3}{4}\left(\frac{1}{1+\exp\left(-\alpha(\theta_i-\gamma_j)\right)}\right),$$

and

$$\Pr^{(FC)}(x_{ij}=0)=\frac{1}{12}+\frac{3}{4}\left(1-\frac{1}{1+\exp\left(-\alpha(\theta_i-\gamma_j)\right)}\right).$$

Let $\Pr_j$ be the probability that a person answers affirmatively to item j. Then, using the c and e notation discussed earlier in this chapter, the Rasch model yields the following likelihood function:

$$L=\prod_{i=1}^{n}\int\prod_{j=1}^{J}[c+e\Pr_j(\theta)]^{x_{ij}}\ [1-(c+e\Pr_j(\theta))]^{1-x_{ij}}f(\theta;\mu,\sigma)d\theta, \tag{30}$$

where $f(\theta;\mu,\sigma)$ is the normal density with mean μ and standard deviation σ. Since the mean μ of the population distribution cannot be estimated independently of the item locations, it is convenient to set $\mu=0$.

5.2 Self-Protective Responses

Despite the fact that the respondents' privacy is protected by the RR design, it is not always perceived as such by the respondents. The notion that aggregate—but not individual—level information can be inferred from randomized responses may neither be intuitive nor obvious to most survey participants. Thus, even when respondents are told that their privacy is protected, not all of them may be convinced that this is indeed the case. Because RR forces respondents to give a self-incriminating answer for something they potentially did not do, it is susceptible to self-protective responses (SP), ie, respondents answer *no* although they should have responded *yes* according to the randomizing device (see for example, Edgell et al., 1982). The online questionnaires that we used were designed in such a way that the outcome of the dice is not recorded and this was mentioned in the instructions given to the respondents. As a result, the respondents were free to give a different answer than the forced *yes* or *no* induced by the dice. Although RR performs relatively well, by eliciting more admittances of fraud than direct-questioning or computer-assisted self-interviews (Lensvelt-Mulders et al., 2005), noncompliance probabilities might still be underestimated if SP is not taken into account.

Recently, several studies have focussed on the detection or estimation of SP in the setting of RR. Clark and Desharnais (1998) showed that by splitting the sample into two groups and assigning each group a different randomization probability, it is possible t of SP responses and to measure its extent (see Ostapczuk et al., 2009; Ostapczuk et al., 2009; and Moshagen and Musch, 2012 for applications). Böckenholt and van der Heijden (2007) use a multivariate approach to estimate SP by using the item randomized-response model. As an extension of this model, the response behavior that does not follow the RR design is approached by introducing mixture components in the IRR models with a first component consisting of respondents who answer truthfully and follow an item response model, and a second component consisting of respondents who systematically say *no* to every item in a subset of items. It is straightforward to account for SP-"No" responses by extending the likelihood function (30) as follows:

$$
\begin{aligned}
L =& \prod_{i=1}^{n}\Big(\pi_i \int \prod_{j=1}^{J}\{[c+e\Pr_j(\theta)]^{x_{ij}}\,[1-(c+e\Pr_j(\theta))]^{1-x_{ij}}\}f(\theta;\mu,\sigma)d\theta \\
&+(1-\pi_i)\prod_{j=1}^{J}\{\Pr(\text{“No”})^{x_{ij}}\,[1-\Pr(\text{“No”})]^{1-x_{ij}}\}\Big),
\end{aligned}
\tag{31}
$$

where π denotes the probability of a randomly sampled person to answer the questions according to the randomization mechanism. By decomposing the "No" responses into self-protective and real ones, the estimates of the underlying noncompliance rates under (31) are higher than under (30).

In the reported application, we specified that participants who decided to give a SP-"No" response, select this response with probability 1. The crucial assumption of the mixture-IRR model (31) is that members of the SP-"No"-group do not provide any information about the items' location and discrimination parameters. This assumption is restrictive and thus easily testable in RR data sets.

5.3 Example

We now discuss analyses reported by Böckenholt and van der Heijden (2007). Table 5 reports the goodness-of-fit statistics obtained from fitting the IRR models to the work and health items for the 2002 and 2004 surveys. We also include the fit statistics obtained when considering the health items only. Because a minimum of three items are needed to identify an IRR model without covariates, no separate IRR models are estimated for the two work items. The first set of fitted models are based on the baseline RR assumptions and serve as a benchmark for the (mixture) IRR models. The second part of Table 5 is obtained by fitting a Rasch version of (30) to the health item set and by fitting a multidimensional Rasch version to both item sets simultaneously. The fit statistics reported in the third part of Table 5 are obtained from model (31) for the health domain and the multidimensional model for the work and health domains.

TABLE 5 Fit Statistics of RR-Models for Work and Health Items

Survey Year	Health Items	Health and Work Items
	G^2 (df)	G^2 (df)
1. Homogeneous compliance		
2002	124.0 (11)	282.4 (57)
2004	56.2 (11)	184.4 (57)
2. Heterogenous compliance		
2002	39.0 (10)	100.6 (54)
2004	23.8 (10)	95.6 (54)
3. Heterogenous compliance and SP-"No"-sayers		
2002	14.9 (9)	54.2 (53)
2004	10.8 (9)	63.5 (53)
2002 and 2004	29.4 (24)	131.0 (114)

The homogeneous-compliance models require the estimation of two and six item location parameters when applied to the four health items and the four health and two work items, respectively. None of the reported fits are satisfactory, indicating that the assumption of no individual differences does not agree with the data. This result is supported by the fit improvement obtained from the IRR models (30) and its multidimensional version that allow for heterogenous compliance behavior without requiring item-specific discrimination parameters. With one additional parameter, the variance of the normal distribution, σ^2, for the health items and three additional parameters for the bivariate covariance matrix of the health and work items, these IRR models provide major fit improvements. However, despite the better fit, these models do not describe the data satisfactorily. As shown by a residual analysis of the data, the main reason for the misfit is that the outcome of consistent "No"-responses to the items is greatly underestimated by these models. Thus, more respondents than expected under the IRR models give exclusively a "No" responses when asked questions about their compliance with the health and work regulations. Models (31) and its multidimensional version can address the problem of extra-"No" responses. The models' parsimonious representation appears to be in good agreement with the 2002 and 2004 survey data as indicated by the fit statistics in Table 5. As a further step, we fitted mixture-IRR models that constrain the model parameters to be the same for both survey years. As indicated by the goodness-of-fit statistics in Table 5, the uni- and two-dimensional IRR models describe the data well. We conclude that there is no significant change in the item parameters, in the association between the work and health domains, and the incidence of SP respondents over the 2 year period.

Table 6 reports the resulting parameter estimates. The second column of Table 6 contains the parameters of the health items which are ordered as expected. The population standard deviation of the person parameters is estimated to be 2.51 (0.52). About $\frac{1}{1+\exp(1.85)} = 14\%$ of the respondents are categorized as SP-"No" sayers. The remaining columns of Table 6 contain the estimates of the two-dimensional IRR mixture model. Although the parameter estimates of the health items are similar to the ones obtained from the univariate IRR model, the large standard errors of the work-item parameter estimates indicate that this model part is only weakly identified. The estimated joint probabilities are $\hat{\pi}_{00}=0.81$, $\hat{\pi}_{01}=0.06$, $\hat{\pi}_{10}=0.02$, and $\hat{\pi}_{11}=0.11$ indicating that about 13% and 17% of the respondents are SP-"No" sayers for the health and work items, respectively.

The estimated correlation between the θ parameters of the two domains is substantial with $\hat{\rho}_{12}=0.58$ (0.08) but still distinguishable from 1. Thus, not surprisingly, a significant fit deterioration is obtained when estimating an one-dimensional model (31) with different discrimination parameters for both item sets with $G^2 = 163.4$ ($df = 118$), demonstrating that individual differences for both domains are separable.

Most importantly, the mixture-IRR models provide more accurate estimates about the noncompliance rate in the population of interest than RR methods that do not allow for response biases. Table 7 contains the

TABLE 6 Parameter Estimates (and Standard Errors) of RR-Models for 2002 and 2004 Health and Work Items

Parameter	Health (h = 1)	Health (h = 1) and Work (h = 2)		
$\hat{\gamma}_{h1}$	5.64 (1.03)	5.55 (1.01)		3.40 (2.56)
$\hat{\gamma}_{h2}$	4.86 (0.92)	4.85 (0.90)		5.09 (3.71)
$\hat{\gamma}_{h3}$	4.09 (0.81)	4.13 (0.81)		–
$\hat{\gamma}_{h4}$	3.33 (0.64)	3.35 (0.64)		–
$\hat{\sigma}_h$	2.51 (0.52)	2.50 (0.52)		3.34 (2.45)
$\ln\frac{\hat{\pi}_{11}}{\hat{\pi}_{00}}$	−1.85 (0.20)		−1.96 (0.18)	
$\ln\frac{\hat{\pi}_{10}}{\hat{\pi}_{00}}$			−3.72 (1.40)	
$\ln\frac{\hat{\pi}_{01}}{\hat{\pi}_{00}}$			−2.67 (1.66)	

TABLE 7 Noncompliance Estimates and 95% Bootstrap Confidence Intervals

Domain	Items	No Bias Correction	Bias Correction Model
Health	1	0.002 (0.000, 0.015)	0.033 (0.010, 0.050)
	2	0.014 (0.000, 0.034)	0.053 (0.033, 0.075)
	3	0.048 (0.027, 0.070)	0.083 (0.055, 0.112)
	4	0.085 (0.063, 0.107)	0.130 (0.087, 0.159)
Work	1	0.030 (0.009, 0.052)	0.074 (0.050, 0.096)
	2	0.110 (0.086, 0.133)	0.159 (0.104, 0.190)

noncompliance estimates and their 95% bootstrap confidence intervals obtained both under the homogeneous-compliance model without a response bias correction and under the multidimensional model with the response bias correction. For example, for the health domain, the bias-uncorrected noncompliance percentages for the four items are estimated as 0.2%, 1.4%, 4.8%, and 8.5% respectively. In contrast, under the multidimensional model the corresponding estimates are 3.3%, 5.3%, 8.3%, and 13.0%. These differences are substantial and demonstrate the value of the proposed approach for the analysis of RR data. By not taking into account possible response biases, the actual incidence of noncompliance is severely underestimated. Equally important, these estimates do not include the mixture component consisting of the SP respondents. Thus, it is possible that even the response-bias corrected estimates are too low if some or all of the SP respondents are noncompliant as well.

Table 8 displays the joint distribution of the number of infringements for the health and work domains estimated under the multidimensional model. About 71% of the respondents report no compliance violations. The estimated levels of compliance for the Health and Work domains are 81% and 82%, respectively. These figures are useful in monitoring the effectiveness of interventions aimed at increasing the overall and domain-specific compliance level in the population of interest.

The IRR model is presented in this manuscript because it also allows to investigate the association of covariates with θ_i. The fit measures in Table 5 demonstrates that respondents differ both in their compliance behavior and in their willingness to follow the randomized response format. It is important to understand the sources of these individual differences. We investigate whether attitudinal, risk return, and demographic items can account for the variability in the IRR person parameters and predict membership in the

TABLE 8 Estimated Compliance Percentages for Health and Work Domains

Counts	Work			
Health	*0*	*1*	*2*	Total
0	71	7	3	81
1	7	2	1	11
2	3	1	1	5
3	1	1	1	2
4	0	0	0	1
Total	82	11	6	100

SP-"No" sayer class for extended versions of the earlier models that allow for covariates.

Both, θ_{i1} and θ_{i2}, can be expressed as a function of covariates with $\theta_{ih} = \sum_f x_{if}\beta_f + \epsilon_{ih}$; $(h = 1,2)$. Random effects that are not captured by the available covariates are represented by ϵ_{ih} which are assumed to be normally distributed with mean 0. In addition, the probability of being in the group following the RR design can be expressed as a logistic function of the covariates.

Our analyses proceeded thus in two steps. First, we parameterized the θ_i's as a linear function of the nine variables that possibly explain regulatory noncompliance (compare Elffers et al., 2003 and Böckenholt and van der Heijden, 2007) and tested for both time/survey and domain effects. Subsequently, we removed the nonsignificant covariates from the mixture IRR models and estimated a model in which both the θ and π parameters are expressed as a function of covariates. These analyses were conducted for the health, the work, and the combined health and work items for both survey years.

Modeling individual differences in noncompliance: The year difference is not significant and we estimate the regression model (included in the Rasch-type model with the SP parameters) in the model where parameters over years are set equal to each other. The estimated regression effects of this final model as well as descriptive statistics of the covariates are listed in Table 9. Four of the nine items predict individual variation in θ: Acceptance, Benefits, Law Abidance, and Social Control. Thus, only cost-benefit considerations and social norms appear to play a significant role in a person's decision to act in accordance with the regulatory health and work laws.

Modeling response biases: In the analysis of the covariates' effects on the individual-specific π parameters, we conjectured that the choice of a person to select the "No"-response category is determined by both a person's perception that the randomization scheme protects one's privacy (Trust) and the

TABLE 9 Effects of Covariates (and their Standard Errors) on θ Parameters and Descriptive Statistics for the Combined 2002 and 2004 Survey Data

Items	Work and Health		Means	Std.
Acceptance	−0.32	(0.12)	2.2	0.6
Clarity	−0.01	(0.12)	2.6	1.0
Benefits	0.80	(0.14)	2.6	1.1
Social control	−0.39	(0.12)	3.6	1.0
Law abidance	−0.41	(0.12)	2.5	0.9
Control	−0.18	(0.11)	2.9	1.1
Detection	−0.15	(0.13)	2.7	1.0
Sanction certainty	−0.15	(0.11)	3.4	0.9
Sanction severity	−0.21	(0.11)	2.4	1.1

perceived clarity of the RR instructions (Instructions). Respondents who did not feel protected or did not understand the instructions were expected to give a SP-"No" response more frequently. Because we found in preliminary analyses that the perceived clarity of the instructions covaried negatively with the educational level of a person, we included "Education" as a third variable. Finally, we added the remaining attitudinal and risk-return variables to test whether they can predict a person's decision to respond to the questions truthfully. Table 10 displays the results obtained under a joint analysis of the item sets for the two survey years.

As expected, both "Instruction" and "Education" are significant predictors of self-protective responses. Respondents who found the instructions to be clear are more likely to state the truth. Out of the set of attitudinal and risk-return factors, an understanding of the regulations (Clarity) and the subjective probability that a case will be prosecuted and sanctioned when detected (Sanction Certainty) as well as the severity of the sanction are significant.

5.4 Remaining Issues

In Böckenholt et al. (2009) alternatives are investigated for the simple SP parameter, such as SP parameters that are item specific. These models do not have a better fit in the data of the earlier section. There are also other models that make use of SP parameters. Cruyff et al. (2007) present log linear models with SP parameters. Another interesting and potentially important model for data collection in a design where RR is compared to, for example, asking questions directly, is Van den Hout et al. (2010). Interestingly,

TABLE 10 Effects of Covariates (and their Standard Errors) on θ and π Parameters for the Combined 2002 and 2004 Survey Data

Items	Work and Health	
θ Effects; noncompliance with IIA regulations		
Acceptance	−0.31	(0.11)
Benefits	0.76	(0.13)
Law abidance	−0.39	(0.11)
Social control	−0.40	(0.13)
π Effects; adherence to RR instructions		
Trust	0.19	(0.11)
Instruction	0.46	(0.09)
Education	0.54	(0.14)
Clarity	0.24	(0.11)
Detection	−0.20	(0.11)
Control	−0.21	(0.12)
Sanction certainty	−0.27	(0.11)
Sanction severity	−0.24	(0.12)

they present an example where the SP parameter estimated under RR is 0.16 whereas under direct questions it is 0.54. For this example, RR still leaves some bias in the data, that is eliminated by using the SP parameter, where the comparable bias under direct questions is so large that data may be considered useless.

Recently Cruyff et al. (2015) published work where an item is measured twice, once with a dichotomous RR question and once with a categorized quantitative RR variable. We used the first work item on moonlighting (with answer yes/no) together with a question on how much they earned with it (measured with six categories). Whereas the $2 \times 6 = 12$ observed answers could be found, on the true (latent) level only six answers are possible (namely no-0 and yes with five options for ranges of amount of money involved). Since only 6 parameters are needed in the model whereas 12 counts are observed, it is possible to include SP parameters in the model to investigate biases. It turned out that, compared to a model for a single dichotomous RR question, the model yielded surprisingly efficient parameter estimates.

It should be noted that it is not possible to estimate SP from univariate RR data, simply because multivariate RR data are needed to estimate the probability of SP. If we want to the existence of SP into account in the analysis of univariate RR data, we have to "borrow" information from the SP parameters estimated in multivariate models. The basic idea is to estimate the SP parameter and adapt equations so that part of the sample with no-answers is eliminated. An example of such an approach can be found in Frank et al. (2009).

REFERENCES

Agresti, A., 2002. Categorical Data Analysis. John Wiley & Sons, Hoboken, NJ.

Ahangar, A., Wang, R., Perez, J., Signh, S., 2012. Extensive study of logistic regression using randomized response sampling. Association for the Advancement of Modelling and Simulation. https://www.researchgate.net/publication/260192741_Extensive_Study_of_Logistic_Regression_using_Randomized_Response_Sampling.

Böckenholt, U., van der Heijden, P.G.M., 2004. Measuring noncompliance in insurance benefit regulations with randomized response methods for multiple items. In: Biggeri, A., Dreassi, E., Lagazio, C., Marchi, M. (Eds.), 19th International Workshop on Statistical Modelling, Florence, Italy, pp. 106–110.

Böckenholt, U., van der Heijden, P.G.M., 2007. Item randomized-response models for measuring noncompliance: risk-return perceptions, social influences, and self-protective responses. Psychometrika 72 (2), 245–262.

Böckenholt, U., Barlas, S., van der Heijden, P.G.M., 2009. Do randomized-response designs eliminate response biases? An empirical study of non-compliance. J. Appl. Econometr. 24, 377–392.

Boruch, R.F., 1971. Assuring confidentiality of responses in social research: a note on strategies. Am. Sociol. 6, 308–311.

Chaudhuri, A., Mukerjee, R., 1988. Randomized Response: Theory and Techniques. Marcel Dekker, New York, NY.

Chaudhuri, A., Saha, A., 2004. Utilizing covariates in logistic regression modelling in improved estimation of population proportions bearing stigmatizing features through randomized responses in complex surveys. J. Ind. Soc. Agric. Stat. 58, 190–211.

Clark, S.J., Desharnais, R.A., 1998. Honest answers to embarissing questions: detecting cheating in the randomized response model. Psychol. Meth. 3, 160–168.

Cruyff, M.J.L.F., van den Hout, A., van der Heijden, P.G.M., Böckenholt, U., 2007. Log-linear randomized-response models taking self-protective response behavior into account. Sociol. Methods Res. 36, 266–282.

Cruyff, M.J.L.F., van den Hout, A., van der Heijden, P.G.M., Böckenholt, U., 2008. The analysis of randomized-response sum score variables. J. R. Stat. Soc. B 71, 21–30.

Cruyff, M.J.L.F., Böckenholt, U., van der Heijden, P.G.M., 2015. The multidimensional randomized response design: estimating estimating different aspects of the same sensitive behavior. Behav. Res. Meth. http://dx.doi.org/10.3758/s13428-015-0583-2.

Edgell, S.E., Himmelfarb, S., Duncan, K.L., 1982. Validity of forced response in a randomized response model. Sociol. Methods Res. 11, 89–110.

Elffers, H., van der Heijden, P.G.M., Hezemans, M., 2003. Explaining regulatory non-compliance. A survey study of rule transgression for two dutch instrumental laws, applying the randomized response method. J. Quant. Criminol. 19, 409–439.

Fox, J.P., 2005. Randomized item response theory models. J. Educ. Behav. Stat. 30, 1–24.

Frank, L.E., van den Hout, A., van der Heijden, P.G.M., 2009. A logit model for repeated cross-sectional reandomized response data. Methodology 5, 145–152.

Glonek, G.F.V., McCullagh, P., 1995. Multivariate logistic models. J. R. Stat. Soc. B 57, 533–546.

Hsieh, S.H., Lee, S.M., Shen, S.H., 2009. Semiparametric analysis of randomized response data with missing covariates in logistic regression. Comput. Stat. Data Anal. 53, 2673–2692.

Jann, B., 2005. rrlogit: stata module to estimate logistic regression for randomized response data. http:/ideas.repec.org/c/boc/bocode/s456203.html.

Krumpal, I., 2012. Estimating the prevalence of xenophobia and anti-semitism in germany: a comparison of randomized response and direct questioning. Soc. Sci. Res. 41, 1387–1403.

Kuha, J., Skinner, C., 1997. Categorical data analysis and misclassification. In: Lyberg, L. (Ed.), Survey Measurement and Process Quality. Wiley, New York, NY, pp. 633–670.

Kuk, A.Y.C., 1990. Asking sensitive questions indirectly. Biometrika 77, 436–438.

Landsheer, J.A., van der Heijden, P.G.M., van Gils, G., 1999. Trust and understanding, two psychological aspects of randomized response. A study of a method for improving the estimate of social security fraud. Qual. Quant. 33, 1–12.

Lensvelt-Mulders, G.J.L.M., Hox, J.J., van der Heijden, P.G.M., Maas, C.J.M., 2005. Meta-analysis of randomized response research: thirty-five years of validation. Sociol. Methods Res. 33, 319–348.

Lensvelt-Mulders, G.J.L.M., van der Heijden, P.G.M., Laudy, O., 2006. A validation of a computer-assisted randomized response survey to estimate the prevalence of fraud in social security. J. R. Stat. Soc. A 169, 305–318.

Maddala, G.S., 1983. Limited Dependent and Qualitative Variables in Econometrics. Cambridge University Press, Cambridge.

McCullagh, P., 1980. Regression models for ordinal data (with discussion). J. R. Stat. Soc. B 42, 109–142.

Moshagen, M., Musch, J., 2012. Surveying multiple sensitive attributes using an extension of the randomized-response technique. Int. J. Public Opin. Res. 24, 508–523.

Ostapczuk, M., Musch, J., Moshagen, M., 2009. A randomized-response investigation of the education effect in attitudes towards foreigners. Eur. J. Soc. Psychol. 39, 920–931.

Rasch, G., 1980. Probabilistic Models for Some Intelligence and Attainment Tests. The University of Chicago Press, Chicago, IL.

Ronning, G., 2005. Randomized rsponse and the binary probit model. Econ. Lett. 86, 221–228.

Scheers, N.J., Dayton, C.M., 1988. Covariate randomized response models. J. Am. Stat. Assoc. 83, 969–974.

St John, F.A.V., Keane, A.M., Edwards-Jones, G., Jones, L., Yarnell, R.W., Jones, J.P.G., 2011. Identifying indicators of illegal behaviour: carnivore killing in human-managed landscapes. Proc. R. Soc. Biol. Sci. 279, 804–812.

van den Hout, A., Kooiman, P., 2007. Estimating the linear regression model with categorical covariates subject to randomized response. Comput. Stat. Data Anal. 50, 3311–3323.

Van den Hout, A., van der Heijden, P.G.M., 2002. Randomized response, statistical disclosure control and misclassification: a review. Int. Stat. Rev. 70, 269–288.

van den Hout, A., van der Heijden, P.G.M., Gilchrist, R., 2007. The logistic regression model with response variables subject to randomized response. Comput. Stat. Data Anal. 51, 6060–6069.

van den Hout, A., Bockenholt, U., van der Heijden, P.G.M., 2010. Estimating the prevalence of sensitive behaviour and cheating with a dual design for direct questioning and randomized response. J. R. Stat. Soc. C 59, 723–736.

van der Heijden, P.G.M., Böckenholt, U., 2008. Applications of randomized response methodology in e-commerc. In: Jank, W., Shmueli, G. (Eds.), Statistical Methods in e-commerce. Wiley, New York, NY, pp. 401–416.

van der Heijden, P.G.M., van Gils, G., 1996. Some logistic regression models for randomized response data. In: Forcina, A., Marchetti, G.M., Hatzinger, R. (Eds.), Statistical Modelling. Proceedings of the 11th International Workshop on Statistical Modelling, Orvieto, Italy, pp. 341–348.

van der Heijden, P.G.M., van Gils, G., Bouts, J., Hox, J.J., 2000. A comparison of randomized response, computer-assisted self-interview, and face-to-face direct questioning. eliciting sensitive information in the context of welfare and unemployment benefit. Sociol. Methods Res. 28, 505–537.

Warner, S.L., 1965. Randomized response: A survey technique for eliminating answer bias. J. Am. Soc. Assoc. 60, 63–69.

Chapter 19

Eliciting Information on Sensitive Features: Block Total Response Technique and Related Inference

K. Nandy*, M. Marcovitz† and B.K. Sinha‡,1,2

*School of Nursing and Department of Biostatistics, Fielding School of Public Health, University of California, Los Angeles, CA, United States

†Baylor University, Waco, TX, United States

‡Retired Professor, Indian Statistical Institute, Kolkata, India

[2]Corresponding author: e-mail: bikassinha1946@gmail.com

ABSTRACT

Randomized Response Technique was first introduced and popularized by Warner in 1965. Since then, survey sampling theoreticians and practitioners have contributed significantly in this area of survey methodological research. The idea is to be able to elicit a "truthful" response on *sensitive* feature(s) from the sampled respondents (of a finite labeled population of respondents), so that eventually the population mean of the sensitive feature can be unbiasedly estimated. Toward this, a novel technique was introduced by Raghavarao and Federer (1979) and it was termed "Block Total Response" (BTR) technique. We undertake various meaningful versions/generalizations of the BTR technique, after a brief review of the literature in this direction. In the process, we also introduce empirical Bayes estimators.

Keywords: Sensitive feature, Regular feature, Randomized response technique, Block total response technique, Balanced incomplete block design, Binary proper equireplicate block design, Empirical Bayes estimators

1 RANDOMIZED RESPONSE TECHNIQUE

Recent textbooks on survey sampling provide a chapter on Randomized Response Techniques (RRTs). There are considerable details given in some

[1]Retired Professor

Handbook of Statistics, Vol. 34. http://dx.doi.org/10.1016/bs.host.2016.01.017

texts, viz., Hedayat and Sinha (1991), chapter 11. We refer to an excellent expository early book by Chaudhuri and Mukerjee (1988). Two recent books (Chaudhuri, 2011; Chaudhuri and Christofides, 2013) are also worth mentioning. In this chapter, we confine the discussion to one or more Sensitive Qualitative Features (SQlFs) denoted by Q^*'s, each with a binary response (yes/no) and we denote by P^*'s, the corresponding population proportion(s) of "yes" response(s). The problem is to provide (i) a method of ascertaining "truthful" response to the Q^*'s from each respondent facing the SQlFs in the surveyed population and (ii) to provide unbiased estimator(s) of P^*'s. Generally, simple random sampling with replacement of respondents is contemplated.

With reference to a single SQlF, Q^*, the simplest unrelated-question technique refers to one arbitrarily chosen Regular Qualitative Feature (RQlF), Q, with associated incidence proportion, P, which is assumed to be known. A Randomization Device (RD) is developed such that the respondent can make a choice of Q^* with probability, θ, or, a choice of Q with probability, $1-\theta$, $0<\theta<1$, $\theta \neq 0.50$, being known beforehand. It is believed that this randomization mechanism will convince the respondent about retaining the confidentiality of the response (yes/no) provided by him/her, without disclosing the choice of Q or Q^* to the interviewer!

Routine formulae are there to work out the details of estimation, etc. There are many versions of the RRT, covering SQlFs as well as SQnFs. Here, "Ql" refers to "Qualitative" and "Qn" refers to "Quantitative." We do not enter into the details. We confine to SQlFs in this study.

We will discuss some aspects of Block Total Response Technique (BTRT) as introduced in the literature and provide some extensions. Next, we will also introduce an empirical Bayes (EB) estimator in this context.

2 BLOCK TOTAL RESPONSE TECHNIQUE

It is truly amazing to note that an altogether different thinking and data collection mechanism emerged in a 1979 paper by Raghavarao and Federer. Subsequently, there have been studies along this direction. A precursor to this study was undertaken by Smith et al. (1974). We refer to the book by Aslam and Yab (2011).

We describe the basic idea first. Consider a collection of v RQlFs $[Q1,Q2,\ldots,Qv]$ and only one single SQlF, Q^*. We thus have a total collection of $(v+1)$ QlFs. A respondent is presented with a "block" of questions, i.e. a "questionnaire" involving some k of the RSlFs and the SQlF and he/she is to provide only the Block Total Response (BTR) in terms of total "score" (i.e. number of "yes" responses) without divulging any separate information as to his/her status in respect of the $k+1$ questions included in the questionnaire. It is believed that for any reasonable choice of k, the respondent is convinced about the honesty and integrity of the data collection agency. This will protect the privacy of the respondent and hence a truthful response to Q^* will emerge! Since only the total score from a block is divulged by the respondent, this scheme is termed Block Total Response Method/Technique (BTRM/BTRT). There are variations of this method. The particular method described above is termed Supplemented Block Total Response Method (SBTRM). In applications, there are a number of blocks of equal/unequal sizes and additionally, there is one block containing all the v RQIFs. Basically, the SQIF acts as a "Supplementary Question" and is attached to every block of RQIFs, except the last one as mentioned above. The blocks to be used may be based on our notion of block designs such as Balanced Incomplete Block Designs (BIBDs). It is to be noted that in the absence of an additional block formed exclusively of all the RQIFs, all other blocks have to be based on unequal sizes, as otherwise, unbiased estimation cannot be ensured (Raghavarao and Federer, 1979).

We may mention in passing that in a thought-provoking paper, Nayak and Adeshiyan (2009) have argued as follows:

If all the proportions corresponding to the v RQIF's are known beforehand, then the respondents in each of the b blocks are presented with a known discrete distribution over the set $[0,1,2,\ldots,k]$ and each one of them is supposed to draw a random observation from the specific discrete distribution in the first place. Next, depending on the nature of true response category of the respondent (either DA or NDA), he/she is supposed to add "1" or "0" to the random number so chosen in the next step. This specifies the final response from the respondent—the details being unnoticed by the investigator. Naturally, in such a case, there is no need for the additional block containing all the v RQIF's.

Denoting this procedure as [2 to k], they have also argued that it is possible to generate the above data starting with a suitably defined [2 to 2] based on DA/NDA classification.

However, the following points may be noted in this context.

(a) The BTR procedure, as described by Raghavarao and Federer (1979), is likely to provide more "perceived" protection and confidentiality to the respondents. The BTR approach is clear and concise to explain and adopt.

(b) In case two known discrete probability distributions are handed over to the respondents and each respondent is asked to utilize the right one and generate an observation to be treated as his/her recorded response, their immediate reaction may not be very favorable!

(c) On the other hand, the BTR procedure handles the case (of unknown proportions for the RQIF's) in a natural manner and it also provides unbiased estimates of all these proportions. This is a routine exercise involving linear models. We only discuss the primary problem of estimation of P^* corresponding to the SQIF.

We should also mention in passing that while dealing with the SQIF, Q^*, the respondent may use the RRT, if he/she so desires. For simplicity, it is advisable to use the same randomization mechanism for every block in the SBTRM—with the same choice of Q and θ.

In early papers on the topic, Supplemented BIBDs were used and methodologies and formulae were derived according to those designs. Supplemented BIBDs are also known in the design literature as Balanced Treatment Incomplete Block Designs (BTIBDs). To reinforce the use of this fascinating technique, we will entertain other variations of this SBTRM.

3 SBTRM : USE OF BIBD AND COMPLIMENTARY BIBD

We start with a BIBD $[b,v,r,k,\lambda]$ which consists of b blocks, each of size k. We represent the v treatments by v RQIF's, so that each block contains exactly k of these Fs. We now reinforce use of the RRT in the sense that in every block, we utilize the remaining $v - k$ RQIF's along with the SQIF, Q^*, in the formation of the RRT having $(v - k) + 1$ options for the respondent to choose from. As before, we associate a chance θ to Q^* and each of the remaining questions in the RRT has an associated chance of $(1 - \theta)/(v - k)$. The message is very clear and quite appealing. Each respondent has k clearly stated RQIF's to respond to and on top of that, he/she has to respond to exactly one more question—chosen according to the RRT described above. Finally, he/she has to report only the total score and not any details! For example, with $v = 9$ and $k = 4$, the RRT will involve $5+1 = 6$ questions altogether, and if $\theta = 0.20$, then the respondent has an 80% chance of picking up any one of the five RQIF's in the complementary block, along with the four in the original block. Naturally, a total score of 3, for example, provided by a respondent in any of these b blocks, may arise in so many ways that it is virtually impossible to track down the actual "response category" of the respondent(s) in respect of the SQIF!

To top it all, we need not start with a BIBD; any Binary Proper Equireplicate Block Design (BPEBD) with parameters $[b,v,r,k]$ would suffice. A proof of this fact is given in the Appendix. Of course, in all of the above, it is implicitly assumed that there is an additional block of all the v RSIFs for a group of respondents to attend to and provide the score in the form of BTR.

4 RELATIVE COMPARISON OF BIBD-BASED SBTRMS

Let us start with a stimulating example. There are $v = 5$ RQIFs and one single SQIF, Q^*, with associated incidence parameter, P^*. We have a total of 11 blocks to implement and there are two options.

Option I : Use five blocks—each repeated twice and an additional block (11th block) of all five RQIFs. The former set of blocks involve four of the five RQIFs, taking turns by skipping one-by-one, being a BIBD $[b = v = 5, r = k = 4, \lambda = 3]$. The left-over RQIF and Q^* form a pair of Fs for an RRT with chance, θ, for Q^*. For example, block one has as its first component [1, 2, 3, 4] and as its second component, an RRT with two options given by $[(5,1 - \theta), (Q^*,\theta)]$. A respondent is asked to answer all the first four RQIFs and adopt a RRT for the second part in each of these five blocks of both the sets, and of course, only to report the BTR from the entire exercise. Under this option,

$$\hat{P}* = [\Sigma_T - (5 - \theta)\Sigma_0]/5\theta,$$

where,

Σ_T = BTRs added over all 10 blocks of the plan divided by 2; Σ_0 = BTR for the 11th block.

It turns out that,

$$V(\hat{P}^*) = [5/2 + (5 - \theta)^2]/25\theta^2.$$

Option II : Use 10 blocks based on the BIBD [$b = 10$, $v = 5$, $r = 6$, $k = 3$, $\lambda = 2$] and the 11th block as before. For each block of the BIBD, there is an RRT counter-part comprising of Q^* with chance, θ, and the complimentary pair of treatments, each with chance, $(1 - \theta)/2$. It follows that,

$$\hat{P}^* = [\Sigma_{\text{BTR}} - (8 - 2\theta)\Sigma_0]/10\theta.$$

Further,

$$V(\hat{P}^*) = [10 + (8 - 2\theta)^2]/100\theta^2.$$

It now turns out that $V_{II} < V_I$, uniformly in θ!

It is interesting to note that seemingly Option II gives better protection of privacy and hence increased confidence to the participants and it also provides increased efficiency toward estimation of P^*! One may argue that the fact that each participant under Option I is addressing five Fs (four RQIFs and one in the form of an RRT as a combination of only one RQIF and the SQIF) is possibly going against it.

5 DERIVING THE EB ESTIMATORS

First introduced by Robbins (1956), EB analysis has been used in many settings to strengthen estimators by borrowing information across a data set. This idea of borrowing strength from the data lends itself to our situation in which the traditional least squares estimators may not be suitable for estimating the population proportion for sensitive behaviors due to participants who do not feel comfortable answering survey questions. EB analysis utilizes observed data to estimate a prior distribution, which is then used to produce a posterior distribution. This is useful when the prior cannot be completely specified beforehand. In this section, we propose using the EB approach to estimate prior probabilities of belonging to the v categories from the data collected in a survey. Using the prior distribution estimated from the data and a least squares estimator, also from the data, we arrive at a posterior distribution that does well at estimating the true proportions of respondents who belong to the v categories. Such an approach is taken up in Marcovitz et al. (2014) in a recent study.

Our intuition is that if participants are uncomfortable with the study questions, the least squares estimator may change because block totals might be affected. To account for this potential deviation in the least squares estimator, we utilize the block totals in other possible ways, by incorporating suitable permutation subgroups. Our EB estimator, $\tilde{p_l}$, is a shrinkage estimator of the type:

$$\tilde{p_l} = a\hat{p}_l + b, \tag{1}$$

where a scale factor, *a,* and a shrinkage factor, *b,* will shrink the least squares estimator, $\hat{p}_l$, toward the true value. These factors are suitably derived from the survey data.

It is assumed that the proportions of respondents belonging to the sensitive categories addressed by the survey questions have the same underlying prior distribution. The same is also assumed of the nonsensitive questions in the list. This latter assumption may be ensured by a proper selection of the nonsensitive questions at our disposal. However, the two distributions will be taken to be different. We take an EB approach to finding the underlying parameters by evaluating the prior distribution from the data. Note that the least squares estimator, $\hat{p}_l$, is a linear combination of observed block means. We now elaborate on the details.

The optimal BIBD for use in BTR has parameters, $v = 4t - 1 = b, r = 2t = k$, and $\lambda = t$ for a choice of t (vide Smith et al., 1974). If there are l sensitive questions in a study, then the design should have $4t - 1 > l$ total questions and $v - l$ nonsensitive questions. We work with the BIBD $[b = v = 11, r = k = 6, \lambda = 3]$ and $l = 3$ in the next section.

In general terms, the method requires a sample of n participants with subsets of the participants randomized to respond to one of the b questionnaires. Let $n = mv$, so that the sample size is a multiple of the v questionnaires with m participants randomized to each questionnaire. For illustration, we take $v =$ 11. Thus, if there are $v = 11$ individual questionnaires and $n = 33$ participants, three participants will be randomized to answer the $k = 6$ questions in each of the 11 questionnaires.

The procedure assumes that $p_1, \ldots, p_{4t-1}$ are the true probabilities of belonging to the $4t - 1$ categories. Let Y_{ij} be the j-th individual's total score for the i-th block of questions. Then $E(Y_{ij}) = \sum_{l \in B_i} p_l$. Let $Y_i = \frac{1}{m}\sum Y_{ij}$ so that,

$$E(\bar{Y}_i) = \sum_{l \in B_i} p_l, l = 1, \ldots, 4t - 1. \tag{2}$$

The p_l's can be unbiasedly estimated using the $4t - 1$ equations given by Formula (2). Specifically, we have,

$$\hat{p}_l = \frac{\sum_{i : l \in B_i} \bar{Y}_i - \sum_{u : l \notin B_u} \bar{Y}_u}{2t}, l = 1, \ldots, 4t - 1. \tag{3}$$

It is evident that $\hat{p}_l$ is unbiased for p_l with variance, σ^2, estimated by,

$$\hat{\sigma^2} = \frac{\sum_{i : l \in B_i} V(\bar{Y}_i) + \sum_{u : l \notin B_u} V(\bar{Y}_u)}{2t} = \frac{\sum_{l=1}^{4t-1} s_l^2}{(2t)^2 m}, \tag{4}$$

where s_i^2, an estimate of the variance of the total response for the ith block of questions, is given by

$$s_i^2 = \frac{1}{m-1}\sum (Y_{ij} - \bar{Y}_i)^2, i = 1, \ldots, 4t-1. \tag{5}$$

Interestingly, for all $\hat{p}_l$'s, we have the same variance σ^2.

Mathematical details for the BTR method can be found in Raghavarao and Federer (1979). An application can be found in Smith et al. (1974), who estimated the proportions for three sensitive questions.

We assume the prior distribution of p_l's to be the empirical distribution of the $\hat{p}_l$'s under suitably defined subgroups of $b!$ permutations of the $\bar{Y}_l$'s.

We now elaborate on this with reference to an example derived from the BIBD [$b = v = 11$, $r = k = 6$, $\lambda = 3$]. There are 11 questions in all–8 are nonsensitive [Q1 to Q8] and 3 are sensitive [Q9, Q10, Q11]. The block specifications of the BIBD are given below.

$$[6,7,8;9,10,11];[3,4,5;9,10,11];[2,4,5,7,8;11];$$
$$[2,3,5,6,8;10];[2,3,4,6,7;9];[1,4,5,6,7;10];[1,3,5,7,8;9];$$
$$[1,3,4,6,8;11];[1,2,5,6;9,11];[1,2,4,8;9,10];[1,2,3,7;10,11]$$

Note that the first two blocks are of one type: three nonsensitive questions followed by all the three sensitive questions in each; the second lot of six blocks are of another type: five nonsensitive questions and one sensitive question in each; lastly, there are three blocks of a third type: four are nonsensitive questions and two are sensitive questions. In view of this, and the assumptions about the nature of the prior distributions of the eight nonsensitive and three sensitive questions, we derive the empirical distribution by permuting the BTRs (or, equivalently, $\bar{Y}_l$'s) for the first lot of two blocks, second lot of six blocks, and the last lot of three blocks—all independent of each other. Therefore, we have a total of $2! \times 6! \times 3!$ permutations to deal with.

We have already stipulated the same prior for all eight p_l's, $l = 1,\ldots,8$ and also the same prior for all three p_l's, $l = 9,10,11$. Let us assume the prior distributions of the p_l's to have mean $= v$ and variance $= \tau^2$, $l = 1,\ldots,8$, and similarly, for the other 3, mean $= \theta$ and variance $= \delta^2$.

Because this is an EB analysis, no distributional assumptions are put on v, θ, τ^2, and δ^2. Instead we estimate them using the $\hat{p}_l$'s based upon all $2! \times 6! \times 3!$ permutations. Naturally, v and τ^2 are to be based on $\hat{p}_l$'s for $l = 1,2,\ldots,8$. Likewise, for the other three parameters.

The posterior mean of p_l, $\widetilde{p_l}$, is the EB estimator of p_l. It is well known that $\widetilde{p_l}$ is linearly related to $\hat{p}_l$ and shifts the scale and origin of the $\hat{p}_l$s:

$$\widetilde{p_l} = \frac{\tau^2\hat{p}_l + \hat{\sigma}^2 v}{\tau^2 + \hat{\sigma}^2}, l = 1,\ldots,8; \tag{6}$$

$$\widetilde{p_l} = \frac{\delta^2 \hat{p}_l + \hat{\sigma}^2 \theta}{\delta^2 + \hat{\sigma}^2}, l = 9, 10, 11. \quad (7)$$

Note that the formula for $\hat{\sigma}^2$ is readily available.

A simple proof of (6) and (7) can be found in Lehmann and Casella (1998), chapter 4, assuming normality. We measure the accuracy of the EB estimator under squared error loss, $E(p_l \sim -p_l)^2$, where the expectation is conditioned only on the observed data. For fixed p_l,

$$\begin{aligned} E(\widetilde{p_l} - p_l)^2 &= E\left(\frac{\tau^2 \hat{p}_l + \hat{\sigma}^2 v}{\tau^2 + \hat{\sigma}^2} - p_l\right)^2 \\ &= \frac{\tau^4 \hat{\sigma}^2}{(\tau^2 + \hat{\sigma}^2)^2} + \frac{\hat{\sigma}^4 (v - p_l)^2}{(\tau^2 + \hat{\sigma}^2)^2} \\ &= \hat{\sigma}^2 \left[\frac{\tau^4 + \hat{\sigma}^2 (v - p_l)^2}{(\tau^2 + \hat{\sigma}^2)^2}\right]. \end{aligned} \quad (8)$$

The above formula applies to l = 1,…,8. Similarly, we can compute (8) for the last three $\widetilde{p}_l$'s as well. The EB estimator will either be less efficient, as efficient, or more efficient than the least squares estimator if $\left[\frac{\tau^4 + \hat{\sigma}^2 (v - p_l)^2}{(\tau^2 + \hat{\sigma}^2)^2}\right]$ is >, =, or < 1.

6 ILLUSTRATIVE EXAMPLE

We carry out an example of a study using the above BIBD[$b = v = 11$, $r = k = 6$, $\lambda = 3$] to compare the performance of the EB estimator with the standard least squares estimator. For this example, as explained before, we create an 11-question survey with a mixture of eight nonsensitive questions (RQIFs) labeled 1 to 8 and three sensitive questions (SQIFs) labeled 9, 10, and 11. We replicate each block of questions, m = 3 times. We simulate the data using known p_l's as indicated in Table 2. Table 1 shows the block composition of the BIBD and the simulated Y_{ij}. Average block totals, $\bar{Y}_i$, and variances, s_i^2, for i = 1,…,11 are also given in Table 1. Note that the variance computation based on the block totals is done by the usual formula given above.

The mean of the average block totals is computed as $\bar{\bar{Y}} = 3.24$. Table 2 shows the known proportions, p_l, and estimators, $\hat{p}_l$ and $\widetilde{p}_l$, obtained through equations (2), (6) and (7). We compute the value of $\hat{\sigma}^2$ using formula (4) as

$$\begin{aligned} \hat{\sigma}^2 &= \frac{0.89 + 0.22 + 1.56 + 0.67 + 0.67 + 1.56 + 0.89 + 0.67 + 0.22 + 1.56 + 0.22}{(6)^2 \times 3} \\ &= 0.0845. \end{aligned}$$

TABLE 1 Totals, Means, and Variances of Blocks

Block Questions	Y_{ij}	$\overline{Y_i}$	s_i^2
6, 7, 8, 9, 10, 11	4, 2, 2	2.67	0.89
3, 4, 5, 9, 10, 11	2, 2, 3	2.33	0.22
2, 4, 5, 7, 8, 11	5, 3, 2	3.33	1.56
2, 3, 5, 6, 8, 10	3, 4, 5	4.00	0.67
2, 3, 4, 6, 7, 9	3, 5, 4	4.00	0.67
1, 4, 5, 6, 7, 10	3, 4, 1	2.67	1.56
1, 3, 5, 7, 8, 9	2, 4, 4	3.33	0.89
1, 3, 4, 6, 8, 11	3, 4, 5	4.00	0.67
1, 2, 5, 6, 9, 11	3, 3, 2	2.67	0.22
1, 2, 4, 8, 9, 10	3, 5, 2	3.33	1.56
1, 2, 3, 7, 10, 11	3, 3, 4	3.33	0.22

TABLE 2 p_I, $\hat{p}_I$, and $\tilde{p}_I$ and Performance Ratio

Question	p_I	$\hat{p}_I$	$\tilde{p}_I$	$E(\tilde{p}_I - p_I)^2/\hat{\sigma}^2$
1	0.55	0.50	0.5927	0.1540
2	0.58	0.95	0.7304	0.4631
3	0.60	1.00[a]	0.6960	0.2813
4	0.62	0.61	0.4141	1.7777
5	0.64	0.17	0.6960	0.2813
6	0.65	0.72	0.5927	0.1540
7	0.70	0.50	0.6615	0.1787
8	0.72	0.95	0.7303	0.4631
9	0.20	0.17	0.1668	0.0730
10	0.25	0.17	0.1068	0.2497
11	0.30	0.17	0.2268	0.2507

[a]*Actual value is more than 1.*

Using the block total means, the calculated v and τ^2 can be obtained through usual computations: $v = 0.6343$ and $\tau^2 = 0.06652$. Likewise, $\theta = 0.1667$ and $\delta^2 = 0.05486$.

Table 2 shows that for all the questions, except Q4, the EB estimators have smaller mean squared error loss (column 5) than $\hat{\sigma}^2$, indicating that the EB estimator is preferred over the least squares estimator for estimating p_l for all the remaining 10 questions, including the sensitive questions.

7 CONCLUDING REMARKS

In this chapter we discussed some variations of the BTR technique as is available in the literature. Next, following Marcovitz et al. (2014), we used an EB approach to estimate the incidence proportions in a BTR experimental setting that protects the anonymity of respondents. Compared against the least squares estimators, the performances of the EB estimators seem to be encouraging. The technique explored here of creating an empirical Bayes estimator by suitably permuting responses in groups can possibly be applied in a variety of other settings.

ACKNOWLEDGMENT

We are thankful to Professor Tapan K. Nayak of George Washington University, USA for exchanging some of his ideas with one of the coauthors in a recent meeting.

APPENDIX

Consider a BPEBD consisting of b blocks, each of size $k < v$, and involving a set of v RQIFs in the entire design. In each block, we assemble k RQIFs in the main part of the block design and the remaining $(v - k)$ RQIFs are assembled in the complimentary part of the design. The SQIF, Q^*, is then included in the complimentary part of each block as a supplementary question. Next, a respondent is supposed to entertain all the k RQIFs in the main part and only one of the $(v - k + 1)$ Fs in the complimentary part. This latter task involves RRT with a chance of θ for selection of the SQIF and a chance of $(1 - \theta)/(v - k)$ for selection of any one of the other RQIFs. For the entire exercise, one is to report only the BTR and no details. Lastly, apart from the BPEBD, there is one "master block," say B_0, comprising all the v RQIFs and it is also a part of the data-gathering exercise. There are $b_0 > 0$ respondents to respond to B_0 and provide BTR independently of one another. It is believed that the respondents in the main part will cooperate with the understanding that their privacy is totally protected under this data collection plan.

We now proceed to carry out a data analysis—with particular emphasis on unbiased estimation of P^* along with an expression for its variance.

Notations:

(i) B_t = t-th block in the BPEBD;
(ii) B_t^c = compliment of B_t;
(iii) $\Sigma_t = \sum_{s \in B_t} P_s$;
(iv) $\Sigma_P = \sum_{s=1}^{v} P_s$.

It follows that,

$$E[\text{BTR}_t] = \Sigma_t + \theta P^* + (1-\theta)[\Sigma_P - \Sigma_t]/(v-k).$$

Further,

$$E[\text{BTR}_{B_0,u}] = \Sigma_P.$$

From the above, we deduce,

$$\begin{aligned} b\theta \hat{P^*} &= \Sigma_T - [r + (1-\theta)(b-r)/(v-k)] \sum_u \text{BTR}_{B_0,u}/b_0 \\ &= \Sigma_T - [r(k+1-\theta)/k] \sum_u \text{BTR}_{B_0,u}/b_0, \end{aligned}$$

where $\Sigma_T = \sum_t \text{BTR}_t$ and $\text{BTR}_{B_0,u}$ is the BTR accrued from the u-th respondent of B_0;u = 1,2,…,b_0. It turns out that, dropping the constant error variance, σ^2, which may be taken to be unity,

$$b^2\theta^2 V(\hat{P^*}) = b + [r(k+1-\theta)/k]^2/b_0.$$

Remark. In case two Fs are selected from the complimentary part involving $(v - k + 1)$ Fs by adopting the PPSWOR sampling procedure, then the estimating equation for P^* is to be based on the defining relations, like

$$E[\text{BTR}_t] = \Sigma_t + \pi^* P^* + (2-\pi^*)[\Sigma_P - \Sigma_t]/(v-k), t = 1,2,\ldots,b$$

where $\pi^* = \theta[(v - k - 1)(1 - \theta) + (v - k)]/[\theta + (v - k - 1)]$.

It is possible to describe and implement other sampling procedures in the complimentary part. We can adopt a πPS design, in which case π^* in the above expression will assume the value 2θ, and the Midzuno scheme can be employed to implement this, provided our choice of θ satisfies $2(v - k)\theta > 1$. We can also extend this method for sample sizes larger than two from the complimentary part, yet again using the BTR on the whole from each block.

A.1 ILLUSTRATIVE EXAMPLES

A.1.1 Illustrative Example 1

Consider a BPEBD with $b = v = 7, r = k = 3$ and with the block design having a cyclic pattern of the treatments starting with [1,2,3] in Block 1. This means that for the Block 2, the treatments are [2,3,4] and so on. It follows that,

$$7^2\theta^2 V(\hat{P^*}) = 7 + (4-\theta)^2/b_0.$$

Had we used the complimentary design in the main part and shifted the other to the RRT part, using a different RRT parameter, δ, we would land into,

$$7^2\delta^2 V(\hat{P^*})_{II} = 7 + (5-\delta)^2/b_0.$$

At this stage, we can also examine relative preference between two choices of the RRT parameter: $\theta = 0.4$ vs $\delta = 0.5$. It turns out that $V_{II} < V_I$, whenever $b_0 > 0$! The message is again clear. A higher chance of selection of the SQIF in an RRT, will result in better precision of the estimator for P^*. Interestingly, a comparison between RRT$_I$ based on $\theta = 0.45$ and RRT$_{II}$ based on $\delta = 0.5$ reveals that if $b_0 = 1$ or 2; we will opt for RRT$_I$ and otherwise, we will opt for RRT$_{II}$.

It also follows from the above that $V(\hat{P^*})$ has a nonattainable lower bound $1/b\theta^2$. For a margin of accuracy given by variance, V_0, and a comfortable value of θ (from the point of view of perceived protection of privacy of the respondents), we may solve for b,v,k,b_0 such that $1/b\theta^2 < V_0 \leq 1/b\theta^2 + b + [r(k+1-\theta)/k]^2/b_0 b^2\theta^2$.

A.1.2 Illustrative Example 2

Consider $\theta = 0.20$ and $V_0 = 1.00$. Then we must necessarily have $b > 25$. Taking $b = 30$, $v = 30$, $k = 3$, $r = 3$, we solve for b_0 as $6(3.8)^2 = 87$. Again, for the choice $V_0 = 2.00$, we must have $b > 12$. Taking $b = 20 = v$, $r = k = 7$, we solve for $b_0 = 5$. The message is clear in both the situations.

A.1.3 Illustrative Example 3

Consider a BIBD $[b,v,r,k,\lambda]$ with RRT parameter, θ, and the Complimentary BIBD $[b^* = b,\ v^* = v,\ r^* = b-r,\ k^* = v-k,\ \lambda^* = b-2r+\lambda]$ with RRT parameter, δ, and the same number, b_0, of additional blocks formed exclusively of the RQIFs. Under the assumption of $\theta = \delta$, it follows that $V_{II} \geq V_I$ iff $b - 2r \geq 0$, independent of the value of $b_0 > 0$.

Note that for $b > 2r$, ie, $v > 2k$, in the case when the BIBD is used, there are more RQIFs along with the single SQIF in the formation of the RRT. This, in a way, gives more privacy to the respondent. It is interesting to see that this also gives more efficiency to the estimator of P^*.

REFERENCES

Aslam, A., Yab, M.Z., 2011. Comparative Study of Balanced Incomplete Block Designs: For Block Total Response Technique. LAP Lambert Academic Publishing.

Chaudhuri, A., 2011. Randomized Response and Indirect Questioning Techniques in Surveys. CRC Press, Chapman and Hall, Taylor and Francis Group, Boca Raton, FL.

Chaudhuri, A., Christofides, T.C., 2013. Indirect Questioning in Sample Surveys. Springer, Heidelberg.

Chaudhuri, A., Mukerjee, R., 1988. Randomized Response : Theory and Applications. Marcel and Dekker, New York, NY.

Hedayat, A.S., Sinha, B.K., 1991. Design and Inference in Finite Population sampling. Wiley, New York, NY.

Lehmann, E.L., Casella, G., 1998. Theory of Point Estimation. Springer, New York, NY.

Marcovitz, M., Nandy, K., Sinha, B.K., 2014. Use of empirical bayes techniques for obtaining sensitive information in block total response designs. The American Statistician (Submitted).

Nayak, T.K., Adeshiyan, S.A., 2009. A unified framework for analysis and comparison of randomized response surveys of binary characteristics. J. Stat. Plan. Inference 139, 2757–2766.

Raghavarao, D., Federer, W.T., 1979. Block total response as an alternative to the randomized response method in surveys. J. R. Stat. Soc. B Methodol. 41, 40–45.

Robbins, H., 1956. An empirical bayes approach to statistics. In: Proc. Third Berkeley Symp, University of California Press, pp. 152–163.

Smith, L., Federer, W., Raghavarao, D., 1974. A comparison of three techniques for eliciting truthful answers to sensitive questions. In: Proceedings of the Social Statistics Association, American Statistical Association. American Statistical Association, Baltimore, pp. 447–452.

Warner, S.L., 1965. "Randomized response: a survey technique for eliminating evasive answer bias" J. Am. Stat. Assoc. 60, 63–69.

Chapter 20

Optional Randomized Response Revisited

R. Mukerjee[1]
Indian Institute of Management Calcutta, Kolkata, India
[1]*Corresponding author: e-mail: rmuk0902@gmail.com*

ABSTRACT

In a randomized response (RR) survey on a sensitive character, the perception of sensitivity can vary from person to person and a section of the respondents may even be willing to disclose their true states rather than providing responses via the use of a randomization device. This opens up the prospect of allowing optional, rather than compulsory, RRs, leading to enhanced estimation efficiency. Research in this direction has received significant attention and the present chapter reviews these developments. We begin with the early work on qualitative characters and then pass on to more recent literature pertaining to quantitative characters. Results under both simple random sampling with replacement and general sampling designs are touched upon.

Keywords: Estimation efficiency, Finite population, Qualitative character, Quantitative character, Respondent cooperation, Sensitive character, Unbiased estimator

1 INTRODUCTION

Over the last 50 years since the pioneering work of Warner (1965), randomized response (RR) has emerged as an extremely powerful technique for eliciting dependable information on sensitive or stigmatizing characters, with applications to such diverse fields as sociology, psephology, public health, and so on. Chaudhuri and Mukerjee (1988) made a comprehensive survey of the developments in RR till about the mid 1980s. This was preceded by another account of more applied nature by Fox and Tracy (1986). More recent in-depth reviews of various aspects of RR are available in Chaudhuri (2011a) and Chaudhuri and Christofides (2013). All these books come up with detailed lists of further references.

While RR aims at drawing trustworthy inference on sensitive characters, the perception of such sensitivity is often subjective. Indeed, even in an RR survey, the subject of the study may not be considered sensitive by some of

Handbook of Statistics, Vol. 34. http://dx.doi.org/10.1016/bs.host.2016.01.018

the respondents and they may wish to provide direct responses (DRs). For example, some of the people included in a psephological investigation may not mind disclosing their true electoral preferences to the interviewer. One then wonders if, instead of forcing these people to go through RR, one can effectively utilize the DRs that they are willing to provide. This calls for devising statistical methodology that allows for optional, rather than compulsory, RR and appropriately combines the resulting RRs and DRs with a view to arriving at statistical inference which can be more efficient than its counterpart under compulsory RR.

Chaudhuri and Mukerjee (1985) initiated work on optional RR from the above perspective. The topic then remained dormant for some time, barring the studies by Mangat (1991), Mangat and Singh (1994), and Singh and Joarder (1997), before undergoing a significant revival in the early 2000s. Since then, there has been a vigorous growth in the literature on optional RR, with attention focused not only on estimating the proportion, mean, or total of a sensitive character but also in assessing the level of sensitivity itself. We aim at reviewing these developments.

2 EARLY WORK

We begin by revisiting the work of Chaudhuri and Mukerjee (1985) on optional RR in the context of a qualitative sensitive character with multiple classes. In order to give a flavor of their ideas without making the notation too heavy, we consider the dichotomous case under the framework of the Warner (1965) model. Suppose interest lies in estimating the proportion, π, of individuals belonging to a sensitive group A in a population. Let $\bar{A}$ denote the complement of A. Under simple random sampling with replacement (SRSWR), each respondent in a sample of size n is asked to choose, using a randomization device, one of the two questions "Do you belong to A?" or "Do you belong to $\bar{A}$?" with respective probabilities p and $1-p$, and say only "yes" or "no" in response to the question chosen, without disclosing the question to which his response corresponds. Clearly, then the probability of a "yes" response is $\lambda = p\pi + (1-p)(1-\pi)$. As such, under compulsory RR, if $\hat{\lambda}$ be the sample proportion of "yes" responses and $p \neq 1/2$, then

$$\hat{\pi} = \frac{\left\{\hat{\lambda} - (1-p)\right\}}{(2p-1)}$$

emerges as an unbiased estimator (UE) of π, with

$$\mathrm{var}(\hat{\pi}) = \frac{\pi(1-\pi)}{n} + \frac{p(1-p)}{n(2p-1)^2}, \tag{1}$$

on simplification. The first term on the right-hand side of (1) is the variance of the usual estimator of π in a direct survey and hence the second term represents the contribution of RR to $\mathrm{var}(\hat{\pi})$.

In the above setup, now suppose DRs are also allowed from respondents who are willing to disclose their true states to the interviewer. Then the optional RR procedure due to Chaudhuri and Mukerjee (1985) entertains four types of responses, namely, (i) DR stating A, (ii) DR stating $\bar{A}$, (iii) RR stating "yes," and (iv) RR stating "no." In a random sample of size n, let $n_1,\ldots,n_4$ be frequencies of responses of types (i)–(iv), respectively. The joint distribution of $(n_1,\ldots,n_4)$ is clearly four-class multinomial having parameters n and $\lambda_1,\ldots,\lambda_4$, with

$$\lambda_1=\pi_1,\ \ \lambda_2=\pi_2,\ \ \lambda_3=p\pi_3+(1-p)\pi_4,\ \ \lambda_4=(1-p)\pi_3+p\pi_4 \qquad (2)$$

Here $\pi_1,\ldots,\pi_4$ are the population proportions of people in the four categories (a) A and willing to disclose the true state, (b) $\bar{A}$ and willing to disclose the true state, (c) A and unwilling to disclose the true state, and (d) $\bar{A}$ and unwilling to disclose the true state, respectively. By (2), under optional RR, a UE of $\pi=\pi_1+\pi_3$ is obtained as

$$\tilde{\pi}=\frac{1}{n}\left[n_1+\frac{pn_3-(1-p)n_4}{2p-1}\right],$$

with

$$\mathrm{var}(\tilde{\pi})=\frac{\pi(1-\pi)}{n}+\frac{p(1-p)}{n(2p-1)^2}(\pi_3+\pi_4), \qquad (3)$$

after some simplification. Since $\pi_3+\pi_4\leq 1$, the inequality being typically strict, a comparison between (1) and (3) shows that the $\tilde{\pi}$ estimates π more efficiently than $\hat{\pi}$.

The gain in efficiency, however, comes with a caveat. If A and $\bar{A}$ are equally stigmatizing, then the above optional RR procedure is expected to entail as much respondent cooperation as compulsory RR. But, if only A is stigmatizing and not $\bar{A}$, then it may prompt some of the respondents in group A to willfully report $\bar{A}$ in the form of a DR, thus defeating the very purpose of the survey. This happens because here the pattern of responses under DR, namely A or $\bar{A}$, grossly differs from that under RR, namely "yes" or "no."

Continuing with the above setup, Mangat and Singh (1994) proposed an optional RR technique where both DRs and RRs are of the form "yes" or "no." This is achieved by asking each sampled individual to respond "yes" or "no" directly either to the question "Do you belong to A?" or to the question generated via randomization, without telling the interviewer whether the response is direct or randomized. This idea was followed up in Singh and Mathur (2003), while Singh and Tarray (2014) extended it to stratified populations. The estimators of π obtained by all these authors are, however, biased. Gupta (2001), on the other hand, showed that a two-sample version of Mangat and Singh's (1994) approach admits a UE of π.

As seen above, in order to facilitate generation of truthful responses in optional RR, the set of possible responses under DR should be about the same as that under RR, thus assuring the respondents that the interviewer would get no idea as to whether any response is direct or randomized. We will see in the next section how more recent work on optional RR, focusing on quantitative sensitive characters, meets this stipulation while enhancing estimation efficiency at the same time.

3 SCRAMBLED RESPONSE

We first outline Eichhorn and Hayre's (1983) multiplicative scrambling approach which forms the basis of quite a few optional RR techniques that have been proposed since the early 2000s. With a sensitive quantitative character Y, this approach asks each respondent to generate a scrambling variable S using a randomization device and then report the scrambled response $Z=SY$. Here S is independent of Y and so chosen that $E(S)=1$. It is not hard to see that

$$E(Z)=\mu_Y, \quad \text{var}(Z)=\sigma_Y^2+\sigma_S^2\left(\sigma_Y^2+\mu_Y^2\right),$$

where μ_Y and σ_Y^2 are the population mean and variance of Y, respectively, and $\sigma_S^2=\text{var}(S)$. As a result, the mean scrambled response $\overline{Z}$, in an SRSWR of size n, is a UE of μ_Y, with

$$\text{var}\left(\overline{Z}\right)=\frac{\left\{\sigma_Y^2+\sigma_S^2\left(\sigma_Y^2+\mu_Y^2\right)\right\}}{n}. \tag{4}$$

Gupta et al. (2002) explored optional RR in the above setup allowing each respondent to report either Y or the scrambled response SY as he likes, without revealing what is being reported. The scrambling variable S, with $E(S)=1$, is so chosen that the ranges of Y and SY are about the same. The response Z now equals $Z=(1-X)Y+XSY$, where X is an indicator taking values 1 or 0 according as the response is SY or Y, respectively. Hence the mean response $\overline{Z}$, in an SRSWR of size n, remains a UE of μ_Y. Furthermore, assuming that Y, X, and S are independent,

$$\text{var}\left(\overline{Z}\right)=\frac{\left\{\sigma_Y^2+w\sigma_S^2\left(\sigma_Y^2+\mu_Y^2\right)\right\}}{n}, \tag{5}$$

where $w=P(X=1)$ represents the sensitivity level of Y. Since $w\leq 1$, the inequality being typically strict, a comparison between (4) and (5) reveals the gain in efficiency through optional RR. As usual, a UE of $\text{var}\left(\overline{Z}\right)$ in (5) is given by $\Sigma_{i=1}^n\left(Z_i-\overline{Z}\right)^2/\{n(n-1)\}$, where $Z_1,\ldots,Z_n$ are the individual responses.

Turning to the estimation of the sensitivity level w, note that $Z=(1-X)Y+XSY$ can also be expressed as $Z=S^XY$, because X is binary. Therefore, if S and Y are both positive, then

$$w = E(X) = E\left[\frac{\log(Z) - \log(Y)}{\log S}\right],$$

which prompted Gupta et al. (2002) to propose the estimator

$$\widetilde{w} = \frac{\left[n^{-1}\sum_{i=1}^{n}\log(Z_i) - \{\log(\overline{Z})\}\right]}{E\{\log(S)\}}$$

of w. As noted by Chaudhuri and Christofides (2013, Chapter 5), this estimator is biased and its mean squared error is hard to obtain. At the same time, it may be unreasonable to expect a single sample to yield UEs of both μ_Y and w and, intuitively, it appears that two independent samples would be needed for this purpose. We next turn to these developments but, before doing so, comment that a subsequent work by Bar-Lev et al. (2004), which does not really dwell on optional RR, has a formal similarity with the ideas in Gupta et al. (2002).

Huang (2010) proposed drawing two independent samples of sizes n_1 and n_2 by SRSWR such that each respondent in the jth sample ($j = 1, 2$) comes up with either a DR Y or a scrambled response $S_{1j}Y + S_{2j}$ as he may wish, without revealing what is being reported. Here Y is again the sensitive study variable with population mean and variance μ_Y and σ_Y^2, respectively. On the other hand, S_{1j} and S_{2j} are scrambling variables so chosen that $E(S_{1j}) = 1$ and the direct and scrambled responses have about the same range. As before, a response $Z^{(j)}$ in the jth sample equals $(1-X)Y + X(S_{1j}Y + S_{2j})$, where X is an indicator taking values 1 or 0 according as the response is scrambled or direct, respectively. Assuming independence of Y, X, S_{1j}, and S_{2j},

$$E\left\{Z^{(j)}\right\} = \mu_Y + w\mu_{2j}, \quad \text{var}\left\{Z^{(j)}\right\} = \sigma_Y^2 + w\sigma_{1j}^2\left(\sigma_Y^2 + \mu_Y^2\right) + w\sigma_{2j}^2 + w(1-w)\mu_{2j}^2, \tag{6}$$

where $\mu_{2j} = E(S_{2j})$, $\sigma_{1j}^2 = \text{var}(S_{1j})$, $\sigma_{2j}^2 = \text{var}(S_{2j})$, and $w = P(X = 1)$; note that μ_{2j}, σ_{1j}^2, and σ_{2j}^2 are known from the choice of S_{1j} and S_{2j}.

If we now write $\overline{Z}^{(j)}$ and s_j^2 for the mean and variance of the n_j responses from the jth sample, with $n_j - 1$ used as the divisor in s_j^2, then the first equation in (6) yields

$$\hat{\mu}_Y = \frac{\mu_{21}\overline{Z}^{(2)} - \mu_{22}\overline{Z}^{(1)}}{\mu_{21} - \mu_{22}} \quad \text{and} \quad \hat{w} = \frac{\overline{Z}^{(1)} - \overline{Z}^{(2)}}{\mu_{21} - \mu_{22}} \tag{7}$$

as UEs of μ_Y and w, respectively, provided $\mu_{21} \neq \mu_{22}$. Furthermore, it is easily seen that

$$\hat{\text{var}}(\hat{\mu}_Y) = \frac{\mu_{21}^2 (s_2^2/n_2) + \mu_{22}^2 (s_1^2/n_1)}{(\mu_{21} - \mu_{22})^2} \quad \text{and} \quad \hat{\text{var}}(\hat{w}) = \frac{(s_1^2/n_1) + (s_2^2/n_2)}{(\mu_{21} - \mu_{22})^2} \tag{8}$$

are UEs of $\text{var}(\hat{\mu}_Y)$ and $\text{var}(\hat{w})$, respectively. Indeed, s_j^2 is a UE of $\text{var}\{Z^{(j)}\}$ in (6), whereas $\hat{w}(1-\hat{w}) + \hat{\text{var}}(\hat{w})$ estimates $w(1-w)$ unbiasedly. Hence, if $\sigma_{11}^2 \neq \sigma_{12}^2$, then from the second equation in (6), a UE of σ_Y^2 is obtained as

$$\frac{\sigma_{12}^2 s_1^2 - \sigma_{11}^2 s_2^2 - \hat{w}(\sigma_{12}^2 \sigma_{21}^2 - \sigma_{11}^2 \sigma_{22}^2) - \{\hat{w}(1-\hat{w}) + \hat{\text{var}}(\hat{w})\}(\sigma_{12}^2 \mu_{21}^2 - \sigma_{11}^2 \mu_{22}^2)}{\sigma_{12}^2 - \sigma_{11}^2}.$$

Thus the two-sample approach is capable of yielding UEs of not only μ_Y and w but also σ_Y^2.

With a view to estimating both μ_Y and w unbiasedly, Gupta et al. (2010) proposed another procedure based on two independent samples drawn by SRSWR. Using a randomization device, each respondent in the jth sample ($j=1, 2$) reports (i) with probability p, his Y-value, and (ii) with probability $1-p$, either his Y-value or a scrambled response $S_j + Y$ as per his will, without disclosing what is being reported. Then, for a response $Z^{(j)}$ in the jth sample, one can check that $E\left\{Z^{(j)}\right\} = \mu_Y + w(1-p)\mu_j$, where μ_j is the known expectation of the scrambling variable S_j. Hence, denoting the mean response from the jth sample by $\overline{Z}^{(j)}$, analogous to (7),

$$\hat{\mu}_Y = \frac{\mu_1 \overline{Z}^{(2)} - \mu_2 \overline{Z}^{(1)}}{\mu_1 - \mu_2} \quad \text{and} \quad \hat{w} = \frac{\overline{Z}^{(1)} - \overline{Z}^{(2)}}{(1-p)(\mu_1 - \mu_2)}$$

are UEs of μ_Y and w, respectively, provided $\mu_1 \neq \mu_2$. Along the lines of (8), one can find UEs of $\text{var}(\hat{\mu}_Y)$ and $\text{var}(\hat{w})$ in a straightforward manner. As in Huang's (2010) approach, here too it is possible to estimate σ_Y^2 unbiasedly from the responses arising from the two samples; we refer to Chaudhuri and Christofides (2013, Chapter 5) for details.

The developments in Huang (2010) and Gupta et al. (2010) were preceded by some other related work. Gupta and Shabbir (2004) considered a two-sample procedure where each respondent in the jth sample ($j=1, 2$) reports either his Y-value or a scrambled response S_jY as he may wish. This was seen to yield a UE of μ_Y and a biased ratio-type estimator of w. Gupta et al. (2006) proposed an approach which amounts to taking S_{11} and S_{12} identically equal to 1 in Huang (2010) and discussed how this entails unbiased estimation of μ_Y and w. Ryu et al. (2006) studied two-stage optional RR but assumed w to be known and dealt only with the estimation of μ_Y. Gupta and Shabbir (2007) suggested a one-sample precursor of the approach in Gupta et al. (2010) with a multiplicative scrambling variable. They examined how this leads to a UE of μ_Y along with a biased estimator of w. Taking cognizance of the potential difference between subpopulations, Huang (2008) developed optional RR procedures for unbiased estimation of μ_Y and σ_Y^2, while also addressing the issue of estimating w. We refer to Sehra (2008, Chapter 3) for an account of some of these results.

Optional RR, allowing direct or scrambled responses, continues to remain quite an active area of research. In continuation of the work of Gupta et al. (2010) as reviewed above, Gupta et al. (2011) studied the issue of choosing p optimally and Mehta (2012) investigated the impact of possible noncooperation from a section of the respondents. With a view to achieving enhanced respondent cooperation, Mehta et al. (2012) proposed a technique where each respondent reports his Y-value with probability p_1, a scrambled response with probability p_2, and either his Y-value or a scrambled response as per his will with probability $1-p_1-p_2$. They showed that this procedure compares favorably with the one in Gupta et al. (2010) in terms of estimation efficiency. Gupta et al. (2013) compared the findings in Huang (2010) with those in Gupta et al. (2010). Saha (2011) examined the situation where the population lacks homogeneity with regard to the probability of choosing between direct and scrambled responses. Hussain et al. (2014) explored efficiencies of estimators arising from optional RR with additive as well as subtractive scrambling. We conclude this section with a reference to Gupta et al. (2014, 2015) and Kalucha et al. (2015) for some recent results on optional RR techniques which employ ratio or regression estimators, under simple random sampling without replacement, in order to utilize information on an innocuous auxiliary variable.

4 GENERAL SAMPLING DESIGNS

So far in this chapter, we have reviewed work on optional RR almost entirely in the framework of SRSWR. In keeping with the literature on mainstream finite population sampling and under the leadership of A. Chaudhuri, there has also been significant progress since the early 2000s on optional RR procedures for general sampling designs. The technicalities underlying this body of work are, however, deeper than those under SRSWR and the notation is also heavier. In a limited space, therefore, an attempt to cover such details will be unrealistic. So, we present here only a general account of this work while citing relevant references for the interested reader.

A brief but very informative account of the key ideas is available in Chaudhuri (2011b). Following him, consider a finite population of N persons identified by labels $i=1, \ldots, N$. Here N is known. Let Y be the study variable which can be binary, i.e., qualitative, or quantitative potentially assuming any real value. Let y_i be the unknown value of Y for the ith person. Interest lies in unbiased estimation of the population total $T=y_1+\cdots+y_N$ of Y from a sample s drawn from the population with probability $p(s)$ according to a sampling design. For Y of innocuous nature, one commonly considers a linear UE of T as given by

$$t(y)=\Sigma_{i\in s}b_{si}y_i, \tag{9}$$

where the sum extends over i such that $i\in s$ and, in order to ensure unbiasedness, the constants b_{si} are so chosen that $\Sigma_{s\ni i}b_{si}p(s)=1,\ \ 1\leq i\leq N$, this latter sum $\Sigma_{s \ni i}$ being over all samples s containing i. The form (9) covers the well-known Horvitz and Thompson (1952) estimator as a special case.

In case Y is sensitive, the sampled persons may be unwilling to divulge their y_i values and the estimator $t(y)$ becomes infeasible. In this situation, one can employ an optional RR technique where for each $i \in s$, the ith person is asked to report, as he may wish, either y_i or an RR r_i such that $E_R(r_i) = y_i$, without disclosing what is being reported. Here E_R denotes expectation with respect to the randomization device. For instance, with quantitative Y, in the spirit of Eichhorn and Hayre (1983), one may consider $r_i = Sy_i$, where S is a scrambling variable so chosen that $E(S) = 1$. For $i \in s$, denote the response from the ith person by $\widetilde{r}_i$ which equals y_i or r_i. Clearly, $E_R(\widetilde{r}_i) = y_i$, and hence from (9), a conditioning argument shows that $t(\widetilde{r}) = \Sigma_{i \in s} b_{si} \widetilde{r}_i$ is a UE of T. Furthermore, again via a conditioning argument, one can check that the variance of this optional RR estimator $t(\widetilde{r})$ cannot exceed its counterpart under compulsory RR where each respondent must report r_i. Thus even under general sampling designs, optional RR has an advantage over compulsory RR in terms of estimation efficiency. We refer to Chaudhuri (2011b) and Chaudhuri and Saha (2005) for more details including those on unbiased estimation of $\text{var}\{t(\widetilde{r})\}$ (see also Arnab, 2004; Pal and Chakraborty, 2006 for some earlier results).

With reference to a general sampling design and a binary study variable Y, Chaudhuri and Dihidar (2009) proposed an alternative optional RR procedure where every sampled person generates two independent binary responses, each either direct or randomized as he likes. These two responses are linearly combined to obtain a counterpart of $\widetilde{r}_i$ and then unbiased estimation of the population total or mean of Y is possible as in the last paragraph. Chaudhuri and Dihidar (2009) also discussed an extension of this two-response approach to the case of quantitative characters and obtained results which supplement an earlier work of Pal (2008) along similar lines. Dihidar (2009), on the other hand, continued with binary Y and studied the properties of a shrinkage estimator which is meaningful when one can make a reasonable guess about the value of the estimand. Dihidar (2011) reported further results based on modification of some classical RR techniques.

5 CONCLUDING REMARKS

From the discussion in the preceding sections, it is clear that optional RR has been a very active area of research, particularly over the last 15 years, and that it continues to evoke significant interest. Even though there has been major progress on various aspects of optional RR, there are several issues that deserve attention. For example, more work, under either SRSWR or more general sampling designs, is needed on estimating proportions associated with qualitative characters with multiple classes, ensuring that the sets of direct and RRs are identical. Another important problem concerns use of Bayesian methodology to incorporate prior knowledge about parameters relating to the sensitive character or the sensitivity level itself. The choice of prior can be also an issue here and one may also wish to explore default priors in this connection. We conclude with the hope that the present review will generate interest in these and related problems, leading to further advancement in the theory and practice of optional RR.

ACKNOWLEDGMENTS

This work was supported by the J.C. Bose National Fellowship of the Government of India and a grant from the Indian Institute of Management Calcutta.

REFERENCES

Arnab, R., 2004. Optional randomized response techniques for complex survey designs. Biom. J. 46, 114–124.

Bar-Lev, S.K., Bobovitch, E., Boukai, B., 2004. A note on randomized response models for quantitative data. Metrika 60, 255–260.

Chaudhuri, A., 2011a. Randomized Response and Indirect Questioning Techniques in Surveys. Chapman and Hall/CRC, Boca Raton.

Chaudhuri, A., 2011b. Analyzing optional randomized response on qualitative and quantitative variables bearing social stigma. In: Int. Statistical Inst. Proceedings of the 58th World Statistical Congress, Dublin, pp. 1939–1947.

Chaudhuri, A., Christofides, T., 2013. Indirect Questioning in Sample Surveys. Springer, Berlin.

Chaudhuri, A., Mukerjee, R., 1985. Optionally randomized response techniques. Calcutta Stat. Assoc. Bull. 34, 225–229.

Chaudhuri, A., Mukerjee, R., 1988. Randomized Response: Theory and Techniques. Marcel Dekker, New York.

Chaudhuri, A., Saha, A., 2005. Optional versus compulsory randomized response techniques in complex surveys. J. Stat. Plann. Inference 135, 516–527.

Chaudhuri, A., Dihidar, K., 2009. Estimating means of stigmatizing qualitative and quantitative variables from discretionary responses randomized or direct. Sankhyā B71, 123–136.

Dihidar, K., 2009. On shrinkage estimation procedure combining direct and randomized responses in unrelated question model. J. Ind. Soc. Agric. Stat. 63, 283–296.

Dihidar, K., 2011. Modifying classical randomized response techniques with provision for the true response. Calcutta Stat. Assoc. Bull. 63, 223–240.

Eichhorn, B.H., Hayre, L.S., 1983. Scrambled randomized response methods for obtaining quantitative data. J. Stat. Plann. Inference 7, 306–316.

Fox, J.A., Tracy, P.E., 1986. Randomized Response: A Method for Sensitive Surveys. Sage, Beverly Hills.

Gupta, S., 2001. Qualifying the sensitivity level of binary response personal interview survey questions. J. Combin. Inform. Syst. Sci. 26, 101–109.

Gupta, S., Shabbir, J., 2004. Sensitivity estimation for personal interview survey questions. Statistica 64, 643–653.

Gupta, S., Shabbir, J., 2007. On the estimation of population mean and sensitivity in two-stage optional randomized response model. J. Indian Soc. Agric. Stat. 61, 164–168.

Gupta, S., Gupta, B., Singh, S., 2002. Estimation of sensitivity level of personal interview survey question. J. Stat. Plann. Inference 100, 239–247.

Gupta, S., Thornton, B., Shabbir, J., Singhal, S., 2006. A comparison of multiplicative and additive optional RRT models. J. Stat. Theory Appl. 5, 226–239.

Gupta, S., Shabbir, J., Sehra, S., 2010. Mean and sensitivity estimation in optional randomized response models. J. Stat. Plann. Inference 140, 2870–2874.

Gupta, S., Mehta, S., Shabbir, J., Dass, B.K., 2011. Some optimality issues in estimating two-stage optional randomized response models. Am. J. Math. Mgmt. Sci. 31, 1–12.

Gupta, S., Mehta, S., Shabbir, J., Dass, B.K., 2013. Generalized scrambling in quantitative optional randomized response models. Commun. Stat. Theory Meth. 42, 4034–4042.

Gupta, S., Kalucha, G., Shabbir, J., Dass, B.K., 2014. Estimation of finite population mean using optional RRT models in the presence of nonsensitive auxiliary information. Am. J. Math. Mgmt. Sci. 33, 147–159.

Gupta, S., Kalucha, G., Shabbir, J., 2015. A regression estimator for finite population mean of a sensitive variable using an optional randomized response model. Commun. Stat. Simul. Comput. http://dx.doi.org/10.1080/03610918.2015.1044614.

Horvitz, D.G., Thompson, D.J., 1952. A generalization of sampling without replacement from finite universe. J. Am. Stat. Assoc. 47, 663–685.

Huang, K.C., 2008. Estimation for sensitive characteristics using optional randomized response technique. Qual. Quant. 42, 679–686.

Huang, K.C., 2010. Unbiased estimators of mean, variance and sensitivity level for quantitative characteristics in finite population sampling. Metrika 71, 341–352.

Hussain, Z., Al-Sobhi, M.M., Al-Zahrani, B., 2014. Additive and subtractive scrambling in optional randomized response modeling. PLoS One 9 (1), e83557.

Kalucha, G., Gupta, S., Dass, B.K., 2015. Ratio estimation of finite population mean using optional randomized response models. J. Stat. Theory Pract. 9, 633–645.

Mangat, N.S., 1991. An optional randomized response sampling technique using nonstigmatized attribute. Statistica 51, 595–602.

Mangat, N.S., Singh, S., 1994. An optional randomized response sampling technique. J. Indian Stat. Assoc. 32, 71–75.

Mehta, S., 2012. Impact of defying instructions by the respondents in a two-stage optional randomized response model. J. Stat. Theory Appl. 11, 311–322.

Mehta, S., Dass, B.K., Shabbir, J., Gupta, S., 2012. A three-stage optional randomized response model. J. Stat. Theory Pract. 6, 417–427.

Pal, S., 2008. Unbiasedly estimating the total of a stigmatizing variable from a complex survey on permitting options for direct or randomized responses. Stat. Papers 49, 157–164.

Pal, S., Chakraborty, S., 2006. Improvement upon Warner's model with an optional randomized response technique. Model Asst. Stat. Appl. 1, 299–304.

Ryu, J.B., Kim, J.M., Heo, T.Y., Park, C.G., 2006. On stratified randomized response sampling. Model Asst. Stat. Appl. 1, 31–36.

Saha, A., 2011. An optional scrambled randomized response technique for practical surveys. Metrika 73, 139–149.

Sehra, S., 2008. Two-Stage Optional Randomized Response Models. Ph.D. Thesis. University of North Carolina at Greensboro, North Carolina.

Singh, S., Joarder, A.H., 1997. Optional randomized response technique for sensitive quantitative variable. Metron 55, 151–157.

Singh, H.P., Mathur, N., 2003. Modified optional randomized response sampling techniques. J. Indian Soc. Agric. Stat. 56, 199–206.

Singh, H.P., Tarray, T.A., 2014. A stratified Mangat and Singh's optional randomized response model using proportional and optimal allocation. Statistica 73, 65–83.

Warner, S.L., 1965. Randomized response: a survey technique for eliminating evasive answer bias. J. Am. Stat. Assoc. 60, 63–69.

Chapter 21

Measures of Respondent Privacy in Randomized Response Surveys

M. Bose*,[1]
*Applied Statistics Unit, Indian Statistical Institute, Kolkata, India
[1]Corresponding author: e-mail: mausumi.bose@gmail.com

ABSTRACT
Socioeconomic surveys are often done on highly personal and sensitive or stigmatizing issues. In such surveys randomized response techniques are popular, as opposed to direct questions, since these techniques do not require a respondent to disclose his true status. Though it is evident that these techniques protect the privacy of a respondent in general, the quantum of protection available from such techniques, vis-á-vis the estimation efficiency, is also of interest. Some researchers have focused on this aspect and measures of jeopardy are available for situations where the sensitive variable is qualitative and the objective is to estimate the proportion of persons in the population who belong to the sensitive group. Measures of privacy protection are also available for cases where the variable of interest is quantitative, both for discrete and continuous variables. In this chapter, some of the available measures of privacy protection will be described.

Keywords: Jeopardy measure, Protection of privacy, Revealing probability, Sensitive issue, Socioeconomic surveys, Stigmatizing variable

MSC classification: 62D05

1 INTRODUCTION

When a survey is carried out on issues which are perceived to be sensitive, incriminating, or stigmatizing for the respondents, most respondents either will not agree to answer direct questions or, even if they do, will give untruthful responses. It is human nature to try to avoid revealing the truth about one's highly personal or stigmatizing issues and protect one's privacy. So, in

Handbook of Statistics, Vol. 34. http://dx.doi.org/10.1016/bs.host.2016.01.019

surveys on such issues, for example, rape, abortion, drug addiction, tax evasion, and criminal conviction, the randomized response technique of data collection is useful. With such a technique, each respondent uses a randomization device to generate his randomized response, without revealing his true status. So, it is expected that this will lead to better participation and truthful answers in surveys on sensitive issues.

Warner (1965) pioneered the randomized response technique for estimating the proportion of persons bearing a sensitive qualitative characteristic in a dichotomous population. Such qualitative characteristics could be, for example, alcoholism, drug addition, or fraud. For details on results available on this technique we refer to the books by Chaudhuri and Mukerjee (1988), Chaudhuri (2011), and Chaudhuri and Christofides (2013).

Although it is well known that randomized response techniques do protect the privacy of a respondent, it is not immediately evident as to exactly how much protection is available to a respondent and how this relates to the efficiency of estimation. A systematic study of the efficiency versus privacy protection was initiated by Lanke (1975, 1976), and Leysieffer and Warner (1976) for dichotomous populations. Since then, this dichotomous case has been widely studied. Loynes (1976) extended the jeopardy measure of Leysieffer and Warner (1976) to polychotomous populations. Ljungqvist (1993) gave a unified and utilitarian approach to measures of privacy for the dichotomous case. This measure was further developed by Nayak and Adeshiyan (2009) who gave an approach for comparing the available randomization procedures based on a measure of jeopardy. Recently, Giordano and Perri (2012) compared some randomization models from the point of view of efficiency and privacy protection. Guerriero and Sandri (2007), Quatember (2009, 2012), and Hong et al. (2010) have continued study in this area. These results are all based on samples drawn by simple random sampling with replacement(SRSWR). Chaudhuri et al. (2009) gave a jeopardy measure to cover general probability sampling schemes.

All the references on privacy protection cited earlier are for sensitive variables which are categorical or qualitative and the objective of the survey is to estimate the proportion of persons in the population who bear the sensitive characteristic. However, in socioeconomic surveys it is quite common to have situations where the sensitive variable under study is quantitative. In particular, the sensitive variable could be continuous, for example, monthly expenditure on alcohol, amount of gambling debts, amount of underreported income for tax evasion, and duration of using illegal drugs, or it could be discrete, for example, the number of criminal convictions, the number of induced abortions, and the number of times one has used illegal drugs. For continuous sensitive variables, Anderson (1977) measured the amount of information provided by the randomized responses and gave measures for privacy. Diana and Perri (2011) also studied continuous variables, and for estimating the mean, they used auxiliary information at the estimation stage and compared different models from the efficiency and privacy protection aspects. Recently, Bose (2014, 2015) gave measures of privacy protection separately for the cases of discrete and continuous sensitive variables and showed how one can choose the randomization device parameter so as to guarantee a certain prespecified level of respondent protection and also maximize the efficiency of estimation. Incidentally, her results for the discrete case may also be easily adapted for use with qualitative variables too.

In this chapter, we review some of the available measures for respondent privacy. Other results and details are available in the cited references. In Section 2, we give measures for the situation when the sensitive variable is qualitative, both for dichotomous and polychotomous cases. In Section 3, we consider measures for the case of quantitative variables, both for discrete and continuous cases. We conclude with some remarks in Section 4.

2 QUALITATIVE STIGMATIZING VARIABLE

We begin with the most popularly studied case where the population of interest is divided into two classes on the basis of a characteristic which is stigmatizing or sensitive. Let A denote a sensitive group and A^c its complement. A^c may be perceived to be nonsensitive or sensitive, depending on the situation. In Warner's model, with the aid of a randomizing device, a respondent is asked to select one of two questions: "Do you belong to A?" or "Do you belong to A^c?", with probabilities p and $1 - p$. The interviewer observes only the answer "yes" or "no" and we assume that the respondent now answers truthfully. In the Simmons' unrelated questions model (Horvitz et al., 1967), the second question is replaced by "Do you belong to B?", where B is an innocuous group unrelated to A. The efficiency of estimation and privacy protection depend on this p, the device parameter. Let π_A be the proportion of persons bearing A and so the proportion bearing A^c is $\pi_{A^c} = 1 - \pi_A$. These π_A and π_{A^c} are sometimes referred to as the *true* probabilities of A and A^c, respectively. The conditional probabilities of belonging to the group A or A^c, for a given response r (either "yes" or "no"), can be expressed as

$$\text{Prob}(A|r) = \frac{\pi_A \text{Prob}(r|A)}{\pi_A \text{Prob}(r|A) + (1-\pi_A)\text{Prob}(r|A^c)},$$

$$\text{Prob}(A^c|r) = \frac{(1-\pi_A)\text{Prob}(r|A^c)}{\pi_A \text{Prob}(r|A) + (1-\pi_A)\text{Prob}(r|A^c)}.$$

These are called the *revealing* probabilities as they give the probabilities of a respondent being perceived as belonging to A or A^c based on his response r. The privacy of a respondent will be protected if his "revealing probability" of belonging to A (or A^c) is not more than the corresponding true probability π_A (or $1 - \pi_A$).

Leysieffer and Warner (1976) called the response r as *jeopardizing* with respect to A or A^c if

$$\text{Prob}(A|r) > \pi_A \text{ or } \text{Prob}(A^c|r) > 1 - \pi_A,$$

and hence they proposed measures of jeopardy as

$$g(r|A) = \frac{\text{Prob}(A|r)/\pi_A}{\text{Prob}(A^c|r)/(1-\pi_A)} = \frac{\text{Prob}(r|A)}{\text{Prob}(r|A^c)} \text{ and } g(r|A^c) = \frac{\text{Prob}(r|A^c)}{\text{Prob}(r|A)}. \quad (1)$$

The higher the value of $g(\text{yes}|A)$ (or $g(\text{no}|A^c)$), the more jeopardizing is a "yes" (or "no") answer with respect to membership in A (or A^c), and consequently, the less is the protection in privacy. They also showed that the variance of the estimate can only be decreased by increasing one or both of these jeopardy measures. They further obtained optimum model designs when there is no jeopardy in a "no" answer, ie, when $g(\text{no}|A^c)$ can be very large. Loynes (1976) extended (1) to polychotomous populations which have more than two sensitive groups and the objective is to estimate the proportions of individuals in these groups.

When A^c is not stigmatizing, Lanke (1976) argued that the respondents would want to hide their membership in A but not membership in A^c. So, based on this focus on "maximum suspicion of belonging to A," he proposed a measure of protection as

$$\max\{\text{Prob}(A|\text{yes}), \text{Prob}(A|\text{no})\}. \quad (2)$$

Thus, the smaller the value of (2), the more is the protection offered. He also obtained conditions under which the estimation is most efficient keeping the value of (2) fixed at a certain level, both for Warner's model and Simmons' model.

Giordano and Perri (2012) compared the estimation efficiency of the Simmons' model and three other alternative unrelated question models for fixed levels of privacy protection as measured by the measures in (1) and (2). Other related measures were given by Fligner et al. (1977). Greenberg et al. (1977) considered the risk to respondents who belong to A as well as those who belong to A^c and defined the "hazard" for an individual in A as

$$\text{Prob}(\text{yes}|A)\text{Prob}(A|\text{yes}) + \text{Prob}(\text{no}|A)\text{Prob}(A|\text{no}),$$

which is the probability that he or she is perceived as belonging to A. Similarly, the "hazard" for an individual belonging to A^c can be obtained. Another measure, the "limited hazard" for an individual in A, was defined as $\text{Prob}(\text{yes}|A)\text{Prob}(A|\text{yes})$.

Ljungqvist (1993) observed that respondents will answer truthfully as long as they do not feel that they are perceived to belong to A. So he defined a respondent's utility function and used this to define a "truth border" to divide the space of revealing probabilities $\text{Prob}(A|\text{yes})$ and $\text{Prob}(A|\text{no})$ into two regions: "truth telling" and "lying," the idea being that enough privacy protection will generate a truthful answer and low protection will lead to a lie. He evaluated some available privacy measures by this method and concluded that one should choose Prob(A|yes) and Prob(A|no) sufficiently close to each other, as had earlier been also proposed by Fligner et al. (1977).

Nayak (1994) studied dichotomous populations with sampling by SRSWR and compared models based on the revealing probabilities. This was extended by Nayak and Adeshiyan (2009) who considered dichotomous populations and polychotomous response variables, there being $k(\geq 2)$ response categories and possible responses being $c_1,\ldots,c_k$, $c_i \neq c_j$ for $i \neq j$. By presenting an unified framework, they used Fisher information to compare efficiencies of models. Extending the ideas of measures of Fligner et al. (1977) and Leysieffer and Warner (1976), they proposed a measure of privacy given by

$$R(\alpha,\beta) = \max\left\{\frac{\alpha_1}{\beta_1},\ldots,\frac{\alpha_k}{\beta_k}\right\}, \tag{3}$$

where Prob$(R = c_i|A) = \alpha_i$, Prob$(R = c_i|A^c) = \beta_i$, $i = 1,\ldots,k$. They then compared the estimation efficiencies of different models, with the value of (3) for all these models fixed at a certain level. They also concluded that use of polychotomous response is not helpful for sampling dichotomous populations and surveys of dichotomous populations should use only binary response variables. However, in this context, the discussion in Christofides (2010) is enlightening. Chaudhuri et al. (2009) also gave an approach to cover models which allow for multiple answers.

Quatember (2009) gave a method for comparing a number of randomized response techniques by first expressing these techniques in a unified manner, response being always "yes" or "no." Consider a variable r_i which takes the value unity if the ith unit of the sample answers "yes" and is equal to zero, otherwise. His proposed measures of loss of privacy are then given by

$$\lambda_j = \frac{\max\{\text{Prob}(r_i = j|i \in A), \text{Prob}(r_i = j|i \in A^c)\}}{\min\{\text{Prob}(r_i = j|i \in A), \text{Prob}(r_i = j|i \in A^c)\}}, \quad j = 0, 1. \tag{4}$$

The larger the values of these measures, the less is the privacy protection and the more is the estimation efficiency. One can express λ_0 and λ_1 in terms of device parameters and so, the device parameters may be chosen so as to achieve stipulated values of these measures. Quatember (2012) extended this approach to the multistage setup so that such techniques may be compared for their protection and efficiencies.

It may be remarked that since Warner (1965), SRSWR has been the preferred scheme used by most of the researchers in this area. The primary advantage of SRSWR is that it ensures that the probability of a "yes" response remains a constant for all respondents. This restriction was removed by Chaudhuri et al. (2009) to cover the practically useful situation where respondents are selected by a general sampling scheme. Following their approach, let L_i denote the probability that unit i of the population belongs to A and $L_i(r)$ denote the conditional probability that unit i belongs to A, given that his response is r. Then, Chaudhuri et al. (2009) proposed a jeopardy measure given by

$$\bar{J}_i = \frac{1}{M}\sum_r \frac{L_i(r)/L_i}{[1 - L_i(r)]/(1 - L_i)}, \tag{5}$$

where M is the number of possible responses of the technique used and the summation is taken over all possible responses. Values of (5) close to unity are considered to guarantee privacy protection of the i-th participant. The measure (5) may be computed for various available techniques and device parameters chosen so that efficiency and protection are at acceptable levels.

3 QUANTITATIVE STIGMATIZING VARIABLE

Now we consider privacy protection measures for the case where the stigmatizing variable under study is quantitative. Discrete and continuous variables are discussed separately. It can be seen that the measures for the discrete case also cover the case of dichotomous or polychotomous populations as special cases.

3.1 Continuous Stigmatizing Variable

Let X be a sensitive variable which is continuous. Let R be the randomized response variable, preferably with the same support as X. By using the randomization device, each respondent reports a value of R, without disclosing his value of X. Thus, R is obtained from X through a scrambling variable whose distribution is known. Let f denote the unknown density of X and g denote the conditional density of X given R. So, f may be called the "true" density, while g may be called the "revealing" density.

Anderson (1977) used Fisher's information to study the amount of information about the true population parameter(s) provided by the randomized responses. He showed that the decrease of Fisher's information due to randomized response is the conditional information from X given R, denoted by $I(X|R)$, say. As a general measure of protection provided by the scheme, he proposed

$$E\{V(X|R)\} \tag{6}$$

and for ensuring more privacy he advocated taking (6) as large as possible, then a response will not specifically point to any particular X. He also gave alternative (relative) measures such as

$$V(X|R)/V(X) \quad \text{and} \quad E\{V(X|R)\}/V(X).$$

When $X \sim N(\theta,1)$ and $R = X + Z$ where the scrambling variable $Z \sim N(0,\sigma)$, σ known, it can be shown that $I(X|R) = E\{V(X|R)\}$ and thus, the loss of information equals the protection as measured by (6). Anderson (1977) also proposed measures of the type $\min_R V(X|R)$ where the respondents may demand a minimal protection from every possible response. In such cases, response densities may be chosen which minimize the loss of information $I(X|R)$ subject to the condition that $\min_R V(X|R) \geq T$ where T is a known constant.

Bose (2015) used a scrambling variable Z to get $R = ZX$ and argued that the degree of disagreement between the true and revealing densities could indicate how much a respondent's privacy has been revealed or compromised by his response. So, her proposed measure for privacy protection is the expectation of the Kullback–Leibler distance between these two densities, as given by:

$$D = \int \left\{ \log_e \left(\frac{f}{g} \right) \right\} f dx. \tag{7}$$

For given forms of f and g, D can be calculated and it can be shown that for a given upper bound of D, one may find the parameter of the distribution of Z so that the efficiency of estimation is maximized.

3.2 Discrete Stigmatizing Variable

It seems that the only study in the context of discrete stigmatizing variables focusing on measures for privacy protection together with efficiency maximization is due to Bose (2014). In this section we present the salient features of this work.

Let X denote a discrete sensitive variable and suppose X may take a finite number of known values $x_1,\ldots,x_m$; the proportions of individuals in the population with these specific values are unknown. The objective of the survey is to estimate the unknown population mean of X. For this, as commonly considered (cf. Nayak and Adeshiyan, 2009; Warner, 1965; and many others), a sample of n individuals is drawn from the population by simple random sampling with replacement.

The randomization device is a box containing a large number of cards, the cards being of $(m + 1)$ types, the ith type of card being marked "Report x_i as your response," $1 \leq i \leq m$, while the $(m + 1)$th type of card is marked: "Report your true value of X as your response." The box has Mp cards of type $(m + 1)$ and $M\frac{1-p}{m}$ cards of each of the types i, $1 \leq i \leq m$, $0 < p < 1$, with M a large integer. Each sampled individual is asked to draw a card at random from the box and give a truthful response according to the card drawn by him, without disclosing the label on the card to the investigator. The n responses received comprise the data and the respondent's privacy is assured since the actual card drawn is not disclosed.

Let the "true" probabilities be

$$\text{Prob}(X = x_i) = \pi_i, \quad 1 \leq i \leq m, \quad \text{where } \pi_i \geq 0, \ \sum_{i=1}^{m} \pi_i = 1, \quad 1 \leq i \leq m. \tag{8}$$

The survey designer has to choose p so as to achieve maximum estimation efficiency while guaranteeing a prespecified level of respondent privacy. Let R denote the randomized response variable. The randomization device leads to

$$\text{Prob}(R=x_i|X=x_j)=(1-p)/m,\quad \text{Prob}(R=x_j|X=x_j)=p+(1-p)/m,\\ 1\le i\ne j\le m.$$

Now, the "true" probabilities are as in (8), while the revealing probabilities are given by the conditional probabilities Prob($X = x_i|R = x_j$), $1 \le i,j \le m$. As usual, for a high degree of privacy protection, it is required that these true and revealing probabilities are as close as possible. The notion of privacy will vary depending on whether all values of X are sensitive or not.

(I) First, suppose that all values $x_1,\ldots,x_m$ are sensitive. For example, if X is the number of times a person has voted for a certain political party in last 6 elections, then $X = 0,1,\ldots,6$ and all these values of X could be considered sensitive. In this case, a respondent will feel assured about his privacy if the perception of his having a value $X = x_i$ is not much altered after knowing his response, $i = 1,\ldots,m$. Consequently, the measure of protection proposed by Bose (2014) in this case is

$$\alpha=\max_{1\le i,j\le m}\frac{p\pi_i|\pi_j-\delta_{ij}|}{p\pi_j+\dfrac{1-p}{m}},\tag{9}$$

where $\delta_{ij}=1$ if $i=j$, and equals zero otherwise. A randomization device with a privacy protection value $\alpha = \alpha_0$ would guarantee that the discrepancies between the true and revealing probabilities will be at most α_0 for all respondents, *irrespective of their true values.*

Now, for a particular survey, suppose the survey designer wants a certain target level of protection to be accorded to all respondents, ie, it is stipulated that $\alpha \le T$ should hold, for some preassigned T, where $0 < T < 1$. Then, for given T, this protection can be accorded to respondents, *irrespective of the values* of $\pi,\ldots,\pi_m$, *if and only if* the device parameter p satisfies

$$p\le\left\{1+\frac{m}{T}\left(\frac{1-T}{2}\right)^2\right\}^{-1}=p_{01}.$$

In order to maximize the efficiency of estimation while guaranteeing this protection, the designer should prepare the randomization device with $p = p_{01}$.

(II) Now suppose that *not all values of X are stigmatizing*. For example, if X is the number of past convictions, then the value $X = 0$ is not stigmatizing. Again, if X is the number of traffic violations in the past year, then values $X = 0$ or $X = 1$ might not be considered stigmatizing while larger values of X may well be perceived to be so. To begin, without loss of generality, suppose the value x_1 of X is not stigmatizing, while all other values are. Here, respondents will feel comfortable with a randomization device for which the "revealing" probability of their having a true value x_1 is large. So, the following measure of privacy was proposed in this case

$$\beta = \min_{1 \le j \le m} P(X = x_1 | R = x_j) = \min_{1 \le j \le m} \frac{(p\delta_{1j} + \frac{1-p}{m})\pi_1}{p\pi_j + \frac{1-p}{m}}, \tag{10}$$

where $\delta_{1j} = 1$ if $j = 1$, and equals zero, otherwise. A device leading to a larger value of β will ensure greater privacy to respondents than one with a smaller β.

Now, for a preassigned level of respondents' privacy, say T, $0 < T < 1$, suppose one requires that $\beta \ge T$, *irrespective of the values of* $\pi_1,\ldots,\pi_m$. It can be shown that this cannot be achieved if $\pi_1 \to 0$. However, in most populations, π_1 will be positive and some lower bound to it can be set. So, if it is assumed that $\pi_1 \ge c > 0$, one should work with $T < c$ as it is impractical to require that Prob($X = x_1 | R = x_j) \ge T(\ge c)$ for all j. Hence, it can be shown that for a given $0 < T < c$, $\beta \ge T$ will hold *irrespective of the values* of $\pi_1,\ldots,\pi_m$, *if and only if*,

$$p \le \frac{c-T}{m}\left\{\frac{c-T}{m} + T(1-c)\right\}^{-1} = p_{02}. \tag{11}$$

Again, to maximize the efficiency of estimation while guaranteeing this protection, the designer should prepare the randomization device with $p = p_{02}$.

The measure in (10) can be extended to the general case where X has t nonstigmatizing values while its remaining $m - t$ values are stigmatizing, $1 < t < m$. We refer to Bose (2014) for details.

4 CONCLUDING REMARKS

In the preceding sections, various measures for privacy protection have been described. A survey designer may use his judgment to choose one of these measures for his design in order to ensure a suitable level of privacy protection according to his chosen measure. However, the question remains as to whether the respondents, who are not conversant with the intricacies of probability or statistics, can fully comprehend these measures and whether they will feel confident enough to participate truthfully in the survey. They may still perceive a loss of privacy protection, for example, some may still believe that giving a "yes" answer will imply that they have an increased chance of belonging to a sensitive group, or giving a large value of the response variable R will mean that their true value of X is large too. So, in order to ensure that the "perceived" protection of privacy is commensurate with the actual protection of privacy as quantified by these measures, the surveyor needs to take steps to clarify and enlighten the respondents as to the level of protection accorded to them and assuage the respondents' fears that their true state/value will be easily ascertained.

REFERENCES

Anderson, H., 1977. Efficiency versus protection in a general randomized response model. Scand. J. Stat. 4, 11–19.

Bose, M., 2014. Respondent privacy and estimation efficiency in randomized response surveys for discrete-valued sensitive variables. Stat. Pap. 56, 1055–1069.

Bose, M., 2015. Respondent privacy in randomized response surveys for continuous sensitive variables. (preprint).

Chaudhuri, A., 2011. Randomized Response and Indirect Questioning Techniques in Surveys. CRC Press, Taylor and Francis Group, Boca Raton.

Chaudhuri, A., Christofides, T.C., 2013. Indirect Questioning Techniques in Sample Surveys. Springer, Heidelberg.

Chaudhuri, A., Mukerjee, R., 1988. Randomized Responses: Theory and Techniques. Marcel Dekker, New York.

Chaudhuri, A., Christofides, T.C., Saha, A., 2009. Protection of privacy in efficient application of randomized response techniques. Stat. Methods Appl. 18, 389–418.

Christofides, T.C., 2010. Comments on a method of comparison of randomized response techniques. J. Stat. Plan. Inference 140, 574–575.

Diana, G., Perri, P.F., 2011. A class of estimators for quantitative sensitive data. Stat. Pap. 52, 633–650.

Fligner, M.A., Policello, G.E., Singh, J., 1977. A comparison of two randomized response survey methods with consideration for the level of respondent protection. Commun. Stat. Theory Methods 6, 1511–1524.

Giordano, S., Perri, P.F., 2012. Efficiency comparison of unrelated question models based on same privacy protection degree. Stat. Pap. 53, 987–999.

Greenberg, B.G., Kuebler, R.R., Abernathy, J.R., Horvitz, D.G., 1977. Respondent hazards in the unrelated question randomized response model. J. Stat. Plan. Inference 1, 53–60.

Guerriero, M., Sandri, M.F., 2007. A note on the comparison of some randomized response procedures. J. Stat. Plan. Inference 137, 2184–2190.

Hong, Z., Yan, Z., Wei, L., 2010. A note on proposed privacy measures in randomized response models. Adv. Intell. Soft Comput. 77, 635–642.

Horvitz, D.G., Shah, B., Simmons, W.R., 1967. The unrelated question RR model. In: Proceedings of the Social Statistics Section of the American Statistical Association. ASA, Alexandria, VA, pp. 65–72.

Lanke, J., 1975. On the choice of the unrelated question in Simmons' version of randomized response. J. Am. Stat. Assoc. 70, 80–83.

Lanke, J., 1976. On the degree of protection in randomized interviews. Int. Stat. Rev. 44, 197–203.

Leysieffer, R.W., Warner, S.L., 1976. Respondent jeopardy and optimal designs in randomized response models. J. Am. Stat. Assoc. 71, 649–656.

Ljungqvist, L., 1993. A unified approach to measures of privacy in randomized response models: a utilitarian perspective. J. Am. Stat. Assoc. 88, 97–103.

Loynes, R.M., 1976. Asymptotically optimal randomized response procedures. J. Am. Stat. Assoc. 71, 924–928.

Nayak, T.K., 1994. On randomized response surveys for estimating a proportion. Commun. Stat. Theory Methods 23, 3303–3321.

Nayak, T.K., Adeshiyan, S.A., 2009. A unified framework for analysis and comparison of randomized response surveys of binary characteristics. J. Stat. Plan. Inference 139, 2757–2766.

Quatember, A., 2009. A standardization of randomized response strategies. Surv. Methodol. 35, 143–152.

Quatember, A., 2012. An extension of the standardized randomized response technique to a multi-stage setup. Stat. Methods Appl. 21, 475–484.

Warner, S.L., 1965. Randomized response: a survey technique for eliminating evasive answer bias. J. Am. Stat. Assoc. 60, 63–69.

Chapter 22

Cramer–Rao Lower Bounds of Variance for Estimating Two Proportions and Their Overlap by Using Two Decks of Cards

C.-S. Lee*,[1], S.A. Sedory[†] and S. Singh[†]
*Coastal Bend College, Beeville, TX, United States
[†]Texas A&M University-Kingsville, Kingsville, TX, United States
[1]Corresponding author: e-mail: cheonsig@coastalbend.edu

ABSTRACT

In this chapter, we show that the three estimators proposed by Lee et al. (2013) attain the lower bounds of variance at equal protection of respondents. A comparison of a range restricted estimate with an unrestricted estimate is also given. The findings are supported with an example using real data.

Keywords: Randomized response techniques, Estimation of two sensitive characteristics, Minimum variance, Maximum likelihood estimates

1 INTRODUCTION

Lee et al. (2013) considered a new methodology for estimating the proportions of persons in a population possessing each of two sensitive characteristics, say A and B along with the proportion possessing both, $A \cap B$ by using two different randomized response models: simple model and crossed model. There are many situations where their proposed models could be implemented in real practice. For example, (1) assume A is a group of smokers, B a group of drinkers, then $A \cap B$ will be a group of both smokers and drinkers; (2) assume A is a group of smack users, B is a group of criminally involved people, then $A \cap B$ will be a group of both smack users and criminally active people; and (3) assume A represents hidden membership in a terrorist group I, B represents a hidden membership in a terrorist group II, then $A \cap B$ will be a hidden membership in both terrorist groups. Their models also allow one to estimate several other parameters such as correlation coefficient, conditional

Handbook of Statistics, Vol. 34. http://dx.doi.org/10.1016/bs.host.2016.01.020

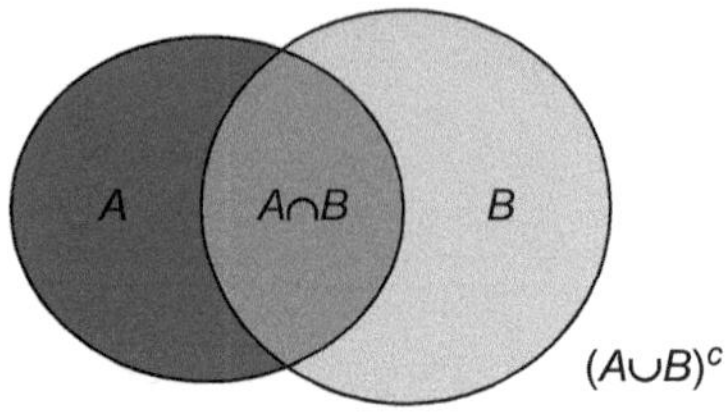

FIG. 1 Population under study.

proportions, and relative risk. A pictorial representation of such a population is shown in Fig. 1. Let π_A, π_B, and $\pi_{A\cap B}$ be the true proportions of respondents possessing the sensitive characteristics A, B, and both $A\cap B$. Also note that $\pi_{A\cap B} \leq \min(\pi_A, \pi_B)$.

Assume a simple random and with replacement sample of n respondents is selected from the population Ω. The authors suggest above two different randomized response models, which are described in brief in the following subsections named as simple model and crossed model.

1.1 Simple Model

In the simple model proposed by Lee et al. (2013), they suggest using a pair of decks of cards in order: say Deck I and Deck II. Each respondent, selected in a simple random with replacement sample of size n, is requested to draw two cards, one card from each deck of cards and keep the response in order from Deck I and Deck II, respectively. Deck I consists of cards, each bearing one of two mutually exclusive statements: "I belong to the sensitive group A," with probability P and "I belong to the nonsensitive group A^c," with probability $(1-P)$. Deck II also consists of cards, each bearing one of two mutually exclusive statements: "I belong to the sensitive group B," with probability T and "I belong to the nonsensitive group B^c," with probability $(1-T)$. Following the notation of Lee et al. (2013) for the simple model (abbreviated hereafter as LSS-SM), the probabilities of obtaining, from a given respondent, each of the following responses: (Yes, Yes), (Yes, No), (No, Yes), and (No, No) are, respectively, given by:

$$\begin{aligned}\theta_{11} = {} & (2P-1)(2T-1)\pi_{AB} + (2P-1)(1-T)\pi_A \\ & + (1-P)(2T-1)\pi_B + (1-P)(1-T),\end{aligned} \tag{1}$$

$$\theta_{10} = -(2P-1)(2T-1)\pi_{AB} + (2P-1)T\pi_A - (1-P)(2T-1)\pi_B + (1-P)T, \tag{2}$$

$$\theta_{01} = -(2P-1)(2T-1)\pi_{AB} - (2P-1)(1-T)\pi_A + P(2T-1)\pi_B + P(1-T), \tag{3}$$

and

$$\theta_{00} = (2P-1)(2T-1)\pi_{AB} - T(2P-1)\pi_A - P(2T-1)\pi_B + PT. \tag{4}$$

Let $\hat{\theta}_{11} = n_{11}/n$, $\hat{\theta}_{10} = n_{10}/n$, $\hat{\theta}_{01} = n_{01}/n$, and $\hat{\theta}_{00} = n_{00}/n$ be the observed proportions of (Yes, Yes), (Yes, No), (No, Yes), and (No, No) responses, so that $n_{11} + n_{10} + n_{01} + n_{00} = n$. LSS-SM considered minimizing a squared distance between the observed proportions and the true proportions, defined as:

$$D_1 = \frac{1}{2}\sum_{i=0}^{1}\sum_{j=0}^{1}\left(\theta_{ij} - \hat{\theta}_{ij}\right)^2. \tag{5}$$

On solving a system of linear equations, obtained by setting: $\frac{\partial D_1}{\partial \pi_A} = 0$, $\frac{\partial D_1}{\partial \pi_B} = 0$, and $\frac{\partial D_1}{\partial \pi_{AB}} = 0$, they obtained unbiased estimators of π_A, π_B, and π_{AB}, given by:

$$\hat{\pi}_A = \frac{\hat{\theta}_{11} + \hat{\theta}_{10} - \hat{\theta}_{01} - \hat{\theta}_{00} + (2P-1)}{2(2P-1)}, \tag{6}$$

$$\hat{\pi}_B = \frac{\hat{\theta}_{11} - \hat{\theta}_{10} + \hat{\theta}_{01} - \hat{\theta}_{00} + (2T-1)}{2(2T-1)}, \tag{7}$$

and

$$\hat{\pi}_{AB} = \frac{(P+T)\hat{\theta}_{11} + (T-P)\hat{\theta}_{10} + (P-T)\hat{\theta}_{01} + (2-P-T)\hat{\theta}_{00} - T(1-P) - P(1-T)}{2(2P-1)(2T-1)}, \tag{8}$$

for $P \neq 0.5$ and $T \neq 0.5$.

The variances of the estimators $\hat{\pi}_A$, $\hat{\pi}_B$, and $\hat{\pi}_{AB}$ are, respectively, given by:

$$V(\hat{\pi}_A) = \frac{\pi_A(1-\pi_A)}{n} + \frac{P(1-P)}{n(2P-1)^2}, \tag{9}$$

$$V(\hat{\pi}_B) = \frac{\pi_B(1-\pi_B)}{n} + \frac{T(1-T)}{n(2T-1)^2}, \tag{10}$$

and

$$V(\hat{\pi}_{AB}) = \frac{\pi_{AB}(1-\pi_{AB})}{n} + \frac{(2P-1)^2 T(1-T)\pi_A + P(1-P)(2T-1)^2\pi_B + PT(1-P)(1-T)}{n(2P-1)^2(2T-1)^2}. \tag{11}$$

The estimators $\hat{V}(\hat{\pi}_A)$, $\hat{V}(\hat{\pi}_B)$, and $V(\hat{\pi}_{AB}\hat{)}$ of the variances $V(\hat{\pi}_A)$, $V(\hat{\pi}_B)$, and $V(\hat{\pi}_{AB})$ can be obtained by replacing π_A with $\hat{\pi}_A$, π_B with $\hat{\pi}_B$, and π_{AB} with $\hat{\pi}_{AB}$ and bias can be adjusted by replacing n by $(n-1)$.

1.2 Crossed Model

In the crossed model, while the rest of the procedure remains the same as for the simple model but the composition of the decks is different. Deck I consists of cards, each bearing one of two mutually exclusive statements: "I belong to the sensitive group A," with probability P and "I belong to the nonsensitive group B^c," with probability $(1-P)$, respectively. Deck II also consists of cards, each bearing one of two mutually exclusive statements: "I belong to the sensitive group B" with probability T and "I belong to the nonsensitive group A^c" with probability $(1-T)$, respectively. Following the notation of Lee et al. (2013) for the crossed model (abbreviated hereafter as LSS-CM), the probabilities of obtaining from a given respondent each of the following responses: (Yes, Yes), (Yes, No), (No, Yes), and (No, No) are, respectively, given by:

$$\begin{aligned}\theta_{11}^* = \pi_{AB}\{PT+(1-P)(1-T)\} - \pi_A(1-P)(1-T)\\ - \pi_B(1-P)(1-T) + (1-P)(1-T),\end{aligned} \tag{12}$$

$$\begin{aligned}\theta_{10}^* = -\pi_{AB}\{PT+(1-P)(1-T)\} - \pi_A\{(1-P)T-1\}\\ - \pi_B(1-P)T + (1-P)T,\end{aligned} \tag{13}$$

$$\begin{aligned}\theta_{01}^* = -\pi_{AB}\{PT+(1-P)(1-T)\} - \pi_A P(1-T)\\ - \pi_B\{P(1-T)-1\} + P(1-T),\end{aligned} \tag{14}$$

and

$$\theta_{00}^* = \pi_{AB}\{PT+(1-P)(1-T)\}^{AB} - \pi_A PT - \pi_B PT + PT. \tag{15}$$

Let $\hat{\theta}_{11}^* = n_{11}^*/n$, $\hat{\theta}_{10}^* = n_{10}^*/n$, $\hat{\theta}_{01}^* = n_{01}^*/n$, and $\hat{\theta}_{00}^* = n_{00}^*/n$ be the observed proportions of (Yes, Yes), (Yes, No), (No, Yes), and (No, No) responses (so that $n_{11}^* + n_{10}^* + n_{01}^* + n_{00}^* = n$). LSS-CM also considered minimizing a squared distance between the observed proportions and the true proportions, defined as:

$$D_1^* = \frac{1}{2}\sum_{i=0}^{1}\sum_{j=0}^{1}\left(\theta_{ij}^* - \hat{\theta}_{ij}^*\right)^2. \tag{16}$$

On solving a system of linear equations, obtained by setting: $\frac{\partial D_1^*}{\partial \pi_A} = 0$, $\frac{\partial D_1^*}{\partial \pi_B} = 0$, and $\frac{\partial D_1^*}{\partial \pi_{AB}} = 0$, they obtained unbiased estimators of π_A, π_B, and π_{AB}, given by:

$$\hat{\pi}_A^* = \frac{1}{2} + \frac{(T-P+1)\left(\hat{\theta}_{11}^* - \hat{\theta}_{00}^*\right) + (P+T-1)\left(\hat{\theta}_{10}^* - \hat{\theta}_{01}^*\right)}{2(P+T-1)}, \tag{17}$$

$$\hat{\pi}_B^* = \frac{1}{2} + \frac{(P-T+1)\left(\hat{\theta}_{11}^* - \hat{\theta}_{00}^*\right) + (P+T-1)\left(\hat{\theta}_{01}^* - \hat{\theta}_{10}^*\right)}{2(P+T-1)}, \tag{18}$$

and

$$\hat{\pi}^*_{AB} = \frac{PT\hat{\theta}^*_{11} - (1-P)(1-T)\hat{\theta}^*_{00}}{\{PT + (1-P)(1-T)\}(P+T-1)}, \tag{19}$$

with $P+T \neq 1$.

The variances of the estimators $\hat{\pi}^*_A$, $\hat{\pi}^*_B$, and $\hat{\pi}^*_{AB}$ are, respectively, given by:

$$V(\hat{\pi}^*_A) = \frac{\pi_A(1-\pi_A)}{n} + \frac{(1-P)T\{PT+(1-P)(1-T)\}(1-\pi_A-\pi_B+2\pi_{AB})}{n(P+T-1)^2}, \tag{20}$$

$$V(\hat{\pi}^*_B) = \frac{\pi_B(1-\pi_B)}{n} + \frac{(1-T)P\{PT+(1-P)(1-T)\}(1-\pi_A-\pi_B+2\pi_{AB})}{n(P+T-1)^2}, \tag{21}$$

and

$$V\left(\hat{\pi}^*_{AB}\right) = \frac{\pi_{AB}\left(1-\pi_{AB}\right)}{n} + \frac{\pi_{AB}\left[P^2T^2+(1-P)^2(1-T)^2-\{PT+(1-P)(1-T)\}(P+T-1)^2\right] + PT(1-P)(1-T)\left(1-\pi_A-\pi_B\right)}{n\{PT+(1-P)(1-T)\}(P+T-1)^2}. \tag{22}$$

The estimators $\hat{V}(\hat{\pi}^*_A)$, $\hat{V}(\hat{\pi}^*_B)$, and $V(\hat{\pi}^*_{AB}\hat{)}$ of the variances $V(\hat{\pi}^*_A)$, $V(\hat{\pi}^*_B)$, and $V(\hat{\pi}^*_{AB})$ can be obtained by replacing π_A with $\hat{\pi}^*_A$, π_B with $\hat{\pi}^*_B$, and π_{AB} with $\hat{\pi}^*_{AB}$ and bias can be adjusted by replacing n by $(n-1)$.

The crossed model has recently been cited in Chaudhuri and Christofides (2013) and Fox (2016). In the following sections, we consider the construction of Cramer–Rao lower bounds of variances for the simple model and then for the crossed model.

2 CRAMER–RAO LOWER BOUNDS OF VARIANCES FOR THE SIMPLE MODEL

Theorem 1. *The maximum likelihood estimates* $\hat{\pi}^{(mle)}_{AB}$, $\hat{\pi}^{(mle)}_{A}$, *and* $\hat{\pi}^{(mle)}_{B}$ *of* π_{AB}, π_A, *and* π_B *are a solution to the three nonlinear equations given as:*

$$\frac{\hat{\theta}_{11}}{\theta_{11}} - \frac{\hat{\theta}_{10}}{\theta_{10}} - \frac{\hat{\theta}_{01}}{\theta_{01}} + \frac{\hat{\theta}_{00}}{\theta_{00}} = 0, \tag{23}$$

$$\frac{\hat{\theta}_{11}}{\theta_{11}}(1-T) + \frac{\hat{\theta}_{10}}{\theta_{10}}T - \frac{\hat{\theta}_{01}}{\theta_{01}}(1-T) - \frac{\hat{\theta}_{00}}{\theta_{00}}T = 0, \tag{24}$$

and

$$\frac{\hat{\theta}_{11}}{\theta_{11}}(1-P)-\frac{\hat{\theta}_{10}}{\theta_{10}}(1-P)+\frac{\hat{\theta}_{01}}{\theta_{01}}P-\frac{\hat{\theta}_{00}}{\theta_{00}}P=0. \tag{25}$$

Proof. In the case of simple model, we have the likelihood function given by:

$$L=\binom{n}{n_{11}n_{10}n_{01}n_{00}}(\theta_{11})^{n_{11}}(\theta_{10})^{n_{10}}(\theta_{01})^{n_{01}}(\theta_{00})^{n_{00}}.$$

Taking log and substituting (1) to (4), we get:

$$\begin{aligned}\log L &= \log C+n_{11}\log\theta_{11}+n_{10}\log\theta_{10}+n_{01}\log\theta_{01}+n_{00}\log\theta_{00}\\ &= \log C+n_{11}\log[(2P-1)(2T-1)\pi_{AB}+(2P-1)(1-T)\pi_A+(1-P)(2T-1)\pi_B+(1-P)(1-T)]\\ &\quad +n_{10}\log[-(2P-1)(2T-1)\pi_{AB}+(2P-1)T\pi_A-(1-P)(2T-1)\pi_B+(1-P)T]\\ &\quad +n_{01}\log[-(2P-1)(2T-1)\pi_{AB}-(2P-1)(1-T)\pi_A+P(2T-1)\pi_B+P(1-T)]\\ &\quad +n_{00}\log[(2P-1)(2T-1)\pi_{AB}-T(2P-1)\pi_A-P(2T-1)\pi_B+PT].\end{aligned}$$

On taking the first-order partial derivates of $\log L$ with respect to π_{AB}, π_A, and π_B, respectively, we obtain

$$\frac{\partial \log L}{\partial \pi_{AB}}=n(2P-1)(2T-1)\left\{\frac{\hat{\theta}_{11}}{\theta_{11}}-\frac{\hat{\theta}_{10}}{\theta_{10}}-\frac{\hat{\theta}_{01}}{\theta_{01}}+\frac{\hat{\theta}_{00}}{\theta_{00}}\right\},$$

$$=n(2P-1)\left\{\frac{\hat{\theta}_{11}}{\theta_{11}}(1-T)+\frac{\hat{\theta}_{10}}{\theta_{10}}T-\frac{\hat{\theta}_{01}}{\theta_{01}}(1-T)-\frac{\hat{\theta}_{00}}{\theta_{00}}T\right\},$$

and

$$\frac{\partial \log L}{\partial \pi_B}=n(2T-1)\left\{\frac{\hat{\theta}_{11}}{\theta_{11}}(1-P)-\frac{\hat{\theta}_{10}}{\theta_{10}}(1-P)+\frac{\hat{\theta}_{01}}{\theta_{01}}P-\frac{\hat{\theta}_{00}}{\theta_{00}}P\right\}.$$

On setting $\frac{\partial \log L}{\partial \pi_{AB}}=0$, $\frac{\partial \log L}{\partial \pi_A}=0$, and $\frac{\partial \log L}{\partial \pi_B}=0$, we have the theorem. □

Theorem 2. *The lower bounds of variance–covariance of the maximum likelihood estimates* $\hat{\pi}_{AB}^{mle}$, $\hat{\pi}_A^{mle}$, *and* $\hat{\pi}_B^{mle}$ *of* π_{AB}, π_A, *and* π_B *for the simple model are, respectively, given by:*

$$V\begin{bmatrix}\hat{\pi}_{AB}^{mle}\\ \hat{\pi}_A^{mle}\\ \hat{\pi}_B^{mle}\end{bmatrix}\geq\frac{1}{\Delta}\begin{bmatrix}a_{11}&a_{12}&a_{13}\\ a_{21}&a_{22}&a_{23}\\ a_{31}&a_{32}&a_{33}\end{bmatrix}, \tag{26}$$

where the inequality sign (≥) has its usual meaning, and Δ *and* a_{ij} *are given by*

$$\Delta=n\left(gb^2-2bcf+dc^2+af^2-adg\right), a_{11}=g^2-df,\quad a_{22}=c^2-ag,$$
$$a_{33}=b^2-ad, a_{12}=a_{21}=bf-cg,\quad a_{13}=a_{31}=cd-bg,$$
$$a_{23}=a_{32}=ag-bc$$

and $a,b,c,d,f,$ *and* g *are given next.*

Proof. By the definition of Cramer–Rao lower bound of variance, we have

$$V\begin{bmatrix}\hat{\pi}_{AB}^{(mle)}\\ \hat{\pi}_{A}^{(mle)}\\ \hat{\pi}_{B}^{(mle)}\end{bmatrix} \geq -\begin{bmatrix} E\left(\frac{\partial^2 \log L}{\partial \pi_{AB}^2}\right) & E\left(\frac{\partial^2 \log L}{\partial \pi_A \partial \pi_{AB}}\right) & E\left(\frac{\partial^2 \log L}{\partial \pi_B \partial \pi_{AB}}\right)\\ E\left(\frac{\partial^2 \log L}{\partial \pi_A \partial \pi_{AB}}\right) & E\left(\frac{\partial^2 \log L}{\partial \pi_A^2}\right) & E\left(\frac{\partial^2 \log L}{\partial \pi_B \partial \pi_A}\right)\\ E\left(\frac{\partial^2 \log L}{\partial \pi_B \partial \pi_{AB}}\right) & E\left(\frac{\partial^2 \log L}{\partial \pi_B \partial \pi_A}\right) & E\left(\frac{\partial^2 \log L}{\partial \pi_B^2}\right)\end{bmatrix}^{-1} = \begin{bmatrix} a & b & c\\ b & d & f\\ c & f & g\end{bmatrix}^{-1},$$

where

$$a = -E\left[\frac{\partial^2 \log L}{\partial \pi_{AB}^2}\right] = n(2P-1)^2(2T-1)^2\left[\frac{1}{\theta_{11}} + \frac{1}{\theta_{10}} + \frac{1}{\theta_{01}} + \frac{1}{\theta_{00}}\right],$$

$$b = -E\left[\frac{\partial^2 \log L}{\partial \pi_A \partial \pi_{AB}}\right] = n(2P-1)^2(2T-1)\left[\frac{(1-T)}{\theta_{11}} - \frac{T}{\theta_{10}} + \frac{(1-T)}{\theta_{01}} - \frac{T}{\theta_{00}}\right],$$

$$c = -E\left[\frac{\partial^2 \log L}{\partial \pi_B \partial \pi_{AB}}\right] = n(2P-1)(2T-1)^2\left\{\frac{(1-P)}{\theta_{11}} + \frac{(1-P)}{\theta_{10}} - \frac{P}{\theta_{01}} - \frac{P}{\theta_{00}}\right\},$$

$$d = -E\left[\frac{\partial^2 \log L}{\partial \pi_A^2}\right] = n(2P-1)^2\left[\frac{(1-T)^2}{\theta_{11}} + \frac{T^2}{\theta_{10}} + \frac{(1-T)^2}{\theta_{01}} + \frac{T^2}{\theta_{00}}\right],$$

$$\begin{aligned} f &= -E\left[\frac{\partial^2 \log L}{\partial \pi_B \partial \pi_A}\right]\\ &= n(2P-1)(2T-1)\left\{\frac{(1-P)(1-T)}{\theta_{11}} - \frac{T(1-P)}{\theta_{10}} - \frac{P(1-T)}{\theta_{01}} + \frac{PT}{\theta_{00}}\right\},\end{aligned}$$

and

$$g = -E\left[\frac{\partial^2 \log L}{\partial \pi_B^2}\right] = n(2T-1)^2\left[\frac{(1-P)^2}{\theta_{11}} + \frac{(1-P)^2}{\theta_{10}} + \frac{P^2}{\theta_{01}} + \frac{P^2}{\theta_{00}}\right]$$

which proves the theorem. □

3 CRAMER–RAO LOWER BOUNDS OF VARIANCES FOR THE CROSSED MODEL

Theorem 3. *The maximum likelihood estimates* $\hat{\pi}_{AB}^{*(mle)}$, $\hat{\pi}_{A}^{*(mle)}$, *and* $\hat{\pi}_{B}^{*(mle)}$ *of* π_{AB}, π_A, *and* π_B *are a solution to the three nonlinear equations given as:*

$$\frac{\hat{\theta}_{11}^*}{\theta_{11}^*} - \frac{\hat{\theta}_{10}^*}{\theta_{10}^*} - \frac{\hat{\theta}_{01}^*}{\theta_{01}^*} + \frac{\hat{\theta}_{00}^*}{\theta_{00}^*} = 0, \tag{27}$$

$$\frac{\hat{\theta}_{11}^*}{\theta_{11}^*}(1-P)(1-T)+\frac{\hat{\theta}_{10}^*}{\theta_{10}^*}\{(1-P)T-1\}+\frac{\hat{\theta}_{01}^*}{\theta_{01}^*}P(1-T)+\frac{\hat{\theta}_{00}^*}{\theta_{00}^*}PT=0, \quad (28)$$

and

$$\frac{\hat{\theta}_{11}^*}{\theta_{11}^*}(1-P)(1-T)+\frac{\hat{\theta}_{10}^*}{\theta_{10}^*}(1-P)T+\frac{\hat{\theta}_{01}^*}{\theta_{01}^*}\{P(1-T)-1\}+\frac{\hat{\theta}_{00}^*}{\theta_{00}^*}PT=0. \quad (29)$$

Proof. In case of the crossed model, the likelihood function is given by:

$$L^*=\binom{n}{n_{11}^* n_{10}^* n_{01}^* n_{00}^*}(\theta_{11}^*)^{n_{11}^*}(\theta_{10}^*)^{n_{10}^*}(\theta_{01}^*)^{n_{01}^*}(\theta_{00}^*)^{n_{00}^*}.$$

Taking log on both sides, we have:

$$\begin{aligned}\log L^* &= \log C+n_{11}^*\log\theta_{11}^*+n_{10}^*\log\theta_{10}^*+n_{01}^*\log\theta_{01}^*+n_{00}^*\log\theta_{00}^*=\log C\\ &+n_{11}^*\log[\pi_{AB}\{PT+(1-P)(1-T)\}-\pi_A(1-P)(1-T)-\pi_B(1-P)(1-T)+(1-P)(1-T)]\\ &+n_{10}^*\log[-\pi_{AB}\{PT+(1-P)(1-T)\}-\pi_A\{(1-P)T-1\}-\pi_B(1-P)T+(1-P)T]\\ &+n_{01}^*\log[-\pi_{AB}\{PT+(1-P)(1-T)\}-\pi_A P(1-T)-\pi_B\{P(1-T)-1\}+P(1-T)]\\ &+n_{00}^*\log[\pi_{AB}\{PT+(1-P)(1-T)\}-\pi_A PT-\pi_B PT+PT].\end{aligned}$$

On taking the first-order partial derivates of log L^* with respect to π_{AB}, π_A, and π_B, respectively, we have

$$\frac{\partial\log L^*}{\partial\pi_{AB}}=n\{PT+(1-P)(1-T)\}\left[\frac{\hat{\theta}_{11}^*}{\theta_{11}^*}-\frac{\hat{\theta}_{10}^*}{\theta_{10}^*}-\frac{\hat{\theta}_{01}^*}{\theta_{01}^*}+\frac{\hat{\theta}_{00}^*}{\theta_{00}^*}\right],$$

$$\frac{\partial\log L^*}{\partial\pi_A}=-n\left[\frac{\hat{\theta}_{11}^*}{\theta_{11}^*}(1-P)(1-T)+\frac{\hat{\theta}_{10}^*}{\theta_{10}^*}\{(1-P)T-1\}+\frac{\hat{\theta}_{01}^*}{\theta_{01}^*}P(1-T)+\frac{\hat{\theta}_{00}^*}{\theta_{00}^*}PT\right],$$

and

$$\frac{\partial\log L^*}{\partial\pi_B}=-n\left[\frac{\hat{\theta}_{11}^*}{\theta_{11}^*}(1-P)(1-T)+\frac{\hat{\theta}_{10}^*}{\theta_{10}^*}(1-P)T+\frac{\hat{\theta}_{01}^*}{\theta_{01}^*}\{P(1-T)-1\}+\frac{\hat{\theta}_{00}^*}{\theta_{00}^*}PT\right].$$

On setting $\frac{\partial\log L^*}{\partial\pi_{AB}}=0$, $\frac{\partial\log L^*}{\partial\pi_A}=0$, and $\frac{\partial\log L^*}{\partial\pi_B}=0$, we have the theorem. □

Theorem 4. *The lower bounds of variance–covariance of the maximum likelihood estimates* $\hat{\pi}_{AB}^{*(mle)}$, $\hat{\pi}_A^{*(mle)}$, *and* $\hat{\pi}_B^{*(mle)}$ *of* π_{AB}, π_A, *and* π_B *for the crossed model are, respectively, given by:*

$$V\begin{bmatrix}\hat{\pi}_{AB}^{*(mle)}\\ \hat{\pi}_A^{*(mle)}\\ \hat{\pi}_B^{*(mle)}\end{bmatrix}\geq\frac{1}{\Delta^*}\begin{bmatrix}a_{11}^* & a_{12}^* & a_{13}^*\\ a_{21}^* & a_{22}^* & a_{23}^*\\ a_{31}^* & a_{32}^* & a_{33}^*\end{bmatrix}, \quad (30)$$

where Δ^* and $a_{ij}{}^*$ are given by

$$\Delta^* = n\left[g^*(b*)^2 - 2b^*c^*f^* + d^*\left(c^*\right)^2 + a^*\left(f^*\right)^2 - a^*d^*g^*\right],$$

$$a_{11}^* = \left(g^*\right)^2 - d^*f^*,\ a_{22}^* = \left(c^*\right)^2 - a^*c^*,\ a_{33}^* = \left(b^*\right)^2 - a^*d^*,$$

$$a_{12}^* = a_{21}^* = b^*f^* - c^*g^*,\ a_{13}^* = a_{31}^* = c^*d^* - b^*g^*,\ a_{23}^* = a_{32}^* = a^*g^* - b^*c^*,$$

$$a^* = n\{PT + (1-P)(1-T)\}^2\left(\frac{1}{\theta_{11}^*} + \frac{1}{\theta_{10}^*} + \frac{1}{\theta_{01}^*} + \frac{1}{\theta_{00}^*}\right),$$

$$\begin{aligned} b^* = &\ -n\{PT + (1-P)(1-T)\} \\ &\times\left(\frac{(1-P)(1-T)}{\theta_{11}^*} - \frac{(1-P)T-1}{\theta_{10}^*} - \frac{P(1-T)}{\theta_{01}^*} + \frac{PT}{\theta_{00}^*}\right), \end{aligned}$$

$$\begin{aligned} c^* = &\ -n\{PT + (1-P)(1-T)\} \\ &\times\left(\frac{(1-P)(1-T)}{\theta_{11}^*} - \frac{(1-P)T}{\theta_{10}^*} - \frac{P(1-T)-1}{\theta_{01}^*} + \frac{PT}{\theta_{00}^*}\right), \end{aligned}$$

$$d^* = n\left(\frac{(1-P)^2(1-T)^2}{\theta_{11}^*} + \frac{\{(1-P)T-1\}^2}{\theta_{10}^*} + \frac{P^2(1-T)^2}{\theta_{01}^*} + \frac{P^2T^2}{\theta_{00}^*}\right),$$

$$f^* = n\left[\frac{(1-P)^2(1-T)^2}{\theta_{11}^*} + \frac{(1-P)T\{(1-P)T-1\}}{\theta_{10}^*} + \frac{P(1-T)\{P(1-T)-1\}}{\theta_{01}^*} + \frac{P^2T^2}{\theta_{00}^*}\right],$$

and

$$g^* = n\left[\frac{(1-P)^2(1-T)^2}{\theta_{11}^*} + \frac{(1-P)^2T^2}{\theta_{10}^*} + \frac{\{(1-P)T-1\}^2}{\theta_{01}^*} + \frac{P^2T^2}{\theta_{00}^*}\right].$$

Proof. It follows from the proof of Theorem 2. □

In the next section, we consider a comparison of the variances of the estimators obtained by minimizing squared distance with the maximum likelihood estimates for both the simple and the crossed models.

4 COMPARISON OF THE VARIANCES AND LOWER BOUNDS

Let $\hat{\theta}_1 = \hat{\pi}_A$, $\hat{\theta}_2 = \hat{\pi}_B$, $\hat{\theta}_3 = \hat{\pi}_{AB}$, $\hat{\theta}_1^{(mle)} = \hat{\pi}_A^{(mle)}$, $\hat{\theta}_2^{(mle)} = \hat{\pi}_B^{(mle)}$, and $\hat{\theta}_3^{(mle)} = \hat{\pi}_{AB}^{(mle)}$. We define the percent relative efficiency of the proposed maximum likelihood estimators of π_A, π_B, and π_{AB} by comparing the Cramer–Rao lower bounds with the variances of the LSS-SM estimators, respectively, as:

$$\mathrm{RE}(j) = \frac{V\left(\hat{\theta}_j\right)}{V\left(\hat{\theta}_j^{(mle)}\right)} \times 100\%,\quad j = 1,2,3. \tag{31}$$

TABLE 1 The Ratios of Variances to the Cramer–Rao Lower Bounds for Different Parameters for the Simple Model

π_A	π_B	π_{AB}	P	T	RE(A)	RE(B)	RE(AB)
0.25	0.2	0.05	0.7	0.60	100	100	100
0.25	0.2	0.10	0.7	0.60	100	100	100
0.25	0.2	0.15	0.7	0.60	100	100	100
0.25	0.2	0.20	0.7	0.60	100	100	100
0.25	0.2	0.05	0.7	0.65	100	100	100
0.25	0.2	0.10	0.7	0.65	100	100	100
0.25	0.2	0.15	0.7	0.65	100	100	100
0.25	0.2	0.20	0.7	0.65	100	100	100
0.25	0.2	0.05	0.7	0.70	100	100	100
0.25	0.2	0.10	0.7	0.70	100	100	100
0.25	0.2	0.15	0.7	0.70	100	100	100
0.25	0.2	0.20	0.7	0.70	100	100	100
0.25	0.2	0.05	0.7	0.75	100	100	100
0.25	0.2	0.10	0.7	0.75	100	100	100
0.25	0.2	0.15	0.7	0.75	100	100	100
0.25	0.2	0.20	0.7	0.75	100	100	100

Note carefully that the percent relative efficiency values defined in (31) are free from the value of the sample size n. For demonstration purposes, we wrote FORTRAN codes to compute the values RE(1) = RE(A), RE(2) = RE(B), and RE(3) = RE(AB) for different values of the parameters P, T, π_A, π_B, and π_{AB}. Please refer to PROGRAM_1.F95 (see Appendix B) for detailed coding used in producing the results in Table 1. The results so obtained are presented in Table 1, which clearly show that the three LSS-SM estimators attain the Cramer–Rao lower bounds of variances.

Next, let $\hat{\theta}_1^* = \hat{\pi}_A^*$, $\hat{\theta}_2^* = \hat{\pi}_B^*$, $\hat{\theta}_3^* = \hat{\pi}_{AB}^*$, $\hat{\theta}_1^{*(mle)} = \hat{\pi}_A^{*(mle)}$, $\hat{\theta}_2^{*(mle)} = \hat{\pi}_B^{*(mle)}$, and $\hat{\theta}_3^{*(mle)} = \hat{\pi}_{AB}^{*(mle)}$. Again we define the percent relative efficiency of the proposed maximum likelihood estimators of π_A, π_B, and π_{AB} by comparing the Cramer–Rao lower bounds with the variances of the LSS-CM estimators, respectively, as:

$$\text{RE}^*(j) = \frac{V\left(\hat{\theta}_j^*\right)}{V\left(\hat{\theta}_j^{*(mle)}\right)} \times 100\%, \quad j = 1,2,3. \tag{32}$$

Again note carefully that the percent relative efficiency values defined in (32) are free from the value of the sample size n. For demonstration purposes, again we wrote FORTRAN codes to compute the values RE*(1) = RE*(A), RE*(2) = RE*(B), and RE*(3) = RE*(AB) for different values of the parameters P, T, π_A, π_B, and π_{AB}. Refer to PROGRAM_2.F95 (see Appendix B) for details of the coding used to produce results listed in Table 2. The results so obtained are presented in Table 2, which clearly show that the three LSS-CM estimators attain the Cramer–Rao lower bounds of variances.

Lee et al. (2013) have shown that the crossed model is more efficient than the simple model. We solved the three nonlinear equations in (27)–(29) by using SAS codes to obtain the maximum likelihood estimates from the same data set used in Lee et al. (2013). Please refer to PROGRAM_3.SAS (see Appendix B) used for producing these findings. Note carefully that these findings are not the same as listed in Singh and Sedory (2012). It is interesting to conclude that if one wishes to estimate single parameter, then the maximum

TABLE 2 The Ratios of Variances to the Cramer–Rao Lower Bounds for Different Parameters for the Crossed Model

π_A	π_B	π_{AB}	P	T	RE*(A)	RE*(B)	RE*(AB)
0.25	0.2	0.05	0.7	0.60	100	100	100
0.25	0.2	0.10	0.7	0.60	100	100	100
0.25	0.2	0.15	0.7	0.60	100	100	100
0.25	0.2	0.20	0.7	0.60	100	100	100
0.25	0.2	0.05	0.7	0.65	100	100	100
0.25	0.2	0.10	0.7	0.65	100	100	100
0.25	0.2	0.15	0.7	0.65	100	100	100
0.25	0.2	0.20	0.7	0.65	100	100	100
0.25	0.2	0.05	0.7	0.70	100	100	100
0.25	0.2	0.10	0.7	0.70	100	100	100
0.25	0.2	0.15	0.7	0.70	100	100	100
0.25	0.2	0.20	0.7	0.70	100	100	100
0.25	0.2	0.05	0.7	0.75	100	100	100
0.25	0.2	0.10	0.7	0.75	100	100	100
0.25	0.2	0.15	0.7	0.75	100	100	100
0.25	0.2	0.20	0.7	0.75	100	100	100

likelihood estimates are different from the least square estimates, but if one estimates three parameters, then the maximum likelihood estimates and the least square estimates are equally efficient. One can easily check that the output of this SAS code gives us: $X = \hat{\pi}_{AB} = 0.23678$, $Y = \hat{\pi}_A = 0.240$, and $Z = \hat{\pi}_B = 0.360$, which are exactly the same estimates reported by Lee et al. (2013) based on least squared distance estimators for a real data set on smoking and drinking. Thus these estimates also confirm that the Cramer–Rao lower bounds of variances are also attained be the squared distance estimators.

5 UNIQUE ESTIMATES

Note that although it is natural to set a constraint on the three proportions being estimated as: $\pi_{A\cap B} \leq \min(\pi_A, \pi_B)$ along with the two constraints $0 \leq \pi_A \leq 1$ and $0 \leq \pi_B \leq 1$. It is very interesting to note that the estimates of π_A, π_B, and $\pi_{A\cap B}$ are observed to be unique obtained by solving the nonlinear equations either by using SAS programming or by using FORTRAN programming. The estimates obtained by using the least square method and maximum likelihood method always match with each other irrespective of satisfying boundary conditions. We took the extreme situation of estimating $\pi_A = 0.1$, $\pi_B = 0.1$, and $\pi_{A\cap B} = 0.05$ with different sample sizes in the range 1000–1500 with a step of 100. No doubt the estimate of $\pi_{A\cap B} = 0.05$ may lead to a negative value very frequently. Interestingly, this is not the case if sample size becomes sufficiently large, say $n \geq 1100$. We wrote FORTRAN code in finding the minimum sample size required to estimate tiny proportions such as $\pi_A = 0.1$, $\pi_B = 0.1$, and $\pi_{A\cap B} = 0.05$. Note that if these proportions are not so tiny, then a moderate sample size is sufficient to estimate the required proportions either by using the least square method or by using the maximum likelihood method without imposing any constraint. These finding can be easily verified by executing the FORTRAN codes PROGRAM_4.F95 (see Appendix B). The estimates based on various sample sizes ($n = 1000$, 1100, 1200, and 1500) are presented in Appendix A. The results in Table A1 show that the same estimates of $\pi_A = 0.1$, $\pi_B = 0.1$, and $\pi_{A\cap B} = 0.05$ are obtained by both methods. For example, for $n = 1000$ the estimates $\hat{\pi}_A^{*(mle)}$ and $\hat{\pi}_A^*$ are 0.09825, the estimates $\hat{\pi}_B^{*(mle)}$ and $\hat{\pi}_B^*$ are 0.12925, and the estimates $\hat{\pi}_{AB}^{*(mle)}$ and $\hat{\pi}_{AB}^*$ are 0.05082. It may be worth mentioning that a corresponding SAS program also results in the same estimates by both methods viz. squared distance and maximum likelihood method. We modified the above FORTRAN codes to compute the proportion of times the values of the estimates $\hat{\pi}_{AB}^*$ (or $\hat{\pi}_{AB}^{*(mle)}$), $\hat{\pi}_A^*$ (or $\hat{\pi}_A^{*(mle)}$), and $\hat{\pi}_B^*$ (or $\hat{\pi}_B^{*(mle)}$) take on negative values for various sample sizes between 1000 and 5000 with a step of 100. We set NOS = 2000, so that every time we had 2000 estimates of each proportion while computing the proportion of negative estimates. A graphical representation of such

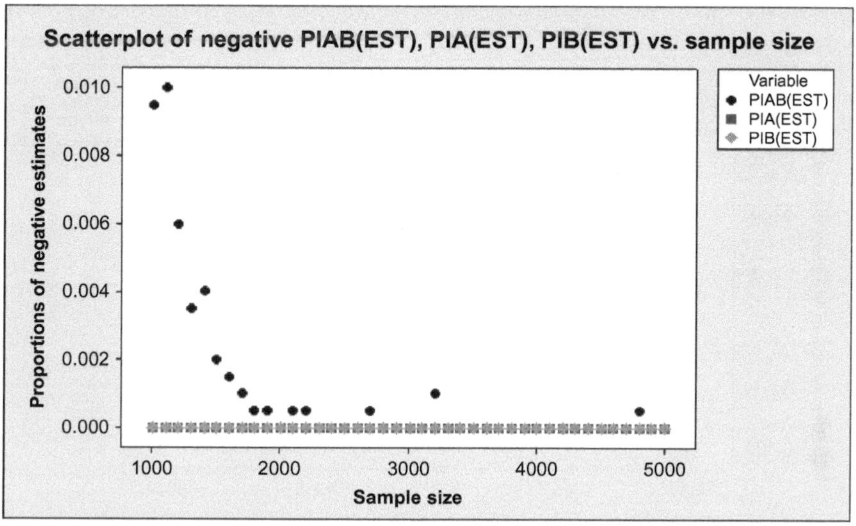

FIG. 2 Proportions of negative estimates.

negative estimates is shown in Fig. 2. A careful examination of Fig. 2 shows that none of the estimates $\hat{\pi}_A^*$ and $\hat{\pi}_B^*$ took negative value beginning with a sample size of $n=1000$ up to a sample size $n=5000$ with a step of 100. However, a maximum of 1% of the estimates $\hat{\pi}_{AB}^*$ took negative values when the sample size was close to $n=1000$. As soon as the sample size became more than $n=3000$, the proportion of negative estimates $\hat{\pi}_{AB}^*$ becomes almost negligible. Thus when estimating tiny proportions such as $\pi_{A\cap B}=0.05$, a large sample size is recommended.

We again modified the above FORTRAN codes to compute the proportion of times the values of estimates $\hat{\pi}_{AB}^*$ (or $\hat{\pi}_{AB}^{*(mle)}$) are greater than $\pi_A=\pi_B=0.1$, the estimate $\hat{\pi}_A^*$ (or $\hat{\pi}_A^{*(mle)}$) is less than the tiny proportion $\pi_{AB}=0.05$ and greater than zero, and also the estimate $\hat{\pi}_B^*$ (or $\hat{\pi}_B^{*(mle)}$) is less than the tiny proportion $\pi_{AB}=0.05$ and greater than zero. Maximum value of inadmissible estimates is around 3% in the case of estimating $\pi_A=\pi_B=0.1$, and the maximum of such proportion is 1.5% in case of estimating $\pi_{AB}=0.05$ with a sample of 1000 units. A pictorial representation of such inadmissible estimates is shown in Fig. 3.

Again a careful examination of the results presented in Fig. 3 shows that the chance of obtaining inadmissible estimates of proportions becomes almost equal to zero as the sample size increases beyond $n=4000$. Thus, when estimating tiny proportions a large sample size is required, as otherwise one may obtain an estimate such that $\hat{\pi}_{AB}^* > \hat{\pi}_A^*$, or $\hat{\pi}_{AB}^* > \hat{\pi}_B^*$ and vice versa. To overcome this limitation, we devote then next section to proposing range restricted maximum likelihood estimates.

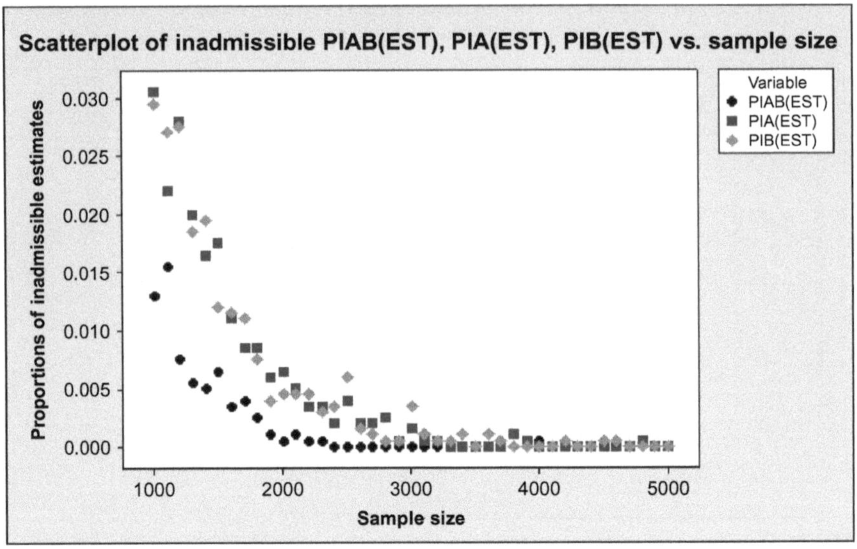

FIG. 3 Proportions of inadmissible estimates of three proportions.

6 RANGE RESTRICTED MAXIMUM LIKELIHOOD ESTIMATES

We propose using the following range restricted maximum likelihood estimates:

$$\hat{\pi}_{AB}^{*(mle)(r)} = \begin{bmatrix} \hat{\pi}_{AB}^{*(mle)} & \text{if } 0 \leq \hat{\pi}_{AB}^{*(mle)} \leq \min\left(\hat{\pi}_{A}^{*(mle)}, \hat{\pi}_{B}^{*(mle)}\right) \\ 0 & \text{if } \hat{\pi}_{AB}^{*(mle)} < 0 \\ \min\left(\hat{\pi}_{A}^{*(mle)}, \hat{\pi}_{B}^{*(mle)}\right) & \text{if } \hat{\pi}_{AB}^{*(mle)} > \min\left(\hat{\pi}_{A}^{*(mle)}, \pi_{B}^{*(mle)}\right) \geq 0 \end{bmatrix} \tag{33}$$

$$\hat{\pi}_{A}^{*(mle)(r)} = \begin{bmatrix} \hat{\pi}_{A}^{*(mle)} & \text{if } 0 \leq \hat{\pi}_{AB}^{*(mle)} \leq \hat{\pi}_{A}^{*(mle)} \leq 1 \\ \hat{\pi}_{AB}^{*(mle)} & \text{if } \hat{\pi}_{A}^{*(mle)} < \hat{\pi}_{AB}^{*(mle)} > 0 \\ 0 & \text{if } \hat{\pi}_{A}^{*(mle)} < 0 \\ 1 & \text{if } \hat{\pi}_{A}^{*(mle)} > 1 \end{bmatrix} \tag{34}$$

and

$$\hat{\pi}_{B}^{*(mle)(r)} = \begin{bmatrix} \hat{\pi}_{B}^{*(mle)} & \text{if } 0 \leq \hat{\pi}_{AB}^{*(mle)} \leq \hat{\pi}_{B}^{*(mle)} \leq 1 \\ \hat{\pi}_{AB}^{*(mle)} & \text{if } \hat{\pi}_{B}^{*(mle)} < \hat{\pi}_{AB}^{*(mle)} > 0 \\ 0 & \text{if } \hat{\pi}_{B}^{*(mle)} < 0 \\ 1 & \text{if } \hat{\pi}_{B}^{*(mle)} > 1 \end{bmatrix} \tag{35}$$

Note that such restriction may lead to biased estimates. We investigated the relative bias and relative efficiency of such range restricted estimates with respect to the least square distance estimators. For clarification, again let $\hat{\theta}_1^{*(mle)} = \hat{\pi}_A^{*(mle)}$, $\hat{\theta}_2^{*(mle)} = \hat{\pi}_B^{*(mle)}$, $\hat{\theta}_3^{*(mle)} = \hat{\pi}_{AB}^{*(mle)}$, $\hat{\theta}_1^{*(r)} = \hat{\pi}_A^{*(mle)(r)}$, $\hat{\theta}_2^{*(r)} = \hat{\pi}_B^{*(mle)(r)}$, and $\hat{\theta}_3^{*(r)} = \hat{\pi}_{AB}^{*(mle)(r)}$, where the superscript (r) indicates range restricted estimators. Again we define the percent relative efficiency of the proposed range restricted maximum likelihood estimates of π_A, π_B, and π_{AB} by comparing the empirical ratio of the LSS-CM estimators, respectively, defined as:

$$\mathrm{RE}^{*(r)}(j) = \frac{\sum_{t=1}^{\Theta}\left[\left(\hat{\theta}_j^{*(mle)}\right)_t - \theta_j\right]^2}{\sum_{t=1}^{\Theta}\left[\left(\hat{\theta}_j^{*(r)}\right)_t - \theta_j\right]^2} \times 100\%, \quad j = 1,2,3. \tag{36}$$

where $\theta_1 = \pi_A$, $\theta_2 = \pi_B$, and $\theta_3 = \pi_{AB}$, and $\Theta = 2000$ is the number of simulated samples. The percent relative bias in the range restricted estimators is computed as:

$$\mathrm{RB}\left(\hat{\theta}_j^{*(r)}\right) = \frac{\frac{1}{\Theta}\sum_{t=1}^{\Theta}\left(\hat{\theta}_j^{*(r)}\right)_t - \theta_j}{\theta_j} \times 100\%, \quad \text{for } j = 1,2,3. \tag{37}$$

The FORTRAN codes used to find the percent relative efficiency values and the percent relative bias are given in PROGRAM_5.F95. In the case of tiny proportions $\pi_A = 0.1$, $\pi_B = 0.1$, and $\pi_{AB} = 0.05$, the results so obtained are presented in Table 3.

TABLE 3 Sample Size vs Recent Relative Bias and Percent Relative Efficiency of the Restricted Maximum Likelihood Estimates

n	$\mathrm{RB}\left(\hat{\theta}_1^{*(r)}\right)$	$\mathrm{RB}\left(\hat{\theta}_2^{*(r)}\right)$	$\mathrm{RB}\left(\hat{\theta}_3^{*(r)}\right)$	$\mathrm{RE}^{*(r)}(1)$	$\mathrm{RE}^{*(r)}(2)$	$\mathrm{RE}^{*(r)}(3)$
700	−2.561	−2.301	−0.522	86.3	85.3	105.1
750	2.108	1.851	2.009	105.7	103.9	105.5
800	2.012	2.026	2.065	103.7	103.1	102.9
850	0.039	−0.018	−1.703	107.7	106.5	103.9
900	1.118	1.879	0.024	102.5	103.6	102.6
950	0.423	0.185	−0.011	103.3	104.4	101.5
1000	0.370	0.001	−0.405	104.7	103.8	102.2

Continued

TABLE 3 Sample Size vs Recent Relative Bias and Percent Relative Efficiency of the Restricted Maximum Likelihood Estimates—cont'd

n	$RB\left(\hat{\theta}_1^{*(r)}\right)$	$RB\left(\hat{\theta}_2^{*(r)}\right)$	$RB\left(\hat{\theta}_3^{*(r)}\right)$	$RE^{*(r)}(1)$	$RE^{*(r)}(2)$	$RE^{*(r)}(3)$
1050	0.067	−0.099	−1.602	104.2	103.7	101.4
1100	−0.171	−0.311	−0.398	103.0	102.0	100.7
1150	0.581	1.258	0.878	104.6	103.4	101.6
1200	0.129	0.126	0.726	101.7	101.2	101.4
1250	−0.551	−0.413	−1.249	104.2	103.8	101.3
1300	0.555	0.952	1.205	101.7	101.4	100.6
1350	0.830	0.231	0.999	102.8	102.4	101.0
1400	0.222	0.114	0.075	102.4	102.4	100.5
1450	−0.219	0.165	−0.741	102.3	101.9	100.5
1500	−0.016	−0.151	−1.104	101.1	101.1	100.1
1550	1.037	−0.007	0.930	101.5	101.9	100.6
1600	−0.046	−0.248	−0.098	102.1	102.5	100.3
1650	0.405	0.289	0.407	100.0	100.0	100.0
1700	0.083	−0.181	−0.379	100.4	100.6	100.3
1750	0.558	0.627	0.945	100.9	100.7	100.1
1800	0.133	0.316	−0.263	100.5	100.9	100.2
1850	−0.399	−0.276	−0.285	100.5	100.4	100.1
1900	0.714	0.708	1.214	100.4	100.6	100.0
1950	0.171	−0.073	−0.336	100.7	100.5	100.1
2000	0.172	0.342	0.243	100.0	100.0	100.0
2050	0.482	0.350	0.704	100.3	100.1	100.1
2100	1.271	0.931	1.880	100.5	100.2	100.1
2150	0.436	−0.252	0.196	101.5	100.8	100.5
2200	0.152	−0.305	−0.107	100.0	100.0	100.0
2250	0.403	0.072	0.469	100.0	100.0	100.0
2300	−0.659	−0.944	−1.082	100.0	100.0	100.0
2350	−0.500	−0.119	−0.428	100.0	100.0	100.0

Continued

TABLE 3 Sample Size vs Recent Relative Bias and Percent Relative Efficiency of the Restricted Maximum Likelihood Estimates—cont'd

n	$RB\left(\hat{\theta}_1^{*(r)}\right)$	$RB\left(\hat{\theta}_2^{*(r)}\right)$	$RB\left(\hat{\theta}_3^{*(r)}\right)$	$RE^{*(r)}(1)$	$RE^{*(r)}(2)$	$RE^{*(r)}(3)$
2400	0.182	−0.034	0.672	100.0	100.0	100.0
2450	0.137	0.498	1.329	100.0	100.0	100.0
2500	−0.480	−0.504	−1.087	100.0	100.0	100.0
2550	−0.165	−0.194	−0.206	100.2	100.5	100.1
2600	0.501	0.607	0.271	100.0	100.0	100.0
2650	−0.310	−0.321	−0.364	100.0	100.0	100.0
2700	0.030	−0.103	−0.186	100.0	100.0	100.0
2750	−0.132	0.091	0.064	100.0	100.0	100.0
2800	−0.324	−0.338	−0.217	100.0	100.0	100.0
2850	−0.008	−0.257	−0.514	100.0	100.0	100.0
2900	0.312	0.140	0.601	100.0	100.0	100.0
2950	0.586	0.414	0.593	100.0	100.0	100.0
3000	0.040	−0.003	−0.477	100.0	100.0	100.0

The results presented in Table 3 show that the percent relative bias value remains negligible in the case of restricted maximum likelihood estimates, and there is a gain in relative efficiency of at least 103% for a sample size in the range of 750 and 1150, but as soon as the sample size becomes 2000, then both the restricted and unrestricted maximum likelihood estimates become equally efficient. If the sample size is less than 700, then the first two restricted maximum likelihood estimators remain less efficient than the unrestricted maximum likelihood estimators, whereas the third is shown to be a more efficient estimator. Note that in Table 3, we have considered a very extreme situation with tiny proportions. Next we change the true proportions to be estimated to: $\pi_A = 0.3$, $\pi_B = 0.4$, and $\pi_{AB} = 0.2$; the results so obtained are presented in Table 4.

The results in Table 4 suggest that if the true proportions are not too small, then a sample size of 300 units is sufficient to show that the proposed restricted and unrestricted maximum likelihood estimates are equally efficient. Thus based on our simulation study, we conclude that in randomized

TABLE 4 Sample Size vs Percent Relative Bias and Percent Relative Efficiency of the Restricted Maximum Likelihood Estimates

n	$RB\left(\hat{\theta}_1^{*(r)}\right)$	$RB\left(\hat{\theta}_2^{*(r)}\right)$	$RB\left(\hat{\theta}_3^{*(r)}\right)$	$RE^{*(r)}(1)$	$RE^{*(r)}(2)$	$RE^{*(r)}(3)$
150	−0.1906	−0.1114	−0.2626	100.3	100.0	100.2
200	0.4208	0.0715	0.0496	100.2	100.0	100.2
250	−0.4472	−0.2841	−0.3883	100.0	100.0	100.2
300	0.3583	0.0884	0.4729	100.0	100.0	100.0
350	0.1649	0.1924	0.0562	100.0	100.0	100.0
400	0.0133	0.0753	0.2905	100.0	100.0	100.0
450	−0.5400	−0.2844	−0.2173	100.0	100.0	100.0
500	0.3740	0.1294	0.4702	100.0	100.0	100.0

response sampling, based on a prior good guess of the true proportions being estimated, an adjustment of the requirement of sample size is necessary when using the unrestricted maximum likelihood estimates; otherwise, the use of the restricted maximum likelihood estimates is recommended. We remark that although the range restricted maximum likelihood estimates is not always the more efficient method compared to the unrestricted maximum likelihood estimate, they do guarantee admissible estimates from a given sample and hence should be preferred over the unrestricted maximum likelihood estimator.

ACKNOWLEDGMENT

The authors are thankful to Prof. Arijit Chaudhuri, Purnima Shaw, and a referee for valuable comments on the original version of this chapter.

TABLE A1 Unique Estimates of $\pi_{AB} = 0.05$, $\pi_A = 0.1$, and $\pi_B = 0.1$ Obtained by Least Square and Maximum Likelihood Methods

$\hat{\pi}_{AB}^{*(mle)}$	$\hat{\pi}_{A}^{*(mle)}$	$\hat{\pi}_{B}^{*(mle)}$	$\hat{\pi}_{AB}^{*}$	$\hat{\pi}_{A}^{*}$	$\hat{\pi}_{B}^{*}$	$\hat{\pi}_{AB}^{*(mle)}$	$\hat{\pi}_{A}^{*(mle)}$	$\hat{\pi}_{B}^{*(mle)}$	$\hat{\pi}_{AB}^{*}$	$\hat{\pi}_{A}^{*}$	$\hat{\pi}_{B}^{*}$
	$n = 1000$						$n = 1200$				
0.05082	0.09825	0.12925	0.05082	0.09825	0.12925	0.07209	0.10187	0.11271	0.07209	0.10187	0.11271
0.02517	0.06500	0.07500	0.02517	0.06500	0.07500	0.03764	0.07083	0.07917	0.03764	0.07083	0.07917
0.04716	0.12050	0.09450	0.04716	0.12050	0.09450	0.05517	0.10250	0.11417	0.05517	0.10250	0.11417
0.08233	0.12550	0.14950	0.08233	0.12550	0.14950	0.06850	0.12104	0.13521	0.06850	0.12104	0.13521
0.05806	0.10625	0.10125	0.05806	0.10625	0.10125	0.04515	0.08229	0.06979	0.04515	0.08229	0.06979
0.03289	0.08725	0.08025	0.03289	0.08725	0.08025	0.04415	0.11146	0.09896	0.04415	0.11146	0.09896
0.07789	0.12725	0.13025	0.07789	0.12725	0.13025	0.01501	0.04958	0.03792	0.01501	0.04958	0.03792
0.03194	0.07175	0.10075	0.03194	0.07175	0.10075	0.04713	0.14500	0.12167	0.04713	0.14500	0.12167
0.09116	0.13025	0.15725	0.09116	0.13025	0.15725	0.05255	0.09771	0.08354	0.05255	0.09771	0.08354
0.07616	0.14125	0.11625	0.07616	0.14125	0.11625	0.04777	0.07875	0.10875	0.04777	0.07875	0.10875
0.06789	0.11325	0.12425	0.06789	0.11325	0.12425	0.04749	0.10375	0.10042	0.04749	0.10375	0.10042
0.02655	0.10800	0.05200	0.02655	0.10800	0.05200	0.04364	0.08771	0.11021	0.04364	0.08771	0.11021
0.02056	0.06675	0.06575	0.02056	0.06675	0.06575	0.06674	0.13208	0.12208	0.06674	0.13208	0.12208
−0.01168	0.04875	0.05375	−0.01168	0.04875	0.05375	0.01645	0.06542	0.10542	0.01645	0.06542	0.10542
0.06091	0.10075	0.09175	0.06091	0.10075	0.09175	0.03279	0.09729	0.10479	0.03279	0.09729	0.10479
0.02940	0.07550	0.06950	0.02940	0.07550	0.06950	0.06171	0.09792	0.11458	0.06171	0.09792	0.11458
0.06582	0.13825	0.11925	0.06582	0.13825	0.11925	0.06767	0.12083	0.12083	0.06767	0.12083	0.12083
0.04108	0.08075	0.06175	0.04108	0.08075	0.06175	0.06965	0.12687	0.14604	0.06965	0.12687	0.14604
0.03720	0.09925	0.11825	0.03720	0.09925	0.11825	0.04619	0.09375	0.10208	0.04619	0.09375	0.10208
0.01849	0.07275	0.07975	0.01849	0.07275	0.07975	0.05639	0.10875	0.07875	0.05639	0.10875	0.07875

Continued

TABLE A1 Unique Estimates of $\pi AB=0.05$, $\pi A=0.1$, and $\pi B=0.1$ Obtained by Least Square and Maximum Likelihood Methods—cont'd

	$n=1100$						$n=1500$				
0.05404	0.09386	0.11295	0.05404	0.09386	0.11295	0.04029	0.08633	0.07700	0.04029	0.08633	0.07700
0.07563	0.11318	0.09136	0.07563	0.11318	0.09136	0.06399	0.10950	0.09550	0.06399	0.10950	0.09550
0.04804	0.08500	0.08318	0.04804	0.08500	0.08318	0.05991	0.10783	0.10050	0.05991	0.10783	0.10050
0.04879	0.11614	0.07705	0.04879	0.11614	0.07705	0.04980	0.10117	0.09383	0.04980	0.10117	0.09383
0.05827	0.12295	0.11114	0.05827	0.12295	0.11114	0.05842	0.11383	0.11450	0.05842	0.11383	0.11450
0.07394	0.10114	0.13295	0.07394	0.10114	0.13295	0.03152	0.09317	0.09517	0.03152	0.09317	0.09517
0.09557	0.19023	0.15295	0.09557	0.19023	0.15295	0.05615	0.09100	0.08567	0.05615	0.09100	0.08567
0.03777	0.08727	0.08545	0.03777	0.08727	0.08545	0.03184	0.06667	0.08667	0.03184	0.06667	0.08667
0.06266	0.13386	0.11841	0.06266	0.13386	0.11841	0.05799	0.12100	0.08233	0.05799	0.12100	0.08233
0.04706	0.11159	0.09068	0.04706	0.11159	0.09068	0.04770	0.09000	0.07667	0.04770	0.09000	0.07667
0.07778	0.09977	0.13886	0.07778	0.09977	0.13886	0.06420	0.11633	0.12700	0.06420	0.11633	0.12700
0.04835	0.06591	0.08409	0.04835	0.06591	0.08409	0.05216	0.10083	0.12417	0.05216	0.10083	0.12417
0.03738	0.06318	0.08682	0.03738	0.06318	0.08682	0.04253	0.06400	0.06933	0.04253	0.06400	0.06933
0.03656	0.08795	0.08705	0.03656	0.08795	0.08705	0.03681	0.07850	0.06983	0.03681	0.07850	0.06983
0.04495	0.10659	0.08205	0.04495	0.10659	0.08205	0.04977	0.11267	0.06733	0.04977	0.11267	0.06733
0.04953	0.10273	0.11545	0.04953	0.10273	0.11545	0.01057	0.06200	0.05800	0.01057	0.06200	0.05800
0.03632	0.06568	0.07750	0.03632	0.06568	0.07750	0.04917	0.09750	0.10083	0.04917	0.09750	0.10083
0.02876	0.10591	0.08955	0.02876	0.10591	0.08955	0.06325	0.14517	0.13650	0.06325	0.14517	0.13650
0.07100	0.10909	0.13636	0.07100	0.10909	0.13636	0.05359	0.07717	0.09783	0.05359	0.07717	0.09783
0.06132	0.08523	0.10795	0.06132	0.08523	0.10795	0.05491	0.08683	0.11150	0.05491	0.08683	0.11150

APPENDIX B CODES USED IN SIMULATION STUDIES

!PROGRAM_1.F95	!PROGRAM_2.F95
!FORTRAN CODES USED FOR SIMPLE MODEL	!FORTRAN CODES USED FOR CROSSED MODEL
USE NUMERICAL_LIBRARIES	USE NUMERICAL_LIBRARIES
IMPLICIT NONE	IMPLICIT NONE
REAL P, T, PIA,PIB,PIAB,TH11,TH10,TH01,TH00	REAL P, T, PIA,PIB,PIAB,TH11S,TH10S,TH01S,TH00S
REAL SUMTH, A, B, C, D, F, G, VARPIA, VARPIB,	REAL SUMTH, VPIAS,VPIBS, VPIABS
1 VARPIAB, DELTA1,VALOW, VBLOW,VABLOW, REA,	REAL AS, BS, CS, DS, FS, GS
1 REB, REAB	REAL DELTA1, VALOW, REAS, VBLOW, REBS, VABLOW, REABS
CHARACTER*20 OUT_FILE	CHARACTER*20 OUT_FILE
CHARACTER*20 IN_FILE	CHARACTER*20 IN_FILE
WRITE(*,'(A)') 'NAME OF THE OUTPUT FILE'	WRITE(*,'(A)') 'NAME OF THE OUTPUT FILE'
READ(*,'(A20)') OUT_FILE	READ(*,'(A20)') OUT_FILE
OPEN(42, FILE=OUT_FILE, STATUS='UNKNOWN')	OPEN(42, FILE=OUT_FILE, STATUS='UNKNOWN')
P = 0.7	P = 0.7
DO 9999 T = 0.6, 0.8, 0.05	DO 9999 T = 0.6, 0.8, 0.05
PIA = 0.25	PIA = 0.25
PIB = 0.20	PIB = 0.20
DO 8888 PIAB = 0.05, 0.20, 0.05	DO 8888 PIAB = 0.05, 0.20, 0.05
TH11 = (2*P-1)*(2*T-1)*PIAB+(2*P-1)*(1-T)*PIA	TH11S = PIAB*(P*T+(1-P)*(1-T))-PIA*(1-P)*(1-T)
1 + (1-P)*(2*T-1)*PIB+(1-P)*(1-T)	1 -PIB*(1-P)*(1-T)+(1-P)*(1-T)
TH10 = -(2*P-1)*(2*T-1)*PIAB+(2*P-1)*T*PIA	TH10S = -PIAB*(P*T+(1-P)*(1-T))-PIA*((1-P)*T-1)

Continued

```
     1 - (1-P)*(2*T-1)*PIB+(1-P)*T
      TH01 = -(2*P-1)*(2*T-1)*PIAB-(2*P-1)*(1-T)
*PIA
     1 + P*(2*T-1)*PIB+P*(1-T)
      TH00 = (2*P-1)*(2*T-1)*PIAB-T*(2*P-1)*PIA
     1 -P*(2*T-1)*PIB+P*T
      SUMTH = TH11+TH10+TH01+TH00
      VARPIA = PIA*(1-PIA)+P*(1-P)/(2*P-1)**2
      VARPIB = PIB*(1-PIB)+T*(1-T)/(2*T-1)**2
      VARPIAB = PIAB*(1-PIAB)
     1 + ( (2*P-1)**2*T*(1-T)*PIA + P*(1-P)*(2*T-1)
**2*PIB
     1 + P*T*(1-P)*(1-T))/ ( (2*P-1)**2*(2*T-1)**2 )
      A = (2*P-1)**2*(2*T-1)**2* (1./TH11+ 1./TH10+
1./TH01
     1 +1./TH00)
      B = (2*P-1)**2*(2*T-1)*((1-T)/TH11-T/TH10+(1-
T)/
     1 TH01-T/TH00)
      C = (2*P-1)*(2*T-1)**2*((1-P)/TH11+(1-P)/TH10
     1-P/TH01-P/TH00)
      D =(2*P-1)**2*((1-T)**2/TH11+T**2/TH10+(1-T)
**2
     1 /TH01+T**2/TH00)
     1 -PIB*(1-P)*T+(1-P)*T
      TH01S = -PIAB*(P*T+(1-P)*(1-T))-PIA*P*(1-T)
     1 -PIB*(P*(1-T)-1)+P*(1-T)
      TH00S = PIAB*(P*T+(1-P)*(1-T))-PIA*P*T-PIB*P*T
+P*T
      SUMTH = TH11S+TH10S+TH01S+TH00S
      VPIAS = PIA*(1-PIA)
     1 + (1-P)*T*( P*T+(1-P)*(1-T) )*(1-PIA-PIB+2.
*PIAB)/(P+T-1)**2
      VPIBS = PIB*(1-PIB)
     1 + (1-T)*P*( P*T+(1-P)*(1-T) )*(1-PIA-PIB+2.
*PIAB)/(P+T-1)**2
      VPIABS = PIAB*(1-PIAB)
     1 + ( PIAB*( P**2*T**2 + (1-P)**2*(1-T)**2
     1 - ( P*T+(1-P)*(1-T) )*(P+T-1)**2 )
     1 + P*T*(1-P)*(1-T)*(1-PIA-PIB) )
     1/ (( P*T+(1-P)*(1-T)) * (P+T-1)**2)
      AS = (P*T+(1-P)*(1-T))**2*(1/TH11S+1/TH10S+1/
TH01S+1/TH00S)
      BS = - (P*T+(1-P)*(1-T))*( (1-P)*(1-T)/TH11S
     1 - ((1-P)*T-1)/TH10S -P*(1-T)/TH01S+P*T/TH00S)
      CS = -(P*T+(1-P)*(1-T))*( (1-P)*(1-T)/TH11S
     1 -(1-P)*T/TH10S - (P*(1-T)-1)/TH01S + P*T/TH00S)
```

`F = (2*P-1)*(2*T-1)*((1-P)*(1-T)/TH11-T*(1-P)/TH10`
`1-P*(1-T)/TH01 + P*T/TH00)`
`G =(2*T-1)**2*((1-P)**2/TH11+(1-P)**2/TH10+P**2/TH01`
`1 +P**2/TH00)`
`DELTA1 = G*B**2-2.0*B*C*F+D*C**2+A*F**2-A*D*G`
`VALOW = (C**2-A*G)/DELTA1`
`VBLOW = (B**2-A*D)/DELTA1`
`VABLOW = (F**2-D*G)/DELTA1`
`REA = VARPIA*100.0/VALOW`
`REB = VARPIB*100.0/VBLOW`
`REAB = VARPIAB*100.0/VABLOW`
`WRITE(42,101)PIA, PIB, PIAB, P, T, REA, REB, REAB`
`101  FORMAT(2X, 5(F8.5,1X), 2X, 3(F9.2,1X))`
`8888 CONTINUE`
`9999 CONTINUE`
`STOP`
`END`
	`DS = (1-P)**2*(1-T)**2/TH11S + ((1-P)*T-1)**2/TH10S`
	`1 + P**2*(1-T)**2/TH01S + P**2*T**2/TH00S`
	`FS = (1-P)**2*(1-T)**2/TH11S + (1-P)*T*((1-P)*T-1)/TH10S`
	`1 + P*(1-T)*( P*(1-T)-1)/TH01S + P**2*T**2/TH00S`
	`GS = (1-P)**2*(1-T)**2/TH11S + (1-P)**2*T**2/TH10S`
	`1 + ( (1-T)*P-1)**2/TH01S + P**2* T**2/TH00S`
	`DELTA1 = GS*BS**2-2.*BS*CS*FS+DS*CS**2+AS*FS**2-AS*DS*GS`
	`VALOW = (CS**2-AS*GS)/DELTA1`
	`VBLOW = (BS**2-AS*DS)/DELTA1`
	`VABLOW = (FS**2-DS*GS)/DELTA1`
	`REAS = VPIAS*100/VALOW`
	`REBS = VPIBS*100/VBLOW`
	`REABS = VPIABS*100/VABLOW`
	`WRITE(42,101)PIA, PIB, PIAB, P, T, REAS,REBS,REABS`
	`101  FORMAT(2X,5(F8.5,1X),2X,3(F9.2,1X))`
	`8888 CONTINUE`
	`9999 CONTINUE`
	`END`

Continued

```
!PROGRAM_2.F95
!FORTRAN CODES USED FOR CROSSED MODEL
    USE NUMERICAL_LIBRARIES
    IMPLICIT NONE
    REAL P, T, PIA,PIB,PIAB,TH11S,TH10S,TH01S,
THOOS
    REAL SUMTH, VPIAS,VPIBS, VPIABS
    REAL AS, BS, CS, DS, FS, GS
    REAL DELTA1, VALOW, REAS, VBLOW, REBS, VABLOW,
REABS
    CHARACTER*20 OUT_FILE
    CHARACTER*20 IN_FILE
    WRITE(*,'(A)') 'NAME OF THE OUTPUT FILE'
    READ(*,'(A20)') OUT_FILE
    OPEN(42, FILE=OUT_FILE, STATUS='UNKNOWN')
    P = 0.7
    DO 9999 T = 0.6, 0.8, 0.05
    PIA = 0.25
    PIB = 0.20
    DO 8888 PIAB = 0.05, 0.20, 0.05
    TH11S = PIAB*(P*T+(1-P)*(1-T))-PIA*(1-P)*(1-T)
   1 -PIB*(1-P)*(1-T)+(1-P)*(1-T)
```

```
!PROGRAM_4.F95
! FORTRAN CODES USED FOR SAMPLE SIZE DETERMINATION
    USE NEQNF_INT
    IMPLICIT NONE
    INTEGER N
    PARAMETER (N=3)
    INTEGER K, NOUT, IR(5000,4),M,LDIR,I,J,NOS
    REAL PIABLEE(5000),PIALEE(5000),PIBLEE(5000)
    REAL X(N),FNORM,PN(4)
    REAL XGUESS(N)
    REAL F(N)
    REAL TH11C,TH10C,TH01C,THOOC,PIAB,PIA,PIB
    REAL TH11S, TH10S, TH01S, THOOS
    COMMON /THETAS/TH11S, TH10S, TH01S, THOOS
    REAL P,T, W1
    COMMON /DEVICE/P,T,W1
    REAL TH11H, TH10H, TH01H, THOOH
    COMMON /THETA/TH11H, TH10H, TH01H, THOOH
    REAL PIABE(5000),PIAE(5000),PIBE(5000)
    CHARACTER*20 OUT_FILE
    EXTERNAL FCN
    DATA XGUESS/0.21,0.62,0.11/
```

```
      TH10S = -PIAB*(P*T+(1-P)*(1-T))-PIA*((1-P)
*T-1)
    1 -PIB*(1-P)*T+(1-P)*T
      TH01S = -PIAB*(P*T+(1-P)*(1-T))-PIA*P*(1-T)
    1 -PIB*(P*(1-T)-1)+P*(1-T)
      TH00S = PIAB*(P*T+(1-P)*(1-T))-PIA*P*T-
PIB*P*T+P*T
      SUMTH = TH11S+TH10S+TH01S+TH00S
      VPIAS = PIA*(1-PIA)
    1 + (1-P)*T*( P*T+(1-P)*(1-T) )*(1-PIA-PIB+2.
*PIAB)/(P+T-1)**2
      VPIBS = PIB*(1-PIB)
    1 + (1-T)*P*( P*T+(1-P)*(1-T) )*(1-PIA-PIB+2.
*PIAB)/(P+T-1)**2
      VPIABS = PIAB*(1-PIAB)
    1 + ( PIAB*( P**2*T**2 + (1-P)**2*(1-T)**2
    1 - ( P*T+(1-P)*(1-T) )*(P+T-1)**2 )
    1 + P*T*(1-P)*(1-T)*(1-PIA-PIB) )
    1/ (( P*T+(1-P)*(1-T)) * (P+T-1)**2)
      AS = (P*T+(1-P)*(1-T))**2*(1/TH11S+1/TH10S
+1/TH01S+1/TH00S)
      BS = - (P*T+(1-P)*(1-T))*( (1-P)*(1-T)/
TH11S
```

```
      WRITE(*,'(A)') 'NAME OF THE OUTPUT FILE'
      READ(*,'(A20)')OUT_FILE
      OPEN(42,FILE=OUT_FILE, STATUS='UNKNOWN')
      DO 9999 M = 1000, 1500, 100
      K = 4
      NOS = 20
      P = 0.7
      T = 0.7
      W1 = P*T+(1-P)*(1-T)
      PIAB = 0.05
      PIA = 0.1
      PIB = 0.1
      TH11C = PIAB*W1-PIA*(1.-P)*(1.-T)-PIB*(1.-P)*
(1.-T)
    1 +(1.-P)*(1.-T)
      TH10C = -PIAB*W1-PIA*((1.-P)*T-1.)-PIB*(1.-P)*T
+(1.-P)*T
      TH01C = -PIAB*W1-PIA*P*(1.-T)-PIB*(P*(1.-T)-1)
+P*(1.-T)
      TH00C = PIAB*W1-PIA*P*T-PIB*P*T+P*T
      PN(1) = TH11C
      PN(2) = TH10C
```

Continued

```
     1 - ((1-P)*T-1)/TH10S -P*(1-T)/TH01S+P*T/
THOOS)
      CS = -(P*T+(1-P)*(1-T))*( (1-P)*(1-T)/TH11S
     1 -(1-P)*T/TH10S - (P*(1-T)-1)/TH01S + P*T/
THOOS)
      DS = (1-P)**2*(1-T)**2/TH11S + ((1-P)*T-1)
**2/TH10S
     1 + P**2*(1-T)**2/TH01S + P**2*T**2/THOOS
      FS = (1-P)**2*(1-T)**2/TH11S + (1-P)*T*
((1-P)*T-1)/TH10S
     1 + P*(1-T)*( P*(1-T)-1)/TH01S + P**2*T**2/
THOOS
      GS = (1-P)**2*(1-T)**2/TH11S + (1-P)
**2*T**2/TH10S
     1 + ( (1-T)*P-1)**2/TH01S + P**2* T**2/THOOS
      DELTA1 = GS*BS**2-2.*BS*CS*FS+DS*CS**2
+AS*FS**2-AS*DS*GS
      VALOW = (CS**2-AS*GS)/DELTA1
      VBLOW = (BS**2-AS*DS)/DELTA1
      VABLOW = (FS**2-DS*GS)/DELTA1
      REAS = VPIAS*100/VALOW
      REBS = VPIBS*100/VBLOW
      REABS = VPIABS*100/VABLOW
      PN(3) = TH01C
      PN(4) = THOOC
      LDIR = 5000
      CALL RNMTN(NOS,M,K,PN,IR,LDIR)
      DO 12 I =1, NOS
      TH11H = DBLE(IR(I,1))/DBLE(M)
      TH10H = DBLE(IR(I,2))/DBLE(M)
      TH01H = DBLE(IR(I,3))/DBLE(M)
      THOOH = DBLE(IR(I,4))/DBLE(M)
      PIABLEE(I) = (P*T*TH11H-(1-P)*(1-T)*THOOH)/(W1*(P
+T-1))
      PIALEE(I) = 0.5 + ( (T-P+1)*(TH11H-THOOH)
     1 + (P+T-1)*(TH10H-TH01H) )/ (2*(P+T-1))
      PIBLEE(I) = 0.5 + ( (P-T+1)*(TH11H-THOOH)
     1 + (P+T-1)*(TH01H-TH10H) )/ (2*(P+T-1))
      CALL NEQNF(FCN, X, XGUESS=XGUESS, FNORM=FNORM)
      PIABE(I) = X(1)
      PIAE(I) = X(2)
      PIBE(I) = X(3)
12    CONTINUE
      WRITE(42,*)M
      DO 107 I=1,NOS
```

WRITE(42,101)PIA, PIB, PIAB, P, T, REAS,REBS, REABS 101 FORMAT(2X,5(F8.5,1X),2X,3(F9.2,1X)) 8888 CONTINUE 9999 CONTINUE END *PROGRAM_3.SAS *SAS CODES USED TO OBTAIN MAXIMUM LIKELIHOOD ESTIMATES FROM THE CROSSED MODEL **DATA** DATA1; TH11H = **13/75**; TH10H = **14/75**; TH01H = **23/75**; TH00H = **25/75**; P = **0.7**; T = **0.7**; **RUN**; **PROC MODEL** DATA=DATA1; EQ.A = TH11H/(X*(P*T+(1-P)*(1-T))-Y*(1-P)*(1-T)-Z*(1-P)*(1-T) +(1-P)*(1-T)) - TH10H/(-X*(P*T+(1-P)*(1-T))-Y*((1-P)*T-1)-Z*(1-P)*T +(1-P)*T)- TH01H/(-X*(P*T+(1-P)*(1-T))-Y*P*	WRITE(42,999)PIABE(I),PIAE(I),PIBE(I),PIABLEE(I), PIALEE(I), 1 PIBLEE(I) 999 FORMAT(2X,7(F9.5,1X)) 107 CONTINUE 9999 CONTINUE STOP END SUBROUTINE FCN (X, F, N) INTEGER N REAL X(N), F(N) REAL TH11H, TH10H, TH01H, TH00H COMMON /THETA/TH11H, TH10H, TH01H, TH00H REAL TH11S, TH10S, TH01S, TH00S COMMON /THETAS/TH11S, TH10S, TH01S, TH00S REAL P,T, W1 COMMON /DEVICE/P,T,W1 TH11S= X(1)*W1-X(2)*(1.0-P)*(1.0-T)-X(3)*(1.0-P)* (1.0-T) 1 +(1.0-P)*(1.0-T) TH10S=-X(1)*W1-X(2)*((1.0-P)*T-1)-X(3)*(1.0-P)*T +(1.0-P)*T

Continued

```
(1-T)-Z*(P*(1-T)-1)+P*(1-T) )+ TH00H/( X*(P*T
+(1-P)*(1-T))-Y*P*T-Z*P*T+P*T ) ;

EQ.B = TH11H*(1-P)*(1-T)/( X*(P*T+(1-P)*(1-T))-
Y*(1-P)*(1-T)-Z*(1-P)*(1-T)+(1-P)*(1-T) ) +
TH10H*((1-P)*T-1)/(-X*(P*T+(1-P)*(1-T))-Y*
((1-P)*T-1)-Z*(1-P)*T+(1-P)*T ) + TH01H*P*(1-T)/
( -X* (P*T+(1-P)*(1-T))-Y*P*(1-T)-Z*(P*(1-T)-1)
+P*(1-T) ) + TH00H * P*T/( X*(P*T+(1-P)*(1-T))-
Y*P*T-Z*P*T+P*T ) ;

EQ.C = TH11H*(1-P)*(1-T)/( X*(P*T+(1-P)*(1-T) )-
Y*(1-P)*(1-T)-Z*(1-P)*(1-T)+(1-P)*(1-T) ) +
TH10H*(1-P)*T/( -X*(P*T+(1-P)*(1-T))-Y*((1-P)
*T-1)-Z*(1-P)*T+(1-P)*T ) + TH01H*(P*(1-T)-1)/ (
-X* (P*T+(1-P)*(1-T))-Y*P*(1-T)-Z*(P*(1-T)-1)
+P*(1-T) )+ TH00H* P*T/( X* (P*T+(1-P)*(1-T))-
Y*P*T-Z*P*T+P*T );

SOLVE X Y Z /SOLVEPRINT;

RUN;

      TH01S=-X(1)*W1-X(2)*P*(1.0-T)-X(3)*(P*(1.0-T)-
     1.)+P*(1.0-T)

      TH00S= X(1)*W1-X(2)*P*T-X(3)*P*T+P*T

      F(1) = TH11H/TH11S-TH10H/TH10S-TH01H/TH01S

     1+TH00H/TH00S

      F(2) = TH11H*(1.0-P)*(1.0-T)/TH11S+TH10H*((1.0-P)
     *T-1.0) /

     1 TH10S + TH01H*P*(1.0-T)/TH01S + TH00H*P*T/TH00S

      F(3) = TH11H*(1.0-P)*(1.0-T)/TH11S + TH10H*(1.0-P)
     *T/TH10S

     1 + TH01H*(P*(1.0-T)-1.0)/TH01S + TH00H*P*T/TH00S

      RETURN

      END
```

```
!PROGRAM_5.F95
! PROGRAM USED TO COMPARE RESTRICED AND UNRESTRICED MLEs
      USE NEQNF_INT
      IMPLICIT NONE
      INTEGER N
      REAL PROPAB,PROPA, PROPB
      PARAMETER (N=3)
      INTEGER K, NOUT, IR(5000,4),M,LDIR,I,J,NOS
      REAL PIABLEE(5000),PIALEE(5000),PIBLEE(5000)
      REAL X(N),FNORM,PN(4),PIABR(5000),PIAR(5000)
      REAL PIBR(5000)
      REAL XGUESS(N)
      REAL F(N)
      REAL TH11C,TH10C,TH01C,TH00C,PIAB,PIA,PIB
      REAL TH11S, TH10S, TH01S, TH00S
      COMMON /THETAS/TH11S, TH10S, TH01S, TH00S
      REAL P,T, W1
      COMMON /DEVICE/P,T,W1
      REAL TH11H, TH10H, TH01H, TH00H
      COMMON /THETA/TH11H, TH10H, TH01H, TH00H
      REAL PIABE(5000),PIAE(5000),PIBE(5000)
      REAL SPIABR,SPIAR,SPIBR, RBABR,RBAR,RBBR
      REAL AMABR,AMAR,AMBR,REA,REB,REAB
      REAL AMABML,AMAML,AMBML
      CHARACTER*20 OUT_FILE
      EXTERNAL FCN
      DATA XGUESS/0.21,0.62,0.11/
       WRITE(*,'(A)') 'NAME OF THE OUTPUT FILE'
       READ(*,'(A20)')OUT_FILE
       OPEN(42,FILE = OUT_FILE, STATUS = 'UNKNOWN')
       DO 9999 M = 700, 3000, 50
       K = 4
       NOS = 2000
       P = 0.7
       T = 0.7
      W1 = P*T+(1-P)*(1-T)
       PIAB = 0.05
```

Continued

```
      PIA = 0.1
      PIB = 0.1
     TH11C = PIAB*W1-PIA*(1.-P)*(1.-T)-PIB*(1.-P)*(1.-T)
    1 + (1.-P)*(1.-T)
     TH10C = -PIAB*W1-PIA*((1.-P)*T-1.)-PIB*(1.-P)*T+(1.-P)*T
     TH01C = -PIAB*W1-PIA*P*(1.-T)-PIB*(P*(1.-T)-1) + P*(1.-T)
     TH00C = PIAB*W1-PIA*P*T-PIB*P*T + P*T
     PN(1) = TH11C
     PN(2) = TH10C
     PN(3) = TH01C
     PN(4) = TH00C
     LDIR = 5000
     CALL RNMTN(NOS,M,K,PN,IR,LDIR)
     DO 12 I =1, NOS
     TH11H = DBLE(IR(I,1))/DBLE(M)
     TH10H = DBLE(IR(I,2))/DBLE(M)
     TH01H = DBLE(IR(I,3))/DBLE(M)
     TH00H = DBLE(IR(I,4))/DBLE(M)
     PIABLEE(I) = (P*T*TH11H-(1-P)*(1-T)*TH00H)/(W1*(P + T-1))
     PIALEE(I) = 0.5 + ( (T-P + 1)*(TH11H-TH00H)
    1 + (P + T-1)*(TH10H-TH01H) )/ (2*(P + T-1))
     PIBLEE(I) = 0.5 + ( (P-T + 1)*(TH11H-TH00H)
    1 + (P + T-1)*(TH01H-TH10H) )/ (2*(P + T-1))
     CALL NEQNF(FCN, X, XGUESS = XGUESS, FNORM = FNORM)
     PIABE(I) = X(1)
     PIAE(I) = X(2)
     PIBE(I) = X(3)
12   CONTINUE
     DO 112 I = 1, NOS
     IF( (PIABE(I).GE.0.0).AND.(PIABE(I).LE.(MIN(PIAE(I),PIBE
(I)))))
    1 THEN
     PIABR(I) = PIABE(I)
     ELSE IF((PIABE(I).GT.MIN(PIAE(I),PIBE(I))).AND.
    1 (MIN(PIAE(I),PIBE(I)).GT.0.0)) THEN
     PIABR(I) = MIN(PIAE(I),PIBE(I))
     ELSE IF (PIABE(I).LE.0.0) THEN
```

Continued

```
      PIABR(I) = 0.0
      END IF
112   CONTINUE
      DO 113 I = 1, NOS
      IF( (PIAE(I).GE.PIABE(I)).AND.(PIAE(I).LE.1.0).AND.
     1 (PIABE(I).GE.0.0) ) THEN
      PIAR(I) = PIAE(I)
      ELSE IF ( (PIAE(I).LT.PIABE(I)).AND.(PIABE(I).GE.0.0) )
THEN
      PIAR(I) = PIABE(I)
      ELSE IF( PIAE(I).LT.0.0 ) THEN
      PIAR(I) = 0.0
      ELSE IF(PIAE(I).GT.1.0) THEN
      PIAR(I) = 1.
      END IF
113   CONTINUE
      DO 114 I = 1, NOS
      IF( (PIBE(I).GE.PIABE(I)).AND.(PIBE(I).LE.1.0).AND.
     1 (PIABE(I).GE.0.0) ) THEN
      PIBR(I) = PIBE(I)
      ELSE IF ( (PIBE(I).LT.PIABE(I)).AND.(PIABE(I).GE.0.0) )
THEN
      PIBR(I) = PIABE(I)
      ELSE IF( PIBE(I).LT.0.0 ) THEN
      PIBR(I) = 0.0
      ELSE IF(PIBE(I).GT.1.0) THEN
      PIBR(I) = 1.
      END IF
114   CONTINUE
      SPIABR = 0.0
      SPIAR = 0.0
      SPIBR = 0.0
       DO 116 I = 1, NOS
       SPIABR = SPIABR + PIABR(I)
       SPIAR = SPIAR + PIAR(I)
116    SPIBR = SPIBR + PIBR(I)
       RBABR = (SPIABR/DBLE(NOS)-PIAB)*100/PIAB
```

Continued

```
      RBAR = (SPIAR/DBLE(NOS)-PIA)*100/PIA
      RBBR = (SPIBR/DBLE(NOS)-PIB)*100/PIB
      AMABR = 0.0
      AMAR = 0.0
      AMBR = 0.0
      DO 117 I = 1, NOS
      AMABR = AMABR + (PIABR(I)-PIAB)**2
      AMAR = AMAR + (PIAR(I)-PIA)**2
117   AMBR = AMBR + (PIBR(I)-PIB)**2
       AMABML = 0.0
       AMAML = 0.0
       AMBML = 0.0
       DO 118 I = 1, NOS
       AMABML = AMABML + (PIABE(I)-PIAB)**2
      AMAML = AMAML + (PIAE(I)-PIA)**2
118   AMBML = AMBML + (PIBE(I)-PIB)**2
      REAB = AMABML*100/AMABR
      REA = AMAML*100/AMAR
      REB = AMBML*100/AMBR
      WRITE(42, 105)M,RBAR,RBBR,RBABR,REA,REB,REAB
105    FORMAT(2X,I6,2X,3(F8.4,1X),2X,3(F9.2,1X))
9999   CONTINUE
       STOP
       END
       SUBROUTINE FCN (X, F, N)
       INTEGER N
       REAL X(N), F(N)
       REAL TH11H, TH10H, TH01H, TH00H
      COMMON /THETA/TH11H, TH10H, TH01H, TH00H
      REAL TH11S, TH10S, TH01S, TH00S
      COMMON /THETAS/TH11S, TH10S, TH01S, TH00S
       REAL P,T, W1
       COMMON /DEVICE/P,T,W1
       TH11S = X(1)*W1-X(2)*(1.0-P)*(1.0-T)-X(3)*(1.0-P)*
(1.0-T)
      1 + (1.0-P)*(1.0-T)
       TH10S = -X(1)*W1-X(2)*((1.0-P)*T-1)-X(3)*(1.0-P)*T
+(1.0-P)*T
```

Continued

```
      TH01S = -X(1)*W1-X(2)*P*(1.0-T)-X(3)*(P*(1.0-T)-1.) +
P*(1.0-T)
      TH00S = X(1)*W1-X(2)*P*T-X(3)*P*T + P*T
      F(1) = TH11H/TH11S-TH10H/TH10S-TH01H/TH01S + TH00H/
TH00S
      F(2) = TH11H*(1.0-P)*(1.0-T)/TH11S + TH10H*((1.0-P)*T-
1.0)/TH10S
     1 + TH01H*P*(1.0-T)/TH01S + TH00H*P*T/TH00S
      F(3) = TH11H*(1.0-P)*(1.0-T)/TH11S + TH10H*(1.0-P)*T/
TH10S
     1 + TH01H*(P*(1.0-T)-1.0)/TH01S + TH00H*P*T/TH00S
      RETURN
      END
```

REFERENCES

Chaudhuri, A., Christofides, T.C., 2013. Indirect Questioning in Sample Surveys. Springer, Heidelberg.

Fox, J.A., 2016. Randomized Response and Related Methods, second ed. Sage, Los Angeles. ISBN: 978-1-4833-8103-9.

Lee, C.S., Sedory, S.A., Singh, S., 2013. Estimating at least seven measures for qualitative variables using randomized response sampling. Stat. Prob. Lett. 83, 399–409.

Singh, S., Sedory, S.A., 2012. A true simulation study of three estimators at equal protection of respondents in randomized response sampling. Stat. Neerlandica 66 (4), 442–451.

Chapter 23

Estimating a Finite Population Proportion Bearing a Sensitive Attribute from a Single Probability Sample by Item Count Technique

P. Shaw[☆,1]
Reserve Bank of India, Mumbai, Maharashtra, India
[1]*Corresponding author: e-mail: purnimashaw2011@gmail.com*

ABSTRACT

Randomized Response Techniques (RRTs) initiated by Warner (1965) have some disadvantages. Some participants either do not understand the RRT procedure or suspect revelation of privacy. Sometimes they end up concluding the randomization procedure as a foul trick. To overcome these problems as well as to shape the whole procedure in the form of canvassing a survey questionnaire, Raghavarao and Federer (1979), Miller (1984), and Miller et al. (1986) introduced the Item Count Technique also known as the List Experiment or the Block Total Response or the Unmatched Count Technique which is user friendly. This method was revised by Chaudhuri and Christofides (2007) which was an improvement over the original Item Count Technique in terms of protection of privacy of the respondents drawn from the population using a general sampling design, whereas the original method was restricted to simple random sampling. But a serious disadvantage is that it requires the selection of two independent samples costing more time and money. Also it needs the knowledge of the population proportion of an innocuous characteristic unrelated to the sensitive characteristic whose proportion is to be estimated. This chapter avoids both the problems.

Keywords: Approximately unbiased, Item Count Technique, Protection of privacy

MSC: 62 DO5

[☆]The views expressed in this chapter are of the author alone and not of the institution, author is working.

1 INTRODUCTION

Let $U=(1, 2, \ldots, i, \ldots, N)$ denote a finite population of a known number N of persons. Consider θ_A as the unknown proportion of a sensitive characteristic A in the population and θ_F as the proportion of an innocuous characteristic F unrelated to A. In the revised Item Count Technique by Chaudhuri and Christofides (2007), θ_F is assumed to be known. Two independent samples s_1 and s_2 are taken from U with probabilities $p(s_1)$ and $p(s_2)$, respectively, according to a design p admitting positive first-order and second-order inclusion probabilities $\pi_i=\sum_{s\ni i}p(s), \pi_{ij}=\sum_{s\ni i,j}p(s), i\neq j$. A questionnaire is provided to the participants in s_1 in which there are G innocuous item-statements and the $(G+1)$th item is—"I have characteristic A or F." Each participant in s_1 responds to the investigator the number of statements out of $(G+1)$ that are valid for him/her without revealing the statements which are valid for him/her.

Another questionnaire is provided to the participants in s_2 in which there are the same G innocuous item-statements as were in the first questionnaire and the $(G+1)$th item is—"I do not have characteristic A or I do not have characteristic F." Each participant in s_2 responds to the investigator the number of statements out of $(G+1)$ that are valid for him/her without revealing the statements applicable to him/her. The set of all possible answers for both the questionnaires is $\{0, 1, 2, \ldots, G-1, G, G+1\}$. An unbiased estimator of θ_A is given by

$$\hat{\theta}_A=\frac{1}{N}\sum_{i\in s_1}\frac{y_i}{\pi_i}-\frac{1}{N}\sum_{j\in s_2}\frac{x_j}{\pi_j}+1-\theta_F$$

such that $E_P\left(\hat{\theta}_A\right)=\theta_A$; here y_i=response given by ith sampled person in s_1 and x_j=response given by jth sampled person in s_2

$$V_P\left(\hat{\theta}_A\right)=V_P\left(\frac{1}{N}\sum_{i\in s_1}\frac{y_i}{\pi_i}\right)+V_P\left(\frac{1}{N}\sum_{j\in s_2}\frac{x_j}{\pi_j}\right)$$

$$=\frac{1}{N^2}\left[\sum_{i}^{N}\sum_{<j}^{N}(\pi_i\pi_j-\pi_{ij})\left\{\left(\frac{y_i}{\pi_i}-\frac{y_j}{\pi_j}\right)^2+\left(\frac{x_i}{\pi_i}-\frac{x_j}{\pi_j}\right)^2\right\}+\sum_{i=1}^{N}\frac{\beta_i}{\pi_i}\left({y_i}^2+{x_i}^2\right)\right]$$

$$\hat{V}_P\left(\hat{\theta}_A\right)=\frac{1}{N^2}\left[\sum_{i\in s_1}\sum_{j\in s_1, j>i}\left(\frac{\pi_i\pi_j-\pi_{ij}}{\pi_{ij}}\right)\left(\frac{y_i}{\pi_i}-\frac{y_j}{\pi_j}\right)^2\right.$$

$$\left.+\sum_{i\in s_2}\sum_{j\in s_2, j>i}\left(\frac{\pi_i\pi_j-\pi_{ij}}{\pi_{ij}}\right)\left(\frac{x_i}{\pi_i}-\frac{x_j}{\pi_j}\right)^2+\sum_{i\in s_1}\frac{{y_i}^2}{{\pi_i}^2}\beta_i+\sum_{i\in s_2}\frac{{x_i}^2}{{\pi_i}^2}\beta_i\right]$$

where $\beta_i=1+\frac{1}{\pi_i}\sum_{j\neq i}^{N}\pi_{ij}-\sum_{i=1}^{N}\pi_i$ such that $E_P\left(\hat{V}_P\left(\hat{\theta}_A\right)\right)=V_P\left(\hat{\theta}_A\right)$

The disadvantages of the above procedure are that it needs the selection of two independent random samples from the population and θ_F needs to be known. Hence, removing the above two restrictions, a further revised Item Count Technique is presented in Section 2.

2 ITEM COUNT TECHNIQUE USING A SINGLE SAMPLE

A sample s is taken from U with probability $p(s)$ according to a design admitting positive first-order and second-order inclusion probabilities $\pi_i = \sum_{s \ni i} p(s)$, $\pi_{ij} = \sum_{s \ni i,j} p(s), \mathrm{i} \neq \mathrm{j}$. Two questionnaires are provided to each participant. The first questionnaire contains G innocuous item-statements and the $(G+1)$th statement is—"I have characteristic A or F." The second questionnaire contains the same G innocuous item-statements with the $(G+1)$th statement being—"I have characteristic F" and the $(G+2)$th statement being—"I do not have characteristic A or I do not have characteristic F." Three boxes named box-1, box-2, and box-3 are provided at a distance from the investigators. Each respondent is provided with three empty identical slips of papers. The ith respondent is requested to give six answers. The first two answers correspond to the number of statements in the first questionnaire (which are applicable to him/her) and the value (corresponding to him/her) of the variable which is taken as size measure, respectively. The next two answers correspond to the number of statements in the second questionnaire (which are applicable to the respondent) and the value (corresponding to him/her) of the variable which is taken as size measure, respectively. The last two answers refer to the sum of the number of statements valid for him/her with respect to both the questionnaires and the value (corresponding to him/her) of the variable which is taken as size measure, respectively. These six answers are to be written in the three slips of paper. The first slip containing the first two answers is to be dropped in box-1, second slip containing the second two answers in box-2 and third slip containing the last two answers in box-3 by the respondent without revealing the answers or the statements to the investigator. The respondents are allowed to shuffle the slips within each box after dropping their answers but they should be clearly instructed not to shuffle the slips among the boxes. The set of all possible answers for the first questionnaire is $\{0, 1, 2, \ldots, G-1, G, G+1\}$ and for the second questionnaire is $\{0, 1, 2, \ldots, G, G+1, G+2\}$. Using the Horvitz–Thompson estimator, an unbiased estimator of θ_A is given by:

$$\hat{\theta}_A = \frac{1}{N}\sum_{i \in s} \frac{y_i}{\pi_i} - \frac{1}{N}\sum_{i \in s} \frac{x_i}{\pi_i} + \frac{1}{N}\sum_{i \in s} \frac{1}{\pi_i}$$

where y_i = number of statements in the first questionnaire valid for the ith individual, x_i = number of statements in the second questionnaire valid for the ith individual and π_i is obtained by performing suitable calculations using the size measure in the slip.

Proof.

$$\begin{aligned}E_P\left(\hat{\theta}_A\right) &= \frac{1}{N}\sum_{i=1}^{N} y_i - \frac{1}{N}\sum_{i=1}^{N} x_i + 1\\ &= \theta_A + \theta_F - \theta_{A\cap F} - \theta_F - \theta_{(A\cap F)^c} + 1\\ &= \theta_A + \theta_F - \theta_{A\cap F} - \theta_F - 1 + \theta_{A\cap F} + 1\\ &= \theta_A\end{aligned}$$

where E_P = expectation operator with respect to sampling, θ_F = population proportion of individuals bearing F, $\theta_{A\cap F}$ = population proportion of individuals bearing A and F, $\theta_{(A\cap F)^c}$ = population proportion of individuals not bearing A and F together.□

$$\begin{aligned}V_P\left(\hat{\theta}_A\right) = {} & V_P\left(\frac{1}{N}\sum_{i\in s}\frac{y_i}{\pi_i}\right) + V_P\left(\frac{1}{N}\sum_{i\in s}\frac{x_i}{\pi_i}\right) + V_P\left(\frac{1}{N}\sum_{i\in s}\frac{1}{\pi_i}\right)\\ & - 2Cov_P\left(\frac{1}{N}\sum_{i\in s}\frac{y_i}{\pi_i}, \frac{1}{N}\sum_{i\in s}\frac{x_i}{\pi_i}\right) + 2Cov_P\left(\frac{1}{N}\sum_{i\in s}\frac{y_i}{\pi_i}, \frac{1}{N}\sum_{i\in s}\frac{1}{\pi_i}\right)\\ & - 2Cov_P\left(\frac{1}{N}\sum_{i\in s}\frac{x_i}{\pi_i}, \frac{1}{N}\sum_{i\in s}\frac{1}{\pi_i}\right)\end{aligned} \tag{1}$$

Now,

$$V_P\left(\frac{1}{N}\sum_{i\in s}\frac{z_i}{\pi_i}\right) = V_P\left(\frac{1}{N}\sum_{i\in s}\frac{y_i}{\pi_i}\right) + V_P\left(\frac{1}{N}\sum_{i\in s}\frac{x_i}{\pi_i}\right) + 2Cov_P\left(\frac{1}{N}\sum_{i\in s}\frac{y_i}{\pi_i}, \frac{1}{N}\sum_{i\in s}\frac{x_i}{\pi_i}\right)$$

where $z_i = y_i + x_i$.

So,

$$2Cov_P\left(\frac{1}{N}\sum_{i\in s}\frac{y_i}{\pi_i}, \frac{1}{N}\sum_{i\in s}\frac{x_i}{\pi_i}\right) = V_P\left(\frac{1}{N}\sum_{i\in s}\frac{z_i}{\pi_i}\right) - V_P\left(\frac{1}{N}\sum_{i\in s}\frac{y_i}{\pi_i}\right) - V_P\left(\frac{1}{N}\sum_{i\in s}\frac{x_i}{\pi_i}\right) \tag{2}$$

Similarly,

$$V_P\left(\frac{1}{N}\sum_{i\in s}\frac{w_i}{\pi_i}\right) = V_P\left(\frac{1}{N}\sum_{i\in s}\frac{y_i}{\pi_i}\right) + V_P\left(\frac{1}{N}\sum_{i\in s}\frac{1}{\pi_i}\right) + 2Cov_P\left(\frac{1}{N}\sum_{i\in s}\frac{y_i}{\pi_i}, \frac{1}{N}\sum_{i\in s}\frac{1}{\pi_i}\right)$$

where $w_i = y_i + 1$.

So,

$$2Cov_P\left(\frac{1}{N}\sum_{i\in s}\frac{y_i}{\pi_i}, \frac{1}{N}\sum_{i\in s}\frac{1}{\pi_i}\right) = V_P\left(\frac{1}{N}\sum_{i\in s}\frac{w_i}{\pi_i}\right) - V_P\left(\frac{1}{N}\sum_{i\in s}\frac{y_i}{\pi_i}\right) - V_P\left(\frac{1}{N}\sum_{i\in s}\frac{1}{\pi_i}\right) \tag{3}$$

Similarly,

$$V_P\left(\frac{1}{N}\sum_{i\in s}\frac{u_i}{\pi_i}\right)=V_P\left(\frac{1}{N}\sum_{i\in s}\frac{x_i}{\pi_i}\right)+V_P\left(\frac{1}{N}\sum_{i\in s}\frac{1}{\pi_i}\right)+2Cov_P\left(\frac{1}{N}\sum_{i\in s}\frac{x_i}{\pi_i},\frac{1}{N}\sum_{i\in s}\frac{1}{\pi_i}\right)$$

where $u_i=x_i+1$

So,

$$2Cov_P\left(\frac{1}{N}\sum_{i\in s}\frac{x_i}{\pi_i},\frac{1}{N}\sum_{i\in s}\frac{1}{\pi_i}\right)=V_P\left(\frac{1}{N}\sum_{i\in s}\frac{u_i}{\pi_i}\right)-V_P\left(\frac{1}{N}\sum_{i\in s}\frac{x_i}{\pi_i}\right)-V_P\left(\frac{1}{N}\sum_{i\in s}\frac{1}{\pi_i}\right) \tag{4}$$

Putting Eqs. (2), (3), and (4) in Eq. (1) and using Chaudhuri and Pal (2002), we get

$$\begin{aligned}
V_P\left(\hat{\theta}_A\right)&=V_P\left(\frac{1}{N}\sum_{i\in s}\frac{y_i}{\pi_i}\right)+3V_P\left(\frac{1}{N}\sum_{i\in s}\frac{x_i}{\pi_i}\right)+V_P\left(\frac{1}{N}\sum_{i\in s}\frac{1}{\pi_i}\right)\\
&\quad -V_P\left(\frac{1}{N}\sum_{i\in s}\frac{z_i}{\pi_i}\right)+V_P\left(\frac{1}{N}\sum_{i\in s}\frac{w_i}{\pi_i}\right)-V_P\left(\frac{1}{N}\sum_{i\in s}\frac{u_i}{\pi_i}\right)\\
&=\frac{1}{N^2}\left[\sum_{i}^{N}\sum_{<j}^{N}(\pi_i\pi_j-\pi_{ij})\left(\frac{y_i}{\pi_i}-\frac{y_j}{\pi_j}\right)^2+\sum_{i=1}^{N}\frac{\beta_i}{\pi_i}{y_i}^2\right]\\
&\quad+\frac{3}{N^2}\left[\sum_{i}^{N}\sum_{<j}^{N}(\pi_i\pi_j-\pi_{ij})\left(\frac{x_i}{\pi_i}-\frac{x_j}{\pi_j}\right)^2+\sum_{i=1}^{N}\frac{\beta_i}{\pi_i}{x_i}^2\right]\\
&\quad+\frac{1}{N^2}\left[\sum_{i}^{N}\sum_{<j}^{N}(\pi_i\pi_j-\pi_{ij})\left(\frac{1}{\pi_i}-\frac{1}{\pi_j}\right)^2+\sum_{i=1}^{N}\frac{\beta_i}{\pi_i}\right]\\
&\quad-\frac{1}{N^2}\left[\sum_{i}^{N}\sum_{<j}^{N}(\pi_i\pi_j-\pi_{ij})\left(\frac{z_i}{\pi_i}-\frac{z_j}{\pi_j}\right)^2+\sum_{i=1}^{N}\frac{\beta_i}{\pi_i}{z_i}^2\right]\\
&\quad+\frac{1}{N^2}\left[\sum_{i}^{N}\sum_{<j}^{N}(\pi_i\pi_j-\pi_{ij})\left(\frac{w_i}{\pi_i}-\frac{w_j}{\pi_j}\right)^2+\sum_{i=1}^{N}\frac{\beta_i}{\pi_i}{w_i}^2\right]\\
&\quad-\frac{1}{N^2}\left[\sum_{i}^{N}\sum_{<j}^{N}(\pi_i\pi_j-\pi_{ij})\left(\frac{u_i}{\pi_i}-\frac{u_j}{\pi_j}\right)^2+\sum_{i=1}^{N}\frac{\beta_i}{\pi_i}{u_i}^2\right]
\end{aligned} \tag{5}$$

where $\beta_i=1+\frac{1}{\pi_i}\sum_{j\neq i}^{N}\pi_{ij}-\sum_{i=1}^{N}\pi_i$.

If every sample s contains a common number of distinct units in it, then, $\beta_i=0\forall i$ throughout in Eq. (5), using Chaudhuri and Pal (2002).

$$\hat{V}_P\left(\hat{\theta}_A\right) = \frac{1}{N^2}\sum_{i\in s}\sum_{j\in s, j>i}\left(\frac{\pi_i\pi_j - \pi_{ij}}{\pi_{ij}}\right)\left(\frac{y_i}{\pi_i} - \frac{y_j}{\pi_j}\right)^2$$
$$+\frac{3}{N^2}\sum_{i\in s}\sum_{j\in s, j>i}\left(\frac{\pi_i\pi_j - \pi_{ij}}{\pi_{ij}}\right)\left(\frac{x_i}{\pi_i} - \frac{x_j}{\pi_j}\right)^2$$
$$+\frac{1}{N^2}\left[\sum_{i\in s}\sum_{j\in s, j>i}\left(\frac{\pi_i\pi_j - \pi_{ij}}{\pi_{ij}}\right)\left(\frac{1}{\pi_i} - \frac{1}{\pi_j}\right)^2\sum_{i\in s}\sum_{j\in s, j>i}\left(\frac{\pi_i\pi_j - \pi_{ij}}{\pi_{ij}}\right)\left(\frac{z_i}{\pi_i} - \frac{z_j}{\pi_j}\right)^2\right.$$
$$+\sum_{i\in s}\sum_{j\in s, j>i}\left(\frac{\pi_i\pi_j - \pi_{ij}}{\pi_{ij}}\right)\left(\frac{w_i}{\pi_i} - \frac{w_j}{\pi_j}\right)^2 - \sum_{i\in s}\sum_{j\in s, j>i}\left(\frac{\pi_i\pi_j - \pi_{ij}}{\pi_{ij}}\right)\left(\frac{u_i}{\pi_i} - \frac{u_j}{\pi_j}\right)^2$$
$$\left.+\sum_{i\in s}\frac{\beta_i}{\pi_i^2}y_i^2 + 3\sum_{i\in s}\frac{\beta_i}{\pi_i^2}x_i^2 + \sum_{i\in s}\frac{\beta_i}{\pi_i^2} - \sum_{i\in s}\frac{\beta_i}{\pi_i^2}z_i^2 + \sum_{i\in s}\frac{\beta_i}{\pi_i^2}w_i^2 - \sum_{i\in s}\frac{\beta_i}{\pi_i^2}u_i^2\right] \tag{6}$$

with $\beta_i = 0 \forall i$ in Eq. (6) when applicable.

So, $E_P\left(\hat{V}_P\left(\hat{\theta}_A\right)\right) = V_P\left(\hat{\theta}_A\right)$.

Instead of the revised Item Count Technique described earlier, one may think of using a single box and asking the respondent to write on a slip of paper whether he/she belongs to the stigmatizing group and the value (corresponding to him/her) of the variable which is taken as size measure and then requesting him/her to drop the slip in that box, which is a much simpler procedure (say Procedure P) in addition to the fact that it will produce a more efficient estimator than any other Randomized Response Technique (RRT) on and from Warner's (1965). But this does not protect the privacy of the respondent in the sense that the respondents may not agree to cooperate with the exercise. But here in the revised Item Count Technique described, the respondent need not answer directly regarding his/her characteristics. The questions regarding A is mixed with other innocuous questions. This makes the respondent rely on the procedure even more than that of the suggested easier procedure using a single box and also in the RRT technique by Warner (1965) and the follow-ups. This was the main reason behind the invention of Item Count Technique. But since the latest procedure of Item Count Technique requires certain restrictions; hence an attempt has been to present in this chapter a revised version of the Item Count Technique which need not be bound to those restrictions.

Now,

$$\hat{\theta}_A{}^P = \frac{1}{N}\sum_{i\in s}\frac{z_i}{\pi_i}$$

where

$$z_i = 1 \text{ if } i\text{th respondent belongs to } A$$
$$= 0 \text{ otherwise,}$$

In other words, z_i is the response from the ith respondent's slip obtained from the box when Procedure P is used, π_i is obtained by performing suitable calculations using the size measure in the slip.

$\hat{\theta}_A^P$ = unbiased estimator of θ_A when Procedure P is used.

Again,

$$\begin{aligned}\hat{\theta}_A^P &= \frac{1}{N}\sum_{i\in s}\frac{z_i}{\pi_i}\\ &= \frac{1}{N}\sum_{i\in s}\frac{y_i - x_i + 1}{\pi_i}\\ &= \frac{1}{N}\sum_{i\in s}\frac{y_i}{\pi_i} - \frac{1}{N}\sum_{i\in s}\frac{x_i}{\pi_i} + \frac{1}{N}\sum_{i\in s}\frac{1}{\pi_i}\\ &= \hat{\theta}_A\end{aligned}$$

Hence, $V\left(\hat{\theta}_A^P\right) = V\left(\hat{\theta}_A\right)$ which shows that the revised Item Count Technique described is equally efficient to the Procedure P.

3 AN ALTERNATIVE ESTIMATOR OF $V_P\left(\hat{\theta}_A\right)$

We may see whether it is possible to find an alternative estimator of $V_P\left(\hat{\theta}_A\right)$ using a procedure described in Chaudhuri (2014).

$$\begin{aligned}V_P\left(\hat{\theta}_A\right) &= V_P\left(\frac{1}{N}\sum_{i\in s}\frac{y_i}{\pi_i}\right) + V_P\left(\frac{1}{N}\sum_{i\in s}\frac{x_i}{\pi_i}\right) + V_P\left(\frac{1}{N}\sum_{i\in s}\frac{1}{\pi_i}\right)\\ &\quad - 2Cov_P\left(\frac{1}{N}\sum_{i\in s}\frac{y_i}{\pi_i}, \frac{1}{N}\sum_{i\in s}\frac{x_i}{\pi_i}\right) + 2Cov_P\left(\frac{1}{N}\sum_{i\in s}\frac{y_i}{\pi_i}, \frac{1}{N}\sum_{i\in s}\frac{1}{\pi_i}\right)\\ &\quad - 2Cov_P\left(\frac{1}{N}\sum_{i\in s}\frac{x_i}{\pi_i}, \frac{1}{N}\sum_{i\in s}\frac{1}{\pi_i}\right)\end{aligned}$$

Now,

$$\begin{aligned}Cov_P\left(\frac{1}{N}\sum_{i\in s}\frac{y_i}{\pi_i}, \frac{1}{N}\sum_{i\in s}\frac{x_i}{\pi_i}\right) &= E_P\left(\frac{1}{N}\sum_{i\in s}\frac{y_i}{\pi_i} - \frac{1}{N}\sum_{i=1}^{N} y_i\right)\left(\frac{1}{N}\sum_{i\in s}\frac{x_i}{\pi_i} - \frac{1}{N}\sum_{i=1}^{N} x_i\right)\\ &= \frac{1}{N^2}\left[E_P\left(\sum_{i\in s}\frac{y_i}{\pi_i}\sum_{i\in s}\frac{x_i}{\pi_i}\right) - \sum_{i=1}^{N} y_i\sum_{i=1}^{N} x_i\right]\\ &= \frac{1}{N^2}\left[E_P\left(\sum_{i\in s}\frac{y_i x_i}{\pi_i^2} + \sum_{i}\sum_{\neq j\in s}\frac{y_i\, x_j}{\pi_i\,\pi_j}\right) - \sum_{i=1}^{N} y_i\sum_{i=1}^{N} x_i\right]\\ &= \frac{1}{\theta_1}\left[\theta_2 + \theta_3 - \theta_4\theta_5\right]\end{aligned}$$

This is a nonlinear function of six population totals, namely:

$$\theta_1 = N^2, \theta_2 = E_P\left(\sum_{i\in s}\frac{y_i x_i}{\pi_i{}^2}\right), \theta_3 = E_P\left(\sum_{i}\sum_{\neq j\in s}\frac{y_i\, x_j}{\pi_i\,\pi_j}\right), \theta_4 = \sum_{i=1}^{N} y_i, \text{ and } \theta_5 = \sum_{i=1}^{N} x_i,$$

each of which is unbiasedly estimable as follows:

$$\begin{aligned} V_P\left(\widehat{N}\right) &= E_P\left(\widehat{N}^2\right) - E_P^2\left(\widehat{N}\right) \\ &= E_P\left(\widehat{N}^2\right) - N^2 \\ \Rightarrow N^2 &= E_P\left(\widehat{N}^2\right) - V_P\left(\widehat{N}\right) \end{aligned}$$

Hence,

$$\hat{\theta}_1 = \widehat{N^2} = \widehat{N}^2 - \widehat{V}_P\left(\widehat{N}\right) = \left(\sum_{i\in s}\frac{1}{\pi_i}\right)^2 - \sum_{i\in s}\sum_{j\in s, j>i}\left(\frac{\pi_i\pi_j - \pi_{ij}}{\pi_{ij}}\right)\left(\frac{1}{\pi_i} - \frac{1}{\pi_j}\right)^2$$

$$-\sum_{i\in s}\frac{\beta_i}{\pi_i{}^2} \text{ with } \beta_i = 0\,\forall i \text{ when applicable.}$$

$$\hat{\theta}_2 = \sum_{i\in s}\frac{y_i x_i}{\pi_i{}^2},\ \hat{\theta}_3 = \sum_{i}\sum_{\neq j\in s}\frac{y_i\, x_j}{\pi_i\,\pi_j},\ \hat{\theta}_4 = \sum_{i\in s}\frac{y_i}{\pi_i} \text{ and}$$

$$\hat{\theta}_5 = \sum_{i\in s}\frac{x_i}{\pi_i}$$

Then an approximately unbiased estimator of $Cov_P\left(\frac{1}{N}\sum_{i\in s}\frac{y_i}{\pi_i}, \frac{1}{N}\sum_{i\in s}\frac{x_i}{\pi_i}\right)$ is given by:

$$\widehat{Cov_P}\left(\frac{1}{N}\sum_{i\in s}\frac{y_i}{\pi_i}, \frac{1}{N}\sum_{i\in s}\frac{x_i}{\pi_i}\right) = \frac{1}{\left(\sum_{i\in s}\frac{1}{\pi_i}\right)^2 - \sum_{i\in s}\sum_{j\in s, j>i}\left(\frac{\pi_i\pi_j - \pi_{ij}}{\pi_{ij}}\right)\left(\frac{1}{\pi_i} - \frac{1}{\pi_j}\right)^2 - \sum_{i\in s}\frac{\beta_i}{\pi_i{}^2}}$$

$$\left[\sum_{i\in s}\frac{y_i x_i}{\pi_i{}^2} + \sum_{i}\sum_{\neq j\in s}\frac{y_i\, x_j}{\pi_i\,\pi_j} - \sum_{i\in s}\frac{y_i}{\pi_i}\sum_{i\in s}\frac{x_i}{\pi_i}\right], \text{with } \beta_i = 0\,\forall i \text{ when applicable.}$$

$$= \frac{1}{\left(\sum_{i\in s}\frac{1}{\pi_i}\right)^2 - \sum_{i\in s}\sum_{j\in s, j>i}\left(\frac{\pi_i\pi_j - \pi_{ij}}{\pi_{ij}}\right)\left(\frac{1}{\pi_i} - \frac{1}{\pi_j}\right)^2 - \sum_{i\in s}\frac{\beta_i}{\pi_i{}^2}}$$

$$\left[\sum_{i\in s}\frac{y_i}{\pi_i}\sum_{i\in s}\frac{x_i}{\pi_i} - \sum_{i\in s}\frac{y_i}{\pi_i}\sum_{i\in s}\frac{x_i}{\pi_i}\right], \text{with } \beta_i = 0\,\forall i \text{ when applicable.} = 0$$

Similarly, approximately,

$$\widehat{Cov}_P\left(\frac{1}{N}\sum_{i\in s}\frac{y_i}{\pi_i},\frac{1}{N}\sum_{i\in s}\frac{1}{\pi_i}\right)=0 \text{ and } \widehat{Cov}_P\left(\frac{1}{N}\sum_{i\in s}\frac{x_i}{\pi_i},\frac{1}{N}\sum_{i\in s}\frac{1}{\pi_i}\right)=0$$

such that

$$E_P\left(\widehat{Cov}_P\left(\frac{1}{N}\sum_{i\in s}\frac{y_i}{\pi_i},\frac{1}{N}\sum_{i\in s}\frac{x_i}{\pi_i}\right)\right)\cong Cov_P\left(\frac{1}{N}\sum_{i\in s}\frac{y_i}{\pi_i},\frac{1}{N}\sum_{i\in s}\frac{x_i}{\pi_i}\right)$$

$$E_P\left(\widehat{Cov}_P\left(\frac{1}{N}\sum_{i\in s}\frac{y_i}{\pi_i},\frac{1}{N}\sum_{i\in s}\frac{1}{\pi_i}\right)\right)\cong Cov_P\left(\frac{1}{N}\sum_{i\in s}\frac{y_i}{\pi_i},\frac{1}{N}\sum_{i\in s}\frac{1}{\pi_i}\right)$$

$$E_P\left(\widehat{Cov}_P\left(\frac{1}{N}\sum_{i\in s}\frac{x_i}{\pi_i},\frac{1}{N}\sum_{i\in s}\frac{1}{\pi_i}\right)\right)\cong Cov_P\left(\frac{1}{N}\sum_{i\in s}\frac{x_i}{\pi_i},\frac{1}{N}\sum_{i\in s}\frac{1}{\pi_i}\right)$$

Hence we can conclude that the approximately unbiased estimator of the variance term is

$$\hat{V}_P(\hat{\theta}_A)=\frac{1}{N^2}\left[\sum_{i\in s}\sum_{j\in s,j>i}\left(\frac{\pi_i\pi_j-\pi_{ij}}{\pi_{ij}}\right)\left(\frac{y_i}{\pi_i}-\frac{y_j}{\pi_j}\right)^2+\frac{1}{N^2}\sum_{i\in s}\sum_{j\in s,j>i}\left(\frac{\pi_i\pi_j-\pi_{ij}}{\pi_{ij}}\right)\left(\frac{x_i}{\pi_i}-\frac{x_j}{\pi_j}\right)^2\right.$$

$$\left.+\sum_{i\in s}\sum_{j\in s,j>i}\left(\frac{\pi_i\pi_j-\pi_{ij}}{\pi_{ij}}\right)\left(\frac{1}{\pi_i}-\frac{1}{\pi_j}\right)^2+\sum_{i\in s}\frac{\beta_i}{\pi_i^2}y_i^2+\sum_{i\in s}\frac{\beta_i}{\pi_i^2}x_i^2+\sum_{i\in s}\frac{\beta_i}{\pi_i^2}\right],$$

$\beta_i=0\forall i$ whenever applicable.
Now,

$$E_P\left(\hat{V}_P\left(\hat{\theta}_A\right)\right)\cong V_P\left(\hat{\theta}_A\right)$$

4 NUMERICAL PRESENTATION

Consider the following imaginary population:

Unit	Items					w	A	F	Unit	Items					w	A	F
	1	*2*	*3*	*4*	*5*					*1*	*2*	*3*	*4*	*5*			
1	0	1	0	0	1	17	1	0	60	0	1	0	1	0	16	0	1
2	1	1	0	0	1	1	1	0	61	1	1	0	1	0	14	0	0
3	1	0	1	1	1	5	0	1	62	0	0	0	1	0	4	0	1
4	1	1	1	0	1	12	0	0	63	1	1	1	0	1	1	0	1
5	1	1	1	0	0	8	0	1	64	1	0	0	1	0	5	1	0
6	1	1	0	0	1	9	1	1	65	1	0	0	1	1	13	0	0
7	0	0	1	0	0	8	1	0	66	1	1	1	1	1	15	0	1
8	0	0	0	0	1	8	1	1	67	1	1	1	0	1	4	1	0
9	1	0	1	1	1	4	1	1	68	0	0	1	1	0	17	0	0
10	1	1	0	0	1	13	1	1	69	1	1	0	0	0	10	1	1
11	1	0	0	0	0	8	0	1	70	0	0	1	0	0	14	1	0
12	0	1	1	0	0	11	1	0	71	0	1	0	1	1	16	1	0
13	0	0	1	1	1	16	0	0	72	0	0	1	1	0	6	1	0
14	1	1	0	0	0	5	1	0	73	1	0	0	0	0	5	1	1
15	1	0	0	0	0	12	0	1	74	1	0	1	1	0	6	0	0
16	0	0	1	1	1	17	1	0	75	1	1	0	1	0	2	1	0
17	1	1	0	1	1	16	1	0	76	1	1	0	1	0	9	1	0
18	0	0	0	1	1	1	0	1	77	0	1	1	1	0	16	1	1

19	0	1	1	0	0	7	0	0	78	0	1	1	1	1	2	1	1
20	0	1	1	1	0	11	0	0	79	0	1	0	1	1	5	0	0
21	1	0	1	1	1	9	0	0	80	0	1	0	0	0	9	0	0
22	1	0	1	0	1	16	1	1	81	1	0	1	1	1	8	0	1
23	0	0	1	1	0	11	1	1	82	1	1	1	1	0	14	0	1
24	0	1	1	1	0	15	0	1	83	1	1	1	1	0	13	0	1
25	0	1	1	1	1	10	0	0	84	0	1	0	0	1	2	1	1
26	0	0	1	0	0	13	1	0	85	1	0	1	1	1	1	1	1
27	1	1	0	1	0	14	1	0	86	1	0	0	1	0	13	0	0
28	1	1	0	1	0	15	1	0	87	0	0	1	0	1	5	0	0
29	0	0	0	0	1	5	0	0	88	0	1	1	1	0	7	1	0
30	1	0	1	0	1	17	0	0	89	0	1	1	0	1	9	0	1
31	1	0	0	0	0	11	0	1	90	0	0	1	1	0	7	1	1
32	1	0	0	0	0	4	1	0	91	0	1	0	0	0	1	0	0
33	1	1	0	0	1	13	0	0	92	0	1	0	1	1	10	0	1
34	0	1	0	0	0	5	0	1	93	0	0	1	1	0	16	0	0
35	0	1	1	0	0	5	0	1	94	0	0	1	1	1	17	1	0
36	1	0	0	1	0	3	0	0	95	1	1	0	0	1	7	0	1
37	0	1	1	0	1	1	0	1	96	0	0	0	1	1	12	1	0
38	1	1	1	1	0	8	1	0	97	0	1	1	1	1	7	1	0

Continued

Unit	Items					w	A	F	Unit	Items					w	A	F
	1	2	3	4	5					1	2	3	4	5			
39	0	1	1	0	0	12	0	0	98	1	1	0	0	0	3	1	1
40	1	0	0	1	0	2	0	0	99	0	0	1	0	0	10	1	0
41	0	1	0	1	0	16	1	1	100	0	1	1	1	1	8	1	1
42	0	1	1	1	1	1	1	0	101	1	0	0	1	0	11	1	1
43	1	1	0	0	1	3	1	1	102	1	0	0	1	0	16	0	0
44	0	1	1	0	1	12	1	0	103	0	1	0	1	1	6	1	0
45	1	1	1	1	1	5	0	0	104	1	0	1	0	0	11	0	0
46	0	1	0	1	0	15	0	1	105	1	1	1	1	1	12	1	1
47	1	1	1	0	0	9	0	0	106	1	1	0	0	0	12	0	0
48	1	0	0	1	1	12	1	0	107	0	0	0	0	1	6	1	1
49	0	1	1	1	0	8	1	0	108	0	0	1	1	1	17	0	0
50	0	0	0	0	1	2	0	0	109	0	1	1	1	0	13	0	1
51	1	1	0	1	0	11	1	0	110	0	1	0	0	1	12	0	0
52	1	1	1	0	1	14	0	1	111	0	0	0	0	0	7	1	0
53	1	0	1	0	1	11	0	0	112	0	1	1	0	0	3	0	1
54	1	0	1	1	1	3	0	0	113	1	1	0	1	0	11	1	1
55	0	1	0	0	1	3	0	0	114	0	0	0	0	0	8	1	0
56	0	1	1	0	0	12	1	0	115	1	0	1	0	1	2	0	0
57	1	1	0	0	1	8	1	1	116	0	1	0	1	0	7	0	0
58	0	0	0	1	0	1	0	1	117	0	1	0	1	1	3	0	1
59	1	0	1	0	1	10	1	1									

Here A = intaker of drugs, F = lover of music, and w = number of family members in the respondent's home. Values of "w" are used as size measures. 1000 samples each of size 37 is chosen from the above population of size $N = 117$ by employing the Hartley and Rao (1962) method in which a systematic sample by Probability Proportional to Size (PPS) method is used after the random arrangement of the population units. We calculate $\hat{\theta}_A$, $\hat{V}_P\left(\hat{\theta}_A\right)$, coefficient of variation (CV) $= 100\frac{\sqrt{\hat{V}_P\left(\hat{\theta}_A\right)}}{\hat{\theta}_A}$ and 95% approximate confidence intervals (CI) for θ_A given by $\left(\hat{\theta}_A - \sqrt{\hat{V}_P\left(\hat{\theta}_A\right)}\tau_{\frac{\alpha}{2}}, \hat{\theta}_A + \sqrt{\hat{V}_P\left(\hat{\theta}_A\right)}\tau_{\frac{\alpha}{2}}\right)$, assuming $\hat{\theta}_A \sim \text{Normal}\left(\theta_A, V_P\left(\hat{\theta}_A\right)\right)$ by using the revised Item Count Technique using single sample by both the estimation methods as discussed in Sections 2 and 3. Then based on the 1000 samples we calculate:

ACV (average coefficient of variation) = the average of the coefficient of variation over the 1000 replicates,

ARB (absolute relative bias) $= \left|\frac{\overline{\hat{\theta}_A} - \theta_A}{\theta_A}\right|$, where $\overline{\hat{\theta}_A} = \frac{1}{1000}\sum_{i=1}^{1000}\hat{\theta}_{Ai}$, $\hat{\theta}_{Ai}$ being the ith sample estimate of θ_A,

ACP (actual coverage proportion) = percentage of replicates out of 1000 for which the CI covers θ_A, and

AL (average length) = average length of the CI over 1000 replicates.

Comparison between the original versus revised method and original versus revised method with approximately unbiased variance estimation denoted with an asterisk are shown in Tables 1–6.

TABLE 1 Comparison of the Performances of the Three Methods

	$\theta_A=0.479$, $V_P\left(\hat{\theta}_A\right)=0.014$		
	Original Method (I)	*Revised Method (II)*	*Revised Method* (III)*
ACV	355.307	20.537	142.755
ARB	1.043	0.00634	0.00634
ACP	97%	89%	100%
AL	3.331	0.402	2.730

TABLE 2 Comparison of CV Between Procedures with a Few Illustrated Replicates

No.	Original Method (I)	Revised Method (II)
1	98.95	14.85
2	64.95	19.11
3	88.76	24.29
4	50.05	16.14
5	42.52	16.68

99% Cases are favorable to the revised method.

TABLE 3 Comparison of Estimated Efficiency Between Procedures with a Few Illustrated Replicates

No.	$\hat{V}_P\left(\hat{\theta}_A\right)$ for the Original Method (I)	$\hat{V}_P\left(\hat{\theta}_A\right)$ for the Revised Method (II)	Efficiency of (II) with Respect to (I)
1	0.312	0.063	495.238
2	0.136	0.058	234.483
3	0.235	0.060	391.667
4	0.362	0.103	351.456
5	0.333	0.067	497.015

100% Cases are favorable to the revised method.

TABLE 4 Comparison of CV Between Procedures with a Few Illustrated Replicates

No.	Original Method (I)	Revised Method* (III)
1	71.78	42.10
2	81.65	45.90
3	42.24	42.04
4	66.31	44.54
5	75.37	49.99

38% Cases are favorable to the revised method*.

TABLE 5 Comparison of Efficiency Between Procedures with a Few Examples

No.	$\hat{V}_P\left(\hat{\theta}_A\right)$ for the Original Method (I)	$\hat{V}_P\left(\hat{\theta}_A\right)$ for the Revised Method* (III)	Efficiency of (III) with Respect to (I)
1	0.430	0.119	361.344
2	0.424	0.111	381.982
3	0.405	0.133	304.511
4	0.412	0.196	210.204
5	0.155	0.094	164.894

64% Cases are favorable to the revised method*.

TABLE 6 A Few Out of the 1000 Sample Estimates of θ_A (=0.479)

Original Method (I)			Revised Method (II)		
0.846	0.203	0.346	0.536	0.448	0.466
0.651	0.918	0.264	0.519	0.350	0.655
0.664	0.704	0.119	0.346	0.334	0.485
0.625	0.860	0.168	0.497	0.607	0.660
0.808	0.637	0.834	0.403	0.421	0.569
0.598	0.149	0.230	0.417	0.573	0.414
0.378	0.430	0.322	0.427	0.471	0.587
0.696	0.160	0.420	0.427	0.497	0.481
0.649	0.363	0.010	0.447	0.474	0.602
0.995	0.253	0.762	0.370	0.537	0.585

In 88% cases, $\hat{\theta}_A$ calculated using the revised method is closer to θ_A than that calculated using the original method.

5 CONCLUSION

The revised Item Count Technique described is not only easy but also produces estimates worthy of attention.

An example of a proper form of the two questionnaires: Here A, use of pirated version of softwares, F, having internet banking. The completed number of years of job experience is taken as the size measure.

Questionnaire 1

Please count the number of statements which are valid for you and write it in one of the three empty slips provided to you. Also write the number of years of your job experience in the same slip. It is advised not to write anything else other than these in the slip. Then drop it in box-1. After this remember to shuffle the slips in box-1. For the statements where there are two options please count the statement if at least one of the options is valid for you. Please retain the questionnaire with you.

1	I am currently doing a job
2	I love surfing the internet when free
3	I have recently developed eye-sight problem
4	Cost of household articles are increasing, or
	I am saving money on a regular basis
5	I hope to get a promotion next year
6	I have recently purchased a new computer for myself
7	I am the only working person in my family
8	I use pirated softwares, or
	I have internet banking
9	I am addicted to computer games
10	I am a specialized in computer programming
11	I have a smart phone

Questionnaire 2

Please count the number of statements which are valid for you and write it in one of remaining two empty slips left with you. Also write the number of years of your job experience in the same slip. It is advised not to write anything else other than these in the slip. Then drop it in box-2. After this remember to shuffle the slips in box-2. For the statements where there are two options please count the statement if at least one of the options is valid for you. Please retain the questionnaire with you. Now add the number of statements valid for you in both the questionnaires, please write it in the last slip left with you. Also write the number of years of your job experience in the same slip. It is advised not to write anything else other than these in the slip. Then drop it in box-3. Kindly shuffle the slips inside the box and take care not to shuffle them among the boxes.

1	I am currently doing a job
2	I love surfing the internet when free
3	I purchase the softwares whichever and whenever required, or
	I don't have internet banking
4	I am addicted to computer games
5	I hope to get a promotion next year
6	I have recently purchased a new computer for myself
7	I am the only working person in my family
8	I have recently developed eye-sight problem
9	Cost of household articles are increasing, or
	I am saving money on a regular basis
10	I am a specialized in computer programming
11	I have a smart phone
12	I have internet banking

ACKNOWLEDGMENTS

The author is grateful to the referee and Dr. Arijit Chaudhuri (Honorary Visiting Professor, Applied Statistics Unit, Indian Statistical Institute, Kolkata) for their helpful comments.

REFERENCES

Chaudhuri, A., 2014. Modern Survey Sampling. CRC Press, Boca Raton, pp. 197–202.

Chaudhuri, A., Christofides, T.C., 2007. Item count technique in estimating the proportion of people with a sensitive feature. J. Stat. Plan. Inference 137, 589–593.

Chaudhuri, A., Pal, S., 2002. On certain alternative mean square error estimators in complex survey sampling. J. Stat. Plan. Inference 104, 363–375.

Hartley, H.O., Rao, J.N.K., 1962. Sampling with unequal probabilities and without replacement. Ann. Math. Stat. 33 (2), 350–374.

Miller, J.D., 1984. A New Survey Technique for Studying Deviant Behavior (Ph.D. Thesis). The George Washington University.

Miller, J.D., Cisin, I.H., Harrel, A.V., 1986. A new technique for surveying deviant behavior: item count estimates of marijuana, cocaine and heroin. In: Paper Presented at the Annual Meeting of the American Association for Public Opinion Research, St. Petersburg, Florida.

Raghavarao, D., Federer, W.F., 1979. Block total response as an alternative to the randomized response method in surveys. J. R. Stat. Soc. Ser. B 41, 40–45.

Warner, S.L., 1965. RR: a survey technique for eliminating evasive answer bias. J. Am. Stat. Assoc. 60, 63–69.

Chapter 24

Surveying a Varying Probability Adaptive Sample to Estimate Cost of Hospital Treatments of Sensitive Diseases by RR Data Gathering

S. Pal*[,1] and S. Roy†
*West Bengal State University, Barasat, India
†Reserve Bank of India, Mumbai, India
[1]Corresponding author: e-mail: mitra_pal@yahoo.com

ABSTRACT
Data gathering on sensitive issues is admittedly difficult. Starting with an initial sample we extend to an adaptive sample of young people with history of in-house hospital treatments over the last 1 year in the same city of inhabitance by each. From such an adaptive sample, randomized response-based costs of treatments of sensitive diseases if any are gathered to estimate the mean cost in the city of interest. A simulation-based study of accuracy in estimation is presented.

Keywords: Hard to get data, Quantitative sensitive issues, Simulation results

MSC: 62D05

1 INTRODUCTION

Accurate estimation of per capita cost of treatment of AIDS in a typical modern city, say, Kolkata is not easy. Those treated in hospitals as in-patients can be surveyed only at homes when released. We intend to capture such people to respond to our questionnaire; consent, however, is hard to come by. So, we decide to go for adaptive sampling so as to enhance an initial sample with a view to capturing persons of relevance to us and then gathering data from sampled respondents by employing a convenient randomized response (RR) technique (RRT). To draw an adaptive sample for each person to start with we define a neighborhood composed of all his/her acquaintances in our city who spent sometime in the year gone by as in-patients in some city hospitals or nursing home or health centers or clinics hoping to include some AIDS patients among them. Each such neighbor is planned to be approached to get his/her neighbors as well with the same mission of enquiry stopping only when no such neighbor can be reached through this process. Given a person the set of all the people covered through this process of approaching the neighbors is a "cluster containing the initial person." Omitting from each such cluster the persons who happen to have not been at all a hospital in-patient in the year gone by the remaining persons in the cluster constitute a "Network" for the initial person. Some network may turn out a singleton "Network." By definition any person who has no neighbors as per the definition above including himself (herself) is also treated as a singleton network as are the genuine networks of sets of persons. So, the entire population concerned is a union of a number of disjoint networks. For details one may consult Chaudhuri's (2014b) monograph, especially its chapters 4 and 5.

2 FORMULATION OF PROBLEM

Let $U = (1, 2, \ldots, N)$ denote the population of a known number N of persons and s be an initial sample from it chosen with probability $p(s)$ according to a design p. Let y denote the amount spent if any on the treatment in hospital of diseases like AIDS with y_i as its value for the ith unit of U. Our mission is to suitably estimate unbiasedly $\theta = \frac{1}{N}\sum_{i=1}^{N} y_i = \frac{Y}{N}$, with $Y = \sum_{i=1}^{N} y_i$ along with an unbiased estimate for the variance of the estimate of θ or equivalently of Y. In practice, often the values of y_i, $i \in U$ are nonascertainable. In that situation here we shall estimate the unknown y_i, $i \in U$ using the following RR device. But in the sample such hospital diseases like AIDS patients are rarely available. Considering the above case as a rare disease case, it is obvious that conventional sampling procedure does not yield fruitful results. Then, we need adaptive sampling to capture more units beyond the initial sample. Some modifications are required to develop RR device in the above situation.

The respondents of the sample s are requested to give out independent responses following the instructions below. For any sampled person labeled i, determination of y_i is deemed unethical by asking the delicate question in a direct manner. So, it is considered useful to adopt the following RR technique pioneered by Warner (1965).

For ready references, monographs on RRTs by Chaudhuri and Mukerjee (1988), Chaudhuri (2011), Chaudhuri and Christofides (2013), and Fox and Tracy (1986) among others may be consulted for clarification of ideas.

3 RR SURVEYS

A sampled person labeled $l\,(\in s)$ is offered by the investigator two boxes containing, respectively, identical looking cards inscribed with $(a_1, a_2, \ldots, a_M)$ as certain real numbers in the first box and with $(b_1, b_2, \ldots, b_T)$ in the second box, M and T moderately large enough. Let $\mu = \frac{1}{N}\sum_{j=1}^{M} a_j \neq 0$, $\sigma^2 = \frac{1}{N}\sum_{j=1}^{M}(a_j - \mu)^2$ and $v = \frac{1}{T}\sum_{k=1}^{T} b_k$, $\tau^2 = \frac{1}{T}\sum_{k=1}^{T}(b_k - v)^2$ each being known to a respondent and to the investigator.

Any person labeled, say l, is on request to give the RR as $z_l = a_j y_l + b_k$ without divulging to the investigator the values of y_l and a_j, b_k, respectively, randomly and independently chosen from the first, second boxes. Writing E_R, V_R as the expectation, variance operators in respect of this RR procedure one gets

$$E_R(z_l) = \mu y_l + v \text{ and } V_R(z_l) = \sigma^2 {y_l}^2 + \tau^2.$$

Let $r_l = \frac{z_l - v}{\mu}$; then $E_R(r_l) = y_l$, $V_R(r_l) = \frac{\sigma^2}{\mu^2} y_l^2 + \frac{\tau^2}{\mu^2} = \varphi y_l^2 + \phi$, say. But, we may not get enough sampled units bearing the sensitive rare diseases. This problem may be solved with an idea from adaptive cluster sampling (ACS) introduced by Thompson (1990) and Thompson and Seber (1996), providing rationale for extending the initial sample to include relevant additional sampled persons without disturbing unbiasedness in estimation.

4 ADAPTIVE CLUSTER SAMPLING

ACS is a modern sampling design and it was motivated by the problem of sampling rare, clustered populations. In ACS, an initial sample of units is selected and whenever the value of the variable of interest satisfies a specified condition, neighboring units are added to the sample. The sampling regions defined as "units" are selected based on values of the variables of interest observed during a sample survey. For example, a region containing endangered species or land contamination would be defined as a unit. Positive identifications meeting a specified criterion within a region are then used for future input. In a conventional sampling design, the selection for a sampling unit does not depend on previous observations made during an initial survey; entire sampling units are selected before any physical observation in the field. Therefore, conventional sampling guarantees that the calculated statistics will be unbiased. To use the adaptive sampling technique, however, different estimators may be employed to guarantee unbiasedness (Thompson, 1992). Thompson defines a "neighborhood" as all the surrounding units (or regions) within a given dimension. See Chaudhuri (2014b) for further elaborations covering unequal probability sampling.

In ACS, we have a population of N units labeled $1,2,\ldots,i,\ldots,N$. Associated with the ith member a characteristic or measurement y_i, for example, the number of animals on the ith plot is attached. Usually, we are interested in estimating such parameters as the mean per unit $\bar{Y}=\frac{1}{N}\sum_{i=1}^{N} y_i$ or the total $N\bar{Y}$. For every unit, we define a neighborhood that consists of the unit and a set of "neighboring units."

If the y value of a sampled unit satisfies a certain condition C, for example $y_i > C$, then the rest of the unit's neighborhood is added to the sample. If any other units in that neighborhood satisfy C, then their neighborhoods are also added to the sample. The process is continued until a cluster of units is obtained that contains a "boundary" of units called edge units that do not satisfy C. The final sample then consists of n (not necessarily distinct) clusters, one for each unit selected in the sample. We note that if a unit selected in the initial sample does not satisfy C, there is no augmentation and we have a cluster of size 1. The neighborhood relationship is assumed to be symmetric in the sense that if unit j is in the neighborhood of unit i, then i is also in the neighborhood of unit j. These neighborhoods do not depend on the population of y values. Although the cluster is the natural sample group, it is not useful because of the double role that edge units can play.

If an edge unit is selected in the initial sample, it forms a cluster of size 1. If it is not selected in the initial sample, then it can still be selected by being a member of any cluster for which it is an edge unit. Then the idea of "network" has been developed.

4.1 Network

A network A_i for unit i is defined as a cluster generated by unit i but with its edge units removed. A selection of any unit in A_i leads to the selection of all of A_i. If unit i is the only unit in a cluster satisfying C, then A_i consists of just unit i and forms a network of size 1. If any unit does not satisfy C, then it is also a network of size 1.

Chaudhuri (2000), Chaudhuri and Pal (2004), and Chaudhuri, Bose and Ghosh (2004) simplified the estimation dealing with ACS. From the idea of ACS, Revised Adaptive Randomized Response Surveys is introduced to estimate amount spent if any on the treatment of hospital diseases like AIDS. Chaudhuri et al. (2005) discuss steps to keep the size of adaptive sample in check of it goes on growing.

5 REVISED ADAPTIVE RANDOMIZED RESPONSE SURVEYS

In Revised Adaptive Randomized Response Surveys, a network may be assumed for every unit in the population. Let $A(i)$ denote the Network of the ith person ($i=1, 2, \ldots, U$). The unit i ($i=1, 2, \ldots, U$) is related with m_i persons where m_i denotes the cardinality of $A(i)$, which means the total number of persons in $A(i)$. We will get enough disease-related information if the ith unit possesses the sensitive diseases and more information may be obtained from his/her network $A(i)$.

Let $A(s) = \bigcup_{i\in s} A(i)$ be the Network sample derived from s.

Defining $t_i = \frac{1}{m_i}\sum_{l\in A(i)} y_l$, it may be observed that $T = \sum_1^N t_i = \sum_1^N y_i = Y$ (Chaudhuri, 2014b).

It is clear that to estimate Y is same as to estimate T. But y_i's are unknown. They may be estimated through an RR device.

Let $e_i = \frac{1}{m_i}\sum_{l\in A(i)} r_l$ be the adaptive RR for ith person $i=1,2,\ldots,N$. Choosing the initial sampling design p admitting positive values for $\pi_i = \sum_{s\ni i} p(s)$, it is convenient to employ $t = \sum_{i\in s}\frac{e_i}{\pi_i}$ to unbiasedly estimate θ (following Horvitz–Thompson estimator, (Horvitz and Thompson, 1952)).

Let E_p, V_p denote design-based expectation, variance operators, and $E = E_P E_R = E_R E_P, V = E_P V_R + V_p E_R = E_R V_P + V_R E_P$, the overall expectation, variance operators. Then, we get

$$E = E_P E_R(t) = E_P\left(\sum_{i\in s}\frac{E_R(e_i)}{\pi_i}\right)$$

$$= \sum_1^N\left[\frac{1}{m_i}\sum_{i\in A(i)} y_l\right] = \sum_1^N y_l\left[\frac{1}{m_i}\sum_{A(i)\ni l} 1\right] = \sum_{l=1}^N y_l = Y$$

since by definition $m_i = \sum_{A(i)\ni l} 1 =$ the cardinality of $A(i)$.

Thus, t is an unbiased estimator for $Y = \sum_1^N y_i$. Next, vide Chaudhuri (2014a)

$$V(t) = E_p V_R(t) + V_p E_R(t) = E_p\left[\sum_{i\in s}\frac{V_R(e_i)}{\pi_i^2}\right] + V_p\left[\sum_{i\in s}\frac{t_i}{\pi_i}\right]$$

$$= \sum_1^N \frac{V_R(e_i)}{\pi_i} + \sum_{i<j}\sum (\pi_i\pi_j - \pi_{ij})\left(\frac{t_i}{\pi_i} - \frac{t_j}{\pi_j}\right)^2$$

assuming every sample s has a common number of units each distinct.

Alternatively, we may also write

$$V'(t) = E_R V_p(t) + V_R E_p(t) = E_R V_p(t) + V_R E_p(t)$$

$$= E_R\left[\sum\sum_{i<j}(\pi_i\pi_j - \pi_{ij})\left(\frac{e_i}{\pi_i} - \frac{e_j}{\pi_j}\right)^2\right] + V_R\left[\sum_1^N e_i\right]$$

$$= \sum\sum_{i<j}(\pi_i\pi_j - \pi_{ij})\left[\left(\frac{t_i}{\pi_i} - \frac{t_j}{\pi_j}\right)^2 + \frac{V_R(e_i)}{{\pi_i}^2} + \frac{V_R(e_j)}{\pi_j^2}\right]$$

$$+ V_R\left[\sum_1^N e_i\right].$$

The variance $V(t)$ may be unbiasedly estimated by

$$v(t) = \sum_{i\in s}\frac{v_R(e_i)}{\pi_i^2} + \sum\sum_{i<j\in s}\frac{(\pi_i\pi_j - \pi_{ij})}{\pi_{ij}}\left(\frac{e_i}{\pi_i} - \frac{e_j}{\pi_j}\right)^{2.}$$

$$- \sum\sum_{i<j\in s}\frac{(\pi_i\pi_j - \pi_{ij})}{\pi_{ij}}\left(\frac{v_R(e_i)}{\pi_i^2} + \frac{v_R(e_j)}{\pi_j^2}\right).$$

Similarly $V'(t)$ may be unbiasedly estimated by

$$v'(t) = \sum_{i\in s}\frac{v_R(e_i)}{\pi_i} + \sum\sum_{i<j\in s}\frac{(\pi_i\pi_j - \pi_{ij})}{\pi_{ij}}\left(\frac{e_i}{\pi_i} - \frac{e_j}{\pi_j}\right)^2.$$

writing

$$v_R(e_i) = \frac{\sigma_a^2}{\mu_a^2}\frac{1}{m_i^2}\sum_{l\in A(i)} h_l^2 + \frac{\sigma_b^2}{\mu_a^2}\left(\frac{1}{m_i}\right) \quad \text{and} \quad h_l^2 = \frac{z_l^2 - \mu_b^2 - \sigma_b^2 - 2\mu_a\mu_b r_i}{\mu_a^2 + \sigma_a^2}.$$

So $\theta = \frac{1}{N}\sum_1^N y_i$ may be unbiasedly estimated by $\hat{\theta} = \frac{t}{N}$ and its variance is $V\left(\hat{\theta}\right) = \frac{V(t)}{N^2}$ or $\frac{V'(t)}{N^2}$. $V\left(\hat{\theta}\right)$ may be estimated by $\frac{v(t)}{N^2}$ or $\frac{v'(t)}{N^2}$.

6 SIMULATION STUDY

We use the data with $N = 113$ households for which we are interested to estimate average cost of hospital treatments of AIDS patients over the last 1 year in Kolkata. Our problem is to estimate $\theta = \frac{1}{N}\sum_1^N y_i$, where y_i is the treatment cost for patient, $i \in U$, the population. A sample of $n = 33$ households is selected by Hartley and Rao (1962) scheme using household income as the size measure.

Hartley and Rao (1962) have given a modified CSSPPS (circular systematic sampling with probability proportional to size) scheme for which the units of the population U are first randomly permuted and the CSSPPS scheme is applied on the realized permutation of the elements of U. For this scheme, sample size is fixed, $\pi_i = np_i$ and $\pi_{ij} > 0 \ \forall i, j, i \neq j$. One may refer to Murthy (1967) for ordinary CSSPPS scheme.

We have used the network for each sampled patient. The patients will give out their responses according to the RR procedure. The performances of the estimators based on RRs and adaptive RRs are judged by the following criteria.

In ACS, sample size restriction is a vital issue. The network size may be very large. Sometimes $A(s)$ is big and too costly to survey $A(s)$. To tackle that issue, Chaudhuri et al. (2005) introduced "Constrained Adaptive" sample in which subsamples are taken from each $A(i)$ where $i \in s$. But in our case we have skipped that issue as m_i's are moderate. In Table 1, the sizes of $A(s)$ are shown for different replicated samples (described later).

To examine the performance of any pair of an estimator e for a parameter admitting a positive variance estimator v, we take $R = 1000$ independently replicated samples (Table 2) and calculate the "Actual coverage percentage," ie,

TABLE 1 Estimates of θ and Estimated Coefficient of Variation for Simulated Samples ($\theta = 0.167$)

Iteration Number	Randomized Response *Estimate(t)/CV*	Adaptive Randomized Response *Estimate(t)/CV*	Size in ACS *Size of A(s)*
5000th	0.456/33.2	0.157/8.5	59
10,000th	0.456/33.2	0.174/11.89	78

TABLE 2 Overall Performances of RR Versus Adaptive RR Based on 10,000 Simulated Samples

Scheme	ACP RR/Adaptive RR	ACV RR/Adaptive RR	AL RR/Adaptive RR
(1) With $v(t)$	64.8/94.6	37.8/13.4	3452.89/1458.76
(2) With $v'(t)$	57.3/92.3	32.7/18.9	3998.5/1078.25

(a) ACP = the percentage of the replicates for which the interval (CI) $\left(e-1.96\sqrt{v}, e+1.96\sqrt{v}\right)$ covers the parameter θ—the closer it is numerically to 95 the better;

(b) the average coefficient of variation.
ACV = the average, over the R replicates, of the values of $100\frac{\sqrt{v}}{|e|}$ this reflects the length of CI—the smaller it is the better.

(c) Average length (AL) of the calculated CI—the smaller it is the better.
In this problem, $\theta=0.167$.

7 CONCLUDING REMARKS

In our sample, we do not get enough units with hospital treatments of sensitive diseases. In such a case, RRT is not enough to get the information. We have introduced the idea of ACS here. We have captured enough units in our sample with sensitive diseases.

From our simulation study, we may conclude that Revised Adaptive RR Surveys give better ACVs and the lengths of the CIs and a desirable magnitude of the ACP values.

REFERENCES

Chaudhuri, A., 2000. Network and adaptive sampling for unequal probabilities. Calcutta. Stat. Assoc. Bull. 50, 237–253.

Chaudhuri, A., 2011. Randomized Response and Indirect Questioning Techniques in Surveys. Chapman and Hall, CRC Press, Boca Raton, FL.

Chaudhuri, A., 2014a. Modern Survey Sampling. Chapman and Hall, CRC Press, Boca Raton, FL.

Chaudhuri, A., 2014b. Network and Adaptive Sampling. CRC Press, Boca Raton, FL.

Chaudhuri, A., Christofides, T., 2013. Indirect Questioning in Sample Surveys. Springer-Verlag, Berlin, Heidelberg.

Chaudhuri, A., Pal, S., 2004. Estimating domain-wise distribution of scarce objects by adaptive sampling and model-based borrowing of strength. J. Indian Soc. Agric. Stat. 58 (1), 136–145.

Chaudhuri, A., Bose, M., Ghosh, J.K., 2004. An application of adaptive sampling to estimate highly localized population segments. J. Stat. Plan. Inference 121, 175–189.

Chaudhuri, A., Mukherjee, R., 1988. Randomized responses: Theory and Techniques. Marcel Dekker, New York, NY.

Chaudhuri, A., Bose, M., Dihidar, K., 2005. Sample-size-restrictive adaptive sampling: an application in estimating localized elements. J. Stat. Plan. Inference 134 (1), 254–267.

Fox, J.A., Tracy, P.E., 1986. Randomized Response: A Method for Sensitive Surveys. Sage, London.

Hartley, H.O., Rao, J.N.K., 1962. Sampling with unequal probabilities and without replacement. Ann. Math. Stat. 33, 350–374.

Horvitz, D.G., Thompson, D.J., 1952. A generalization of sampling without replacement from a finite universe. J. Am. Stat. Assoc. 47, 663–685.

Murthy, M.N., 1967. Sampling Theory and Methods. Statistical Publishing Society, Calcutta.

Thompson, S.K., 1990. Adaptive cluster sampling. J. Am. Stat. Assoc. 85, 1050–1059.

Thompson, S.K., 1992. Sampling. John Wiley & Sons, New York, NY.

Thompson, S.K., Seber, G.A.F., 1996. Adaptive Sampling. John Wiley & Sons, New York, NY.

Warner, S.L., 1965. RR: a survey technique for eliminating evasive answer bias. J. Am. Stat. Assoc. 60, 63–69.

Chapter 25

Estimation of Means of Two Rare Sensitive Characteristics: Cramer–Rao Lower Bound of Variances

S.-C. Su*, C.-S. Lee†, S.A. Sedory‡ and S. Singh‡,1
*Freer Independent School District, Freer, TX, United States
†Coastal Bend College, Beeville, TX, United States
‡Texas A&M University-Kingsville, Kingsville, TX, United States
1Corresponding author: e-mail: sarjinder.sing@tamuk.edu

ABSTRACT

In this chapter, the problem of estimating the means of two rare sensitive characteristics has been considered by following the pioneer work of Land et al. (2012). The proposed estimators will be investigated analytically as well as empirically.

Keywords: Poisson distribution, Two or more rare sensitive characteristics, Cramer–Rao lower bound of variance

1 INTRODUCTION

The problem of estimating the proportion π of a sensitive characteristic in a population was first dealt with Warner (1965). He considers the case of taking a simple random and with replacement sample (SRSWR) of n respondents from a target population. He proposed a randomization device which works as a black box between an interviewer and an interviewee. An interviewer knows only the output from the response of an interviewer, but has no clue what happened between the interviewee and the randomization device. For example, one such device could be deck of cards with each card having one of the two statements: (i) "I belong to group A" and (ii) "I do not belong to group A." The statements on the cards in the deck occur with probabilities P and $(1-P)$, respectively. Each individual in the sample is asked to draw one card at random from the well-shuffled deck. Based on the statement

Handbook of Statistics, Vol. 34. http://dx.doi.org/10.1016/bs.host.2016.01.023

written on the card drawn by an interviewee, and without presenting the card to the interviewer, the interviewee responds to the question. Obviously, the number of interviewees, n_1, that answer "yes" is binomially distributed with parameters n and $P\pi + (1-P)(1-\pi)$. An unbiased and maximum-likelihood estimator of π exists for $P \neq 0.5$ and is given by:

$$\hat{\pi}_w = \frac{(n_1/n) - (1-P)}{2P - 1} \tag{1}$$

with variance:

$$V(\hat{\pi}_w) = \frac{\pi(1-\pi)}{n} + \frac{P(1-P)}{n(2P-1)^2} \tag{2}$$

In the pioneer Warner's randomization device, the two statements are associated with the same sensitive characteristic, A, or its complement A^c. Greenberg et al. (1969) propose that to protect respondents, it could be better to use two unrelated questions in a randomization device. They suggested an unrelated question randomization device with two questions: (i) Do you possess membership in group A? and (ii) Do you possess membership in group Y?, where membership in the group Y or its complement is innocuous and unrelated to membership in the sensitive group A. For example, the two questions may be: (a) Have you ever used an electric cigarette? and (b) Were you born in winter? Then the second question of being born in winter has nothing to do with use of an electronic cigarette. The Greenberg et al. (1969) model considered two situations involving π_y (the proportion of individuals possessing membership in group, Y): that where π_y is known and that where π_y is unknown. Greenberg et al. (1969) recommended that one of the optimal choices of P_i, $i = 1, 2$, should be close to zero and the other close to one. The choice of unrelated question should be based on a criterion they proposed, namely, that π_y should be close to zero or one according as $\pi < 0.5$ or $\pi > 0.5$.

Land et al. (2012) consider a different and unique problem where the number of persons possessing a rare sensitive attribute is very small, and a huge sample size is required to estimate this number. It is challenging issue to estimate the proportion of a rate sensitive attribute for a survey statistician. They claim that the capacity of our communication systems is increasing rapidly, so it should soon be possible to conduct such large randomized response surveys over the internet, by telephone, etc. They also suggest the need of some sort of conference to be organized by sociologists across the world to teach the general public about the role of randomized response devices in human surveys and to help create trust in its use because some respondents do not believe in the use of randomization devices in surveys. Land et al. (2012) assume π_1 be the true proportion of a rare sensitive attribute A_1 in a population Ω. For example, it may be the proportion of truly innocent women in jails. They suggest selecting a large sample of n persons from the population such that as

$n \to \infty$ and $\pi_1 \to 0$ then $n\pi_1 = \lambda_1$ (finite). Let π_y be the true proportion of the population having the rare unrelated attribute Y such that as $n \to \infty$ and $\pi_y \to 0$ then $n\pi_y = \lambda_y$ (finite and known). For example, π_y might be a proportion of girls born blind. Each respondent selected in the sample is requested to rotate a spinner bearing two types of statements: (a) Do you possess the rare sensitive attribute A_1? and (b) Do you possess the rare unrelated attribute Y? with probabilities P and $(1-P)$, respectively. Thus for Land et al. (2012) model, the probability of a "yes" answer is same as in Greenberg et al. (1969) model and is given by:

$$\theta_0 = P\pi_1 + (1-P)\pi_y \tag{3}$$

Assuming that $n \to \infty$ and $\theta_0 \to 0$ in such a way that $n\theta_0 = \lambda_0$ (finite). Let $y_1, y_2, \ldots, y_n$ be a random sample of n observations from the Poisson distribution with parameter λ_0. Land et al. (2012) developed an unbiased and maximum-likelihood estimator of λ_1 given by:

$$\hat{\lambda}_1 = \frac{1}{P}\left[\frac{1}{n}\sum_{i=1}^{n} y_i - (1-P)\lambda_y\right] \tag{4}$$

The variance of the estimator $\hat{\lambda}_1$ is given by:

$$V\left(\hat{\lambda}_1\right) = \frac{\lambda_1}{nP} + \frac{(1-P)\lambda_y}{nP^2} \tag{5}$$

A review of the literature on randomized response sampling as cited at several places, such as Fox (2016), Chaudhuri and Christofides (2013), Chaudhuri (2011), and Tracy and Mangat (1996), indicates that only a few contributions have been made to estimate sensitive multinomial proportions in a population. Abul-Ela et al. (1967) were the first to modify the Warner (1965) model for the multiproportion case when a population can be considered to be divided into t disjoint classes C_j with unknown proportions π_j ($j = 1,2,\ldots,\ t,\ 0 < \pi_j < 1,\ \sum \pi_j = 1$). They assume that at least one of the classes is sensitive in nature and at least one class is not sensitive or is unrelated to the stigma issues. They consider taking $s(=t-1)$ independent SRSWR samples of sizes n_i ($i = 1,2,\ldots,\ s,\ \sum n_i = n$), and then a randomized response device is employed to each one of the samples. They examined, in detail, the extent of bias and the mean square error of the estimators for a trinomial case with $t = 3$. Bourke and Dalenius (1973, 1974) used a Latin square measurement design to extend Warner's model to the multinomial case. Their design uses t different possible responses and requires only one sample. The respondent is asked to select one of the t-types of cards using a random device. Each of the t-mutually exclusive classes is described on each card, except that the order of the classes is permuted from card to card. The permutations for the t-cards form a Latin square. The respondent reads the cards selected and reports only the position on the card (ie, $t = 1,2,\ \ldots\ (t-1)$ or t)

of the statement describing the class to which he/she belongs. The unrelated question model was also extended by Bourke (1974) to estimate the proportion of a population in each of t-mutually exclusive classes of which $(t-1)$ are sensitive. Only one sample is needed if the distribution of the unrelated character is known; otherwise an additional sample is required. The design uses a deck of cards. Each of these cards contains a number of statements. The arrangement of the statements is a part of the design. Hochberg (1975) outlined an alternative scheme for estimating the t group proportions of which at a most $(t-2)$ are stigmatizing. The realizations for any sampled individuals constitute a two-stage scheme. The second stage is conditional on the random individual's response in the first stage. Drane (1976) used his "forced yes" stochastic model to estimate the proportion of more than one sensitive character. The use of supplemented block, balanced incomplete block, and spring balance weighing designs were introduced by Raghavarao and Federer (1979). Their models allow the surveyor to obtain answers to several sensitive questions. Mukhopadhyay (1980), Mukherjee (1981), Tamhane (1981), Bourke (1981, 1982, 1990), Silva (1983), and Christofides (2003) have also considered the estimation of multiattribute parameters. Chen and Singh (2012) used higher order moments of scrambling variables to estimate multiproportions.

In a very short span of time, the model due to Land et al. (2012) has been extended in several directions by Son and Kim (2015), Lee et al. (2013, 2014), Wakeel and Aslam (2013), and Singh and Tarray (2014) and has been cited by Fox (2016) and Chaudhuri and Christofides (2013). Lee, Sedory, and Singh (2013) consider the problems of two sensitive attributes and their overlap by using two decks of cards.

It is an interesting observation that no one has considered the problem of estimating two (or more) rare sensitive attributes by using a randomization device. This motivated the authors to consider such a situation.

2 ESTIMATION OF TWO RARE SENSITIVE ATTRIBUTES

Consider a situation, where a survey statistician wishes to estimate the prevalence of two rare sensitive characteristics in a population and later wishes to know the difference between the means of such rare sensitive attributes. On obvious method is to apply the Land et al. (2012) methodology twice as follows. Assume π_1 and π_2 be the true proportions of two rare sensitive attributes A_1 and A_2 in a population Ω. Select independently two large samples of sizes n_1 and n_2 persons from the population such that as $n_1 \to \infty$ and $\pi_1 \to 0$ then $n_1\pi_1 = \lambda_1$ (finite) and as $n_2 \to \infty$ and $\pi_2 \to 0$ then $n_2\pi_2 = \lambda_2$ (finite). Let π_y be the true proportion of the population having the rare unrelated attribute Y such that as $n_1 \to \infty$ (or as $n_2 \to \infty$) and $\pi_y \to 0$ then for $k = 1,2$ $n_k\pi_y = \lambda_y$ (finite and known). Each respondent selected in the first sample is requested to rotate a spinner bearing two types of statements: (a) Do you possess the rare sensitive attribute A_1? and (b) Do you possess the rare unrelated attribute Y? with probabilities P and $(1-P)$, respectively. Also each respondent selected in the second sample is requested to rotate a spinner bearing two types of statements: (a) Do you possess the rare sensitive attribute A_2? and (b) Do you possess the rare unrelated attribute Y? with probabilities T and $(1-T)$, respectively. Following Land et al. (2012) model, the probability of a "yes" answer in the first and second samples will be, respectively, given by:

$$\theta_{0(1)} = P\pi_1 + (1-P)\pi_y \tag{6}$$

and

$$\theta_{0(2)} = T\pi_2 + (1-T)\pi_y \tag{7}$$

Assume that as $n_k \to \infty$ and $\theta_{0(k)} \to 0$ we have $n_k\theta_{0(k)} = \lambda_{0(k)}$ (finite); this is not unconsiderable because all the three attributes A_1, A_2, and Y are assumed to be very rare in the population. Let $y_1, y_2, \ldots, y_{n_1}$ be a random sample of n_1 observations from the Poisson distribution with parameter $\lambda_{0(1)} = P\lambda_1 + (1-P)\lambda_y$. Then following Land et al. (2012) an unbiased and the maximum-likelihood estimator of λ_1 given by:

$$\hat{\lambda}_{1(L)} = \frac{1}{P}\left[\frac{1}{n_1}\sum_{i=1}^{n_1} y_i - (1-P)\lambda_y\right] \tag{8}$$

with variance of the estimator $\hat{\lambda}_{1(L)}$ given by:

$$V\left(\hat{\lambda}_{1(L)}\right) = \frac{\lambda_1}{n_1P} + \frac{(1-P)\lambda_y}{n_1P^2} \tag{9}$$

Let $x_1, x_2, \ldots, x_{n_2}$ be a random sample of n_2 observations from the Poisson distribution with parameter $\lambda_{0(2)} = T\lambda_2 + (1-T)\lambda_y$. Then again following Land et al. (2012), an unbiased and the maximum-likelihood estimator of λ_2 given by:

$$\hat{\lambda}_{2(L)} = \frac{1}{T}\left[\frac{1}{n_2}\sum_{i=1}^{n_2} x_i - (1-T)\lambda_y\right] \tag{10}$$

with variance of the estimator $\hat{\lambda}_{2(L)}$ given by:

$$V\left(\hat{\lambda}_{2(L)}\right) = \frac{\lambda_2}{n_2T} + \frac{(1-T)\lambda_y}{n_2T^2} \tag{11}$$

The difference between the population means of the two sensitive characteristics is given by:

$$\lambda_{\text{diff}} = \lambda_1 - \lambda_2 \tag{12}$$

An unbiased estimator of difference between the means of two sensitive characteristics is obviously given by:

$$\hat{\lambda}_{\text{diff}(L)} = \hat{\lambda}_{1(L)} - \hat{\lambda}_{2(L)} \tag{13}$$

Note that both samples are independent, thus the variance of the estimator $\hat{\lambda}_{\text{diff}(L)}$ is given by:

$$\begin{aligned} V\left(\hat{\lambda}_{\text{diff}(L)}\right) &= V\left(\hat{\lambda}_{1(L)}\right) + V\left(\hat{\lambda}_{2(L)}\right) \\ &= \frac{1}{n_1}\left[\frac{\lambda_1}{P} + \frac{(1-P)\lambda_y}{P^2}\right] + \frac{1}{n_2}\left[\frac{\lambda_2}{T} + \frac{(1-T)\lambda_y}{T^2}\right] \end{aligned} \tag{14}$$

The optimum values of n_1 and n_2 which minimize the variance in (14) are given by:

$$n_1 = \frac{n\sqrt{\frac{\lambda_1}{P} + \frac{(1-P)\lambda_y}{P^2}}}{\sqrt{\frac{\lambda_1}{P} + \frac{(1-P)\lambda_y}{P^2}} + \sqrt{\frac{\lambda_2}{T} + \frac{(1-T)\lambda_y}{T^2}}} \tag{15}$$

and

$$n_2 = \frac{n\sqrt{\frac{\lambda_2}{T} + \frac{(1-T)\lambda_y}{T^2}}}{\sqrt{\frac{\lambda_1}{P} + \frac{(1-P)\lambda_y}{P^2}} + \sqrt{\frac{\lambda_2}{T} + \frac{(1-T)\lambda_y}{T^2}}} \tag{16}$$

where $n = n_1 + n_2$, the total sample size. Using these optimum values of the sample sizes, the minimum variance of the estimator $\hat{\lambda}_{\text{diff}(L)}$ of the difference λ_{diff} between two means is give by:

$$\text{Min. } V\left(\hat{\lambda}_{\text{diff}(L)}\right) = \frac{\left[\sqrt{\lambda_1 P + (1-P)\lambda_y} + \sqrt{T\lambda_2 + (1-T)\lambda_y}\right]^2}{nP^2T^2} \tag{17}$$

An unbiased estimator of $V\left(\hat{\lambda}_{\text{diff}(L)}\right)$ in (14) is given by:

$$\hat{V}\left(\hat{\lambda}_{\text{diff}(L)}\right) = \frac{1}{n_1}\left[\frac{\hat{\lambda}_1}{P} + \frac{(1-P)\lambda_y}{P^2}\right] + \frac{1}{n_2}\left[\frac{\hat{\lambda}_2}{T} + \frac{(1-T)\lambda_y}{T^2}\right] \tag{18}$$

An estimator of the minimum variance in (17) is given by:

$$\text{Min. } \hat{V}\left(\hat{\lambda}_{\text{diff}(L)}\right) = \frac{\left[\sqrt{\hat{\lambda}_1 P + (1-P)\lambda_y} + \sqrt{T\hat{\lambda}_2 + (1-T)\lambda_y}\right]^2}{nP^2T^2} \tag{19}$$

Note that unbiased estimator of the minimum variance in (17) cannot be easily found.

In the next section, we propose a new randomization device for estimating prevalence of two or more sensitive characteristics in a population.

3 PROPOSED RANDOMIZED RESPONSE MODEL FOR TWO RARE SENSITIVE ATTRIBUTES

The proposed randomized response model also considers taking two large independent samples s_1 and s_2 consisting of n_1 and n_2 respondents from the population. Each respondent selected in the first sample is asked to use a randomization device, say a spinner, which could produce three types of outcomes: (i) Do you possess the rare sensitive attribute A_1?; (ii) Do you possess the rare sensitive attribute A_2?; and (iii) Do you possess the rare unrelated attribute Y? with probabilities P_1, P_2, and P_3, respectively, such that $P_1+P_2+P_3=1$. Thus for a respondent in the first sample, the probability of "yes" answer is given by:

$$P(\text{Yes})=\theta_1=P_1\pi_1+P_2\pi_2+P_3\pi_y \tag{20}$$

Assuming that $n_1\to\infty$ and $\theta_1\to 0$ such that $\mu_1=n_1\theta_1$ (finite), because all the three attributes A_1, A_2, and Y are very rare in the population. Let $x_{11},x_{12},\ldots,x_{1n_1}$ be a random sample of n_1 observations from the Poisson distribution with parameter

$$\mu_1=\theta_1=P_1(n_1\pi_1)+P_2(n_1\pi_2)+P_3\left(n_1\pi_y\right)$$

or

$$\mu_1=P_1\lambda_1+P_2\lambda_2+P_3\lambda_y \tag{21}$$

Each respondent selected in the second sample is asked to use another randomization device, say a spinner, which could produce three types of outcomes: (i) Do you possess the rare sensitive attribute A_1?; (ii) Do you possess the rare sensitive attribute A_2?; and (iii) Do you possess the rare unrelated attribute Y? with probabilities T_1, T_2, and T_3, respectively, such that $T_1+T_2+T_3=1$. Thus for a respondent in the second sample, the probability of "yes" answer is given by:

$$P(\text{Yes})=\theta_2=T_1\pi_1+T_2\pi_2+T_3\pi_y \tag{22}$$

Assuming that $n_2\to\infty$ and $\theta_2\to 0$ such that $\mu_2=n_2\theta_2$ (finite), because all the three attributes A_1, A_2, and Y are very rare in the population. Let $x_{21},x_{22},\ldots,x_{2n_2}$ be a random sample of n_2 observations from the Poisson distribution with parameter

$$\mu_2=\theta_2=T_1(n_2\pi_2)+T_2(n_2\pi_2)+T_3\left(n_2\pi_y\right)$$

or

$$\mu_2=T_1\lambda_1+T_2\lambda_2+T_3\lambda_y \tag{23}$$

where $P_1T_2\neq T_1P_2$. Note that both samples are independent, the likelihood function of the random sample of $n=n_1+n_2$ observations is given by:

$$L=\left(\prod_{i=1}^{n_1}\frac{\mathrm{e}^{-\mu_1}\mu_1^{x_{1i}}}{x_{1i}!}\right)\left(\prod_{i=1}^{n_2}\frac{\mathrm{e}^{-\mu_2}\mu_1^{x_{2i}}}{x_{2i}!}\right)=\frac{\mathrm{e}^{-(\mu_1+\mu_2)}\cdot\mu_1^{x_1}\cdot\mu_2^{x_2}}{x_1^*x_2^*} \tag{24}$$

where $x_1 = \sum_{i=1}^{n_1} x_{1i}$, $x_2 = \sum_{i=1}^{n_2} x_{2i}$, $x_1^* = \prod_{i=1}^{n_1} (x_{1i}!)$, and $x_2^* = \prod_{i=1}^{n_1} (x_{2i}!)$. Taking natural log on both sides of (24) we have

$$\begin{aligned}
\ln(L) &= \ln \frac{e^{-(\mu_1+\mu_2)} \cdot \mu_1^{x_1} \cdot \mu_2^{x_2}}{x_1^* x_2^*} \\
&= -(\mu_1+\mu_2) + x_1 \ln(\mu_1) + x_2 \ln(\mu_2) - \ln(x_1^*) - \ln(x_2^*) \\
&= -(P_1\lambda_1 + P_2\lambda_2 + P_3\lambda_3 + T_1\lambda_1 + T_2\lambda_2 + T_3\lambda_3) \\
&\quad + x_1 \ln(\mu_1) + x_2 \ln(\mu_2) - \ln(x_1^*) - \ln(x_2^*) \\
&= -(P_1+T_1)\lambda_1 - (P_2+T_2)\lambda_2 - (P_3+T_3)\lambda_3 \\
&\quad + x_1 \ln(\mu_1) + x_2 \ln(\mu_2) - \ln(x_1^*) - \ln(x_2^*)
\end{aligned} \tag{25}$$

For a given value of λ_y, the maximum-likelihood estimates $\hat{\lambda}_{1(\text{new})}$ and $\hat{\lambda}_{2(\text{new})}$ of λ_1 and λ_2 are solutions to the nonlinear equations given by:

$$\frac{\partial \ln(L)}{\partial \lambda_1} = -(P_1+T_1) + \frac{x_1 P_1}{\mu_1} + \frac{x_2 T_1}{\mu_2} = 0 \tag{26}$$

and

$$\frac{\partial \ln(L)}{\partial \lambda_2} = -(P_2+T_2) + \frac{x_1 P_2}{\mu_1} + \frac{x_2 T_2}{\mu_2} = 0 \tag{27}$$

A computer grid search code in any programming language, such as FORTRAN or R, could be written to find estimates a solution to the nonlinear equations in (26) and (27). Alternatively, already available standard SAS procedure PROC MODEL can be used to solve such nonlinear system of equations.

Let $\hat{\mu}_1$ and $\hat{\mu}_2$ be the estimates of μ_1 and μ_2 obtained by solving the Eqs. (26) and (27) by any iterative method, like Newton–Raphson method. Then the estimates of λ_1 and λ_2 will be given by:

$$\hat{\lambda}_{1(\text{new})} = \frac{(T_2\hat{\mu}_1 - P_2\hat{\mu}_2) + \lambda_y(P_2T_3 - T_2P_3)}{(P_1T_2 - T_1P_2)} \tag{28}$$

and

$$\hat{\lambda}_{2(\text{new})} = \frac{(P_1\hat{\mu}_2 - T_1\hat{\mu}_1) + \lambda_y(T_1P_3 - P_1T_3)}{(P_1T_2 - T_1P_2)} \tag{29}$$

Also note that:

$$\frac{\partial^2 \ln(L)}{\partial \lambda_1^2} = -\frac{x_1 P_1^2}{\mu_1^2} - \frac{x_2 T_1^2}{\mu_2^2} \tag{30}$$

$$\frac{\partial^2 \ln(L)}{\partial \lambda_2^2} = -\frac{x_1 P_2^2}{\mu_1^2} - \frac{x_2 T_2^2}{\mu_2^2} \tag{31}$$

and

$$\frac{\partial^2 \ln(L)}{\partial\lambda_1\partial\lambda_2} = \frac{\partial^2 \ln(L)}{\partial\lambda_2\partial\lambda_1} = -\frac{x_1P_1P_2}{\mu_1^2} - \frac{x_2T_1T_2}{\mu_2^2} \tag{32}$$

Using (30), (31), and (32), the Cramer–Rao lower bounds of variance–covariance for the estimators $\hat{\lambda}_{1(\text{new})}$ and $\hat{\lambda}_{2(\text{new})}$ are given by:

$$V\begin{bmatrix}\hat{\lambda}_{1(\text{new})}\\ \hat{\lambda}_{2(\text{new})}\end{bmatrix} \ge -E\begin{bmatrix}\frac{\partial^2 \ln(L)}{\partial\lambda_1^2} & \frac{\partial^2 \ln(L)}{\partial\lambda_1\partial\lambda_2}\\ \frac{\partial^2 \ln(L)}{\partial\lambda_2\partial\lambda_1} & \frac{\partial^2 \ln(L)}{\partial\lambda_2^2}\end{bmatrix}^{-1}$$

$$= \frac{1}{(P_1T_2 - T_1P_2)^2}\begin{bmatrix}\frac{P_2^2\mu_2}{n_2} + \frac{T_2^2\mu_1}{n_1} & -\left(\frac{P_1P_2\mu_2}{n_2} + \frac{T_1T_2\mu_1}{n_1}\right)\\ -\left(\frac{P_1P_2\mu_2}{n_2} + \frac{T_1T_2\mu_1}{n_1}\right) & \frac{P_1^2\mu_2}{n_2} + \frac{T_1^2\mu_1}{n_1}\end{bmatrix} \tag{33}$$

From (33), we have

$$V\left(\hat{\lambda}_{1(\text{new})}\right) = \frac{1}{(P_1T_2 - T_1P_2)^2}\left(\frac{P_2^2\mu_2}{n_2} + \frac{T_2^2\mu_1}{n_1}\right) \tag{34}$$

$$V\left(\hat{\lambda}_{2(\text{new})}\right) = \frac{1}{(P_1T_2 - T_1P_2)^2}\left(\frac{P_1^2\mu_2}{n_2} + \frac{T_1^2\mu_1}{n_1}\right) \tag{35}$$

and

$$\text{Cov}\left(\hat{\lambda}_{1(\text{new})}, \hat{\lambda}_{2(\text{new})}\right) = \frac{-1}{(P_1T_2 - T_1P_2)^2}\left(\frac{P_1P_2\mu_2}{n_2} + \frac{T_1T_2\mu_1}{n_1}\right) \tag{36}$$

Now a new unbiased estimator of the difference between the means of two sensitive characteristics is obviously given by:

$$\hat{\lambda}_{\text{diff(new)}} = \hat{\lambda}_{1(\text{new})} - \hat{\lambda}_{2(\text{new})} \tag{37}$$

Now the variance of the estimator $\hat{\lambda}_{\text{diff(new)}}$ is given by:

$$\begin{aligned}V\left(\hat{\lambda}_{\text{diff(new)}}\right) &= V\left(\hat{\lambda}_{1(\text{new})}\right) + V\left(\hat{\lambda}_{2(\text{new})}\right) - 2\text{Cov}\left(\hat{\lambda}_{1(\text{new})}, \hat{\lambda}_{2(\text{new})}\right)\\ &= \frac{1}{(P_1T_2 - T_1P_2)^2}\left[\frac{P_2^2\mu_2}{n_2} + \frac{T_2^2\mu_1}{n_1} + \frac{P_1^2\mu_2}{n_2} + \frac{T_1^2\mu_1}{n_1} + 2\left(\frac{P_1P_2\mu_2}{n_2} + \frac{T_1T_2\mu_1}{n_1}\right)\right]\\ &= \frac{1}{(P_1T_2 - T_1P_2)^2}\left[\frac{\mu_2(P_2+P_1)^2}{n_2} + \frac{\mu_1(T_2+T_1)^2}{n_1}\right]\end{aligned} \tag{38}$$

The optimum values of n_1 and n_2 which minimize the variance in (38) are given by:

$$n_1 = \frac{n(P_2+P_1)\sqrt{\mu_2}}{(P_2+P_1)\sqrt{\mu_2}+(T_2+T_1)\sqrt{\mu_1}} \tag{39}$$

and

$$n_2 = \frac{n(T_2+T_1)\sqrt{\mu_1}}{(P_2+P_1)\sqrt{\mu_2}+(T_2+T_1)\sqrt{\mu_1}} \tag{40}$$

where $n = n_1 + n_2$, the total sample size. For the optimum values of the sample sizes, the minimum variance of the estimator $\hat{\lambda}_{\text{diff(new)}}$ of the difference λ_{diff} between two means is give by:

$$\text{Min. } V\left(\hat{\lambda}_{\text{diff(new)}}\right) = \frac{\left[(P_2+P_1)\sqrt{\mu_2}+(T_2+T_1)\sqrt{\mu_1}\right]^2}{n(P_1T_2-T_1P_2)^2} \tag{41}$$

An unbiased estimator of the variance, $V\left(\hat{\lambda}_{\text{diff(new)}}\right)$, in (36) is given by:

$$\hat{V}\left(\hat{\lambda}_{\text{diff (new)}}\right) = \frac{1}{(P_1T_2-T_1P_2)^2}\left[\frac{\hat{\mu}_2(P_2+P_1)^2}{n_2}+\frac{\hat{\mu}_1(T_2+T_1)^2}{n_1}\right] \tag{42}$$

An estimator of the minimum variance in (41) is given by:

$$\text{Min. } V\left(\hat{\lambda}_{\text{diff(new)}}\right) = \frac{\left[(P_2+P_1)\sqrt{\mu_2}+(T_2+T_1)\sqrt{\mu_1}\right]^2}{n(P_1T_2-T_1P_2)^2} \tag{43}$$

Note that unbiased estimator of the minimum variance in (43) cannot be easily found.

In the next section, we compare the two methods for estimating the difference between two population means.

4 RELATIVE EFFICIENCY

The percent relative efficiency of the newly proposed randomization device with respect to the Land et al. (2012) model is defined as:

$$\begin{aligned} \text{RE} &= \frac{\text{Min.} V\left(\hat{\lambda}_{\text{diff}(L)}\right)}{\text{Min.} V\left(\hat{\lambda}_{\text{diff(new)}}\right)} \times 100\% \\ &= \frac{\left[\sqrt{\lambda_1 P+(1-P)\lambda_y}+\sqrt{T\lambda_2+(1-T)\lambda_y}\right]^2(P_1T_2-T_1P_2)^2}{P^2T^2\left[(P_2+P_1)\sqrt{\mu_2}+(T_2+T_1)\sqrt{\mu_1}\right]} \times 100\% \end{aligned} \tag{44}$$

Note that the value of percent relative efficiency in (44) is independent of the sample size n. To investigate the performance of the proposed estimator, we fixed the randomization device parameters as $P = T = 0.7$, $P_1 = 0.7$, $P_2 = 0.2$, $P_3 = 0.1$, $T_1 = 0.1$, $T_2 = 0.8$, and $T_3 = 0.1$. Then we varied the values of λ_1, λ_2, and λ_y between 0.5 and 2.0 with a step of 0.5. The results obtained are presented in Table 1. For details, refer to the FORTRAN code CYN10.F95 in the Appendix.

TABLE 1 Relative Efficiency Values for Different Values of λ_1, λ_2, and λ_y

λ_1	λ_2	λ_y	RE	λ_1	λ_2	λ_y	RE
0.5	0.5	0.5	149.9	1.5	0.5	0.5	139.3
0.5	0.5	1.0	177.2	1.5	0.5	1.0	156.9
0.5	0.5	1.5	199.9	1.5	0.5	1.5	172.3
0.5	0.5	2.0	219.1	1.5	0.5	2.0	186.1
0.5	1.0	0.5	134.0	1.5	1.0	0.5	133.2
0.5	1.0	1.0	154.2	1.5	1.0	1.0	146.6
0.5	1.0	1.5	171.9	1.5	1.0	1.5	158.8
0.5	1.0	2.0	187.6	1.5	1.0	2.0	170.1
0.5	1.5	0.5	124.7	1.5	1.5	0.5	128.5
0.5	1.5	1.0	141.2	1.5	1.5	1.0	139.6
0.5	1.5	1.5	156.0	1.5	1.5	1.5	149.9
0.5	1.5	2.0	169.3	1.5	1.5	2.0	159.6
0.5	2.0	0.5	118.2	1.5	2.0	0.5	124.7
0.5	2.0	1.0	132.5	1.5	2.0	1.0	134.3
0.5	2.0	1.5	145.3	1.5	2.0	1.5	143.4
0.5	2.0	2.0	157.0	1.5	2.0	2.0	151.9
1.0	0.5	0.5	143.8	2.0	0.5	0.5	135.8
1.0	0.5	1.0	164.7	2.0	0.5	1.0	151.2
1.0	0.5	1.5	182.9	2.0	0.5	1.5	164.8
1.0	0.5	2.0	198.8	2.0	0.5	2.0	177.1
1.0	1.0	0.5	134.2	2.0	1.0	0.5	132.0
1.0	1.0	1.0	149.9	2.0	1.0	1.0	143.8
1.0	1.0	1.5	164.2	2.0	1.0	1.5	154.6
1.0	1.0	2.0	177.2	2.0	1.0	2.0	164.7
1.0	1.5	0.5	127.6	2.0	1.5	0.5	128.6
1.0	1.5	1.0	140.7	2.0	1.5	1.0	138.4
1.0	1.5	1.5	152.7	2.0	1.5	1.5	147.6
1.0	1.5	2.0	163.7	2.0	1.5	2.0	156.3
1.0	2.0	0.5	122.7	2.0	2.0	0.5	125.6
1.0	2.0	1.0	134.0	2.0	2.0	1.0	134.2
1.0	2.0	1.5	144.5	2.0	2.0	1.5	142.2
1.0	2.0	2.0	154.2	2.0	2.0	2.0	149.9

It has been observed that for all the 64 cases considered in Table 1 the average value of the percent relative efficiency is 152.13% with a standard deviation of 20.62. The minimum value of the percent relative efficiency is 118.24% and maximum is 219.14% with a median value of 149.94. Note that we have only reported these choices of the device parameters, but one could search others if required by using the FORTRAN codes in the Appendix.

ACKNOWLEDGMENTS

The authors are thankful to Prof. Arijit Chaudhuri, Purnima Shaw, and a referee for their very constructive comments on the original version of this chapter.

APPENDIX

```
! FORTRAN CODES CYN10.F95

      USE NUMERICAL_LIBRARIES
      IMPLICIT NONE
      REAL P, T, AL1, AL2, ALY
      REAL P1, P2, P3, T1,T2, T3, AMU1, AMU2
      REAL FACT1, FACT2, FACT3, FACT4
      REAL ANOM, DENO, RE
      CHARACTER*20 OUT_FILE
     CHARACTER*20 IN_FILE
     WRITE(*,'(A)') 'NAME OF THE OUTPUT FILE'
      READ(*,'(A20)') OUT_FILE
      OPEN(42, FILE=OUT_FILE, STATUS='UNKNOWN')
      P = 0.7
     T = 0.7
     DO 10 P1 = 0.1, 0.8, 0.6
     DO 10 P2 = 0.2, 0.8, 0.7
     DO 10 T1 = 0.1, 0.8, 0.7
     DO 10 T2 = 0.2, 0.8, 0.6
     P3 = 1-P1-P2
     T3 = 1-T1-T2
     DO 10 AL1 = 0.5, 2.0, 0.5
     DO 10 AL2 = 0.5, 2.0, 0.5
     DO 10 ALY = 0.5, 2.0, 0.5
     AMU1 = P1*AL1+P2*AL2+P3*ALY
     AMU2 = T1*AL1+T2*AL2+T3*ALY
     FACT1 = AL1*P+(1-P)*ALY
     FACT2 = AL2*T+(1-T)*ALY
     FACT3 = AMU2*(P2+P1)**2
     FACT4 = AMU1*(T2+T1)**2
      ANOM = (SQRT(FACT1)+ SQRT(FACT2))**2*(P1*T2-T1*P2)**2
      DENO = P**2*T**2*(SQRT(FACT3)+SQRT(FACT4))**2
       RE = ANOM*100/DENO
```

```
       IF ((P3.GT.0.0).AND.(T3.GT.0.0).AND.(RE.GT.110))THEN
       WRITE(42,101)AL1,AL2, ALY, P,T,P1,P2,P3,T1,T2,T3,RE
101    FORMAT(2X,12(F7.3,1X))
       ENDIF
10     CONTINUE
      STOP
      END
```

REFERENCES

Abul-Ela, A.L.A., Greenberg, B.G., Horvitz, D.G., 1967. A multi-proportion randomized response model. J. Am. Stat. Assoc. 62, 990–1008.

Bourke, P.D., 1974. Multi-proportions randomized response using the unrelated question technique: Report No. 74 of the Errors in Survey research project. Institute of Statistics, University of Stockholm.

Bourke, P.D., 1981. On the analysis of some multivariate randomized response designs for categorical data. J. Stat. Plan. Inference 5, 165–170.

Bourke, P.D., 1982. RR multivariate designs for categorical data. Commun. Stat. Theor. Meth. 11, 2889–2901.

Bourke, P.D., 1990. Estimating a distribution function for each category of a sensitive variable. Commun. Stat. Theor. Meth. 19 (9), 3233–3241.

Bourke, P.D., Dalenious, T., 1974. RR models with lying: Technical Report-71. Institute of Statistics, University of Stockholm.

Bourke, P.D., Dalenius, T., 1973. Multi-proportions randomized response using a single sample: Report No. 68 of the Errors in Survey research project. Institute of Statistics, University of Stockholm.

Chaudhuri, A., 2011. Randomized Response and Indirect Questioning Techniques in Surveys. Chapman & Hall, CRC Press, Boca Raton, FL.

Chaudhuri, A., Christofides, T.C., 2013. Indirect Questioning in Sample Surveys. Springer, Heidelberg.

Chen, C.C., Singh, S., 2012. Estimation of multinomial proportions using higher order moments of scrambling variables in randomized response sampling. J. Mod. App. Stat. Meth. 11 (1), 106–122.

Christofides, T.C., 2003. A generalized randomized response technique. Metrika 57 (2), 195–200.

Drane, W., 1976. On the theory of randomized responses to two sensitive questions. Commun. Stat. Theor. Meth. A 5, 565–574.

Fox, J.A., 2016. Randomized Response and Related Methods. SAGE, Los Angeles, CA, ISBN: 978-1-4833-8103-9.

Greenberg, B.G., Abul-Ela, A.L.A., Simmons, W.R., Horvitz, D.G., 1969. The unrelated question randomized response model: theoretical framework. J. Am. Stat. Assoc. 64, 520–539.

Hochberg, Y., 1975. Two-stage randomized response scheme for estimating a multinomial. Commun. Stat. Theor. Meth. 4, 1021–1032.

Land, M., Singh, S., Sedory, S.A., 2012. Estimation of a rare sensitive attribute using Poisson distribution. Stat. J. Theor. Appl. Stat. 46 (3), 351–360.

Lee, C.-S., Sedory, S.A., Singh, S., 2013. Estimating at least seven measures for qualitative variables using randomized response sampling. Stat. Prob. Lett. 83, 399–409.

Lee, G.-S., Uhm, D., Kim, J.-M., 2013. Estimation of a rare sensitive attribute in a stratified sample using Poisson distribution. Stat. J. Theor. Appl. Stat. 47 (3), 575–589.

Lee, G.-S., Uhm, D., Kim, J.-M., 2014. Estimation of a rare sensitive attribute in probability proportional to size measures using Poisson distribution. Stat. J. Theor. Appl. Stat. 48 (3), 685–709.

Mukherjee, R., 1981. Inference on confidential characters from survey data. Calcutta Stat. Assoc. Bull. 30, 77–88.

Mukhopadhyay, P., 1980. On the estimation of some confidential characters from survey. Calcutta Stat. Assoc. Bull. 29, 77–88.

Raghavarao, D., Federer, W.T., 1979. Block total response as an alternative to the randomized response method. J. R. Stat. Soc. Ser. B 41, 40–45.

Silva, L.C., 1983. On the generalized randomized response model with polychotomous variables. Rev. Invest. Operac 4 (III), 75–100.

Singh, H.P., Tarray, T.A., 2014. A dexterous randomized response model for estimating a rare sensitive attribute using Poisson distribution. Stat. Probabil. Lett. 90, 42–45.

Son, C.-K., Kim, J.-M., 2015. Calibration estimation of a rare sensitive attribute with Poisson distribution. Commun. Stat. Theor. Meth. 44, 855–871.

Tamhane, A.C., 1981. Randomized response techniques for multiple attributes. J. Am. Stat. Assoc. 76, 916–923.

Tracy, D.S., Mangat, N.S., 1996. Some developments in randomized response sampling during the last decade—a follow up of review by Chaudhuri and Mukerjee. J. Appl. Stat. Sci. 4 (2/3), 147–158.

Wakeel, A., Aslam, I., 2013. Bayesian estimation of rare sensitive attribute. Thailand Stat. 11 (1), 17–29.

Warner, S.L., 1965. Randomized response: a survey technique for eliminating evasive answer bias. J. Am. Stat. Assoc. 60, 63–69.

Chapter 26

Estimating Sensitive Population Proportion by Generating Randomized Response Following Direct and Inverse Hypergeometric Distribution

K. Dihidar[1]

Sampling and Official Statistics Unit, Indian Statistical Institute, Kolkata, West Bengal, India
[1]Corresponding author: e-mail: kajaldihidar@gmail.com; dkajal@isical.ac.in

ABSTRACT

We consider estimating the proportion of people containing sensitive attributes like habitual drunkenness, drug addiction, reckless car driving, evading the tax liabilities, etc., in a given community. Following the pioneering work of Singh and Sedory (2013), we examine the effectiveness of generating randomized responses by negative hypergeometric distribution in respect to generating randomized responses by direct hypergeometric distribution. We consider sampling of respondents by general sampling schemes having the positive inclusion probabilities for single and paired population units. Essential theoretical derivations for unbiased estimator, variance and variance estimators are presented here. We perform a numerical illustration for comparison purpose which support the usefulness of Singh and Sedory's (2013) negative hypergeometric approach.

Keywords: Sensitive population proportion, Randomized response techniques, Hypergeometric distribution, Negative hypergeometric distribution, Unequal probability sampling

1 INTRODUCTION

A difficult problem faced while conducting surveys is to gather reliable data on stigmatizing or sensitive topics such as drug addiction, induced abortion, drunken driving, habitual tax evasion, excessive gambling, etc. This is difficult because many respondents may be reluctant to truthfully answer directly on such topics. To overcome this difficulty, Warner (1965) pioneered a

Handbook of Statistics, Vol. 34. http://dx.doi.org/10.1016/bs.host.2016.01.024

randomized response (RR) technique for estimating the proportion of persons in a given community belonging to some sensitive group A, say, on the basis of simple random sampling with replacement (SRSWR) of respondents. Since then many authors have made significant contributions in this area, for instance, Horvitz et al. (1967, 1976), Greenberg et al. (1969), Mukerjee (1981), Kuk (1990), Mangat et al. (1995), Christofides (2003, 2005), Chaudhuri and Pal (2008), Singh and Grewal (2013), Chaudhuri and Dihidar (2014) among others. For an excellent exposition on different developments in the area of randomized response, we refer to Chaudhuri and Mukerjee (1988) and Chaudhuri (2011). We give some details below.

In Warner's (1965) method, each respondent is provided with a randomization device by which he/she chooses one of two questions "Do you belong to A" or "Do you belong to A^c" with respect to probabilities, say, $p : (1 - p)$, where $p \neq 1/2$. The selected respondent is requested to draw randomly one card from the box and is requested to report the "match" or "nonmatch" of his own characteristic with the question written on the card drawn by him. These RR's gathered from a sample of persons provide an unbiased estimator for the sensitive population proportion, say, θ. The variance of this estimator and an unbiased estimator for that variance are also given by Warner (1965). Kuk (1990) for the same problem proposed a scheme for generating randomized responses and gave an unbiased estimator for θ under SRSWR. In his proposed device, each of the two boxes filled with cards of two types, say, red and blue with their mixing proportions being $p_1 : (1 - p_1), 0 < p_1 < 1$ in one box and $p_2 : (1 - p_2), 0 < p_2 < 1$ in the other; $p_1 \neq p_2$ and $p_1 + p_2 \neq 1$. Each selected person in an SRSWR is given these two boxes and is requested to draw cards K times independently, either from the first box or from the second, according as whether this person bears characteristic A or not. The person selected in every draw is requested to give the number of red cards obtained out of the K cards drawn. Based on these RRs an unbiased estimator for θ, variance and variance estimator are duly available.

In Kuk's (1990) model, the cards are drawn from either of the two boxes by with replacement. A natural question arises what will happen if the cards are drawn without replacement. In this chapter we investigate this matter. Also, we know that while drawing n cards without replacement from a box containing two types of cards, the number of first type of cards obtained follows hypergeometric distribution, the number of trials to obtain a fixed number of first type of cards follows the negative hypergeometric distribution. Following the pioneering work of Al-Sobhi et al. (2014) and Singh and Sedory (2013), we consider estimating the sensitive population proportion by generating RRs following both the direct and negative hypergeometric distribution. Also, keeping in mind that many large scale sample surveys consist of sampling of respondents by unequal probability sampling even without replacement, in this chapter we develop unbiased estimators for sensitive population proportion by general sampling schemes instead of only simple random sampling with replacement scheme of respondents. We organize our findings of this research work in the following sections.

In Section 2, we present the necessary derivations for generating RRs by direct hypergeometric distribution. In Section 3, we present the same by negative hypergeometric distribution. In Section 4, we present the unbiased estimators for θ, variance and variance estimators based on some sampling methods, namely simple random sampling (SRS) both with and without replacement (WR and WOR), and some unequal probability sampling methods, namely probability proportional to size with replacement (PPSWR), Rao, Hartley, and Cochran's (1962) and Midzuno's (1952) sampling schemes. We present the numerical illustration in Section 5 for comparison purpose. Finally, we give some concluding remarks in Section 6.

2 GENERATING RR BY HYPERGEOMETRIC DISTRIBUTION

Let $U = (1,\ldots,i,\ldots,N)$ denote a finite population of N persons labeled 1 through N. Let

$$\begin{aligned} y_i &= 1 \text{ if } i^{th} \text{ person bears the sensitive characteristic } A \\ &= 0, \text{otherwise.} \end{aligned}$$

Our objective is to estimate the population proportion $\theta = \frac{1}{N}\sum_{i=1}^{N} y_i$ bearing the sensitive characteristic A, using randomized response technique (RRT).

We consider a modification of Kuk's (1990) proposed scheme for generating randomized responses in the following way. In this modified scheme, two randomized response boxes are used, and each of the two boxes are filled with cards of two types, say red and blue; say the first box contains total N_1 number of cards, of which r_1 cards are red and the rest $N_1 - r_1$ cards are blue; the second box contains total N_2 number of cards, of which r_2 cards are red and the rest $N_2 - r_2$ cards are blue; and $r_1/N_1 \neq r_2/N_2$. Each respondent in sample s of units collected with a given probability $p(s) > 0$ according to a given sampling design p is given two boxes and requested to draw cards K times without replacement, either from the first box or from the second, according as whether this person bears the sensitive characteristic A or not. The ith selected person is requested to give the randomized response as the number of red cards out of the K cards drawn. Thus if f_i denotes the number of red cards happened to be obtained out of the K trials as reported by the person labeled i, then following the approach of Chaudhuri (2001) we can have

$$E_R(f_i) = K\left[y_i\frac{r_1}{N_1} + (1-y_i)\frac{r_2}{N_2}\right],$$

and

$$V_R(f_i) = K\left[y_i\frac{r_1}{N_1}\frac{N_1-r_1}{N_1}\frac{N_1-K}{N_1-1} + (1-y_i)\frac{r_2}{N_2}\frac{N_2-r_2}{N_2}\frac{N_2-K}{N_2-1}\right].$$

This leads to

$$E_R(f_i) = K\left[y_i\left(\frac{r_1}{N_1} - \frac{r_2}{N_2}\right) + \frac{r_2}{N_2}\right].$$

This implies that

$$E_R\left[\frac{\dfrac{f_i}{K}-\dfrac{r_2}{N_2}}{\dfrac{r_1}{N_1}-\dfrac{r_2}{N_2}}\right]=y_i.$$

So if we call

$$r_i=\frac{\dfrac{f_i}{K}-\dfrac{r_2}{N_2}}{\dfrac{r_1}{N_1}-\dfrac{r_2}{N_2}}\text{, then we have } E_R(r_i)=y_i.$$

And

$$V_R(r_i)=\frac{1}{\left(\frac{r_1}{N_1}-\frac{r_2}{N_2}\right)^2}\left[\frac{1}{K^2}V_R(f_i)\right]$$

$$=\frac{1}{\left(\frac{r_1}{N_1}-\frac{r_2}{N_2}\right)^2}\frac{1}{K^2}K\left[y_i\frac{r_1}{N_1}\frac{N_1-r_1}{N_1}\frac{N_1-K}{N_1-1}+(1-y_i)\frac{r_2}{N_2}\frac{N_2-r_2}{N_2}\frac{N_2-K}{N_2-1}\right]$$

$$=\frac{1}{K\left(\frac{r_1}{N_1}-\frac{r_2}{N_2}\right)^2}\left[y_i\left\{\frac{r_1}{N_1}\frac{N_1-r_1}{N_1}\frac{N_1-K}{N_1-1}-\frac{r_2}{N_2}\frac{N_2-r_2}{N_2}\frac{N_2-K}{N_2-1}\right\}\right.$$

$$\left.+\frac{r_2}{N_2}\frac{N_2-r_2}{N_2}\frac{N_2-K}{N_2-1}\right]=ay_i+b\text{, say, where}$$

$$a=\frac{1}{K\left(\frac{r_1}{N_1}-\frac{r_2}{N_2}\right)^2}\left[\frac{r_1}{N_1}\frac{N_1-r_1}{N_1}\frac{N_1-K}{N_1-1}-\frac{r_2}{N_2}\frac{N_2-r_2}{N_2}\frac{N_2-K}{N_2-1}\right],$$

$$b=\frac{1}{K\left(\frac{r_1}{N_1}-\frac{r_2}{N_2}\right)^2}\left[\frac{r_2}{N_2}\frac{N_2-r_2}{N_2}\frac{N_2-K}{N_2-1}\right].$$

Then an unbiased estimator for $V_i = V_R(r_i)$ is

$$\hat{V}_R(r_i)=v_i=ar_i+b, \quad i\in s.$$

3 GENERATING RR BY NEGATIVE HYPERGEOMETRIC DISTRIBUTION

Following the pioneering approach of Singh and Sedory (2013), we consider that the selected respondent is requested to use Box1 if he belongs to sensitive group A, or to use Box2 if he belongs to group A^c.

If the respondent uses Box1, he is requested to draw balls without replacement until he gets $t_1(< r_1)$ red balls. Then he is requested to report the number of draws, say, it is $G = g$, where G is the random variable denoting the number of draws from Box1.

If the respondent uses Box2, he is requested to draw balls without replacement until he gets $t_2(< r_2)$ red balls. Then he is requested to report the number of draws, say, it is $H = h$, where H is the random variable denoting the number of draws from Box 2.

Then G follows a negative hypergeometric distribution given by:

$$P(G=g|N_1,r_1,t_1)=\frac{\binom{r_1}{t_1-1}\binom{N_1-r_1}{g-t_1}}{\binom{N_1}{g-1}}\times\frac{r_1-t_1+1}{N_1-g+1}$$

$$=\frac{\binom{g-1}{t_1-1}\binom{N_1-g}{r_1-t_1}}{\binom{N_1}{r_1}},\ g=t_1,t_1+1,t_1+2,\ldots,(N_1-r_1+t_1).$$

Similarly, the random variable H follows the negative hypergeometric distribution with different parameters given by:

$$P(H=h|N_2,r_2,t_2)=\frac{\binom{h-1}{t_2-1}\binom{N_2-h}{r_2-t_2}}{\binom{N_2}{r_2}},\ h=t_2,t_2+1,t_2+2,\ldots,(N_2-r_2+t_2).$$

So, if Z_i denotes the randomized response obtained from i^{th} chosen person, then

$$\begin{aligned} Z_i &= G \text{ if } i^{th} \text{ person bears } A \\ &= H, \text{if } i^{th} \text{ person bears } A^c. \end{aligned}$$

Let us denote E_R, V_R, C_R as the expectation, variance and covariance operators with respect to randomized response generation. On noting that the expectation and variance of the negative hypergeometric distribution $G(N_1,r_1,t_1)$ are

$$E(G(N_1,r_1,t_1))=t_1\frac{N_1+1}{r_1+1},$$

$$V(G(N_1,r_1,t_1))=t_1\frac{(N_1-r_1)(N_1+1)(r_1+1-t_1)}{(r_1+1)^2(r_1+2)}$$

and similarly for the random variable H, we have

$$E_R(Z_i) = y_i E_R(G) + (1-y_i)E_R(H) = y_i t_1 \frac{N_1+1}{r_1+1} + (1-y_i)t_2 \frac{N_2+1}{r_2+1}$$
$$= y_i \left[\frac{t_1(N_1+1)(r_2+1) - t_2(N_2+1)(r_1+1)}{(r_1+1)(r_2+1)}\right] + t_2\left(\frac{N_2+1}{r_2+1}\right)$$

This implies that if

$$r_i' = \frac{Z_i - t_2\left(\frac{N_2+1}{r_2+1}\right)}{\frac{t_1(N_1+1)(r_2+1)-t_2(N_2+1)(r_1+1)}{(r_1+1)(r_2+1)}},$$

then

$$E_R(r_i') = y_i.$$

We now note that

$$V_R(r_i') = \frac{V_R(Z_i)}{\left[\frac{t_1(N_1+1)(r_2+1)-t_2(N_2+1)(r_1+1)}{(r_1+1)(r_2+1)}\right]^2},$$

and

$$V_R(Z_i) = E_R(Z_i^2) - (E_R(Z_i))^2$$
$$= y_i E_R(G^2) + (1-y_i)E_R(H^2) - (E_R(Z_i))^2$$
$$= y_i(V_R(G) + (E_R(G))^2 + (1-y_i)(V_R(H) + (E_R(H))^2) - (E_R(Z_i))^2$$
$$= cy_i + d = V_i', \text{ say, where if}$$

$$\phi = \left[\frac{t_1(N_1+1)}{(r_1+1)} - \frac{t_2(N_2+1)}{(r_2+1)}\right]^2,$$

$$c = \frac{\frac{t_1(N_1-r_1)(N_1+1)(r_1+1-t_1)}{(r_1+1)^2)(r_1+2)} - \frac{t_2(N_2-r_2)(N_2+1)(r_2+1-t_2)}{(r_2+1)^2)(r_2+2)}}{\phi},$$

$$d = \frac{t_2(N_2-r_2)(N_2+1)(r_2+1-t_2)}{(r_2+1)^2)(r_2+2)\phi}.$$

An unbiased estimator for $V_i' = V_R(r_i')$ is

$$\hat{V}_R(r_i') = v_i' = cr_i' + d, \; i \in s.$$

4 COMPARATIVE EFFICIENCIES OF INVERSE HYPERGEOMETRIC VS DIRECT HYPERGEOMETRIC RR GENERATION FOR DIFFERENT SAMPLING SCHEMES

We now present a study of the relative efficiencies of the inverse hypergeometric RR generation vs direct hypergeometric RR generation procedure as

$e = 100\frac{V}{V'}$ taking V as the variance of the usual estimator of θ for direct hypergeometric RRT and V' as the variance of the estimator of θ for the inverse hypergeometric RRT in different situations, where the units or individuals are being sampled from a finite population by various sampling schemes. For example, in (1) Simple Random Sampling With Replacement (SRSWR) by n draws and (2) Simple Random Samples Without Replacement (SRSWOR) in n draws, the sample mean of the transformed randomized responses are used to estimate θ. Also (3) Rao, Hartley, and Cochran's (RHC, 1962) sampling scheme using their estimation and variance estimation procedure is also treated. Finally, (4) we consider taking a sample of size n by Midzuno's (1952) scheme and employ Horvitz and Thompson's (HT, 1952) estimation method.

Let us denote E_p, V_p as the expectation and variance operators for design p, then the overall expectation, variance operators denoted by E and V are given as $E = E_pE_R$ and $V = E_pV_R + V_pE_R$. We present below the essential formulation for the estimator considering the direct hypergeometric RR generation, the variance and variance estimators for θ based on the various sampling schemes considered in this chapter. For the inverse hypergeometric counterpart, r_i, V_i and v_i will of course change in the manners described already, replacing them by r_i', V_i' and v_i' respectively.

4.1 SRSWR in *n* Draws

Let us denote y_k as the y-value for a person chosen on the kth draw ($k = 1,\ldots,n$) and r_k as the transformed RR generated by direct hypergeometric RRD from that person. Then an unbiased estimator for $\theta = \frac{\sum_{i=1}^{N} y_i}{N} = \bar{Y}$ is given by $\bar{r} = \frac{1}{n}\sum_{k=1}^{n} r_k$ such that $E_pE_R(\bar{r}) = E_p(\bar{y}) = \bar{Y} = \theta$, and $V(\bar{r}) = V_pE_R(\bar{r}) + E_pV_R(\bar{r}) = V_p(\bar{y}) + E_p(\frac{1}{n^2}\sum_{k=1}^{n} V_R(r_k)) = \frac{1}{n}[\theta(1-\theta)] + \frac{1}{Nn}\sum_{i=1}^{N} V_i$, where $V_i = V_R(r_i)$. $V(\bar{r})$ can be unbiasedly estimated by

$$\hat{V}(\bar{r}) = v(\bar{r}) = \frac{1}{n(n-1)}\sum_{k=1}^{n}(r_k - \bar{r})^2.$$

4.2 SRSWOR in *n* Draws

In this case also an unbiased estimator for θ is $\bar{r} = \frac{1}{n}\sum_{i\in s} r_i$ because $E(\bar{r}) = E_pE_R(\bar{r}) = E_p(\bar{y}) = \bar{Y} = \theta$ and $V(\bar{r}) = \frac{N-n}{Nn}\frac{1}{N-1}\sum_{i\in s}(y_i - \bar{Y})^2 + \frac{1}{Nn}\sum_{i=1}^{N} V_i$. $V(\bar{r})$ is unbiasedly estimated by

$$\hat{V}(\bar{r}) = v(\bar{r}) = \frac{N-n}{Nn}\frac{1}{(n-1)}\sum_{i\in s}(r_i - \bar{r})^2 + \frac{1}{Nn}\sum_{i\in s} v_i.$$

4.3 PPSWR in *n* Draws

Let us consider that for unequal probability sample drawing the normed size measures p_i s are formed from an auxiliary variable z with known $z_i > 0$ for all i having $Z=\sum_{i=1}^{N} z_i$ such that $p_i=\frac{z_i}{Z}$, where $0 < p_i < 1$ and $\sum_{i=1}^{N} p_i=1$. Let us denote p_k as the normed size measure and y_k as the y-value for a person chosen on the kth draw ($k = 1,\ldots,n$) and r_k as the transformed RR generated by direct hypergeometric RRD obtained from that person. Then following Hansen and Hurwitz (1943) an unbiased estimator for θ is given by $e_{PPSWR}=\frac{1}{Nn}\sum_{k=1}^{n}\frac{r_k}{p_k}$ such that $E_pE_R(e_{PPSWR})=E_p\left(\frac{1}{Nn}\sum_{k=1}^{n}\frac{y_k}{p_k}\right)=\bar{Y}=\theta$ and $V(e_{PPSWR})=V_pE_R(e_{PPSWR})+E_pV_R(e_{PPSWR})=V_p\left(\frac{1}{Nn}\sum_{k=1}^{n}\frac{y_k}{p_k}\right)+E_p\left(\frac{1}{N^2n^2}\sum_{k=1}^{n}\frac{V_R(r_k)}{p_k^2}\right)$
$=\frac{1}{N^2}\left[\frac{V}{n}+\frac{1}{n}\sum_{i=1}^{N}\frac{V_i}{p_i}\right]$, where

$V=\sum_{i=1}^{N-1}\sum_{j>i}^{N}p_ip_j\left(\frac{y_i}{p_i}-\frac{y_j}{p_j}\right)^2$ and $V_i = V_R(r_i)$. $V(e_{PPSWR})$ can be unbiasedly estimated by

$$\hat{V}(e_{PPSWR})=v(e_{PPSWR})=\frac{1}{N^2}\left[\frac{1}{2n^2(n-1)}\sum_{k=1}^{n}\sum_{k'=1,k'\neq k}^{n}\left(\frac{r_k}{p_k}-\frac{r_{k'}}{p_{k'}}\right)^2\right].$$

4.4 Rao, Hartley, and Cochran's Sampling of Size *n*

Rao, Hartley, and Cochran's (RHC, 1962) sampling of n persons from N population units consists of making n nonoverlapping random groups of the population units of group sizes being N_i s such that $\sum_{i=1}^{n} N_i=N$. Let Q_i denote the sum of the normed size measures of the N_i units falling in the ith group. Then independently from every group only one unit is selected with probability proportional to the normed size measures, thus yielding a sample of required size n by RHC method. For simplicity in notation, we denote the value obtained from the unit selected from i^{th} group as y_i and its normed size measure as p_i. With this notation, the unbiased estimator for θ is

$$e_{RHC}=\frac{1}{N}\sum_{n}r_i\frac{Q_i}{p_i},\ \text{as}\ E_pE_R(e_{RHC})=E_p\left(\frac{1}{N}\sum_{n}y_i\frac{Q_i}{p_i}\right)=\bar{Y}=\theta.$$

Here, $\sum_n$ means sum over the n disjoint groups into which the population U is divided into random groups. Following Rao et al. (1962), the optimal choices of group sizes N_i s are given by $N_i = [N/n]$ for $i = 1,2,\ldots,k$ and $N_i = [N/n] + 1$ for $i = k + 1,2,\ldots,n$, k being determined by solving $\sum_{i=1}^{n}N_i=N$.

Chaudhuri et al. (2000) have given results using which we may write

$$V(e_{RHC}) = \frac{1}{N^2}\left[C\sum_{i=1}^{N}\frac{V_i}{p_i} + (1-C)\sum_{i=1}^{N}V_i + C\left(\sum_{i=1}^{N}\frac{y_i^2}{p_i} - Y^2\right)\right], \text{ with}$$

$$C = \frac{\sum_n N_i^2 - N}{N(N-1)}.$$

$V\ (e_{RHC})$ is unbiasedly estimated by

$$\hat{V}(e_{RHC}) = v(e_{RHC}) = \frac{1}{N^2}\left[D\Sigma_n\Sigma_{n'}Q_iQ_{i'}\left(\frac{r_i}{p_i} - \frac{r_{i'}}{p_{i'}}\right)^2 + \Sigma_n v_i\frac{Q_i}{p_i}\right], \text{ where}$$

$$D = \frac{\sum_n N_i^2 - N}{N^2 - \sum_n N_i^2}.$$

Here, $\Sigma_n\Sigma_{n'}$ denotes the sum over nonoverlapping pairs of n groups.

4.5 Midzuno's (1952) Sampling of *n* Persons

For our illustrative purpose, we consider the fifth scheme as unequal probability sampling of n units by Midzuno's (1952) scheme. Sampling by this scheme is done first by drawing one unit by probability proportional to the size measure of the auxiliary variable, say, z. Then keeping the selected unit aside, the remaining $(n - 1)$ units are chosen by simple random sampling without replacement (SRSWOR) out of $(N - 1)$ units. Then under this scheme, the first and second order inclusion probabilities, π_i and π_{ij}, $i \neq j$ are given below.

$$\pi_i = \frac{z_i}{Z} + \frac{Z - z_i}{Z}\frac{\binom{N-2}{n-2}}{\binom{N-1}{n-1}} = \frac{z_i}{Z}\frac{N-n}{N-1} + \frac{n-1}{N-1} \quad \forall i = 1,2,\ldots,N,$$

and

$$\begin{aligned}\pi_{ij} &= \frac{z_i}{Z}\frac{\binom{N-2}{n-2}}{\binom{N-1}{n-1}} + \frac{z_j}{Z}\frac{\binom{N-2}{n-2}}{\binom{N-1}{n-1}} + \frac{Z - z_i - z_j}{Z}\frac{\binom{N-3}{n-3}}{\binom{N-1}{n-1}} \\ &= \frac{z_i + z_j}{Z}\frac{(N-n)(n-1)}{(N-1)(N-2)} + \frac{(n-1)(n-2)}{(N-1)(N-2)},\end{aligned}$$

$\forall i \neq j \in U$. For this scheme, $\pi_i\pi_j > \pi_{ij} \forall i \neq j \in U$.

An unbiased estimator for the sensitive population proportion θ is given by Horvitz and Thompson (1952)'s estimator as

$$e_{HT} = \frac{1}{N}\sum_{i\in s}\frac{r_i}{\pi_i},$$

such that $E_pE_R(e_{HT}) = E_p\left(\frac{1}{N}\sum_{i\in s}\frac{y_i}{\pi_i}\right) = \bar{Y} = \theta$. Utilizing Yates and Grundy (1953)'s form of variance of the HT estimator the variance of e_{HT} is given by

$$V(e_{HT}) = \frac{1}{N^2}\left[\sum_{i=1}^{N}\sum_{j=1,j>i}^{N}(\pi_i\pi_j - \pi_{ij})\left(\frac{y_i}{\pi_i} - \frac{y_j}{\pi_j}\right)^2 + \sum_{i=1}^{N}\frac{V_i}{\pi_i}\right].$$

It is unbiasedly estimated by

$$\hat{V}(e_{HT}) = v_{e_{HT}} = \frac{1}{N^2}\left[\sum_{i\in s}\sum_{j\in s,j>i}\frac{\pi_i\pi_j - \pi_{ij}}{\pi_{ij}}\left(\frac{r_i}{\pi_i} - \frac{r_j}{\pi_j}\right)^2 + \sum_{i\in s}\frac{v_i}{\pi_i}\right].$$

4.6 Comparison of the Efficiencies

It is clear from the variance formulae for the unbiased estimators for θ in each case as given in above sections, that in each case the V_pE_R term remains same for the direct hypergeometric RRT and the negative hypergeometric RRT and the difference is only in the E_pV_R term. So, to compare the efficiency of the negative hypergeometric approach of RR generation, we need to examine the relative magnitudes of $V_R(r_i')$ vs $V_R(r_i)$.

We note that the negative hypergeometric RRT will be superior to the direct hypergeometric RRT if

$$V_R(r_i') \leq V_R(r_i)$$

that is if $cy_i + d \leq ay_i + b$,

which may be equivalently expressed by the condition

$$0 \leq \theta \leq \frac{b-d}{c-a},$$

and after further simplification it becomes equivalent to

$$0 \leq \theta \leq \frac{b-d}{(b-d) + \left[\frac{t_1(N_1-r_1)(N_1+1)(r_1+1-t_1)}{(r_1+1)^2(r_1+2)\phi} - \frac{1}{K\left(\frac{r_1}{N_1}-\frac{r_2}{N_2}\right)^2}\left(\frac{r_1}{N_1}\frac{N_1-r_1}{N_1}\frac{N_1-K}{N_1-1}\right)\right]}.$$

Because of the complicated form of the above inequality, it becomes very difficult to have any insightful idea about the situation which may result the better performance or otherwise of the estimators of θ realized by survey data obtained through the revised RRT following Singh and Sedory's (2013) approach rather than the procedure following direct hypergeometric RRT. However, we present below a simulation-based numerical evaluation for efficiency comparison.

5 NUMERICAL ILLUSTRATION SHOWING RELATIVE PERFORMANCES BY SIMULATION

For numerical illustration, the values of y_i's and size measures z_i's for $i = 1,\ldots,117 = N$ are taken from Chaudhuri and Dihidar (2014) which are also used by Chaudhuri et al. (2009) and n is taken throughout as 24. For these data we have $\theta = 0.188$. We use the device parameters as $N_1 = 27$, $r_1 = 17$, $t_1 = 9$, $N_2 = 33$, $r_2 = 15$, $t_2 = 9$, and $K = 12$. We present the relative efficiencies of the inverse hypergeometric RR generation procedures vs the direct hypergeometric RR generation procedure with samples and estimators of θ obtained following the methods mentioned in previous section. We also illustrate below few instances of our findings for the estimated standard error (se) which is the positive square root of $\hat{V}(\hat{\theta})$ and estimated coefficient of variation (cv) which is $cv = 100\frac{se}{\hat{\theta}}$ for various situations (Tables 1–5).

TABLE 1 Relative Performances of Direct Hypergeometric RRT vs Negative Hypergeometric RRT Based on SRSWR

	Method 1: Direct Hypergeometric RRT		Method 2: Negative Hypergeometric RRT	
Serial Number	*se*	*cv*	*se*	*cv*
1	0.187	52.78	0.113	49.64
2	0.180	95.46	0.156	68.88
3	0.216	65.12	0.154	56.91
4	0.205	86.84	0.123	76.20
5	0.137	64.69	0.111	63.49
Efficiency = 100 ×(*V*(Method 1)/*V*(Method 2)) = 128.19				
Out of 100 cases estimated *cv*(Method 1)> estimated *cv*(Method 2) in 57 cases.				

TABLE 2 Relative Performances of Direct Hypergeometric RRT vs Negative Hypergeometric RRT Based on SRSWOR

	Method 1: Direct Hypergeometric RRT		Method 2: Negative Hypergeometric RRT	
Serial Number	*se*	*cv*	*se*	*cv*
1	0.123	57.88	0.123	54.10
2	0.136	52.28	0.137	51.52
3	0.130	78.94	0.130	73.91
4	0.114	97.48	0.114	64.85
5	0.100	60.96	0.100	59.56
Efficiency = 100 ×(*V*(Method 1)/*V*(Method 2)) = 130.20				
Out of 100 cases estimated *cv*(Method 1)> estimated *cv*(Method 2) in 62 cases.				

TABLE 3 Relative Performances of Direct Hypergeometric RRT vs Negative Hypergeometric RRT Based on PPSWR

	Method 1: Direct Hypergeometric RRT		Method 2: Negative Hypergeometric RRT	
Serial number	*se*	*cv*	*se*	*cv*
1	0.117	96.03	0.109	72.25
2	0.124	74.72	0.104	73.30
3	0.235	85.11	0.231	81.42
4	0.176	63.07	0.155	53.55
5	0.348	74.81	0.201	66.34
Efficiency = 100 ×(*V*(Method 1)/*V*(Method 2)) = 123.20				
Out of 100 cases estimated *cv*(Method 1)> estimated *cv*(Method 2) in 54 cases.				

TABLE 4 Relative Performances of Direct Hypergeometric RRT vs Negative Hypergeometric RRT Based on Rao, Hartley, and Cochran's Sampling

	Method 1: Direct Hypergeometric RRT		Method 2: Negative Hypergeometric RRT	
Serial Number	*se*	*cv*	*se*	*cv*
1	0.229	83.08	0.424	61.20
2	0.270	64.88	0.438	61.05
3	0.210	94.30	0.485	83.51
4	0.224	93.64	0.271	78.92
5	0.327	93.77	0.232	81.49

Efficiency = 100 ×(*V*(Method 1)/*V*(Method 2)) = 124.07

Out of 100 cases estimated *cv*(Method 1)> estimated *cv*(Method 2) in 67 cases.

TABLE 5 Relative Performances of Direct Hypergeometric RRT vs Negative Hypergeometric RRT Based on Midzuno's Scheme of Sampling

	Method 1: Direct Hypergeometric RRT		Method 2: Negative Hypergeometric RRT	
Serial Number	*se*	*cv*	*se*	*cv*
1	0.128	58.11	0.111	58.06
2	0.106	90.51	0.110	59.86
3	0.119	73.65	0.110	66.03
4	0.087	79.08	0.144	61.26
5	0.091	64.24	0.093	59.93

Efficiency = 100 ×(*V*(Method 1)/*V*(Method 2)) = 130.15

Out of 100 cases estimated *cv*(Method 1)> estimated *cv*(Method 2) in 53 cases.

6 CONCLUDING REMARKS

Our objective in this work is to examine how Singh and Sedory's (2013) negative hypergeometric RR generation approach may fare relative to the direct hypergeometric RR generation procedure in estimating the sensitive population proportion. Also, instead of considering only the SRSWR, we have considered the general sampling scheme for selecting the respondents. Essential theory is developed as shown in this chapter. However, the relative efficacies are revealed only with numerical studies based on simulations. Although the estimated coefficient of variations are shown to be very large in the numerical presentation, but relatively Singh and Sedory's (2013) negative hypergeometric distributional approach of generating RRs seems to perform better and thus it may establish the justification of our research work.

ACKNOWLEDGMENT

I gratefully acknowledge the constructive and insightful comments received from the referee which led to an improved version of the manuscript.

REFERENCES

Al-Sobhi, M.M., Hussain, Z., Al-Zahrani, B., 2014. General randomized response techniques using Polya's urn process as a randomization device. PLoS One 9 (12), e115612. http://dx.doi.org/10.1371/journal.pone.0115612.

Chaudhuri, A., 2001. Using randomized response from a complex survey to estimate a sensitive proportion in a dichotomous finite population. J. Stat. Plan. Inference 94, 37–42.

Chaudhuri, A., 2011. Randomized response and indirect questioning techniques in surveys. CRC Press, London.

Chaudhuri, A., Adhikary, A., Dihidar, S., 2000. Mean square error estimation in multi-stage sampling. Metrika 52, 115–131.

Chaudhuri, A., Christofides, T.C., Saha, A., 2009. Protection of privacy in efficient application of randomized response techniques. Stat. Methods Appl. 18, 389–418.

Chaudhuri, A., Dihidar, K., 2014. Generating randomized response by inverse mechanism. Model Assist. Stat. Appl. 9, 343–351.

Chaudhuri, A., Mukerjee, R., 1988. Randomized response. In: Theory and Techniques, Marcel Dekker, New York, NY.

Chaudhuri, A., Pal, S., 2008. Estimating sensitive proportions from Warner's randomized responses in alternative ways restricting to only distinct units sampled. Metrika 68, 147–156.

Christofides, T.C., 2003. A generalized randomized response technique. Metrika 57, 195–200.

Christofides, T.C., 2005. Randomized response in stratified sampling. J. Stat. Plan. Inference 128, 303–310.

Greenberg, B.G., Abul-Ela, A.L.A., Simmons, W.R., Horvitz, D.G., 1969. The unrelated question rr model: theoretical framework. J. Am. Stat. Assoc. 64, 520–539.

Hansen, M.H., Hurwitz, W.N., 1943. On the theory of sampling from finite populations. Ann. Math. Stat. 14 (4), 333–362.

Horvitz, D.G., Shah, B.V., Abernathy, J.R., 1976. Randomized response: a data gathering device for sensitive questions. Int. Stat. Rev. 44, 181–196.

Horvitz, D.G., Shah, B.V., Simmons, W.R., 1967. The unrelated question RR model. In: Proceedings of the Social Statistics Section. American Statistical Association, Washington, pp. 65–72.

Horvitz, D.G., Thompson, D.J., 1952. A generalization of sampling without replacement from a finite universe. J. Am. Stat. Assoc. 47, 663–685.

Kuk, A.Y.C., 1990. Asking sensitive questions indirectly. Biometrika 77 (2), 436–438.

Mangat, N.S., Singh, R., Singh, S., Bellhouse, D.R., Kashani, H.B., 1995. On efficiency of estimator using distinct respondents in randomized response survey. Surv. Methodol. 21, 21–23.

Midzuno, H., 1952. On the sampling system with probabilities proportionate to sum of sizes. Ann. Inst. Stat. Math. 3, 99–107.

Mukerjee, R., 1981. Inference on confidential characters from survey data. Bull. Calcutta Stat. Assoc. 30, 77–88.

Rao, J.N.K., Hartley, H.O., Cochran, W.G., 1962. On a simple procedure of unequal probability sampling without replacement. J. R. Stat. Soc. 24, 482–491.

Singh, S., Grewal, I.S., 2013. Geometric distribution as a randomization device: implemented to the Kuks model. Int. J. Contemp. Math. Sci. 8 (5), 243–248.

Singh, S., Sedory, S.A., 2013. A new randomized response device for sensitive characteristics: an application of the negative hypergeometric distribuion. Metron 71, 3–8.

Warner, S.L., 1965. Randomized response: a survey technique for eliminating evasive answer bias. J. Am. Stat. Assoc. 60, 63–69.

Yates, F., Grundy, P.M., 1953. Selection without replacement from within strata with probability proportional to size. J. Am. Stat. Assoc. 75, 206–211.

Chapter 27

Incredibly Efficient Use of a Negative Hypergeometric Distribution in Randomized Response Techniques

M.L. Johnson, S.A. Sedory and S. Singh
Texas A&M University-Kingsville, Kingsville, TX, United States

ABSTRACT

In this chapter, a new randomized response device is proposed based on the use of a pair of negative hypergeometric distribution when estimating the proportion of persons in a population possessing a sensitive characteristic in a population. The proposed resultant estimator is found to be more efficient than the recent model studied by Singh and Sedory (2013) and hence in turn remains more efficient than the Warner (1965), the Kuk (1990), and the Mangat (1994) estimators under the same situations investigated by Singh and Sedory (2013).

Keywords: Randomized response technique, Estimation of proportion, Relative efficiency

1 INTRODUCTION

Warner (1965) proposed an interviewing technique, called randomized response, to protect an interviewee's privacy and to reduce a major source of bias (evasive answers or refusing to respond) in surveys when estimating the prevalence of sensitive characteristics on human populations. Warner (1965) designed a randomization device, for example, a spinner or a deck of cards that consists of two mutually exclusive outcomes. In the case of cards, each card has one of the following statements: (i) I possess attribute A and (ii) I do not possess attribute A. The maximum likelihood estimator of π, the proportion of respondents in the population possessing the attribute A, is given by:

Handbook of Statistics, Vol. 34. http://dx.doi.org/10.1016/bs.host.2016.01.025

$$\hat{\pi}_w = \frac{(n_1/n) - (1-P)}{2P-1} \tag{1}$$

where n_1 is the number of individuals responding "yes," n is the number of respondents selected by a simple random and with replacement sampling (SRSWR), and P is the proportion of cards bearing the statement, "I possess an attribute A." The variance of $\hat{\pi}_w$ is given by:

$$V(\hat{\pi}_w) = \frac{\pi(1-\pi)}{n} + \frac{P(1-P)}{n(2P-1)^2} \tag{2}$$

Kuk (1990) introduced an ingenious randomized response model in which, if a respondent belongs to a sensitive group A, then he/she is instructed to use a deck of cards having the proportion θ_1 of cards with the statement, "I belong to group A," and if the respondent belongs to nonsensitive group A^c, then the respondent is requested to use a different deck of cards having the proportion θ_2 of cards with the statement, "I do not belong to group A." Again let π be the true proportion of persons belonging to the sensitive group A. Then, the probability of a "yes" answer in the Kuk (1990) model is given by:

$$\theta_{kuk} = \pi\,\theta_1 + (1-\pi)\theta_2 \tag{3}$$

Further assume that a simple random with replacement sample of n respondents is selected from the population and that n_1 is the number of observed "yes" answers. The number of people n_1 that answer "yes" is binomially distributed with parameters θ_{kuk} and n. For the Kuk (1990) model, an unbiased estimator of the population proportion π is given by:

$$\hat{\pi}_{kuk} = \frac{n_1/n - \theta_2}{\theta_1 - \theta_2}, \quad \theta_1 \neq \theta_2 \tag{4}$$

with variance, given by:

$$V(\hat{\pi}_{kuk}) = \frac{\theta_{kuk}(1-\theta_{kuk})}{n(\theta_1 - \theta_2)^2} \tag{5}$$

Mangat (1994) suggested a randomized response model in which, if a respondent belongs to a sensitive group A, then he/she is requested to report "yes" without using any randomization device, and if a respondent does not belong to the sensitive group A, then he/she is requested to respond using the Warner (1965) device. The mode of response of the respondent remains unrevealed to the interviewer. Thus the probability of "yes" answer in the Mangat (1994) model is given by:

$$\theta_m = \pi + (1-\pi)(1-P) \tag{6}$$

Further assume that a simple random with replacement sample of n respondents is selected from the population and that n_m is the number of observed "yes" answers. The number of people n_m that answer "yes" is binomially distributed with parameters θ_m and n. For the Mangat (1994) model, an unbiased estimator of the population proportion π is given by:

$$\hat{\pi}_m = \frac{n_m/n - (1-P)}{P}, \tag{7}$$

with variance, given by:

$$V(\hat{\pi}_m) = \frac{\theta_m(1-\theta_m)}{nP^2} \tag{8}$$

Note that in the Mangat (1994) model, the respondents who report "no" are surely members of the nonsensitive group and lose their privacy. Because a few respondents lose their privacy, it is not easy to develop a model which is more efficient than this one. Further note that if $\theta_1 = 1$ and $\theta_2 = (1-P)$, then the Kuk (1990) model reduces to the Mangat (1994) model. Note that we have considered a particular case of the Kuk (1990) model where each respondent is requested to make only one draw from the deck of cards used by him/her based on his/her membership in a group. A more general case of Kuk (1990) model, where each respondent makes more than one trial and is also valid for complex survey designs, can be had from Chaudhuri and Christofides (2013). Discussion of generalized Kuk's model is beyond the scope of this chapter and can be investigated in future studies. Odumade and Singh (2009) suggested use of a pair of decks of cards which could be made more efficient than Mangat (1994), Kuk (1990), and Warner (1965) estimators. They proposed that each respondent in the simple random and with replacement (SRSWR) sample of n be provided with two decks of cards, marked as Deck-I and Deck-II, as shown in Fig. 1.

Each respondent is requested to draw two cards simultaneously, one card from each deck of cards and read the statements in order. The respondent first matches his/her status with the statement written on the first deck of cards, and then he/she matches his/her status with the statement written on the second deck of cards. Let π be the true proportion of respondents in the population that possesses the characteristic A. Consider the situation when the selected respondent belongs to group A: now if he/she draws the first card with statement $I \in A$ having proportion P in the first deck of cards, and a second card with statement $I \in A$ having proportion T in the second deck of cards,

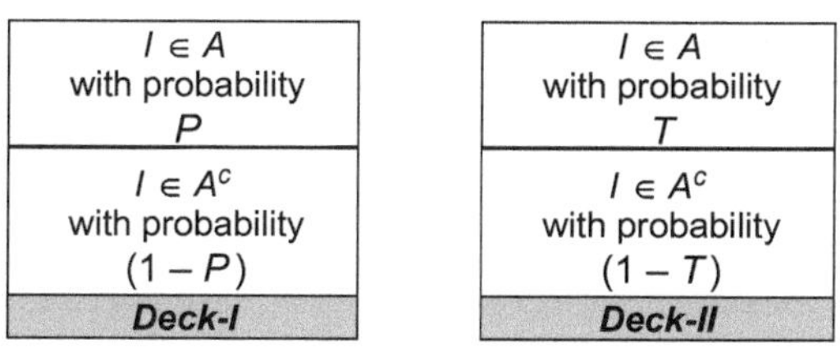

FIG. 1 Two decks of cards.

then he/she is requested to report: (Yes, Yes). Consider another situation when the selected respondent belongs to group A^c: now if he/she draws the first card with statement $I \in A^c$ having proportion $(1-P)$ in the first deck of cards and a second card with statement $I \in A^c$ having proportion $(1-T)$ in the second deck of cards, then he/she is also requested to report: (Yes, Yes). Thus the response (Yes, Yes) can come from either type of respondents, ones belonging to group A or A^c, and hence their privacy will be maintained. The probability of getting (Yes, Yes) response is given by:

$$\begin{aligned} P(\text{Yes, Yes}) = \theta_{11} &= PT\pi + (1-P)(1-T)(1-\pi) \\ &= (P+T-1)\pi + (1-P)(1-T) \end{aligned} \tag{9}$$

Now again consider a situation where the selected respondent belongs to group A: now if he/she draws the first card with statement $I \in A$ from the first deck of cards, and a second card with statement $I \in A^c$ from the second deck of cards, then he/she is requested to report: (Yes, No). Consider another situation where the selected respondent belongs to group A^c: now if he/she draws the first card with statement $I \in A^c$ from the first deck of cards, and a second card with statement $I \in A$ from the second deck of cards, then he/she is also requested to report: (Yes, No). Thus the response (Yes, No) can come from either type of respondents, either belonging to the group A or A^c, and hence their privacy will not be disclosed. The probability of getting (Yes, No) response is given by:

$$P(\text{Yes, No}) = \theta_{10} = P(1-T)\pi + (1-P)T(1-\pi) = (P-T)\pi + T(1-P) \tag{10}$$

Similarly, the probability of obtaining (No, Yes) and (No, No) responses is given by:

$$P(\text{No, Yes}) = \theta_{01} = (1-P)T\pi + P(1-T)(1-\pi) = (T-P)\pi + P(1-T) \tag{11}$$

and

$$P(\text{No, No}) = \theta_{00} = (1-P)(1-T)\pi + PT(1-\pi) = (1-P-T)\pi + PT \tag{12}$$

Let $\hat{\theta}_{11} = \frac{n_{11}}{n}$, $\hat{\theta}_{10} = \frac{n_{10}}{n}$, $\hat{\theta}_{01} = \frac{n_{01}}{n}$, and $\hat{\theta}_{00} = \frac{n_{00}}{n}$ be the observed proportions of (Yes, Yes), (Yes, No), (No, Yes), and (No, No) responses, respectively. Then Odumade and Singh (2009) considered minimizing the squared distance between the observed proportions and the true proportions namely:

$$D = \frac{1}{2}\sum_{i=0}^{1}\sum_{j=0}^{1}\left(\theta_{ij} - \hat{\theta}_{ij}\right)^2 \tag{13}$$

which leads to an unbiased estimator of the population proportion π given by:

$$\hat{\pi}_{os} = \frac{1}{2} + \frac{(P+T-1)\left(\hat{\theta}_{11} - \hat{\theta}_{00}\right) + (P-T)\left(\hat{\theta}_{10} - \hat{\theta}_{01}\right)}{2\left[(P+T-1)^2 + (P-T)^2\right]} \tag{14}$$

The variance of the estimator $\hat{\pi}_{os}$ is given by:

$$V(\hat{\pi}_{os}) = \frac{(P+T-1)^2\{PT+(1-P)(1-T)\}+(P-T)^2\{T(1-P)+P(1-T)\}}{4n\left[(P+T-1)^2+(P-T)^2\right]^2} - \frac{(2\pi-1)^2}{4n} \quad (15)$$

Singh and Grewal (2013) proposed a randomized response model in which each respondent selected in the sample is provided with two decks of cards in the same way as in Kuk (1990) model. In the first deck of cards, let $\theta_1{}^*$ be the proportion of cards with the statement, "I belong to group A" and $\left(1-\theta_1^*\right)$ be the proportion of cards with the statement, "I do not belong to group A." In the second deck of cards, let $\theta_2{}^*$ be the proportion of cards with the statement, "I do not belong to group A" and $\left(1-\theta_2^*\right)$ be the proportion of cards with the statement, "I belong to group A." They suggest if a respondent belongs to group A, he/she is instructed to draw cards, one by one using *with replacement*, from the first deck of cards until he/she gets the first card bearing the statement of his/her own status, and requested to report the total number of cards, say X, drawn by him/her. If a respondent belongs to group A^c, he/she is instructed to draw cards, one by one using with replacement, from the second deck of cards until he/she gets the first card bearing the statement of his/her own status, and requested to report the total number of cards, say Y, drawn. Assuming Z_i to be the number of cards reported by the ith respondent, then an unbiased estimator of the population proportion π is given by:

$$\hat{\pi}_{SG} = \frac{\frac{\theta_1^*\theta_2^*}{n}\sum_{i=1}^{n} Z_i - \theta_1^*}{\theta_2^* - \theta_1^*}, \quad \theta_1^* \neq \theta_2^* \quad (16)$$

with variance of the estimator $\hat{\pi}_{SG}$ given by:

$$V(\hat{\pi}_{SG}) = \frac{\pi(1-\pi)}{n} + \frac{\theta_2^{*2}\left(1-\theta_1^*\right)\pi + \theta_1^{*2}\left(1-\theta_2^*\right)(1-\pi)}{n\left(\theta_2^*-\theta_1^*\right)^2} \quad (17)$$

In section 2, we discuss a new randomization device recently studied by Singh and Sedory (2013) which is in fact an extension of Singh and Grewal (2013) model.

2 SINGH AND SEDORY RANDOMIZATION DEVICE

Singh and Sedory (2013) proposed a new randomized response device consisting of two urns: Urn-I contains N_1 balls, out of which r_1 balls bear the statement, (i) "I belong to the group A," and the remaining $(N_1 - r_1)$ balls are blank with no statement on them. Urn-II contains N_2 balls, out of which r_2 balls bear the statement, (ii) "I do not belong to the group A," and the remaining $(N_2 - r_2)$ balls are blank with no statement on them. Each respondent selected in the sample is instructed as follows: if he/she belongs to the group A, then he/she draws balls using without replacement sampling from Urn-I until he/she gets t_1 $(< r_1)$ balls bearing the statement (i), and reports the total number of balls, say X, drawn by him/her. Thus X follows a negative hypergeometric distribution. If he/she does not belongs to the group A, then he/she draws balls using without replacement sampling from Urn-II until he/she gets t_2 $(< r_2)$ balls bearing the statement (ii), and reports the total number of balls, say Y, drawn by him/her. Thus Y also follows a negative hypergeometric distribution, but with different parameters. Obviously, the distribution of the observed responses Z_i is given by:

$$Z_i = \begin{cases} X & \text{if } i \in A \\ Y & \text{if } i \in A^c \end{cases} \tag{18}$$

Then, we have the following theorems:

Theorem 1. *An unbiased estimator of the population proportion π is given by:*

$$\hat{\pi}_{ss} = \frac{\frac{(r_1+1)(r_2+1)}{n}\sum_{i=1}^{n} Z_i - t_2(r_1+1)(N_2+1)}{t_1(N_1+1)(r_2+1) - t_2(N_2+1)(r_1+1)} \tag{19}$$

Proof. Obvious because the expected value of Z_i is given by:

$$E(Z_i) = \frac{1}{(r_1+1)(r_2+1)}[\pi\, t_1(N_1+1)(r_2+1) + (1-\pi)t_2(N_2+1)(r_1+1)] \quad \square \tag{20}$$

Theorem 2. *The variance of the estimator $\hat{\pi}_{ss}$ is given by:*

$$V(\hat{\pi}_{ss}) = \frac{\pi(1-\pi)}{n} + \frac{\pi t_1(N_1+1)(N_1-r_1)(r_1+1-t_1)(r_2+1)^2(r_2+2) + (1-\pi)(N_2+1)(N_2-r_2)(r_2+1-t_2)(r_1+1)^2(r_1+2)}{n(r_1+2)(r_2+2)\{t_1(r_2+1)(N_1+1) - t_2(r_1+1)(N_2+1)\}^2} \tag{21}$$

Proof. The expected value of Z_i^2 over the randomization device is given by:

$$E(Z_i^2) = \frac{\pi}{(r_1+1)^2(r_1+2)}\left\{t_1(N_1+1)(N_1-r_1)(r_1+1-t_1) + t_1^2(N_1+1)^2(r_1+2)\right\} + \frac{(1-\pi)}{(r_2+1)^2(r_2+2)}\left\{t_2(N_2+1)(N_2-r_2)(r_2+1-t_2) + t_2^2(N_2+1)^2(r_2+2)\right\} \tag{22}$$

Because the responses Z_i are independent, so by the definition of variance, we have

$$V(\hat{\pi}_{ss}) = V\left[\frac{\frac{(r_1+1)(r_2+1)}{n}\sum_{i=1}^{n} Z_i - t_2(r_1+1)(N_2+1)}{t_1(N_1+1)(r_2+1) - t_2(N_2+1)(r_1+1)}\right]$$

$$= \frac{\frac{(r_1+1)^2(r_2+1)^2}{n^2}\sum_{i=1}^{n} V(Z_i)}{\{t_1(N_1+1)(r_2+1) - t_2(N_2+1)(r_1+1)\}^2} \tag{23}$$

$$= \frac{\frac{(r_1+1)^2(r_2+1)^2}{n}\sigma_Z^2}{\{t_1(N_1+1)(r_2+1) - t_2(N_2+1)(r_1+1)\}^2}$$

where

$$\sigma_Z^2 = V(Z_i) = E\left(Z_i^2\right) - \{E(Z_i)\}^2 \tag{24}$$

On using (20) and (22) in (24), and later using it in (23), we get (21). Hence the theorem. □

Singh and Sedory (2013) investigated the relative efficiency of their proposed randomization device over the Warner (1965) and Kuk (1990) estimators. The percent relative efficiency of the estimator $\hat{\pi}_{ss}$ over the Warner (1965) and the Kuk (1990) estimators is, respectively, defined as:

$$\text{RE(W)} = \frac{V(\hat{\pi}_w)}{V(\hat{\pi}_{ss})} \times 100\% \tag{25}$$

and

$$\text{RE(K)} = \frac{V(\hat{\pi}_{kuk})}{V(\hat{\pi}_{ss})} \times 100\% \tag{26}$$

They claimed that the balls in the two urns can be adjusted in many ways to get efficient results and cooperation with respondents and they considered comparison with Mangat (1994) model. Singh and Sedory (2013) suggested choosing a randomization device in which Urn-I consists of $N_1 = 10$ identical balls (say, Ping pong balls), out of which $r_1 = 7$ bear the statement, "I belong to the sensitive group A," and the remaining three balls are blank. Urn-II consists of $N_2 = 12$ identical balls, out of which $r_2 = 6$ bear the statement, "I do not belong to statement A," and the remaining six balls are blank. Each respondent selected in the sample is requested to report the number of balls drawn by him/her once he/she collects $t_1 = t_2 = 5$ balls of the appropriate type. The percent relative efficiency of the Singh and Sedory (2013) model is shown in Table 1.

Singh and Sedory (2013) provided only Table 2 which reports the relative efficiencies of their estimator with respect to the Kuk model with parameters $\theta_1 = 0.8$, $\theta_2 = 0.2$ (or $\theta_1 = 0.2$, $\theta_2 = 0.8$).

TABLE 1 RE of a Version of the New Estimator with Respect to the Warner (1965), the Kuk (1990), and the Mangat (1994) ($\theta_1=1, \theta_2=1-P$) Estimators

θ_1	θ_2	P	π	RE(W)	RE(K)	θ_1	θ_2	P	π	RE(W)	RE(K)
0.7	0.2	0.7	0.05	357.74	183.47	0.8	0.3	0.8	0.50	140.25	199.95
			0.10	339.64	181.63				0.55	143.29	202.42
			0.15	326.75	180.96				0.60	146.98	206.15
			0.20	317.70	181.24				0.65	151.53	211.41
			0.25	311.68	182.33				0.70	157.23	218.63
			0.30	308.18	184.20				0.75	164.57	228.51
			0.35	306.89	186.82				0.80	174.30	242.23
			0.40	307.68	190.26				0.85	187.80	261.86
			0.45	310.54	194.58				0.90	207.69	291.46
			0.50	315.57	199.95	1.0	0.3	0.7	0.05	357.74	119.59
			0.55	323.05	206.56				0.10	339.64	115.20
			0.60	333.38	214.74				0.15	326.75	111.59
			0.65	347.27	224.94				0.20	317.70	108.49
			0.70	365.79	237.85				0.25	311.68	105.75

			0.75	390.62	254.56				0.30	308.18	103.23
			0.80	424.63	276.84				0.35	306.89	100.85
			0.85	472.83	307.83				0.40	307.68	98.53
			0.90	545.03	353.64				0.45	310.54	96.19
0.8	0.3	0.8	0.05	129.40	230.82				0.50	315.57	93.77
			0.10	129.43	220.37				0.55	323.05	91.19
			0.15	129.78	212.72				0.60	333.38	88.35
			0.20	130.41	207.13				0.65	347.27	85.13
			0.25	131.31	203.11				0.70	365.79	81.34
			0.30	132.47	200.39				0.75	390.62	76.73
			0.35	133.90	198.78				0.80	424.63	70.86
			0.40	135.65	198.18				0.85	472.83	62.97
			0.45	137.74	198.56				0.90	545.03	51.63

TABLE 2 Relative Efficiency with Respect to Kuk's Model for θ_1=0.8 and θ_2=0.2

π	0.05	0.10	0.15	0.20	0.25	0.30	0.35	0.40	0.45
RE(K)	129.40	129.43	129.78	130.41	131.31	132.47	133.90	135.65	137.74
π	0.50	0.55	0.60	0.65	0.70	0.75	0.80	0.85	0.90
RE(K)	140.25	143.29	146.98	151.53	157.23	164.57	174.30	187.80	207.69

They noted that they are in the same boat as with the Mangat (1994) model as far as the degree of protection of respondents is concerned, because if a respondent reports more than 10 balls, that is, either 11 balls or 12 balls, then this respondent surly belongs to the nonsensitive group. There is a gain over the Mangat (1994) model for $P=0.7$ and the value of π less than or equal to 0.35. In practice, it is usually the case that the proportion of those having the sensitive characteristic is small, so their model seems to perform better than the Mangat (1994) model in these realistic situations. As it can be seen, it is always more efficient than the Warner (1965) and Kuk (1990) models for the situations listed in Table 1. It can be verified that for a choice of $\theta_1=0.7$ and $\theta_2=0.2$ in the Kuk (1990) model, the average value of RE(K) is 219.0% with a standard deviation on 49.8% and median value of 197.3% as the value of π changes from 0.05 to 0.90 with a step of 0.05. For $P=0.7$ in the Warner (1965) model, the average value of RE(W) is 355.8% with a standard deviation of 65.2% and median value of 330.1%. For the choice $\theta_1=0.8$, $\theta_2=0.2$ in Kuk's model, the relative efficiency changes from 129.40% to 207.69% as the value of π changes from 0.05 to 0.90. Note that the relative efficiency values remain the same for $\theta_1=0.2$, $\theta_2=0.8$ as for $\theta_1=0.8$, $\theta_2=0.2$. Thus Singh and Sedory (2013) conclude that their proposed randomization device can be efficiently and cooperatively used in real practice to estimate the proportion of a sensitive characteristic.

In the next section, we propose a new incredibly efficient use of a randomization device by making use of two pairs of hypergeometric distribution. We demonstrate that the proposed randomization device can perform incredibly more efficiently than the Singh and Sedory (2013) model under similar assumptions of losing privacy by the respondents belong to the nonsensitive group.

3 PROPOSED INCREDIBLY EFFICIENT RANDOMIZATION DEVICE

The proposed randomization device consists of two pairs of urns. The first pair consists of Urn-I and Urn-II. Urn-I contains N_1 balls, out of which r_1 balls bear the statement, (i) "I belong to the group A," and the remaining (N_1-r_1)

balls are blank with no statement on them. Indeed it is the same urn as suggested by Singh and Sedory (2013). Urn-II of the first pair contains $N_1{}^*$ balls, out of which $r_1{}^*$ balls bear the statement, (ii) "I belong to the group A," and the remaining $\left(N_1^* - r_1^*\right)$ balls are blank with no statement on them.

The second pair of urns consists of Urn-III and Urn-IV. The first Urn-III contains N_2 balls, out of which r_2 balls bear the statement, (i) "I do not belong to the group A," and the remaining $(N_2 - r_2)$ balls are blank, with no statement on them. Indeed the first urn in the second pair is the same as proposed by Singh and Sedory (2013). The second, Urn-IV contains $N_2{}^*$ balls, out of which $r_2{}^*$ balls bear the statement, (ii) "I do not belong to the group A," and the remaining $\left(N_2^* - r_2^*\right)$ balls are blank, with no statement on them.

A graphical presentation of the proposed randomization device is given in Figs. 2 and 3. The first pair, Urn-I and Urn-II, is shown in Fig. 2.

The second pair, Urn-III and Urn-IV, is shown in Fig. 3.

Each respondent selected in an SRSWR is instructed as follows: if he/she belongs to the group A, then he/she is instructed to use first pair of urns as follows: he/she draws balls using without replacement sampling from Urn-I until he/she gets t_1 $(< r_1)$ balls bearing the statement (i), say the total number of balls drawn by him/her is X. Then at the same time he/she draws balls, using without replacement sampling, from Urn-II until he/she gets t_1^* $\left(< r_1^*\right)$ balls bearing the statement (i), say the total number of balls drawn by him/her is X^*. Then he/she is asked to report the sum $(X + X^*)$. The distributions

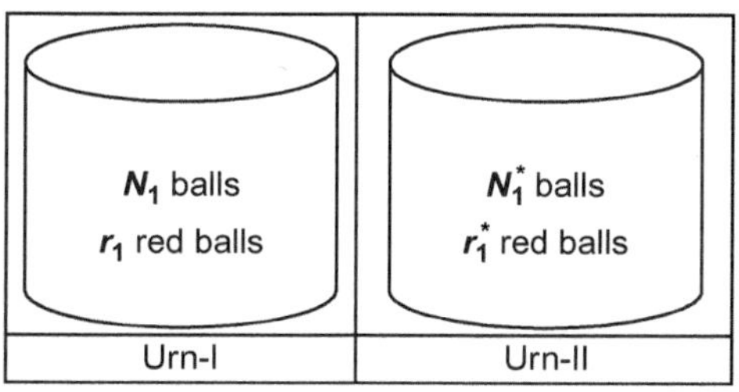

FIG. 2 First pair of urns: Urn-I and Urn-II.

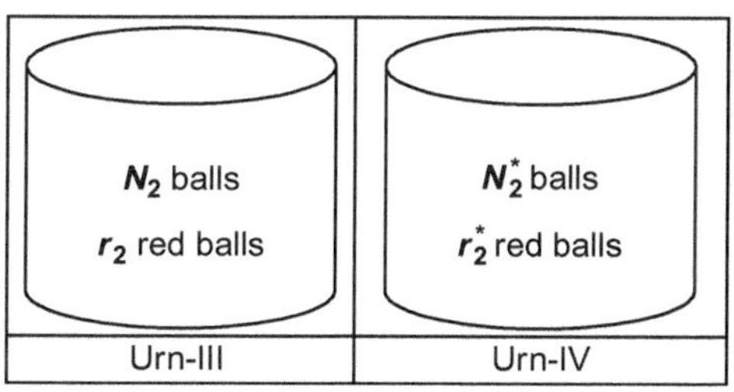

FIG. 3 Second pair of urns: Urn-III and Urn-IV.

of the variables X and X^* are each negative hypergeometric with probability mass functions given by:

$$P(X=x|N_1,r_1,t_1)=\frac{\binom{x-1}{t_1-1}\binom{N_1-x}{r_1-t_1}}{\binom{N_1}{r_1}},\quad x=t_1,(t_1+1),\ldots,(N_1-r_1+t_1) \tag{27}$$

and

$$P\left(X^*=x^*|N_1^*,r_1^*,t_1^*\right)=\frac{\binom{x^*-1}{t_1^*-1}\binom{N_1^*-x^*}{r_1^*-t_1^*}}{\binom{N_1^*}{r_1^*}},\quad x^*=t_1^*,\left(t_1^*+1\right),\ldots,\left(N_1^*-r_1^*+t_1^*\right) \tag{28}$$

If he/she does not belong to the group A, then he/she is instructed to use the second pair of urns. He/she draws balls, using without replacement sampling, from Urn-III until he/she gets t_2 $(<r_2)$ balls bearing the statement (ii), say the total number of balls drawn by him/her is Y. Then at the same time he/she draws balls, using without replacement sampling from Urn-IV until he/she gets t_2^* $\left(<r_2^*\right)$ balls bearing the statement (ii), say the total number of balls drawn by him/her is Y^*. Then he/she is asked to report the sum $(Y+Y^*)$. The distributions of the variables Y and Y^* are each negative hypergeometric with probability mass functions given by:

$$P(Y=y|N_2,r_2,t_2)=\frac{\binom{y-1}{t_2-1}\binom{N_2-y}{r_2-t_2}}{\binom{N_2}{r_2}},\quad y=t_2,(t_2+1),\ldots,(N_2-r_2+t_2) \tag{29}$$

and

$$P\left(Y^*=y^*|N_2^*,r_2^*,t_2^*\right)=\frac{\binom{y^*-1}{t_2^*-1}\binom{N_2^*-y^*}{r_2^*-t_2^*}}{\binom{N_2^*}{r_2^*}},\quad y^*=t_2^*,\left(t_2^*+1\right),\ldots,\left(N_2^*-r_2^*+t_2^*\right) \tag{30}$$

Let W_i, $i=1,2,3,\ldots,n$, be the observed response from the ith respondent in a simple random with replacement sample of n units. Obviously, the distribution of the observed responses W_i is given by:

$$W_i=\begin{cases} X+X^* & \text{if } \ i\in A \\ Y+Y^* & \text{if } \ i\in A^c \end{cases} \tag{31}$$

Then, we have the following theorems:

Theorem 3. *An unbiased estimator of the population proportion π is given by:*

$$\hat{\pi}_{JSS} = \frac{\frac{\Delta_1}{n}\sum_{i=1}^{n} W_i - \Delta_2}{\Delta_3} \tag{32}$$

where

$$\Delta_1 = (r_1+1)(r_1^*+1)(r_2+1)(r_2^*+1)$$

$$\Delta_2 = t_2(N_2+1)(r_1+1)(r_1^*+1)(r_2^*+1) + t_2^*(N_2^*+1)(r_1+1)(r_1^*+1)(r_2+1)$$

and

$$\begin{aligned}\Delta_3 &= t_1(N_1+1)(r_1^*+1)(r_2+1)(r_2^*+1) + t_1^*(N_1^*+1)(r_1+1)(r_2^*+1)\\ &\quad - t_2(N_2+1)(r_1+1)(r_1^*+1)(r_2^*+1) - t_2^*(N_2^*+1)(r_1+1)(r_1^*+1)(r_2+1)\end{aligned}$$

Proof. Given $X \sim \text{NH}(N_1, r_1, t_1)$, $X^* \sim \text{NH}(N_1^*, r_1^*, t_1^*)$, $Y \sim \text{NH}(N_2, r_2, t_2)$, and $Y^* \sim \text{NH}(N_2^*, r_2^*, t_2^*)$, where NH stands for a negative hypergeometric distribution, we have

$$E(X) = \frac{t_1(N_1+1)}{(r_1+1)},\quad E(X^*) = \frac{t_1^*(N_1^*+1)}{(r_1^*+1)},\quad E(Y) = \frac{t_2(N_2+1)}{(r_2+1)},\quad \text{and}$$

$$E(Y^*) = \frac{t_2^*(N_2^*+1)}{(r_2^*+1)}$$

By the definition of expected value, we have

$$\begin{aligned}E(W_i) &= \pi[E(X)+E(X^*)] + (1-\pi)[E(Y)+E(Y^*)]\\ &= \pi\left[\frac{t_1(N_1+1)}{(r_1+1)} + \frac{t_1^*(N_1^*+1)}{(r_1^*+1)}\right] + (1-\pi)\left[\frac{t_2(N_2+1)}{(r_2+1)} + \frac{t_2^*(N_2^*+1)}{(r_2^*+1)}\right]\end{aligned} \tag{33}$$

Taking expected value on both sides of (32), we have

$$E(\hat{\pi}_{JSS}) = \frac{\frac{\Delta_1}{n}\sum_{i=1}^{n} E(W_i) - \Delta_2}{\Delta_3} = \pi$$

which proves the theorem. □

Theorem 4. *The variance of the proposed estimator $\hat{\pi}_{JSS}$ is given by:*

$$V(\hat{\pi}_{JSS}) = \frac{\pi(1-\pi)}{n} + \frac{\pi(\alpha_1+\alpha_2-\alpha_3-\alpha_4) + (r_1+2)(r_1^*+2)(\alpha_3+\alpha_4)}{n\Delta_4} \tag{34}$$

where

$$\begin{aligned}\alpha_1 &= t_1(N_1+1)(N_1-r_1)(r_1-1-t_1)(r_1^*+1)^2(r_2+1)^2(r_2^*+1)^2(r_1^*+2)\\ &\quad \times (r_2+2)(r_2^*+2)\end{aligned}$$

$$\alpha_2 = t_1^*(N_1^*+1)(N_1^*-r_1^*)(r_1^*-1-t_1^*)(r_1^*+1)^2(r_2+1)^2(r_2^*+1)^2(r_1+2) \times (r_2+2)(r_2^*+2)$$

$$\alpha_3 = t_2(N_2+1)(N_2-r_2)(r_2-1-t_2)(r_2^*+1)^2(r_1+1)^2(r_1^*+1)^2(r_1+2) \times (r_1^*+2)(r_2^*+2)$$

$$\alpha_4 = t_2^*(N_2^*+1)(N_2^*-r_2^*)(r_2^*-1-t_2^*)(r_1+1)^2(r_1^*+1)^2(r_2+1)^2(r_1+2) \times (r_1^*+2)(r_2+2)$$

and

$$\Delta_4 = (r_1+2)(r_1^*+2)(r_2+2)(r_2^*+2)\Delta_3^2$$

Proof. Note that all the responses are independent, thus by definition the variance of the proposed estimator $\hat{\pi}_{JSS}$ is given by:

$$V(\hat{\pi}_{JSS}) = V\left[\frac{\frac{\Delta_1}{n}\sum_{i=1}^{n} W_i - \Delta_2}{\Delta_3}\right] = \frac{\Delta_1^2 \sum_{i=1}^{n} V(W_i)}{n^2 \Delta_3^2} = \frac{\Delta_1^2 \sigma_w^2}{n \Delta_3^2} \tag{35}$$

The value of σ_w^2 is given by:

$$\sigma_w^2 = E(W_i^2) - \{E(W_i)\}^2 \tag{36}$$

The value of $E(W_i)$ is given in (33) and the value of $E(W_i^2)$ is given by:

$$\begin{aligned}
E(W_i^2) &= \pi\, E(X+X^*)^2 + (1-\pi)E(Y+Y^*)^2 \\
&= \pi\, E[X^2 + X^{*2} + 2XX^*] + (1-\pi)E[Y^2 + Y^{*2} + 2YY^*] \\
&= \pi\, [E(X^2) + E(X^{*2}) + 2E(X)E(X^*)] + (1-\pi)[E(Y^2) + E(Y^{*2}) + 2E(Y)E(Y^*)] \\
&= \pi\left[\frac{t_1(N_1+1)(N_1-r_1)(r_1+1-t_1) + t_1^2(N_1+1)^2(r_1+2)}{(r_1+1)^2(r_1+2)}\right. \\
&\quad \left. + \frac{t_1^*(N_1^*+1)(N_1^*-r_2^*)(r_1^*+1-t_1^*) + t_1^{*2}(N_1^*+1)^2(r_1^*+2)}{(r_1^*+1)^2(r_1^*+2)} + \frac{2t_1(N_1+1)t_1^*(N_1^*+1)}{(r_1+1)(r_1^*+1)}\right] \\
&\quad + (1-\pi)\left[\frac{t_2(N_2+1)(N_2-r_2)(r_2+1-t_2) + t_2^2(N_2+1)^2(r_2+1)}{(r_2+1)^2(r_2+2)}\right. \\
&\quad \left. + \frac{t_2^*(N_2^*+1)(N_2^*-r_2^*)(r_2^*+1-t_2^*) + t_2^{*2}(N_2^*+1)^2(r_2^*+2)}{(r_2^*+1)^2(r_2^*+2)} + \frac{2t_2(N_2+1)t_2^*(N_2^*+1)}{(r_2+1)(r_2^*+1)}\right]
\end{aligned} \tag{37}$$

On using (33) and (37) in (36), we get the value of σ_w^2 which on substitution into (35) proves the theorem. □

Theorem 5. *An estimator of the variance of proposed estimator $\hat{\pi}_{JSS}$ is given by:*

$$\hat{V}(\hat{\pi}_{JSS}) = \frac{\hat{\pi}_{JSS}(1-\hat{\pi}_{JSS})}{n-1} + \frac{\hat{\pi}_{JSS}(\alpha_1+\alpha_2-\alpha_3-\alpha_4)+(r_1+2)(r_1^*+2)(\alpha_3+\alpha_4)}{n\Delta_4} \tag{38}$$

Note that other estimators are a special case of this estimator, so other estimator of variance can be derived from (38), if required.

In the next section, we investigate the percent relative efficiency of the proposed new randomized response model, with four urns, over the randomization device due to Singh and Sedory (2013), with two urns.

4 EFFICIENCY COMPARISON

The percent relative efficiency of the newly proposed estimator over the Singh and Sedory (2013) estimator is defined as:

$$\text{RE} = \frac{V(\hat{\pi}_{ss})}{V(\hat{\pi}_{JSS})} \times 100\% \tag{39}$$

Clearly, the percent relative efficiency (RE) value in (39) is free from the sample size. Note that the balls in the four urns can be adjusted in many ways to get efficient results and cooperation with respondents. In this illustration, the aim is to beat the Singh and Sedory (2013) model. We consider using the same randomization device as used by Singh and Sedory (2013) by making a randomization device in which Urn-I consists of $N_1 = 10$ identical balls (say, Ping pong balls), out of which $r_1 = 7$ bear the statement, "I belong to the sensitive group A," and the remaining three balls are blank. For simplicity, we let Urn-II be the same as Urn-I both of which are the same as Urn-I in Singh and Sedory (2013).

Similarly, we let Urn-III be the same as Urn-II used by Singh and Sedory (2013); that is, Urn-III consists of $N_2 = 12$ identical balls, out of which $r_2 = 6$ bear the statement, "I do not belong to statement A," and the rest of the six balls are blank. Again for simplicity, we let Urn-IV be the same as Urn-II.

If a respondent belongs to the sensitive group A, then he/she is instructed to draw a number of balls without replacement, X, from the Urn-I until he/she obtains $t_1 = 5$ balls bearing the statement "I belong to the sensitive group A." He/she is also instructed to follow the same procedure for Urn-II. In other words, he/she is also instructed to draw a number of balls, X^*, from Urn-II until he/she obtains $t_1^* = 5$ balls bearing the statement "I belong to the sensitive group A." He/she is then instructed to report the total number of balls $(X + X^*)$ from both urns, viz. Urn-I and Urn-II.

If a respondent belongs to the nonsensitive group A^c, then he/she is instructed to draw a number of balls, Y, from the Urn-III until he/she obtains $t_2 = 5$ balls bearing the statement "I belong to the nonsensitive group A^c."

He/she is also instructed to follow the same procedure for Urn-IV. In other words, he/she is also instructed to draw a number of balls Y^* from Urn-IV until he/she obtains $t_2^* = 5$ balls bearing the statement "I belong to the nonsensitive group A^c." He/she is then instructed to report the total number of balls, $(Y + Y^*)$, from both urns, viz. Urn-III and Urn-IV. The pair of urns used by a respondent in an interview remains unknown to the interviewer; thus the respondent's privacy is protected just as in the Singh and Sedory (2013) model. Basically, in this comparison, in the newly proposed model each respondent is using twice the Singh and Sedory (2013) model and hence a doubling of the efficiency is expected. We wrote the FORTRAN codes PROGRAM-I, given in the Appendix, to compare the proposed new model with the Singh and Sedory (2013) model. The percent RE of such a newly proposed randomization device with respect to the Singh and Sedory (2013) model is shown in Table 3.

A graphical representation of the percent relative efficiency vs the value of π is shown in Fig. 4.

We called the percent relative efficiency of the proposed new randomized response model with four urns "incredible" because it is almost four times higher than the Singh and Sedory (2013) model with two urns for the same choice of device parameters. The gain is higher when the value of π is close to zero. The maximum percent RE value is found to be 456.8% when π is 0.1, and it reduces to a minimum of 409.1% when π is 0.9 with a median value of 424.5 when π is 0.5. The substantial gain is only due to using totals of the

TABLE 3 Percent RE of the Proposed New Estimator

π	0.1	0.2	0.3	0.4	0.5	0.6	0.7	0.8	0.9
RE	456.8	443.7	435.2	429.1	424.5	420.6	417.2	413.6	409.1

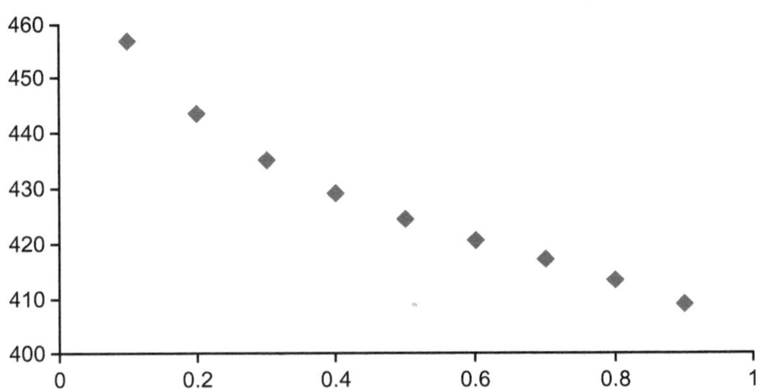

FIG. 4 RE value vs value of π.

balls drawn from two urns by each respondent and reporting back to the interviewer. Obtaining totals instead of individual responses may increase respondent's cooperation and protection. Thus the proposed randomized response model is found to be an incredibly more efficient model than the Singh and Sedory (2013) model.

In the next section, we consider a "limiting case" of the proposed randomization device with four decks of cards and compare it to the model due to Odumade and Singh (2009) which used two decks of cards.

5 LIMITING CASE WITH FOUR DECKS OF CARDS

We consider making two pairs of decks of cards. In the first pair, the first deck D-I consists of cards, each having one of two types of statements: (i) "I belong to the sensitive group A," with proportion P_1 and (ii) "I belong to the nonsensitive group A^c," with proportion $(1-P_1)$. The second deck, D-II, also consists of cards, each bearing one of two types of statements: (i) "I belong to the sensitive group A," with proportion T_1 and (ii) "I belong to the nonsensitive group A^c," with proportion $(1-T_1)$.

In the second pair of decks of cards, the first deck D-III consists of cards, each bearing one of the statements: (i) "I belong to nonsensitive group A^c," with proportion P_2 and (ii) "I belong to the sensitive group A," with proportion $(1-P_2)$. The second deck, D-IV, again consists of cards each of which have one of the statements: (i) "I belong to nonsensitive group A^c," with proportion T_2 and (ii) "I belong to the sensitive group A," with proportion $(1-T_2)$.

If a respondent belongs to the sensitive group A, then he/she is instructed to use the first pair of decks, unobserved by the interviewer. He/she is instructed to make X draws with replacement from deck D-I until he/she draws r_1 cards bearing the statement: (i) "I belong to the sensitive group A." Then he/she is instructed to make X^* draws with replacement from deck D-II until he/she draws $r_1{}^*$ cards bearing the statement: (i) "I belong to the sensitive group A." Finally, he/she is asked to report the total number $(X+X^*)$ of cards drawn.

If a respondent belongs to the nonsensitive group A^c, then he/she is instructed to use the second pair of decks, unobserved by the interviewer. He/she is instructed to make Y draws with replacement from deck D-III until he/she draws r_2 cards bearing the statement: (i) "I belong to the nonsensitive group A^c." Then he/she is instructed to make Y^* draws with replacement from the deck D-IV until he/she draws $r_2{}^*$ cards bearing the statement: (i) "I belong to the nonsensitive group A^c." Finally, he/she is asked to report the total number $(Y+Y^*)$ of cards drawn.

Thus $X \sim NB(r_1, P_1)$, that is, X follows a negative binomial distribution with parameters r_1 and P_1, and its probability mass function is given by:

$$P(X=x)=\binom{x-1}{r_1-1}P_1^{r_1}(1-P_1)^{(x-r_1)}, \quad x=r_1, r_1+1, \ldots, \infty \tag{40}$$

Similarly, $X^* \sim NB(r_1^*, T_1)$, that is, X^* follows negative binomial distribution with parameters $r_1{}^*$ and T_1, and its probability mass function is given by:

$$P(X^* = x^*) = \binom{x^* - 1}{r_1^* - 1} T_1^{r_1^*} (1 - T_1)^{(x^* - r_1^*)}, \quad x^* = r_1^*, r_1^* + 1, \ldots, \infty \tag{41}$$

Also $Y \sim NB(r_2, P_2)$, that is, Y follows negative binomial distribution with parameters r_2 and P_2, and its probability mass function is given by:

$$P(Y = y) = \binom{y - 1}{r_2 - 1} P_2^{r_2} (1 - P_2)^{(y - r_2)}, \quad y = r_2, r_2 + 1, \ldots, \infty \tag{42}$$

Finally $Y^* \sim NB(r_2^*, T_2)$, that is, Y^* follows negative binomial distribution with parameters $r_2{}^*$ and T_2, and its probability mass function is given by:

$$P(Y^* = y^*) = \binom{y^* - 1}{r_2^* - 1} T_2^{r_2^*} (1 - T_2)^{(y^* - r_2^*)}, \quad y^* = r_2^*, r_2^* + 1, \ldots, \infty \tag{43}$$

Thus the observed response Z_i, $i = 1, 2, \ldots, n$, from the ith respondent in a sample is given by:

$$Z_i = \begin{cases} (X + X^*) & \text{with probability } \pi \\ (Y + Y^*) & \text{with probability } (1 - \pi) \end{cases} \tag{44}$$

Now we have the following theorems:

Theorem 6. *An unbiased estimator of the population proportion π is given by:*

$$\hat{\pi}_{JSS(1)} = \frac{\frac{P_1 P_2 T_1 T_2}{n} \sum_{i=1}^{n} Z_i - \left(r_2 P_1 T_1 T_2 + r_2^* P_1 T_1 P_2\right)}{\left(r_1 T_1 P_2 T_2 + r_1^* P_1 P_2 T_2\right) - \left(r_2 P_1 T_1 T_2 + r_2^* P_1 P_2 T_1\right)} \tag{45}$$

Proof. Given $X \sim NB(r_1, P_1)$, $X^* \sim NB(r_1^*, T_1)$, $Y \sim NB(r_2, T_1)$, and $Y^* \sim NB(r_2^*, T_2)$, thus we have:

$$E(X) = \frac{r_1}{P_1}, \quad E(X^*) = \frac{r_1^*}{T_1}, \quad E(Y) = \frac{r_2}{P_2}, \quad \text{and} \quad E(Y^*) = \frac{r_2^*}{T_2}$$

The expected value of the observed response Z_i is given by:

$$\begin{aligned} E(Z_i) &= \pi\, E(X + X^*) + (1 - \pi) E(Y + Y^*) \\ &= \pi\, (E(X) + E(X^*)) + (1 - \pi)(E(Y) + E(Y^*)) \\ &= \pi \left(\frac{r_1}{P_1} + \frac{r_1^*}{T_1}\right) + (1 - \pi)\left(\frac{r_2}{P_2} + \frac{r_2^*}{T_2}\right) \end{aligned} \tag{46}$$

Taking expected value on both sides of (45), we have

$$E\left(\hat{\pi}_{JSS(1)}\right)=\frac{\frac{P_1P_2T_1T_2}{n}\sum_{i=1}^{n}E(Z_i)-\left(r_2P_1T_1T_2+r_2^*P_1T_1P_2\right)}{\left(r_1T_1P_2T_2+r_1^*P_1P_2T_2\right)-\left(r_2P_1T_1T_2+r_2^*P_1P_2T_1\right)} \tag{47}$$

On using (46) in (47), we get

$$E\left(\hat{\pi}_{JSS(1)}\right)$$
$$=\frac{\frac{P_1P_2T_1T_2}{n}\sum_{i=1}^{n}\left[\pi\left(\frac{r_1}{P_1}+\frac{r_1^*}{T_1}\right)+(1-\pi)\left(\frac{r_2}{P_2}+\frac{r_2^*}{T_2}\right)\right]-\left(r_2P_1T_1T_2+r_2^*P_1T_1P_2\right)}{\left(r_1T_1P_2T_2+r_1^*P_1P_2T_2\right)-\left(r_2P_1T_1T_2+r_2^*P_1P_2T_1\right)}$$

which on simplification yields

$$E\left(\hat{\pi}_{JSS(1)}\right)=\pi$$

which proves the theorem. □

Theorem 7. *The variance of the unbiased estimator* $\hat{\pi}_{JSS(1)}$ *of the population proportion* π *is given by:*

$$V\left(\hat{\pi}_{JSS(1)}\right)=\frac{\pi(1-\pi)}{n}$$
$$+\frac{\pi\left[r_1T_1^2P_2^2T_2^2(1-P_1)+r_1^*P_1^2P_2^2T_2^2(1-T_1)-r_2P_1^2T_1^2T_2^2(1-P_2)-r_2^*P_1^2P_2^2T_1^2(1-T_2)\right]}{n\left[\left(r_1T_1P_2T_2+r_1^*P_1P_2T_2\right)-\left(r_2P_1T_1T_2+r_2^*P_1P_2T_1\right)\right]^2}$$
$$+\frac{r_2P_1^2T_1^2T_2^2(1-P_2)+r_2^*P_1^2T_1^2P_2^2(1-T_2)}{n\left[\left(r_1T_1P_2T_2+r_1^*P_1P_2T_2\right)-\left(r_2P_1T_1T_2+r_2^*P_1P_2T_1\right)\right]^2} \tag{48}$$

Proof. Given $X\sim NB(r_1,P_1)$, $X^*\sim NB\left(r_1^*,T_1\right)$, $Y\sim NB(r_2,T_1)$, and $Y^*\sim NB\left(r_2^*,T_2\right)$, thus we have:

$$E(X^2)=\frac{r_1(r_1+1-P_1)}{P_1^2},\quad E(X^{*2})=\frac{r_1^*\left(r_1^*+1-T_1\right)}{T_1^2},$$

$$E(Y^2)=\frac{r_2(r_2+1-T_1)}{T_1^2},\quad E(Y^{*2})=\frac{r_2^*\left(r_2^*+1-T_2\right)}{T_2^2}$$

The expected value of the square of the observed response Z_i^2 is given by:

$$\begin{aligned}
E(Z_i^2) &= \pi\, E(X+X^*)^2 + (1-\pi)E(Y+Y^*)^2 \\
&= \pi\left[E(X^2)+E(X^{*2})+2E\left(XX^*\right)\right] + (1-\pi)\left[E(Y^2)+E(Y^{*2})+2E(YY^*)\right] \\
&= \pi\left[E(X^2)+E(X^{*2})+2E(X)E\left(X^*\right)\right] \\
&\quad + (1-\pi)\left[E(Y^2)+E(Y^{*2})+2E(Y)E(Y^*)\right] \\
&= \pi\left[\frac{r_1(r_1+1-P_1)}{P_1^2}+\frac{r_1^*(r_1^*+1-T_1)}{T_1^2}+\frac{2r_1r_1^*}{P_1T_1}\right] \\
&\quad + (1-\pi)\left[\frac{r_2(r_2+1-P_2)}{P_2^2}+\frac{r_2^*(r_2^*+1-T_2)}{T_2^2}+\frac{2r_2r_2^*}{P_2T_2}\right]
\end{aligned} \tag{49}$$

The variance of the observed response Z_i is given by

$$\begin{aligned}
\sigma_Z^2 = E(Z_i^2)-(E(Z_i))^2 &= \pi\left[\frac{r_1(r_1+1-P_1)}{P_1^2}+\frac{r_1^*(r_1^*+1-T_1)}{T_1^2}+\frac{2r_1r_1^*}{P_1T_1}\right] \\
&\quad + (1-\pi)\left[\frac{r_2(r_2+1-P_2)}{P_2^2}+\frac{r_2^*(r_2^*+1-T_2)}{T_2^2}+\frac{2r_2r_2^*}{P_2T_2}\right] \\
&\quad - \left[\pi\left(\frac{r_1}{P_1}+\frac{r_1^*}{T_1}\right)+(1-\pi)\left(\frac{r_2}{P_2}+\frac{r_2^*}{T_2}\right)\right]^2
\end{aligned} \tag{50}$$

Note that all the responses are independent, thus by the definition of variance, we have

$$\begin{aligned}
V\left(\hat{\pi}_{JSS(1)}\right) &= V\left[\frac{\frac{P_1P_2T_1T_2}{n}\sum_{i=1}^{n} Z_i - \left(r_2P_1T_1T_2+r_2^*P_1T_1P_2\right)}{\left(r_1T_1P_2T_2+r_1^*P_1P_2T_2\right)-\left(r_2P_1T_1T_2+r_2^*P_1P_2T_1\right)}\right] \\
&= \frac{\left(\frac{P_1P_2T_1T_2}{n}\right)^2\sum_{i=1}^{n} V(Z_i)}{\left[\left(r_1T_1P_2T_2+r_1^*P_1P_2T_2\right)-\left(r_2P_1T_1T_2+r_2^*P_1P_2T_1\right)\right]^2} \\
&= \frac{(P_1P_2T_1T_2)^2\left(\frac{\sigma_Z^2}{n}\right)}{\left[\left(r_1T_1P_2T_2+r_1^*P_1P_2T_2\right)-\left(r_2P_1T_1T_2+r_2^*P_1P_2T_1\right)\right]^2}
\end{aligned} \tag{51}$$

On using (50) into (51), we have the theorem. □

6 RELATIVE EFFICIENCY OF THE LIMITING CASE

In this section, we consider a study of percent relative efficiency of the limiting case with respect to the Odumade and Singh (2009) two decks of cards methods. The percent relative efficiency of the newly proposed estimator $\hat{\pi}_{JSS(1)}$ with respect to the Odumade and Singh (2009) model is defined as:

$$\text{RE(OS)} = \frac{V(\hat{\pi}_{os})}{V(\hat{\pi}_{JSS(1)})} \times 100\% \tag{52}$$

where $V(\hat{\pi}_{os})$ stands for the variance of the Odumade and Singh (2009) estimator. The percent relative efficiency of the newly proposed estimator $\hat{\pi}_{JSS(1)}$ with respect to the Singh and Grewal (2013) model is defined as:

$$\text{RE(SG)} = \frac{V(\hat{\pi}_{SG})}{V(\hat{\pi}_{JSS(1)})} \times 100\% \tag{53}$$

where $V(\hat{\pi}_{SG})$ stands for the variance of the Singh and Grewal (2013) estimator. In the numerical comparisons, we consider $P=0.7$, $T=0.6$, $P_1=0.7$, $P_2=0.6$, $T_1=0.7$, $T_2=0.6$, $\theta_1^*=0.7$, and $\theta_2^*=0.6$. We consider all possible combinations of $r_1=2$ or 3, $r_1^*=2$ or 3, $t_1=2$ or 3, and $t_1^*=2$ or 3. Then we wrote the FORTRAN codes PROGRAM-II given in the Appendix to compare the proposed limiting case with the Odumade and Singh (2009) and Singh and Grewal (2013). For π between 0.1 and 0.9, with a step of 0.1, the situations where the proposed estimator in the limiting case can perform better than the Odumade and Singh (2009) and Singh and Grewal (2013) model are listed in Table 4.

TABLE 4 Percent RE(OS) and RE(SG) Values

r_1	r_1^*	r_2	r_2^*	π	RE(OS)	RE(SG)
2	2	2	3	0.1	137	2201
2	2	2	3	0.2	141	2047
2	2	2	3	0.3	146	1942
2	2	2	3	0.4	152	1879
2	2	2	3	0.5	160	1854
2	2	2	3	0.6	170	1871
2	2	2	3	0.7	184	1942
2	2	2	3	0.8	204	2093
2	2	2	3	0.9	238	2391

Continued

TABLE 4 Percent RE(OS) and RE(SG) Values—Cont'd

r_1	r_1^*	r_2	r_2^*	π	RE(OS)	RE(SG)
2	2	3	2	0.1	137	2201
2	2	3	2	0.2	141	2047
2	2	3	2	0.3	146	1942
2	2	3	2	0.4	152	1879
2	2	3	2	0.5	160	1854
2	2	3	2	0.6	170	1871
2	2	3	2	0.7	184	1942
2	2	3	2	0.8	204	2093
2	2	3	2	0.9	238	2391
2	2	3	3	0.1	272	4374
2	2	3	3	0.2	260	3773
2	2	3	3	0.3	256	3406
2	2	3	3	0.4	258	3193
2	2	3	3	0.5	267	3101
2	2	3	3	0.6	284	3130
2	2	3	3	0.7	313	3311
2	2	3	3	0.8	366	3749
2	2	3	3	0.9	475	4779
2	3	3	3	0.1	136	2181
2	3	3	3	0.2	140	2026
2	3	3	3	0.3	144	1920
2	3	3	3	0.4	150	1855
2	3	3	3	0.5	157	1827
2	3	3	3	0.6	167	1839
2	3	3	3	0.7	180	1902
2	3	3	3	0.8	199	2041
2	3	3	3	0.9	230	2312
3	2	3	3	0.1	136	2181

Continued

TABLE 4 Percent RE(OS) and RE(SG) Values—Cont'd

r_1	r_1^*	r_2	r_2^*	π	RE(OS)	RE(SG)
3	2	3	3	0.2	140	2026
3	2	3	3	0.3	144	1920
3	2	3	3	0.4	150	1855
3	2	3	3	0.5	157	1827
3	2	3	3	0.6	167	1839
3	2	3	3	0.7	180	1902
3	2	3	3	0.8	199	2041
3	2	3	3	0.9	230	2312

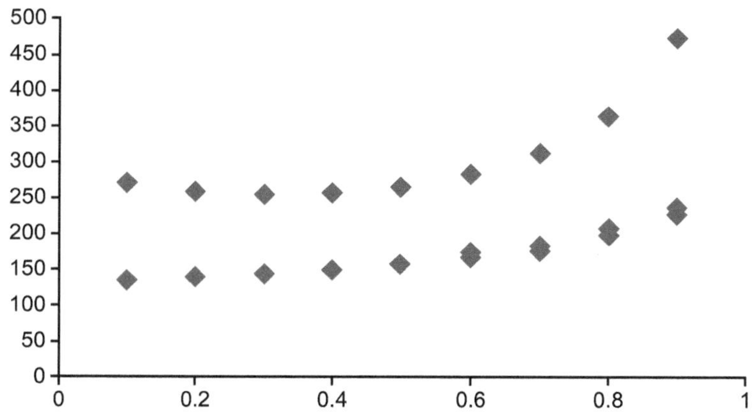

FIG. 5 RE(OS) value for different values of π.

The value of relative efficiency RE(SG) remains much higher than RE(OS) estimators for the situations considered in Table 4. The RE(OS) and RE(SG) vs the value of π are presented in Figs. 5 and 6, respectively.

From the graphs, it seems clear that the proposed model shows higher relative efficiency RE(OS) when the value of π is close to one, while the relative efficiency RE(SG) is higher when the value of π is either close to zero or close to one. A more critical look at both Figs. 5 and 6 shows that the proposed model can also be efficiently used when the value of π is close to zero, which may be a situation more common in real practice.

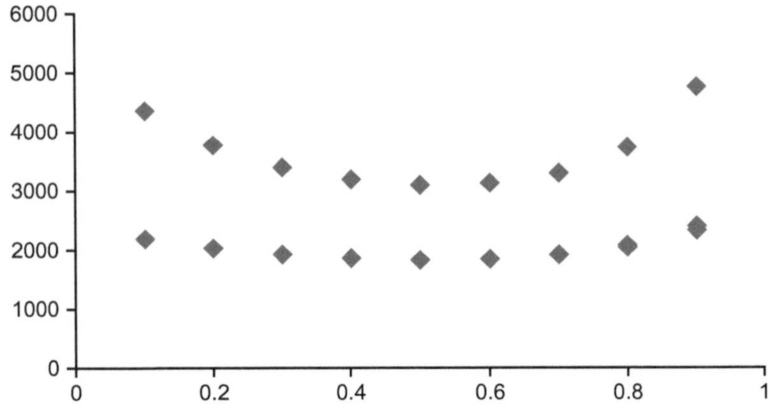

FIG. 6 RE(SG) value for different values of π.

ACKNOWLEDGMENTS

The authors are thankful to Prof. Arijit Chaudhuri, Purnima Shaw, and a referee for their very constructive comments on the original version of this chapter.

APPENDIX

PROGRAM-I

```
! FORTRAN CODED USED IN JSS ESTIMATOR
        USE NUMERICAL_LIBRARIES
        IMPLICIT NONE
        INTEGER IR1, IR2, N1, N2, IT1, IT2
        INTEGER IR1P, IR2P, N1P, N2P, IT1P, IT2P
        REAL PI, AT1, AT2, AN1, AN2, AR1, AR2
        REAL AT1P, AT2P,AN1P, AN2P, AR1P, AR2P
        REAL EX1, EY1, EX1SQ, EY1SQ
        REAL EX1P, EY1P, EX1PSQ, EY1PSQ
        REAL DENOSS, EZ, EZSQ, VARSS
        REAL DENOM, EZP, EZPSQ, VARMICH,RE
        CHARACTER*20 OUT_FILE
        CHARACTER*20 IN_FILE
        WRITE(*,'(A)') 'NAME OF THE OUTPUT FILE'
        READ(*,'(A20)') OUT_FILE
        OPEN(42, FILE=OUT_FILE, STATUS='UNKNOWN')
        IT1 = 5
        IT2 = 5
        N1 = 10
        N2 = 12
        IR1 = 7
        IR2 = 6
```

```
      IT1P = 5
      IT2P = 5
      N1P = 10
      N2P = 12
      IR1P = 7
      IR2P = 6
      AT1 = IT1
      AT2 = IT2
      AN1 = N1
      AN2 = N2
      AR1 = IR1
      AR2 = IR2
      AT1P = IT1P
      AT2P = IT2P
      AN1P = N1P
      AN2P = N2P
      AR1P = IR1P
      AR2P = IR2P
      EX1 = AT1*(AN1+1)/(AR1+1)
      EY1 = AT2*(AN2+1)/(AR2+1)
      EX1SQ  =  AT1*(AN1+1)*(AN1-AR1)*(AR1+1-AT1)/((AR1+1)**2*
(AR1+2))
     1 + EX1**2
      EY1SQ  =  AT2*(AN2+1)*(AN2-AR2)*(AR2+1-AT2)/((AR2+1)**2*
(AR2+2))
     1 + EY1**2
      DENOSS = AT1*(AN1+1)*(AR2+1)-AT2*(AN2+1)*(AR1+1)
      EX1P = AT1P*(AN1P+1)/(AR1P+1)
      EY1P = AT2P*(AN2P+1)/(AR2P+1)
      EX1PSQ = AT1P*(AN1P+1)*(AN1P-AR1P)*(AR1P+1-AT1P)/
     1((AR1P+1)**2*(AR1P+2)) + EX1P**2
      EY1PSQ = AT2P*(AN2P+1)*(AN2P-AR1P)*(AR2P+1-AT2P)/
     1((AR2P+1)**2*(AR2P+2)) + EY1P**2
      DENOM = AT1*(AN1+1)*(AR1P+1)*(AR2+1)*(AR2P+1)
     1+ AT1P*(AN1P+1)*(AR1+1)*(AR2+1)*(AR2P+1)
     1- AT2*(AN2+1)*(AR1+1)*(AR1P+1)*(AR2P+1)
     1- AT2P*(AN2P+1)*(AR1+1)*(AR1P+1)*(AR2+1)
      DO 9999 PI = 0.1, 0.9, 0.1
      EZ = PI*EX1 + (1-PI)* EY1
      EZSQ = PI*EX1SQ + (1-PI)*EY1SQ
      VARSS = (EZSQ-EZ**2)*(AR1+1)**2*(AR2+1)**2/DENOSS**2
      EZP = PI*EX1P + (1-PI)*EY1P
      EZPSQ = PI * EX1PSQ + (1-PI) * EY1PSQ
      VARMICH=((AR1+1)*(AR1P+1)*(AR2+1)*(AR2P+1))**2*(EZPSQ-
EZP**2)/
     1DENOM**2
      RE = VARSS*100/VARMICH
```

```
      WRITE(42,101)N1,IR1,IT1, N2, IR2,IT2,N1P,IR1P,IT1P,
 N2P,IR2P,
     1IT2P, PI,RE
101   FORMAT(1X,12(I3,1X),2X,F7.3,2X,F9.2)
9999 CONTINUE
      STOP
      END
!REMARK: X1 used for X; Y1 used for Y; X1P used for X*; and Y1P used
 for Y*
```

PROGRAM-II

```
! JSS ( 1 ) ESTIMATOR
        USE NUMERICAL_LIBRARIES
        IMPLICIT NONE
        INTEGER IR1, IR2, IR1S, IR2S
        REAL P, T, PI
        REAL TERM1, TERM2, TERM3, VAROS
        REAL P1, T1, P2, T2
        REAL R1, R1S, R2, R2S
        REAL EZ, EZSQ, VARZ,DELT, VARMICH4
        REAL VARSG, TH1S, TH2S, RE1, RE2
        CHARACTER*20 OUT_FILE
        CHARACTER*20 IN_FILE
        WRITE(*,'(A)') 'NAME OF THE OUTPUT FILE'
        READ(*,'(A20)') OUT_FILE
        OPEN(42, FILE=OUT_FILE, STATUS='UNKNOWN')
        P = 0.7
        T = 0.6
        DO 9999 R1 = 2, 3, 1
        DO 9999 R2 = 2, 3, 1
        DO 9999 R1S = 2, 3, 1
        DO 9999 R2S = 2, 3,1
        P1 = 0.7
        P2 = 0.6
        T1 = 0.7
        T2 = 0.6
        TH1S = P
        TH2S = T
        DO 9999 PI = 0.1, 0.9, 0.1
        TERM1 = (P+T-1)**2*(P*T+(1-P)*(1-T))
        TERM2 = (P-T)**2*(T*(1-P)+P*(1-T))
        TERM3 = (P+T-1)**2 + (P-T)**2
        VAROS = (TERM1+TERM2)/(4.*TERM3**2)-(2.*PI-1)**2/4.
         EZ = PI*(R1/P1+R1S/T1)+(1-PI)*(R2/P2+R2S/T2)
```

```
        EZSQ = PI*( R1*(R1+1-P1)/P1**2 + R1S*(R1S+1-T1)/T1**2
       1 + 2*R1*R1S/(P1*T1))
       1 + (1-PI)*( R2*(R2+1-P2)/P2**2 + R2S*(R2S+1-T2)/T2**2
       1 + 2*R2*R2S/(P2*T2))
        VARZ = EZSQ - EZ**2
        DELT = (R1*T1*P2*T2+R1S*P1*P2*T2)-(R2*P1*T1*T2+R2S*P1*P2*T1)
        VARMICH4 = (P1*T1*P2*T2)**2*VARZ/DELT**2
        VARSG = PI*(1-PI)
       1 + (TH2S**2*(1-TH1S)*PI + TH1S**2*(1-TH2S)*(1-PI))/
       1 (TH2S-TH1S)**2
         RE1 = VAROS*100.0/VARMICH4
         RE2 = VARSG*100/VARMICH4
          IR1 = R1
          IR1S = R1S
          IR2 = R2
          IR2S = R2S
          IF((RE1.GT.100).AND.(RE2.GE.100))THEN
         WRITE(42,101)IR1,IR1S,IR2,IR2S,P,T,P1,T1,P2,T2,PI,
RE1,RE2
   101   FORMAT(1X,4(I3,1X),7(F6.3,1X),1X,F7.2,1X,F7.2)
         ENDIF
   9999  CONTINUE
         STOP
         END
```

REFERENCES

Chaudhuri, A., Christofides, T.C., 2013. Indirect Questioning in Sample Surveys. Springer, Heidelberg.

Kuk, A.Y.C., 1990. Asking sensitive questions indirectly. Biomerika 77 (2), 436–438.

Mangat, N.S., 1994. An improved randomized response strategy. J. R. Stat. Soc., Series B Stat. Methodol. 56, 93–95.

Odumade, O., Singh, S., 2009. Efficient use of two decks of cards in randomized response sampling. Commun. Stat. Theory Meth. 38, 439–446.

Singh, S., Grewal, I.S., 2013. Geometric distribution as a randomization device implemented to the Kuk's model. Int. J. Contemp. Math. Sci. 8 (5), 243–248.

Singh, S., Sedory, S.A., 2013. A new randomized response device for sensitive characteristics: an application of the negative hypergeometric distribution. Metron 71, 3–8.

Warner, S.L., 1965. Randomized response: a survey technique for eliminating evasive answer bias. J. Am. Stat. Assoc. 60, 63–69.

Chapter 28

Comparison of Different Imputing Methods for Scrambled Responses

C. Mohamed, S.A. Sedory and S. Singh[1]
Texas A&M University-Kingsville, Kingsville, TX, United States
[1]*Corresponding author: e-mail: Sarjinder.Singh@tamuk.edu*

ABSTRACT
In comparison to follow up a survey, the method of imputing nonresponses has been found to be much cheaper from the cost point of views. The problem of nonresponse in survey sampling has been widely discussed in direct question survey techniques. A very limited effort has been made if the responses are collected by using a randomized response device. In this paper, the problem of imputing scrambled responses will be considered by following Singh et al. (2008) and Zaizai et al. (2008).

Keywords: Randomized response sampling, Imputation, Missing data, Sensitive characteristics

1 INTRODUCTION

Asking sensitive questions about such things as induced abortion, drug abuse, and family income remains a problematic issue for collecting data in survey sampling. Warner (1965) came up with a strategy to deal with asking such questions. His idea was to consider the respondents from a population consisting of two mutually exclusive groups: one group with the sensitive or stigmatizing characteristic, and the other group without it. For estimating π which represents the proportion of respondents in the population belonging to the sensitive group G, a simple random sample of n respondents selected with replacement from the population. Warner (1965) used a randomization device to collect information on the sensitive characteristic. For instance, a deck of cards can be used as a randomization device where each card has one of the following two statements: (i) "I belong to group G" and (ii) "I do not belong to group G." The statements occur with relative frequencies P and $(1-P)$, respectively in the deck, and each respondent from the sample is asked to

Handbook of Statistics, Vol. 34. http://dx.doi.org/10.1016/bs.host.2016.01.026

select a card. The respondent then answers the question "Is the statement true for you?" Assume n_1 representing the number of respondents answering "yes," is binomially distributed with parameters $\theta_w = P\pi + (1-P)(1-\pi)$ and n. Using the known value of P, the maximum likelihood estimator of π is given by:

$$\hat{\pi}_w = \frac{\hat{\theta}_w - (1-P)}{2P-1} \tag{1}$$

where $P \neq 0.5$ and $\hat{\theta}_w = n_1/n$ is the observed proportion of "yes" answers. The estimator $\hat{\pi}_w$ is unbiased for π, with variance

$$V(\hat{\pi}_w) = \frac{\pi(1-\pi)}{n} + \frac{P(1-P)}{n(2P-1)^2} \tag{2}$$

Horvitz et al. (1967) and Greenberg et al. (1971) continued the work of Warner (1965) by allowing for the responses to the sensitive question to be quantitative instead of the usual "yes" or "no." Using a randomization device, the respondent chooses one of two questions: one the sensitive question and the other nonsensitive and unrelated to the sensitive one. However, using this unrelated question method creates several difficulties: The first one is selecting the unrelated question. According to Greenberg et al. (1971), it is important that the population mean and population variance of responses to the sensitive questions would approximately equal those of the responses to the unrelated question. A second difficulty is when the answers to the unrelated question are rounded while those to the sensitive question are not, making it possible to recognize which question was answered. For instance, Greenberg et al. (1971) considered the sensitive question: "how much money did the head of this household earn last year?" and the unrelated question: "how much money do you think the average head of a household of your size earns in a year?" The authors paired up those two questions and found out that an answer to the unrelated question will look like \$26,350, while an answer to the sensitive question will look more like \$18,618. Lastly, some people are reluctant to answer the sensitive question, even though they know that the interviewer cannot be sure that the sensitive question was selected. However, the scrambled randomized response method introduced by Eichhorn and Hayre (1983) deals with these difficulties. In summary, the method is as follows: The scrambled result $Z = YS$ is revealed to the interviewer, with Y being the true response from each respondent and S being the random scrambling variable. Then from this sample of Z values, the mean of the response, $E(Y)$, is estimated. These scrambled responses are also used in regression analysis by Singh et al. (1996), Rodney et al. (1998), and Singh and King (1999). Chaudhuri and Adhikary (1990) suggested a practicable randomization device, where the ith respondent in the sample is required to choose independently and at random two tickets: one S_{1j} from a box containing tickets labeled $A_1, A_2, \ldots, A_m$ with labels having a known mean θ_1 and known variance

σ_1^2 and the other S_{2j}, from a box containing tickets labeled $B_1,B_2,...,B_t$ with labels having a known mean θ_2 and variance σ_2^2. The respondent should report the response as: $z_i = S_{1j}y_i + S_{2j}$.

A very common problem in survey sampling is nonresponse or underreporting on the sensitive issues due to our social culture. Many instances could be stated such as cases where students are reluctant to disclose their GPA, where physicians are reluctant to disclose the number of patients with AIDS, and where women are reluctant to disclose how many abortions they have. Thus, techniques that would guarantee a respondents' privacy are more than necessary in case of sensitive questions. A detail of importance and need of randomized response sampling can be had from Chaudhuri (2011), Chaudhuri and Christofides (2013), and Fox (2016).

In this section, we consider the situation where some of the respondents when using a randomized response scheme refuse to give scrambled responses related to sensitive questions. In particular, we suppose an investigator is using the method proposed by Singh et al. (2008) and Zaizai et al. (2008). In this setup one begins by selecting a simple random without replacement sample of n respondents from a population consisting of N units. Let y_i be the true response, for example, income, of the ith respondent in the sample. The ith respondent selected in the sample is requested to draw two numbers S_1 and S_2 from two independent randomization devises, say D_1 and D_2, respectively. The ith respondent is requested to report the scrambled response as:

$$z_i = \frac{S_1 y_i + S_2 - \theta_2}{\theta_1} \tag{3}$$

Let E_3 and V_3 denote, respectively, the expected value and the variance over the randomization devices. We assume $E_3(S_1) = \theta_1$, $E_3(S_2) = \theta_2$, $V_3(S_1) = \sigma_1^2$, and $V_3(S_2) = \sigma_2^2$ are known. Obviously, $V_3(z_i) = \frac{1}{\theta_1^2}\left(y_i^2\sigma_1^2 + \sigma_2^2\right) = y_i^2 C_{\theta_1}^2 + \left(\frac{\theta_2}{\theta_1}\right)^2 C_{\theta_2}^2$, where $C_{\theta_1}^2 = \sigma_1^2/\theta_1^2$ and $C_{\theta_2}^2 = \sigma_2^2/\theta_2^2$.

Let r be the number of respondents in the subset A of the sample s who responded to the sensitive question with the help of the above randomization device, then $(n-r)$ respondents belong to A^c the subset of the sample s who refused to respond so $s = A \cup A^c$. Let $\bar{z}_r = \frac{1}{r}\sum_{i=1}^{r} z_i$ be the sample mean of the scrambled responses obtained from the responding units. Then we have the following lemma:

Lemma 1. *The variance of $\bar{z}_r$ is approximately given by:*

$$V(\bar{z}_r) = \left(\frac{1}{n} - \frac{1}{N}\right)S_y^2 + \frac{\vec{Y}^2}{r}\left[C_{\theta_2}^2\left(1 + \frac{(N-1)}{N}C_y^2\right) + \left(\frac{\theta_2}{\theta_1}\right)^2 C_{\theta_1}^2\right] \tag{4}$$

Proof. Let E_1 and E_2 be the expected values for given n and r, respectively. Also let V_1 and V_2 be the variances for the given values of n and r, respectively. Then by the definition of variance, we have

$$
\begin{aligned}
V(\bar{z}_r) &= E_1E_2V_3[\bar{z}_r] + E_1V_2E_3[\bar{z}_r] + V_1E_2E_3[\bar{z}_r] \\
&= E_1E_2V_3\left[\frac{1}{r}\sum_{i=1}^{r} z_i\right] + E_1V_2E_3\left[\frac{1}{r}\sum_{i=1}^{r} z_i\right] + V_1E_2E_3\left[\frac{1}{r}\sum_{i=1}^{r} z_i\right] \\
&= E_1E_2\left[\frac{1}{r^2}\sum_{i=1}^{r} V_3(z_i)\right] + E_1V_2\left[\frac{1}{r}\sum_{i=1}^{r} E_3(z_i)\right] + V_1E_2\left[\frac{1}{r}\sum_{i=1}^{r} E_3(z_i)\right] \\
&= E_1E_2\left[\frac{1}{r^2}\sum_{i=1}^{r} \frac{\sigma_1^2 y_i^2 + \sigma_2^2}{\theta_1^2}\right] + E_1V_2\left[\frac{1}{r}\sum_{i=1}^{r} y_i\right] + V_1E_2\left[\frac{1}{r}\sum_{i=1}^{r} y_i\right] \\
&= E_1\left[\frac{1}{rn}\sum_{i=1}^{r} \frac{\sigma_1^2 y_i^2 + \sigma_2^2}{\theta_1^2}\right] + E_1\left[\left(\frac{1}{r} - \frac{1}{n}\right)s_{y(n)}^2\right] + V_1\left[\frac{1}{n}\sum_{i=1}^{n} y_i\right] \\
&= \left[\frac{1}{rN}\sum_{i=1}^{N} \frac{\sigma_1^2 y_i^2 + \sigma_2^2}{\theta_1^2}\right] + \left(\frac{1}{r} - \frac{1}{n}\right)S_y^2 + \left(\frac{1}{n} - \frac{1}{N}\right)S_y^2 \\
&= \frac{1}{r}\left[\frac{\sigma_1^2}{\theta_1^2}\left(\frac{1}{N}\sum_{i=1}^{N} y_i^2\right) + \frac{\sigma_2^2}{\theta_1^2}\right] + \left(\frac{1}{r} - \frac{1}{n}\right)S_y^2 + \left(\frac{1}{n} - \frac{1}{N}\right)S_y^2 \\
&= \frac{1}{r}\left[\frac{\sigma_1^2}{\theta_1^2}\left(\frac{(N-1)}{N}S_y^2 + \bar{Y}^2\right) + \frac{\sigma_2^2}{\theta_1^2}\right] + \left(\frac{1}{r} - \frac{1}{N}\right)S_y^2 \\
&= \frac{\bar{Y}^2}{r}\left[C_{\theta_1}^2\left(1 + \frac{(N-1)}{N}C_y^2\right) + \left(\frac{\theta_2}{\theta_1}\right)^2 C_{\theta_2}^2\right] + \left(\frac{1}{r} - \frac{1}{N}\right)S_y^2
\end{aligned}
$$

which proves the lemma. Note that this lemma gives a better approximation of the result quoted in Singh et al. (2008).■

2 RATIO METHOD OF IMPUTING SCRAMBLED RESPONSES

Singh et al. (2008) considered imputing the missing scrambled responses as follows:

$$
\hat{z}_i = \begin{cases} z_i, & i \in A \\ \hat{b}x_i, & i \in A^c \end{cases} \tag{5}
$$

where x_i is an auxiliary variable related to the study variable y_i. We follow Singh et al. (1996) by using the models:

$$
z_i = bx_i + \eta_i \quad \text{and} \quad y_i = bx_i + e_i \tag{6}
$$

where $e_i \sim N(0, \sigma^2)$ and the distribution of η_i is unknown. By the first principle of least squares, that is, by setting:

$$\sum_{i=1}^{r} \eta_i = 0, \tag{7}$$

the linear unbiased estimator of the regression coefficient b is given by:

$$\hat{b} = \frac{\bar{z}}{\bar{x}} \tag{8}$$

Following Singh et al. (2008), the point estimator $\bar{y}_s$ of the population mean $\bar{Y}$ is defined as:

$$\bar{y}_s = \frac{1}{n}\sum_{i=1}^{n} \hat{z}_i \tag{9}$$

Under the ratio method of imputation defined in (5), the point estimator $\bar{y}_s$ becomes:

$$\bar{y}_{\mathrm{sr}} = \bar{z}_r \left(\frac{\bar{x}_n}{\bar{x}_r}\right) \tag{10}$$

where $\bar{z}_r = \frac{1}{r}\sum_{i=1}^{r} z_i$, $\bar{x}_r = \frac{1}{r}\sum_{i=1}^{r} x_i$, and $\bar{x}_n = \frac{1}{n}\sum_{i=1}^{n} x_i$. The estimator $\bar{y}_{\mathrm{sr}}$ in (10) is a ratio estimator of the population mean based on scrambled responses (sr).

3 REGRESSION METHOD OF IMPUTING SCRAMBLED RESPONSES

In this section, we consider imputing the missing scrambled responses as follows:

$$\hat{z}_i = \begin{cases} z_i, & i \in A \\ \bar{z}_r + \hat{\beta}_s(x_i - \bar{x}_r), & i \in A^c \end{cases} \tag{11}$$

From the model,

$$z_i = bx_i + \eta_i \quad \text{with} \sum_{i=1}^{r} \eta_i = 0 \tag{12}$$

we have

$$\bar{z}_r = b\,\bar{x}_r \tag{13}$$

From (12) and (13) we have

$$\eta_i = (z_i - \bar{z}_r) - b(x_i - \bar{x}_r) \tag{14}$$

By the second principle of least squares, that is, by

$$\text{Minimize} \sum_{i=1}^{r} \eta_i^2, \tag{15}$$

subject to b. In other words, by setting

$$\frac{\partial}{\partial b}\sum_{i=1}^{r}\eta_i^2=0$$

and then by the method of moments, an unbiased estimator of the regression coefficient b is given by:

$$\hat{\beta}_s=\frac{\sum_{i=1}^{r}(z_i-\bar{z}_r)(x_i-\bar{x}_r)}{\sum_{i=1}^{r}(x_i-\bar{x}_r)^2} \tag{16}$$

Under the proposed regression type method of imputation defined in (11), the point estimator $\bar{y}_s$ becomes:

$$\bar{y}_{\text{slr}}=\bar{z}_r+\hat{\beta}_s(\bar{x}_n-\bar{x}_r) \tag{17}$$

Some related work on imputation in survey sampling can be found in Mohamed (2015), Singh et al. (2009), Singh (2009), Singh and Arnab (2010), Singh and Deo (2003), Singh and Horn (2000), Toutenberg and Srivastava (2003), and Toutenberg et al. (2006), but these methods differ from the method developed in section 4, which makes use of higher order moments of an auxiliary variable at the imputation stage, and are as efficient as a traditional regression method of estimation.

The problem of estimating a population parameter by making use of higher order moments of an auxiliary variables has been considered by Das and Tripathi (1978), Isaki (1983), Srivastava and Jhajj (1981). Later, Singh et al. (1998), Singh et al. (1999), Farrell and Singh (2005), and Sitter and Wu (2002) among others, also considered the use of known higher order moments of an auxiliary variable when calibrating estimators of variance. Thus, it is a well-established result that the uses of higher order moments of an auxiliary variable play an important role in estimating population parameter of a study variable in survey sampling. This motivated the present study to investigate whether the higher order moments of an auxiliary variable can be utilized to impute scrambled responses.

4 IMPUTING SCRAMBLED RESPONSES USING HIGHER ORDER MOMENTS OF AN AUXILIARY VARIABLE

In this section, we propose a new method of imputation which makes use of higher order moments of the auxiliary variable as follows:

$$d_{\bullet i}=\begin{cases} z_i \quad \text{if } i\in A \\ \bar{z}_r+\beta_1(x_i-\bar{x}_r)+\beta_2\left(\frac{n\left(x_i-{}^{i}\bar{x}_n\right)^2}{(n-1)}+\frac{n\sum_{i\in A}(x_i-\bar{x}_n)^2}{(n-r)(n-1)}-\frac{n\sum_{i\in A}(x_i-\bar{x}_r)^2}{(n-r)(r-1)}\right) \quad \text{if } i\in A^c \end{cases} \tag{18}$$

where β_1 and β_2 are assumed to be known constant and are determined such that the variance of the resultant estimator is minimum. Now the point estimator of population mean, under the new method of imputation becomes:

$$\begin{aligned}
\bar{y}_{\text{point}} &= \frac{1}{n}\sum_{i\in s} d_{\bullet i} \\
&= \frac{1}{n}\left[\sum_{i\in A} z_i + (n-r)\bar{z}_r + \beta_1\left(\sum_{i\in A^c} x_i - (n-r)\bar{x}_r\right)\right. \\
&\quad \left. + \beta_2\left\{\frac{n\sum_{i\in A^c}(x_i-\bar{x}_n)^2 + n\sum_{i\in A}(x_i-\bar{x}_n)^2}{n-1} - ns^2_{x(r)}\right\}\right] \qquad (19) \\
&= \frac{1}{n}\left[r\bar{z}_r + (n-r)\bar{z}_r + \beta_1((n\bar{x}_n - r\bar{x}_r) - (n-r)\bar{x}_r) + \beta_2\left\{ns^2_{x(n)} - ns^2_{x(r)}\right\}\right] \\
&= \frac{1}{n}\left[n\bar{z}_r + n\beta_1(\bar{x}_n - \bar{x}_r) + n\beta_2\left(s^2_{x(n)} - s^2_{x(r)}\right)\right] \\
&= \bar{z}_r + \beta_1(\bar{x}_n - \bar{x}_r) + \beta_2\left(s^2_{x(n)} - s^2_{x(r)}\right) = \bar{y}_{\text{snew(ch)}} \quad \text{(say)}
\end{aligned}$$

5 SOME USEFUL RESULTS

In order to study the bias and variance of the estimators considered in the previous sections, let us define:

$$\varepsilon_0 = \frac{\bar{z}_r}{\bar{Y}} - 1,\ \varepsilon_1 = \frac{\bar{x}_r}{\bar{X}} - 1,\ \varepsilon_2 = \frac{\bar{x}_n}{\bar{X}} - 1,\ \varepsilon_3 = \frac{s^2_{x(r)}}{S^2_x} - 1,\ \varepsilon_4 = \frac{s^2_{x(n)}}{S^2_x} - 1, \text{and}\ \varepsilon_5 = \frac{s_{xz(r)}}{S_{xy}} - 1$$

such that:

$$E(\varepsilon_0) = E(\varepsilon_1) = E(\varepsilon_2) = E(\varepsilon_3) = E(\varepsilon_4) = E(\varepsilon_5) = 0$$

and, to the first order of approximation,

$$E(\varepsilon_0^2) = \left(\frac{1}{r} - \frac{1}{N}\right)C_y^2 + \frac{1}{r}\left[C^2_{\theta_1}\left(1 + \frac{(N-1)}{N}C_y^2\right) + \left(\frac{\theta_2}{\theta_1}\right)^2 C^2_{\theta_2}\right]$$

$$E(\varepsilon_1^2) = \left(\frac{1}{r} - \frac{1}{N}\right)C_x^2,\ E(\varepsilon_2^2) = \left(\frac{1}{n} - \frac{1}{N}\right)C_x^2,\ E(\varepsilon_3^2) = \left(\frac{1}{r} - \frac{1}{N}\right)(\lambda_{04} - 1)$$

$$E(\varepsilon_4^2) = \left(\frac{1}{n} - \frac{1}{N}\right)(\lambda_{04} - 1),\ E(\varepsilon_0\varepsilon_1) = \left(\frac{1}{r} - \frac{1}{N}\right)\rho_{xy}C_xC_y,\ E(\varepsilon_0\varepsilon_2) = \left(\frac{1}{n} - \frac{1}{N}\right)\rho_{xy}C_xC_y$$

$$E(\varepsilon_0\varepsilon_3) = \left(\frac{1}{r} - \frac{1}{N}\right)C_y\lambda_{12},\ E(\varepsilon_0\varepsilon_4) = \left(\frac{1}{n} - \frac{1}{N}\right)C_y\lambda_{12},\ E(\varepsilon_1\varepsilon_2) = \left(\frac{1}{n} - \frac{1}{N}\right)C_x^2,$$

$$E(\varepsilon_1\varepsilon_3) = \left(\frac{1}{r} - \frac{1}{N}\right)C_x\lambda_{03},\ E(\varepsilon_1\varepsilon_4) = \left(\frac{1}{n} - \frac{1}{N}\right)C_x\lambda_{03},\ E(\varepsilon_2\varepsilon_3) = \left(\frac{1}{n} - \frac{1}{N}\right)C_x\lambda_{03},$$

$$E(\varepsilon_2\varepsilon_4)=\left(\frac{1}{n}-\frac{1}{N}\right)C_x\lambda_{03},\ E(\varepsilon_3\varepsilon_4)=\left(\frac{1}{n}-\frac{1}{N}\right)(\lambda_{04}-1),$$

$$E(\varepsilon_1\varepsilon_5)=\left(\frac{1}{r}-\frac{1}{N}\right)C_x\frac{\lambda_{12}}{\rho_{xy}}\ \text{and}\ E(\varepsilon_2\varepsilon_5)=\left(\frac{1}{n}-\frac{1}{N}\right)C_x\frac{\lambda_{12}}{\rho_{xy}}$$

where

$$C_x^2=S_x^2/\bar{X}^2,\ C_y^2=S_y^2/\bar{Y}^2,\ \rho_{xy}=S_{xy}/\left(S_xS_y\right),\ \bar{X}=N^{-1}\sum_{i=1}^{N}x_i,\ \bar{Y}=N^{-1}\sum_{i=1}^{N}y_i,$$

$$S_x^2=(N-1)^{-1}\sum_{i=1}^{N}\left(x_i-\bar{X}\right)^2,\ S_y^2=(N-1)^{-1}\sum_{i=1}^{N}\left(y_i-\bar{Y}\right)^2,$$

$$S_{xy}=(N-1)^{-1}\sum_{i=1}^{N}(y_i-\bar{Y})(x_i-\bar{X}),\ \mu_{ab}=(N-1)^{-1}\sum_{i=1}^{N}(y_i-\bar{Y})^a(x_i-\bar{X})^b,$$

$$\lambda_{ab}=\frac{\mu_{ab}}{\mu_{20}^{a/2}\mu_{02}^{b/2}},\ C_{\theta_1}^2=\frac{\sigma_1^2}{\theta_1^2},\ \text{and}\ C_{\theta_2}^2=\frac{\sigma_2^2}{\theta_2^2}$$

6 PROPERTIES OF DIFFERENT ESTIMATORS

The ratio estimator $\bar{y}_{\text{sr}}$ in (10) under the scrambled responses in terms of ε_0, ε_1, and ε_2 can be written as:

$$\bar{y}_{\text{sr}}=\bar{z}_r\left(\frac{\bar{x}_n}{\bar{x}_r}\right)=\bar{Y}(1+\varepsilon_0)\frac{\bar{X}(1+\varepsilon_2)}{\bar{X}(1+\varepsilon_1)}=\bar{Y}(1+\varepsilon_0)(1+\varepsilon_2)\left(1+\varepsilon_1\right)^{-1}$$

Assuming $\varepsilon_1<1$, by using binomial expansion, to the first order of approximation, we have

$$\begin{aligned}\bar{y}_{\text{sr}}&=\bar{Y}(1+\varepsilon_0)(1+\varepsilon_2)\left(1-\varepsilon_1+\varepsilon_1^2+\cdots\right)\\&=\bar{Y}(1+\varepsilon_0+\varepsilon_2+\varepsilon_0\varepsilon_2)\left(1-\varepsilon_1+\varepsilon_1^2+\cdots\right)\\&=\bar{Y}\left[1+\varepsilon_0+\varepsilon_2-\varepsilon_1+\varepsilon_1^2-\varepsilon_0\varepsilon_1-\varepsilon_1\varepsilon_2+\varepsilon_0\varepsilon_2\cdots\right]\end{aligned}\tag{20}$$

Thus, the bias in the scrambled ratio estimator $\bar{y}_{\text{sr}}$ is given by:

$$\begin{aligned}B(\bar{y}_{\text{sr}})&=E(\bar{y}_{\text{sr}})-\bar{Y}\\&=\bar{Y}E\left[\varepsilon_0+\varepsilon_2-\varepsilon_1+\varepsilon_1^2-\varepsilon_0\varepsilon_1-\varepsilon_1\varepsilon_2+\varepsilon_0\varepsilon_2\cdots\right]\\&=\bar{Y}\left[E(\varepsilon_0)+E(\varepsilon_2)-E(\varepsilon_1)+E\left(\varepsilon_1^2\right)-E(\varepsilon_0\varepsilon_1)-E(\varepsilon_1\varepsilon_2)+E(\varepsilon_0\varepsilon_2)\right]\\&=\bar{Y}\left[\left(\frac{1}{r}-\frac{1}{N}\right)C_x^2-\left(\frac{1}{r}-\frac{1}{N}\right)\rho_{xy}C_xC_y-\left(\frac{1}{n}-\frac{1}{N}\right)C_x^2+\left(\frac{1}{n}-\frac{1}{N}\right)\rho_{xy}C_xC_y\right]\\&=\left(\frac{1}{r}-\frac{1}{n}\right)\bar{Y}\left[C_x^2-\rho_{xy}C_xC_y\right]\end{aligned}$$

The bias in the scrambled ratio estimator, $\bar{y}_{\text{sr}}$, to the first order of approximation, is given by:

$$B(\bar{y}_{\text{sr}}) = \left(\frac{1}{r} - \frac{1}{n}\right)\bar{Y}\left[C_x^2 - \rho_{xy}C_xC_y\right] \tag{21}$$

The mean squared error (MSE) of the scrambled ratio estimator $\bar{y}_{\text{sr}}$ is given by:

$$\begin{aligned}
\text{MSE}(\bar{y}_{\text{sr}}) &= E[\bar{y}_{\text{sr}} - \bar{Y}]^2 \approx E[\bar{Y}(\varepsilon_0 + \varepsilon_2 - \varepsilon_1)]^2 \\
&\approx \bar{Y}^2 E\left[\varepsilon_0^2 + \varepsilon_2^2 + \varepsilon_1^2 + 2\varepsilon_0\varepsilon_2 - 2\varepsilon_0\varepsilon_1 - 2\varepsilon_1\varepsilon_2\right] \\
&\approx \bar{Y}^2\left[E\left(\varepsilon_0^2\right) + E\left(\varepsilon_2^2\right) + E\left(\varepsilon_1^2\right) + 2E(\varepsilon_0\varepsilon_2) - 2E(\varepsilon_0\varepsilon_1) - 2E(\varepsilon_1\varepsilon_2)\right] \\
&= \bar{Y}^2\left[\left(\frac{1}{r} - \frac{1}{N}\right)C_y^2 + \frac{1}{r}\left\{C_{\theta_1}^2\left(1 + \frac{(N-1)}{N}C_y^2\right) + (\theta_2/\theta_1)^2 C_{\theta_2}^2 + \left(\frac{1}{n} - \frac{1}{N}\right)C_x^2\right.\right. \\
&\quad + \left(\frac{1}{r} - \frac{1}{N}\right)C_x^2 + 2\left(\frac{1}{n} - \frac{1}{N}\right)\rho_{xy}C_xC_y - 2\left(\frac{1}{r} - \frac{1}{N}\right)\rho_{xy}C_xC_y - 2\left(\frac{1}{n} - \frac{1}{N}\right)C_x^2\Big] \\
&= \bar{Y}^2\left[\left(\frac{1}{r} - \frac{1}{N}\right)C_y^2 + \frac{1}{n}\left\{C_{\theta_1}^2\left(1 + \frac{(N-1)}{N}C_y^2\right) + (\theta_2/\theta_1)^2 C_{\theta_2}^2 + \left(\frac{1}{r} - \frac{1}{n}\right)C_x^2 - 2\left(\frac{1}{r} - \frac{1}{n}\right)\rho_{xy}C_xC_y\right]\right. \\
&= \bar{Y}^2\left[\left(\frac{1}{r} - \frac{1}{n} + \frac{1}{n} - \frac{1}{N}\right)C_y^2 + \frac{1}{r}\left\{C_{\theta_1}^2\left(1 + \frac{(N-1)}{N}C_y^2\right) + (\theta_2/\theta_1)^2 C_{\theta_2}^2 + \left(\frac{1}{r} - \frac{1}{n}\right)\left(C_x^2 - 2\rho_{xy}C_xC_y\right)\right]\right. \\
&= \bar{Y}^2\left[\left(\frac{1}{n} - \frac{1}{N}\right)C_y^2 + \frac{1}{n}\left\{C_{\theta_1}^2\left(1 + \frac{(N-1)}{N}C_y^2\right) + (\theta_2/\theta_1)^2 C_{\theta_2}^2 + \left(\frac{1}{r} - \frac{1}{n}\right)\left(C_y^2 + C_x^2 - 2\rho_{xy}C_xC_y\right)\right]\right. \\
&= \left(\frac{1}{n} - \frac{1}{N}\right)S_y^2 + \left(\frac{1}{r} - \frac{1}{n}\right)\left(S_y^2 + R^2S_x^2 - 2RS_{xy}\right) + \frac{\bar{Y}^2}{r}\left\{C_{\theta_1}^2\left(1 + \frac{N-1}{N}C_y^2\right) + (\theta_2/\theta_1)^2 C_{\theta_2}^2\right\}
\end{aligned} \tag{22}$$

The regression estimator $\bar{y}_{\text{slr}}$ in terms of ε_i, $i = 0, 1, 2, 3, 4, 5$ can be written as:

$$\begin{aligned}
\bar{y}_{\text{slr}} &= \bar{z}_r + \hat{\beta}_s(\bar{x}_n - \bar{x}_r) \\
&= \bar{z}_r + \frac{s_{xz(r)}}{s_{x(r)}^2}(\bar{x}_n - \bar{x}_r) \\
&= \bar{Y}(1 + \varepsilon_0) + \frac{S_{yx}(1 + \varepsilon_5)}{S_x^2(1 + \varepsilon_3)}[\bar{X}(1 + \varepsilon_2) - \bar{X}(1 + \varepsilon_1)] \\
&= \bar{Y}(1 + \varepsilon_0) + \frac{S_{yx}(1 + \varepsilon_5)}{S_x^2}(1 + \varepsilon_3)^{-1}[\bar{X}(1 + \varepsilon_2) - \bar{X}(1 + \varepsilon_1)]
\end{aligned}$$

Assuming $|\varepsilon_3| < 1$, and using the binomial expansion, the estimator $\bar{y}_{\text{slr}}$ can be approximated as:

$$\begin{aligned}
\bar{y}_{\text{slr}} &= \bar{Y}(1 + \varepsilon_0) + \bar{X}\frac{S_{yx}(1 + \varepsilon_5)}{S_x^2}\left(1 - \varepsilon_3 + \varepsilon_3^2 + \cdots\right)[\varepsilon_2 - \varepsilon_1] \\
&= \bar{Y}(1 + \varepsilon_0) + \bar{X}\beta(1 + \varepsilon_5)\left(1 - \varepsilon_3 + \varepsilon_3^2 + \cdots\right)(\varepsilon_2 - \varepsilon_1) \\
&= \bar{Y}(1 + \varepsilon_0) + \bar{X}\beta\left[1 + \varepsilon_5 - \varepsilon_3 + \varepsilon_3^2 - \varepsilon_3\varepsilon_5 + \cdots\right](\varepsilon_2 - \varepsilon_1) \\
&= \bar{Y} + \bar{Y}\varepsilon_0 + \bar{X}\beta[\varepsilon_2 - \varepsilon_1 + \varepsilon_2\varepsilon_5 - \varepsilon_1\varepsilon_5 - \varepsilon_2\varepsilon_3 + \varepsilon_1\varepsilon_3 \cdots]
\end{aligned} \tag{23}$$

Taking expected values on both sides of (23), we have

$$\begin{aligned}
E(\bar{y}_{\text{slr}}) &= \bar{Y} + \beta\bar{X}[E(\varepsilon_2\varepsilon_5) - E(\varepsilon_1\varepsilon_5) - E(\varepsilon_2\varepsilon_3) + E(\varepsilon_1\varepsilon)] \\
&= \bar{Y} + \beta\bar{X}\left[\left(\frac{1}{n} - \frac{1}{N}\right)C_x\frac{\lambda_{12}}{\rho} - \left(\frac{1}{r} - \frac{1}{N}\right)C_x\frac{\lambda_{12}}{\rho} - \left(\frac{1}{n} - \frac{1}{N}\right)C_x\lambda_{03} + \left(\frac{1}{r} - \frac{1}{N}\right)C_x\lambda_{03}\right] \\
&= \bar{Y} + \left(\frac{1}{r} - \frac{1}{n}\right)\beta\bar{X}C_x\left[\lambda_{03} - \frac{\lambda_{12}}{\rho}\right]
\end{aligned}$$

By the definition of bias, the approximate bias in the regression method of estimation is given by:

$$B(\bar{y}_{\text{slr}}) = E(\bar{y}_{\text{slr}}) - \bar{Y} = \left(\frac{1}{r} - \frac{1}{n}\right)\beta\bar{X}C_x\left[\lambda_{03} - \frac{\lambda_{12}}{\rho}\right] \tag{24}$$

The MSE of the regression estimator $\bar{y}_{\text{slr}}$ is given by:

$$\begin{aligned}
\text{MSE}(\bar{y}_{\text{slr}}) &= E[\bar{y}_{\text{slr}} - \bar{Y}]^2 \\
&\approx E[\bar{Y} + \bar{Y}\varepsilon_0 + \beta\bar{X}(\varepsilon_2 - \varepsilon_1 + \cdots) - \bar{Y}]^2 \\
&= E[\bar{Y}\varepsilon_0 + \beta\bar{X}(\varepsilon_2 - \varepsilon_1)]^2 \\
&= E\left[\bar{Y}^2\varepsilon_0^2 + \beta^2\bar{X}^2(\varepsilon_2 - \varepsilon_1)^2 + 2\beta\bar{Y}\bar{X}\varepsilon_0(\varepsilon_2 - \varepsilon_1)\right] \\
&= \bar{Y}^2E(\varepsilon_0^2) + \beta^2\bar{X}^2E(\varepsilon_2^2 + \varepsilon_1^2 - 2\varepsilon_1\varepsilon_2) + 2\beta\bar{Y}\bar{X}E(\varepsilon_0\varepsilon_2 - \varepsilon_0\varepsilon_1) \\
&= \bar{Y}^2E(\varepsilon_0^2) + \beta^2\bar{X}^2\left[E(\varepsilon_2^2) + E(\varepsilon_1^2) - 2E(\varepsilon_1\varepsilon_2)\right] + 2\beta\bar{Y}\bar{X}(E(\varepsilon_0\varepsilon_2) - E(\varepsilon_0\varepsilon_1)) \\
&= \bar{Y}^2\left(\frac{1}{r} - \frac{1}{N}\right)C_y^2 + \frac{1}{r}\left[C_{\theta_1}^2\left(1 + \frac{(N-1)}{N}C_y^2\right) + \left(\frac{\theta_2}{\theta_1}\right)^2 C_{\theta_2}^2\right] \\
&\quad + \beta^2\bar{X}^2\left[\left(\frac{1}{n} - \frac{1}{N}\right)C_x^2 + \left(\frac{1}{r} - \frac{1}{N}\right)C_x^2 - 2\left(\frac{1}{n} - \frac{1}{N}\right)C_x^2\right] \\
&\quad + 2\beta\bar{Y}\bar{X}\left[\left(\frac{1}{n} - \frac{1}{N}\right)\rho C_yC_x - \left(\frac{1}{r} - \frac{1}{N}\right)\rho C_yC_x\right] \\
&= \bar{Y}^2\left(\frac{1}{r} - \frac{1}{N}\right)C_y^2 + \frac{\bar{Y}^2}{r}\left[C_{\theta_1}^2\left(1 + \frac{N-1}{N}C_y^2\right) + \left(\frac{\theta_2}{\theta_1}\right)^2 C_{\theta_2}^2\right] \\
&\quad + \beta^2\bar{X}^2\left(\frac{1}{r} - \frac{1}{n}\right)C_x^2 - 2\beta\bar{Y}\bar{X}\left(\frac{1}{r} - \frac{1}{n}\right)\rho C_yC_x \\
&= \bar{Y}^2\left(\frac{1}{r} - \frac{1}{N}\right)\frac{S_y^2}{\bar{Y}^2} + \frac{\bar{Y}^2}{r}\left[C_{\theta_1}^2\left(1 + \frac{N-1}{N}C_y^2\right) + \left(\frac{\theta_2}{\theta_1}\right)^2 C_{\theta_2}^2\right] \\
&\quad + \left(\frac{1}{r} - \frac{1}{n}\right)\left[\frac{S_{yx}^2}{S_x^4}\bar{X}^2\frac{S_x^2}{\bar{X}^2} - 2\frac{S_{yx}}{S_x^2}\bar{Y}\bar{X}\frac{S_{yx}}{S_xS_y}\frac{S_y}{\bar{Y}}\frac{S_x}{\bar{X}}\right] \\
&= \left(\frac{1}{r} - \frac{1}{N}\right)S_y^2 + \frac{\bar{Y}^2}{r}\left[C_{\theta_1}^2\left(1 + \frac{N-1}{N}C_y^2\right) + \left(\frac{\theta_2}{\theta_1}\right)C_{\theta_2}^2\right] + \left(\frac{1}{r} - \frac{1}{n}\right)\left[\frac{S_{yx}^2}{S_x^2} - 2\frac{S_{yx}^2}{S_x^2}\right] \\
&= \left(\frac{1}{r} - \frac{1}{N}\right)S_y^2 + \frac{\bar{Y}^2}{r}\left[C_{\theta_1}^2\left(1 + \frac{N-1}{N}C_y^2\right) + \left(\frac{\theta_2}{\theta_1}\right)^2 C_{\theta_2}^2\right] - \left(\frac{1}{r} - \frac{1}{n}\right)\frac{S_{yx}^2}{S_x^2} \\
&= \left(\frac{1}{r} - \frac{1}{n} + \frac{1}{n} - \frac{1}{N}\right)S_y^2 - \left(\frac{1}{r} - \frac{1}{n}\right)\frac{S_{yx}^2}{S_x^2} + \frac{\vec{Y}^2}{r}\left[C_{\theta_1}^2\left(1 + \frac{N-1}{N}C_y^2\right) + \left(\frac{\theta_2}{\theta_1}\right)^2 C_{\theta_2}^2\right] \\
&= \left(\frac{1}{n} - \frac{1}{N}\right)S_y^2 + \left(\frac{1}{r} - \frac{1}{n}\right)S_y^2 - \left(\frac{1}{r} - \frac{1}{n}\right)\frac{S_{yx}^2}{S_x^2} + \frac{\bar{Y}^2}{r}\left[C_{\theta_1}^2\left(1 + \frac{N-1}{N}C_y^2\right) + \left(\frac{\theta_2}{\theta_1}\right)^2 C_{\theta_2}^2\right] \\
&= \left(\frac{1}{n} - \frac{1}{N}\right)S_y^2 + \left(\frac{1}{r} - \frac{1}{n}\right)S_y^2\left(1 - \frac{S_{yx}^2}{S_x^2S_y^2}\right) + \frac{\bar{Y}^2}{r}\left[C_{\theta_1}^2\left(1 + \frac{N-1}{N}C_y^2\right) + \left(\frac{\theta_2}{\theta_1}\right)^2 C_{\theta_2}^2\right] \\
&= \left(\frac{1}{n} - \frac{1}{N}\right)S_y^2 + \left(\frac{1}{r} - \frac{1}{n}\right)S_y^2(1 - \rho^2) + \frac{\bar{Y}^2}{r}\left[C_{\theta_1}^2\left(1 + \frac{N-1}{N}C_y^2\right) + \left(\frac{\theta_2}{\theta_1}\right)^2 C_{\theta_2}^2\right]
\end{aligned} \tag{25}$$

The proposed estimator $\bar{y}_{\text{snew(ch)}}$ in (19) in terms of ε_i, $i=0,1,2,3,4$ can be written as:

$$\begin{aligned}\bar{y}_{\text{snew(ch)}} &= \bar{z}_r + \beta_1(\bar{x}_n - \bar{x}_r) + \beta_2\left(s^2_{x(n)} - s^2_{x(r)}\right)\\ &= \bar{Y}(1+\varepsilon_0) + \beta_1[\bar{X}(1+\varepsilon_2) - \bar{X}(1+\varepsilon_1)] + \beta_2\left[S_x^2(1+\varepsilon_4) - S_x^2(1+\varepsilon_3)\right]\\ &= \bar{Y}(1+\varepsilon_0) + \beta_1\bar{X}[\varepsilon_2 - \varepsilon_1] + \beta_2 S_x^2[\varepsilon_4 - \varepsilon_3]\end{aligned} \tag{26}$$

Taking expected value on both sides of (26), we have

$$E\left(\bar{y}_{\text{snew(ch)}}\right) = \bar{Y} + \bar{Y}E(\varepsilon_0) + \beta_1\bar{X}[E(\varepsilon_2) - E(\varepsilon_1)] + \beta_2 S_x^2[E(\varepsilon_4) - E(\varepsilon_3)] = \bar{Y}$$

which shows that the proposed imputing method provides an unbiased estimator of the population mean.

Now by the definition of variance, we have

$$\begin{aligned}V\left(\bar{y}_{\text{snew(ch)}}\right) &= E\left[\bar{y}_{\text{snew(ch)}} - \bar{Y}\right]^2\\ &= E\left[\bar{Y}(1+\varepsilon_0) + \beta_1\bar{X}(\varepsilon_2-\varepsilon_1) + \beta_2 S_x^2(\varepsilon_4-\varepsilon_3) - \bar{Y}\right]^2\\ &= \bar{Y}^2E\left(\varepsilon_0^2\right) + \beta_1^2\bar{X}^2\left[E\left(\varepsilon_2^2\right) + E\left(\varepsilon_1^2\right) - 2E(\varepsilon_1\varepsilon_2)\right] + \beta_2^2S_x^4\left[E\left(\varepsilon_4^2\right) + E\left(\varepsilon_3^2\right) - 2E(\varepsilon_3\varepsilon_4)\right]\\ &\quad + 2\beta_1\bar{Y}\bar{X}[E(\varepsilon_0\varepsilon_2) - E(\varepsilon_0\varepsilon_1)] + 2\beta_2\bar{Y}S_x^2[E(\varepsilon_0\varepsilon_4) - E(\varepsilon_0\varepsilon_3)]\\ &\quad + 2\beta_1\beta_2\bar{X}S_x^2[E(\varepsilon_2\varepsilon_4) - E(\varepsilon_1\varepsilon_4) - E(\varepsilon_2\varepsilon_3) + E(\varepsilon_1\varepsilon_3)]\\ &= \left(\frac{1}{n} - \frac{1}{N}\right)S_y^2 + \frac{\vec{Y}^2}{r}\left[C_{\theta_1}^2\left(1 + \frac{N-1}{N}C_y^2\right) + \left(\frac{\theta_2}{\theta_1}\right)^2 C_{\theta_2}^2\right]\\ &\quad + \left(\frac{1}{r} - \frac{1}{n}\right)\left[S_y^2 + \beta_1^2S_x^2 + \beta_2^2S_x^4(\lambda_{04}-1) - 2\beta_1S_{yx} - 2\beta_2S_yS_x^2\lambda_{12} + 2\beta_1\beta_2S_x^3\lambda_{03}\right]\end{aligned} \tag{27}$$

On setting:

$$\frac{\partial}{\partial\beta_1}V\left(\bar{y}_{\text{snew(ch)}}\right) = 0 \text{ and } \frac{\partial}{\partial\beta_2}V\left(\bar{y}_{\text{snew(ch)}}\right) = 0$$

We get the following two "normal" equations, given by:

$$\beta_1S_x^2 + \beta_2S_x^3\lambda_{03} = S_{yx} \tag{28}$$

and

$$\beta_1S_x\lambda_{03} + \beta_2S_x^2(\lambda_{04}-1) = S_y\lambda_{12} \tag{29}$$

The values of β_1 and β_2 are given by:

$$\beta_1 = \frac{\Delta_1}{\Delta} = \frac{S_{yx}(\lambda_{04}-1) - S_yS_x\lambda_{12}\lambda_{03}}{S_x^2\left(\lambda_{04} - 1 - \lambda_{03}^2\right)} \tag{30}$$

and

$$\beta_2 = \frac{\Delta_2}{\Delta} = \frac{S_y S_x \lambda_{12} - S_{yx}\lambda_{03}}{S_x^3\left(\lambda_{04} - 1 - \lambda_{03}^2\right)} \tag{31}$$

On substituting the values of β_1 and β_2 into (27), we have the minimum variance of the resultant estimator $\bar{y}_{\text{snew(ch)}}$ given by:

$$\text{Min.}\, V\left(\bar{y}_{\text{snew(ch)}}\right) = \left(\frac{1}{n} - \frac{1}{N}\right)S_y^2 + \frac{\bar{Y}^2}{r}\left[C_{\theta_1}^2\left(1 + \frac{N-1}{N}C_y^2\right) + \left(\frac{\theta_2}{\theta_1}\right)^2 C_{\theta_2}^2\right] + \left(\frac{1}{r} - \frac{1}{n}\right)S_y^2\left\{1 - \rho_{xy}^2 - \frac{\left(\lambda_{12} - \rho_{xy}\lambda_{03}\right)^2}{\lambda_{04} - 1 - \lambda_{03}^2}\right\} \tag{32}$$

In the next section, we compare four imputing methods, that is, the mean method of imputation, ratio method of imputation, regression method of imputation, and the proposed method of imputation making use of higher order moments of an auxiliary variable, by using a numerical illustrations applied to a real dataset.

7 APPLICATION TO A REAL DATA SET

We use a dataset, FEV.DAT, available on the CD that accompanies the text by Rosner (2006), that contains data on $N = 654$ children from the Childhood Respiratory Disease Study done in Boston. Among the variables considered are Age, HEIGHT, and FEV. We consider the problem of imputing the log of Age of a child given the interaction between HEIGHT and FEV value of a child. It is assumed here that the AGE of a child regarded as sensitive information whereas HEIGHT and FEV values of a child are known from a hospital file. A scatter plot showing the relationship between Y = log(AGE) and X = HEIGHT*FEV is shown in Fig. 1. It will worth pointing out that although Age is not very sensitive characteristic, but as pointed out by Singh et al. (2008) that one variable could be sensitive in one situation and nonsensitive in another situation.

The value of the Pearson's correlation coefficient between Y = log(AGE) and X = HEIGHT*FEV is 0.763 which is a quite reasonable value being neither too high nor too low in real situation. A few descriptive parameters of the variables Y = LOG(AGE) and X = HEIGHT*FEV are shown in Table 1.

In order to investigate the performance of the ratio, regression, and the proposed imputation method with respect to the mean method of imputation, we compute the following percent relative efficiencies values as:

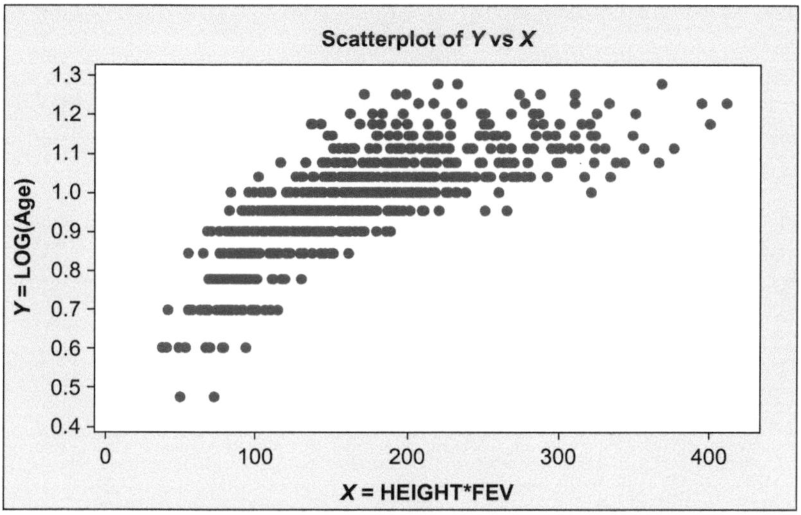

FIG. 1 Scatter plot of Log(Age) vs HEIGHT*FEV.

$$\mathrm{RE}(1) = \frac{V(\bar{z}_r)}{V(\bar{y}_{\mathrm{sr}})} \times 100\% \tag{33}$$

$$\mathrm{RE}(2) = \frac{V(\bar{z}_r)}{V(\bar{y}_{\mathrm{slr}})} \times 100\% \tag{34}$$

$$\mathrm{RE}(3) = \frac{V(\bar{z}_r)}{\mathrm{Min}.V\left(\bar{y}_{\mathrm{snew(ch)}}\right)} \times 100\% \tag{35}$$

We used three values of the ratio of the means of the two scrambling variables (θ_2/θ_1) equal to 1, 0.5, and 1.5, and different situations where C_{θ_1} and C_{θ_2} could be 0.1 and 0.2 in ordered pairs. For a sample of $n = 60$, we considered different values of r between 5 and 50 with a step of 5. This generates situations where response rate (RR) varies from 8.3% to 83.3%. We wrote the following FORTRAN codes to compute the percent relative efficiency values for the various situations.

TABLE 1 Descriptive Parameters of the Variables AGE and HEIGHT

Variable	N	Mean	St. Dev.	Min	Med	Max	Skewness	Kurtosis
Y=Log(Age)	654	0.9768	0.1359	0.4771	1	1.2788	−0.49	0.38
X=HEIGHT*FEV	654	165.51	67.83	37.41	157.65	411.21	0.77	0.43

```
! FORTRAN CODES CHOUKRI12.F95
        USE NUMERICAL_LIBRARIES
        IMPLICIT NONE
        INTEGER NP,I,NS,NR,ID(1000)
        REAL Y1(1000), X1(1000),V1(1000),V2(1000),V3(1000)
        REAL Y(1000),X(1000),ANP,ANS
        REAL SUMY, SUMX, YM, XM, SUMX2, SUMY2, SUMXY
        REAL SX2, SY2, SXY, AMU03, AMU04, AMU12, AL12,AL03,AL04
        REAL CY, CX, THRAT, RHOXY, ANR, FACT, VMEAN, VRAT
        REAL VREG,TERM, VP, RE1,RE2,RE3,R,RR,CTH1,CTH2
        CHARACTER*20 OUT_FILE
        CHARACTER*20 IN_FILE
        WRITE(*,'(A)') 'NAME OF THE INPUT FILE'
        READ(*,'(A20)') IN_FILE
        OPEN(41, FILE =IN_FILE, STATUS='OLD')
        WRITE(*,'(A)') 'NAME OF THE OUTPUT FILE'
        READ(*,'(A20)') OUT_FILE
        OPEN(42, FILE=OUT_FILE, STATUS='UNKNOWN')
        READ(41,*)NP
        DO 10 I =1, NP
10      READ(41,*)ID(I),Y1(I),V1(I),X1(I),V2(I),V3(I)
        ANP = NP
        DO 11 I =1, NP
       Y(I) = LOG(Y1(I))
11     X(I) = X1(I)*V1(I)
       SUMY = 0.0
       SUMX = 0.0
       DO 12 I=1, NP
       SUMY = SUMY + Y(I)
12     SUMX = SUMX + X(I)
       YM = SUMY/ANP
       XM = SUMX/ANP
       SUMY2 = 0.0
       SUMX2 = 0.0
       SUMXY = 0.0
       AMU03 = 0.0
       AMU04 = 0.0
       AMU12 = 0.0
       DO 13 I =1, NP
       SUMY2 = SUMY2 + (Y(I)-YM)**2
       SUMX2 = SUMX2 + (X(I)-XM)**2
       SUMXY = SUMXY + (Y(I)-YM)*(X(I)-XM)
       AMU03 = AMU03 + (X(I)-XM)**3
       AMU04 = AMU04 + (X(I)-XM)**4
13     AMU12 = AMU12 + (Y(I)-YM)*(X(I)-XM)**2
       SY2 = SUMY2/(ANP-1)
        SX2 = SUMX2/(ANP-1)
```

```
        SXY = SUMXY/(ANP-1)
        RHOXY = SXY/SQRT(SY2*SX2)
        AMU03 = AMU03/(ANP-1)
        AMU04 = AMU04/(ANP-1)
        AMU12 = AMU12/(ANP-1)
        AL12 = AMU12/(SQRT(SY2)*SX2)
        AL03 = AMU03/(SX2*SQRT(SX2))
        AL04 = AMU04/SX2**2
        CY = SQRT(SY2)/YM
        CX = SQRT(SX2)/XM
        R = YM/XM
        NS = 60
        ANS = NS
        DO 15 CTH1 = 0.1, 0.21, 0.1
        DO 16 CTH2 = 0.1, 0.21, 0.1
        DO 17 THRAT = 0.5, 1.51, 0.5
        DO 18 NR = 5, 50, 5
        ANR = NR
        FACT = CTH1**2*(1+(ANP-1)*CY**2/ANP)+(THRAT)**2*CTH2**2
        VMEAN = (1/ANR-1/ANP)*SY2+FACT/ANR
        VRAT = (1/ANS-1/ANP)*SY2 + (1/ANR-1/ANS)*(SY2+R**2*SX2-
2*R*SXY)
       1 + FACT/ANR
        VREG = (1/ANS-1/ANP)*SY2 + (1/ANR-1/ANS)*SY2*(1-RHOXY**2)
       1 + FACT/ANR
        TERM = (AL12-RHOXY*AL03)**2/(AL04-1-AL03**2)
        VP = (1/ANS-1/ANP)*SY2 + (1/ANR-1/ANS)*SY2*(1-RHOXY**2-
TERM)
       1 + FACT/ANR
        RE1 = VMEAN*100/VRAT
        RE2 = VMEAN*100/VREG
        RE3 = VMEAN*100/VP
        RR = ANR/ANS
        WRITE(42,101)NP,NS,NR, RR, CTH1, CTH2, THRAT,RE1,RE2,RE3
101     FORMAT(2X,3(I4,1X),2X,4(F7.3,1X),2X,3(F9.2, 2X))
18      CONTINUE
17      CONTINUE
16      CONTINUE
15      CONTINUE
        STOP
        END
```

The results obtained after executing the FORTRAN codes are given in Table 2.

Table 3 gives summary of statistics of RE(j), $j = 1,2,3$ over the 12 combinations of θ_2/θ_1, C_{θ_1}, and C_{θ_2} mentioned above; one set for each of the values of r that were considered.

TABLE 2 RE(1),RE(2), and RE(3) Values for Different Situations

r	RR	C_{θ_1}	C_{θ_2}	θ_2/θ_1	RE(1)	RE(2)	RE(3)
5	0.083	0.1	0.1	0.5	22.6	190.6	221.1
10	0.167	0.1	0.1	0.5	24.2	177.0	200.6
15	0.250	0.1	0.1	0.5	26.1	165.1	183.3
20	0.333	0.1	0.1	0.5	28.3	154.6	168.6
25	0.417	0.1	0.1	0.5	30.9	145.1	155.9
30	0.500	0.1	0.1	0.5	34.1	136.7	144.8
35	0.583	0.1	0.1	0.5	38.2	129.1	135.1
40	0.667	0.1	0.1	0.5	43.4	122.2	126.5
45	0.750	0.1	0.1	0.5	50.3	115.9	118.8
50	0.833	0.1	0.1	0.5	60.2	110.1	111.9
5	0.083	0.1	0.1	1.0	23.8	180.2	205.3
10	0.167	0.1	0.1	1.0	25.5	168.7	188.4
15	0.250	0.1	0.1	1.0	27.4	158.4	173.9
20	0.333	0.1	0.1	1.0	29.6	149.3	161.3
25	0.417	0.1	0.1	1.0	32.4	141.0	150.3
30	0.500	0.1	0.1	1.0	35.7	133.5	140.6
35	0.583	0.1	0.1	1.0	39.8	126.6	132.0
40	0.667	0.1	0.1	1.0	45.1	120.4	124.3
45	0.750	0.1	0.1	1.0	52.1	114.7	117.3
50	0.833	0.1	0.1	1.0	61.8	109.4	111.0
5	0.083	0.1	0.1	1.5	25.7	167.3	186.4
10	0.167	0.1	0.1	1.5	27.4	158.2	173.5
15	0.250	0.1	0.1	1.5	29.5	149.9	162.2
20	0.333	0.1	0.1	1.5	31.8	142.4	152.2
25	0.417	0.1	0.1	1.5	34.7	135.5	143.2
30	0.500	0.1	0.1	1.5	38.1	129.2	135.2
35	0.583	0.1	0.1	1.5	42.3	123.4	127.9
40	0.667	0.1	0.1	1.5	47.7	118.0	121.3
45	0.750	0.1	0.1	1.5	54.7	113.0	115.3
50	0.833	0.1	0.1	1.5	64.3	108.4	109.8

Continued

TABLE 2 RE(1),RE(2), and RE(3) Values for Different Situations—Cont'd

r	RR	C_{θ_1}	C_{θ_2}	θ_2/θ_1	RE(1)	RE(2)	RE(3)
5	0.083	0.1	0.2	0.5	23.8	180.2	205.3
10	0.167	0.1	0.2	0.5	25.5	168.7	188.4
15	0.250	0.1	0.2	0.5	27.4	158.4	173.9
20	0.333	0.1	0.2	0.5	29.6	149.3	161.3
25	0.417	0.1	0.2	0.5	32.4	141.0	150.3
30	0.500	0.1	0.2	0.5	35.7	133.5	140.6
35	0.583	0.1	0.2	0.5	39.8	126.6	132.0
40	0.667	0.1	0.2	0.5	45.1	120.4	124.3
45	0.750	0.1	0.2	0.5	52.1	114.7	117.3
50	0.833	0.1	0.2	0.5	61.8	109.4	111.0
5	0.083	0.1	0.2	1.0	28.2	154.9	169.0
10	0.167	0.1	0.2	1.0	30.0	147.9	159.5
15	0.250	0.1	0.2	1.0	32.2	141.4	151.0
20	0.333	0.1	0.2	1.0	34.7	135.5	143.2
25	0.417	0.1	0.2	1.0	37.6	129.9	136.1
30	0.500	0.1	0.2	1.0	41.2	124.7	129.6
35	0.583	0.1	0.2	1.0	45.6	119.9	123.7
40	0.667	0.1	0.2	1.0	51.0	115.4	118.2
45	0.750	0.1	0.2	1.0	58.0	111.2	113.1
50	0.833	0.1	0.2	1.0	67.3	107.2	108.4
5	0.083	0.1	0.2	1.5	34.4	136.0	143.9
10	0.167	0.1	0.2	1.5	36.5	131.8	138.5
15	0.250	0.1	0.2	1.5	38.9	127.9	133.6
20	0.333	0.1	0.2	1.5	41.7	124.2	128.9
25	0.417	0.1	0.2	1.5	44.8	120.6	124.5
30	0.500	0.1	0.2	1.5	48.6	117.3	120.4
35	0.583	0.1	0.2	1.5	53.0	114.0	116.5
45	0.750	0.1	0.2	1.5	65.1	108.0	109.4
50	0.833	0.1	0.2	1.5	73.6	105.2	106.1
5	0.083	0.2	0.1	0.5	27.2	159.2	175.0

Continued

TABLE 2 RE(1),RE(2), and RE(3) Values for Different Situations—Cont'd

r	RR	C_{θ_1}	C_{θ_2}	θ_2/θ_1	RE(1)	RE(2)	RE(3)
10	0.167	0.2	0.1	0.5	29.0	151.5	164.4
15	0.250	0.2	0.1	0.5	31.1	144.4	154.9
20	0.333	0.2	0.1	0.5	33.6	137.9	146.4
25	0.417	0.2	0.1	0.5	36.5	131.9	138.6
30	0.500	0.2	0.1	0.5	40.0	126.3	131.6
35	0.583	0.2	0.1	0.5	44.3	121.2	125.2
40	0.667	0.2	0.1	0.5	49.7	116.3	119.3
45	0.750	0.2	0.1	0.5	56.8	111.9	113.9
50	0.833	0.2	0.1	0.5	66.2	107.6	108.9
5	0.083	0.2	0.1	1.0	28.3	154.6	168.6
10	0.167	0.2	0.1	1.0	30.1	147.6	159.2
15	0.250	0.2	0.1	1.0	32.3	141.2	150.7
20	0.333	0.2	0.1	1.0	34.8	135.3	142.9
25	0.417	0.2	0.1	1.0	37.7	129.8	135.9
30	0.500	0.2	0.1	1.0	41.3	124.6	129.5
35	0.583	0.2	0.1	1.0	45.7	119.8	123.6
40	0.667	0.2	0.1	1.0	51.1	115.4	118.1
45	0.750	0.2	0.1	1.0	58.1	111.2	113.1
50	0.833	0.2	0.1	1.0	67.4	107.2	108.4
5	0.083	0.2	0.1	1.5	29.9	148.3	160.0
10	0.167	0.2	0.1	1.5	31.9	142.3	152.1
15	0.250	0.2	0.1	1.5	34.1	136.8	144.9
20	0.333	0.2	0.1	1.5	36.7	131.6	138.2
25	0.417	0.2	0.1	1.5	39.7	126.8	132.1
30	0.500	0.2	0.1	1.5	43.3	122.2	126.5
35	0.583	0.2	0.1	1.5	47.7	118.0	121.3
40	0.667	0.2	0.1	1.5	53.2	113.9	116.4
45	0.750	0.2	0.1	1.5	60.1	110.2	111.9
50	0.833	0.2	0.1	1.5	69.2	106.6	107.7
5	0.083	0.2	0.2	0.5	28.3	154.6	168.6

Continued

TABLE 2 RE(1),RE(2), and RE(3) Values for Different Situations—Cont'd

r	RR	C_{θ_1}	C_{θ_2}	θ_2/θ_1	RE(1)	RE(2)	RE(3)
10	0.167	0.2	0.2	0.5	30.1	147.6	159.2
15	0.250	0.2	0.2	0.5	32.3	141.2	150.7
20	0.333	0.2	0.2	0.5	34.8	135.3	142.9
25	0.417	0.2	0.2	0.5	37.7	129.8	135.9
30	0.500	0.2	0.2	0.5	41.3	124.6	129.5
35	0.583	0.2	0.2	0.5	45.7	119.8	123.6
40	0.667	0.2	0.2	0.5	51.1	115.4	118.1
45	0.750	0.2	0.2	0.5	58.1	111.2	113.1
50	0.833	0.2	0.2	0.5	67.4	107.2	108.4
5	0.083	0.2	0.2	1.0	32.1	141.5	151.1
10	0.167	0.2	0.2	1.0	34.2	136.6	144.7
15	0.250	0.2	0.2	1.0	36.5	132.0	138.7
20	0.333	0.2	0.2	1.0	39.1	127.6	133.2
25	0.417	0.2	0.2	1.0	42.3	123.5	128.0
30	0.500	0.2	0.2	1.0	45.9	119.6	123.2
35	0.583	0.2	0.2	1.0	50.4	115.9	118.7
40	0.667	0.2	0.2	1.0	55.8	112.4	114.5
45	0.750	0.2	0.2	1.0	62.7	109.0	110.6
50	0.833	0.2	0.2	1.0	71.5	105.9	106.8
5	0.083	0.2	0.2	1.5	37.8	129.7	135.9
10	0.167	0.2	0.2	1.5	39.9	126.4	131.7
15	0.250	0.2	0.2	1.5	42.4	123.3	127.8
20	0.333	0.2	0.2	1.5	45.2	120.2	124.1
25	0.417	0.2	0.2	1.5	48.5	117.3	120.5
30	0.500	0.2	0.2	1.5	52.2	114.6	117.2
35	0.583	0.2	0.2	1.5	56.7	111.9	114.0
40	0.667	0.2	0.2	1.5	62.0	109.3	110.9
45	0.750	0.2	0.2	1.5	68.4	106.9	108.0
50	0.833	0.2	0.2	1.5	76.4	104.5	105.2

TABLE 3 Descriptive Values of the Results

r	Freq	Mean	St. Dev.	Min	Med	Max
			RE(1)			
5	12	28.51	4.52	22.60	28.25	37.80
10	12	30.36	4.69	24.20	30.05	39.90
20	12	34.99	5.08	28.30	34.75	45.20
25	12	37.93	5.26	30.90	37.65	48.50
30	12	41.45	5.42	34.10	41.25	52.20
35	12	45.77	5.56	38.20	45.65	56.70
40	12	51.13	5.59	43.40	51.05	62.00
45	12	58.04	5.45	50.30	58.05	68.40
50	12	67.26	4.91	60.20	67.35	76.40
			RE(2)			
5	12	158.09	18.64	129.70	154.75	190.60
10	12	150.36	15.49	126.40	147.75	177.00
15	12	143.33	12.78	123.30	141.30	165.10
20	12	136.93	10.50	120.20	135.40	154.60
25	12	131.02	8.49	117.30	129.85	145.10
30	12	125.57	6.75	114.60	124.65	136.70
35	12	120.52	5.24	111.90	119.85	129.10
40	12	115.84	3.92	109.30	115.40	122.20
45	12	111.49	2.76	106.90	111.20	115.90
50	12	107.39	1.72	104.50	107.20	110.10
			RE(3)			
5	12	174.18	26.10	135.90	168.80	221.10
10	12	163.35	21.08	131.70	159.35	200.60
15	12	153.80	16.97	127.80	150.85	183.30
20	12	145.27	13.60	124.10	143.05	168.60
25	12	137.61	10.81	120.50	136.00	155.90
30	12	130.72	8.43	117.20	129.55	144.80
35	12	124.47	6.45	114.00	123.65	135.10
40	12	118.73	4.76	110.90	118.15	126.50
45	12	113.48	3.29	108.00	113.10	118.80
50	12	108.63	2.04	105.20	108.40	111.90

From Table 3, it is clear that out of all the cases considered the maximum value of the percent relative efficiency of the ratio estimator, that is RE(1) value, is 76.40%. Thus, it seems that there is no reason to use the ratio method of imputation in such situations. The proposed method of imputation, using higher order moments of an auxiliary variable, can perform better than both the mean and regression methods of imputations in all the cases considered. If the response rate is as low as 8.3% (5 out of 60 responses), then the average value of RE(3) is 174.18% with a standard deviation of 26.10%. The minimum value of RE(3) is 135.90%, and maximum value is 221.10% with a median value of 168.80%. As the response rate increase to 83.3% (50 out of 60), the average value of RE(3) becomes 108.63% with a standard deviation of 2.04%. The minimum value of RE(3) is 105.20% and a maximum value of 111.90% with a median value of 108.40%. Fig. 2 is devoted to display the values of RE(1), RE(2), and RE(3) as a function of the value of number of responses r considered in this study.

Fig. 3 shows the relationship between the values of RE(3) vs the value of C_{θ_1} (CTH1) and C_{θ_2} (CTH2).

A summary of the results displayed in Fig. 3 are given in Table 4.

From our small-scale study, we conclude that the proposed method of imputations making use of higher order of moments of the auxiliary variable can be used to impute missing values in a more precise way than the ratio or regression method of imputation.

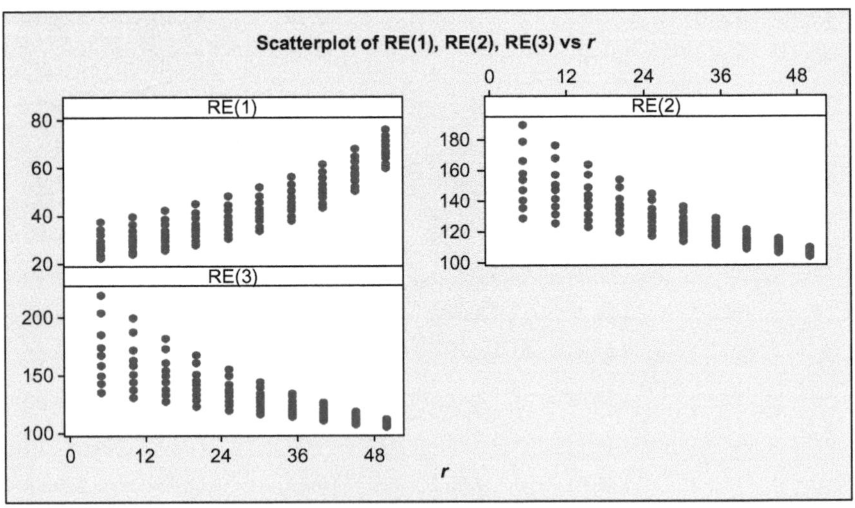

FIG. 2 Pictorial representation of RE(1), RE(2), and RE(3) values.

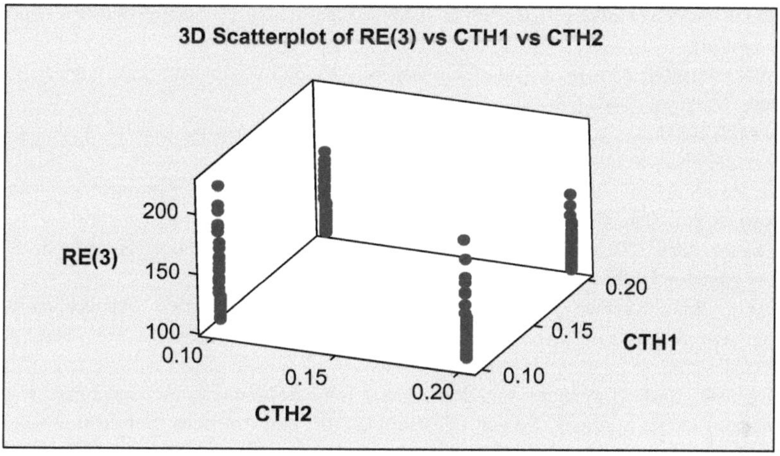

FIG. 3 The value or RE(3) as a function of C_{θ_1} (CTH1) and C_{θ_2} (CTH2).

TABLE 4 RE(3) vs C_{θ_1} (CTH1) and C_{θ_2} (CTH2)

	Freq	Mean	St. Dev.	Min	Med	Max
	$C_{\theta_1}=0.1$					
$C_{\theta_2}=0.1$	30	149.93	30.88	109.80	144.00	221.10
$C_{\theta_2}=0.2$	30	136.36	24.71	106.10	130.80	205.30
	$C_{\theta_1}=0.2$					
$C_{\theta_2}=0.1$	30	134.64	19.53	107.70	131.85	175.00
$C_{\theta_2}=0.2$	30	127.16	16.40	105.20	123.85	168.60

ACKNOWLEDGMENTS

The authors are thankful to Prof. Arijit Chaudhuri, Purnima Shaw, and a referee for their valuable comments on the original version of this chapter.

REFERENCES

Chaudhuri, A., 2011. Randomized Response and Indirect Questioning Techniques in Surveys. Chapman & Hall/CRC, Boca Raton, FL.

Chaudhuri, A., Adhikary, A.K., 1990. Variance estimation with randomized response. Commun. Stat. Theory Methods 19 (3), 1119–1125.

Chaudhuri, A., Christofides, T.C., 2013. Indirect Questioning in Sample Surveys. Springer, New York.

Das, A.K., Tripathi, T.P., 1978. Use of auxilary information in estimating finite population variance. Sankhya, C 40, 139–148.

Eichhorn, B.H., Hayre, L.S., 1983. Scrambled randomized response methods for obtaining sensitive quantitative data. J. Stat. Plan. Inf. 7, 307–316.

Farrell, P.J., Singh, S., 2005. Model-assisted higher order calibration of estimators of variance. Aust. N Z J. Stat. 47 (3), 375–383.

Fox, James Alan, 2016. Randomized Response and Related Methods, second ed. SAGE, Los Angeles, ISBN: 978-1-4833-8103-9.

Greenberg, B.G., Kuebler, R.R., Abernathy, J.R., Horvitz, D.G., 1971. Application of the randomized response technique in obtaining quantitative data. J. Am. Stat. Assoc. 66, 243–250.

Horvitz, D.G., Shah, B.V., Simmons, W.R., 1967. The unrelated question randomized response model. Social Statistics Section, Proceedings of the American Staistical Association 1967, 65–72.

Isaki, C.T., 1983. Variance estimation using auxilary information. J. Am. Stat. Assoc. 78, 117–123.

Mohamed, Choukri, 2015. Improved Imputation Methods in Survey Sampling. Unpublished thesis submitted to Department of Mathematics, Texas A&M University-Kingsville, Kingsville, TX.

Rodney, S., King, M., Singh, S., 1998. Likelihood-based estimation of the regressing model with scrambled responses. Aust. N Z J. Stat. 40 (3), 279–290.

Rosner, B., 2006. Fundamentals of Biostaistics, sixth ed. Belmont, CA, Thomson-Brooks/Cole.

Singh, S., 2009. A new method of imputation in survey sampling. Stat. J. Theor. Appl. Stat. 43 (5), 499–511.

Singh, S., Arnab, R., 2010. Bias-adjustment and calibration of jackknife variance estimator in the presence of non-response. J. Stat. Plan. Inf. 140 (4), 862–871.

Singh, S., Deo, B., 2003. Imputation by power transformation. Stat. Pap. 44, 555–579.

Singh, S., Horn, S., 2000. Compromised imputation in survey sampling. Metrika 51, 267–276.

Singh, S., King, M.L., 1999. Estimation of coefficient of determination using scrambled responses. J. Indian Soc. Agric. Statist. 52 (3), 338–343.

Singh, S., Joarder, A.H., King, M.L., 1996. Regression analysis using scrambled responses. Aust. J. Stat. 38 (2), 201–211.

Singh, S., Horn, S., Yu, F., 1998. Estimation of variance of general regression estimator: higher level calibration approach. Surv. Methodol. 24 (1), 41–50.

Singh, S., Horn, S., Chowdhury, S., Yu, F., 1999. Calibration of the estimator of variance. Aust. N Z J. Stat. 41 (2), 199–212.

Singh, S., Kim, J.-M., Grewal, I.S., 2008. Imputing and jackknifing scrambled responses. Metron LXVI (2), 183–204.

Singh, G.N., Priyanka, K., Kim, J.M., Singh, S., 2009. Estimation of population mean using imputation techniques in sample surveys. J. Korean Stat. Soc. 39 (1), 67–74.

Sitter, R.R., Wu, C., 2002. Efficient estimation of quadratic finite population functions. J. Am. Stat. Assoc. 97, 535–543.

Srivastava, S.K., Jhajj, H.S., 1981. A class of estimators of the population mean in survey sampling using auxiliary information. Biometrika 68, 341–343.

Toutenberg, H., Srivastava, V., 2003. Efficient estimation of population mean using incomplete survey data on study and auxiliary characteristics. Statistica 53, 223–236.

Toutenberg, H., Srivastava, V., Shalabh, 2006. Amputation versus imputation of missing values through ratio method in survey sampling. Stat. Pap. 49, 237–247.

Warner, S.L., 1965. Randomized response: a survey technique for eliminating evasive answer bias. J. Am. Stat. Assoc. 60, 63–69.

Zaizai, Y., Jingyu, W., Junfeng, L., Hua, W., 2008. Ratio imputation method for handling item-nonresponse in Eichhorn model. Model Assist. Stat. Appl. 3 (2), 89–98.

Chapter 29

On an Indirect Response Model

V.R. Padmawar[1]
Stat-Math Unit, Indian Statistical Institute, Bangalore, Karnataka, India
[1]*Corresponding author: e-mail: vrp@isibang.ac.in*

ABSTRACT
It is now part of folklore that when the statistician has to collect data on characteristics that are either too personal, sensitive, or stigmatizing; *randomized response* (RR) techniques are used. There is a large number of such *randomized response models* in the literature. In this article, we propose and study a new indirect response model. We then derive a *method of moments* estimator for the unknown population mean of a stigmatizing variable, under the proposed model. We establish that if we decide to use noise added sensitive and innocuous responses then there is no need to do randomization on the responses, more specifically, it suffices to use just the noise added sensitive response. We then explore the possibility of obtaining an estimator based on maximization of likelihood or *pseudolikelihood* function. We then work in the *Bayesian framework* to explore the possibility of obtaining a *Bayes* or *Bayeslike* estimator based on the *posterior distribution*.

Keywords: Indirect response, Randomized response, Noise, Warner's model, Unrelated question model, Yu–Tian–Tang models, Kim–Warde model, Padmawar–Vijayan model, Pure noise model, Counting model, Poisson model, Bernoulli model

AMS Classification: Primary 62D05

1 PRELIMINARIES

It is now part of folklore that when the statistician has to collect data on characteristics that are either too personal, sensitive or stigmatizing *randomized response(RR)* techniques are used. There is a large number of such *RR models* in the literature. The setbacks that a survey Statistician has to suffer while collecting data on some sensitive issues are well documented in the literature vide (Chaudhuri and Mukerjee, 1987). The book also discusses various *RR* models since the first and now well-known Warner (1965) model. There is now a comprehensive book, Chaudhuri (2010), on *RR*. An expository article, Padmawar (2012) appeared in *Resonance*. In this article, we propose and study an indirect response model that covers a number of models. We,

Handbook of Statistics, Vol. 34. http://dx.doi.org/10.1016/bs.host.2016.01.027

in particular, see its connections with *Padmawar–Vijayan Model* proposed in Padmawar and Vijayan (2000) and *models* proposed or listed in Padmawar (2014, 2015a,b).

Let $U = \{1, 2, \ldots, N\}$ be a finite population under consideration. Let the study variate y take value y_i on unit i, $y_i \in \mathbb{R}$, $1 \le i \le N$. Let x be another variable that takes value x_i on unit i, $x_i \in \mathbb{R}$, $1 \le i \le N$. In particular, x, y may be counting variables taking values 0,1,2,..., or binary variables taking values 0 and 1. Suppose we are interested in estimating the population mean $\overline{Y}$ of variable y given by

$$\overline{Y} = \frac{1}{N}\sum_{i=1}^{N} y_i. \tag{1}$$

Let also

$$\overline{X} = \frac{1}{N}\sum_{i=1}^{N} x_i. \tag{2}$$

Let us also define

$$\sigma_y^2 = \frac{1}{N}\sum_{i=1}^{N}\left(y_i - \overline{Y}\right)^2,\quad \sigma_x^2 = \frac{1}{N}\sum_{i=1}^{N}\left(x_i - \overline{X}\right)^2 \tag{3}$$

and

$$\sigma_{xy} = \frac{1}{N}\sum_{i=1}^{N}\left(x_i - \overline{X}\right)\left(y_i - \overline{Y}\right). \tag{4}$$

If the study variate y happens to be stigmatizing or sensitive like extent of tax evasion or amount of alcohol consumption, etc., the traditional direct response methods are riddled with nonresponse or evasive response. We resort to indirect methods of collecting data. *RR* is an important such method.

A typical *RR* method in the traditional framework may be described as a procedure in which the Statistician chooses x, an innocuous variable whenever the variable y is sensitive. For the method to be effective, x is chosen such that both y and x take values in $\Gamma \subseteq \mathbb{R}$. A respondent reports either the sensitive study variate value y or the innocuous variate value x with some preassigned probabilities. The Statistician, does not, however, know whether the reported value corresponds to the variable x or the variable y. The Statistician has to estimate $\overline{Y}$ based on the *RR*s from the sampled individuals. There are exceptions like *Warner's model* wherein y is a stigmatizing variable and $x = 1 - y$. The model proposed in this article and the techniques therein, however, are valid for any y and x.

In another interesting technique, the respondent adds noise W to her/his true y value and reports $y + W$. The noise variable with known distribution is chosen by the statistician. Padmawar and Vijayan (2000) evolved this idea to estimate $\overline{Y}$ based on the noise added responses $y_i + W_i$ from the sampled individuals.

In this article, we first propose a new indirect response model that combines the above two ideas. The proposed model covers some well-known models. We then derive *method of moments (mom)* estimator for $\overline{Y}$, the unknown population mean of a stigmatizing variable, under the proposed model. As far as *mom* is concerned, we prove an interesting result that once we decide to use noise added responses there is no need to do randomization on noise added sensitive and innocuous variables. In other words, it suffices to use just the noise added sensitive response. We then explore the possibility of obtaining estimators based on maximization of *likelihood* or *pseudolikelihood* functions. We finally work in the *Bayesian framework* to explore the possibility of obtaining *Bayes* or *Bayeslike* estimators based on the *posterior distribution*. We also provide some examples.

2 INTRODUCING THE MODEL

Suppose we draw a sample of size n from the population $U = \{1, 2, ..., N\}$ using, say, *Simple Random Sampling With Replacement(SRSWR)*. In this article, for convenience, we restrict ourselves to *SRSWR*. The treatment, however, can easily be extended to more general sampling designs. Let the respondent i report Z_i. Let $Z_1, Z_2, \ldots, Z_n$ be the n responses from the sampled units. Let us define $\overline{Z}$, the sample mean of the responses, as

$$\overline{Z} = \frac{1}{n}\sum_{i=1}^{n} Z_i \tag{5}$$

and S_Z^2 as the sample variance of the responses

$$S_Z^2 = \frac{1}{n-1}\sum_{i=1}^{n}(Z_i - \overline{Z})^2. \tag{6}$$

Under the proposed model

$$\begin{aligned} Z_i &= Z_{1i} \text{ with probability } p \\ &= Z_{2i} \text{ with probability } 1-p \end{aligned} \tag{7}$$

Z_{1i} and Z_{2i} themselves may be random variables, having for instance, fixed and random components. $p \in [0,1]$, is called mixing or randomizing parameter. The mixing parameter p may be known or unknown. When it is unknown it may even depend on the unknown parameter $\overline{Y}$ that we are actually interested in estimating.

More specifically, we consider the following model for the *RR* of the individual i in the population, $1 \le i \le N$,

$$\begin{aligned} Z_i = Z_{1i} = y_i + W_{1i} &\text{ with probability } p \\ = Z_{2i} = x_i + W_{2i} &\text{ with probability } 1-p \end{aligned} \tag{8}$$

where y_i and x_i are the fixed parts and W_{1i} and W_{2i} are the random parts, $1 \leq i \leq N$. Here, y and x are real valued variables often having the same sample space. y_i is the y-value of the respondent i and x_i is the x-value of the respondent i. $W_{11}, W_{12},\ldots,W_{1N}$ are N *iid* random variables with known common distribution F_1, probability density function (*pdf*) f_1 and $W_{21}, W_{22},\ldots,W_{2N}$ are N *iid* random variables with known common distribution F_2, *pdf* f_2, further the two sets of random variables are independent of each other. Let (ν_1, ξ_1^2) and (ν_2, ξ_2^2) be respectively the mean and variance of F_1 and F_2; $\nu_1, \nu_2 \in (-\infty, \infty)$, and $\xi_1^2, \xi_2^2 \in [0, \infty)$. Typically, y is a stigmatizing variable, x is an innocuous variable and W_1, W_2 are independent noise variables. p in (8), the mixing or randomizing parameter, may be known or unknown.

It is easy to show that many well-known models are particular cases of (7). For instance, Warner's model (Warner, 1965), unrelated question model (Greenberg et al., 1969), Padmawar–Vijayan model (Padmawar and Vijayan, 2000), pure noise models (Padmawar, 2015a,b), Yu-Tian-Tang models (Yu et al., 2008), Bernoulli models, Counting models (Padmawar, 2015b), Kim–Warde models (Kim and Warde, 2005) can all be brought under single umbrella of (7) and some of them also under (8).

2.1 Method of Moments

We first obtain *mom estimator* for $\overline{Y}$.

Let the conditional mean and variance of Z_{ri}, in (8), given i, respectively be ν_{ri} and ξ_r^2; $r = 1,2$. Then $\nu_{1i} = y_i + \nu_1$, $\nu_{2i} = x_i + \nu_2$.

Under *SRSWR* the sample responses $Z_1, Z_2,\ldots,Z_n$ are *iid* and $E(\overline{Z}) = E(Z_1)$. Let their common mean and variance respectively be μ (9) and τ^2 (10). It is then true that the sample mean $\overline{Z}$(5) is unbiased for the population mean μ and sample variance S_Z^2 (6) is unbiased for the population variance τ^2.

We now prove the following results.

Theorem 1. *Under the proposed model (8), under SRSWR, $Z_1, Z_2,\ldots,Z_n$ are iid with common mean μ and common variance τ^2 given by*

$$\mu = p\overline{Y} + (1-p)\overline{X} + p\nu_1 + (1-p)\nu_2 \tag{9}$$

$$\begin{aligned} \textit{and } \tau^2 = &-p^2\left(\left(\overline{Y} + \nu_1\right) - \left(\overline{X} + \nu_2\right)\right)^2 \\ &+ p\left[\left(\left(\sigma_y^2 + \xi_1^2\right) - \left(\sigma_x^2 + \xi_2^2\right)\right) + \left(\left(\overline{Y} + \nu_1\right) - \left(\overline{X} + \nu_2\right)\right)^2\right] \\ &+ \left(\xi_2^2 + \sigma_x^2\right). \end{aligned} \tag{10}$$

Further for p≠0

$$\hat{\overline{Y}} = \frac{\overline{Z} - \left[(1-p)\overline{X} + pv_1 + (1-p)v_2\right]}{p} \tag{11}$$

is an mom estimator for $\overline{Y}$*. It is unbiased for* $\overline{Y}$ *with variance* $\frac{\tau^2}{np^2}$*, that is,*

$$Var\left(\hat{\overline{Y}}\right) = \frac{\tau^2}{np^2}. \tag{12}$$

Further an unbiased estimator for $Var\left(\hat{\overline{Y}}\right)$ *is given by*

$$\widehat{Var}\left(\hat{\overline{Y}}\right) = \frac{S_Z^2}{np^2}. \tag{13}$$

Proof. Note that in our framework there are three stages of randomization. At stage one, we use *SRSWR* to select n respondents. At the next stage, there is randomization to decide whether to report Z_{1i} or Z_{2i}, $1 \le i \le n$. At the third and final stage, the value of a noise variable W is generated. For computational convenience, let subscript 1 signify the first stage of randomization and subscript $2*$ signify the second and third stages of randomization together. Let further E_1, Var_1 denote the operators at stage 1, and E_{2*}, Var_{2*} denote conditional operators at stage $2*$, given stage 1.

It is easy to see that

$$\begin{aligned} E_{2*}(Z_i) &= pv_{1i} + (1-p)v_{2i} \\ &= p(y_i + v_1) + (1-p)(x_i + v_2) \\ Var_{2*}(Z_i) &= p\xi_1^2 + (1-p)\xi_2^2 + p(1-p)(v_{1i} - v_{2i})^2 \\ &= p\xi_1^2 + (1-p)\xi_2^2 + p(1-p)((y_i + v_1) - (x_i + v_2))^2 \end{aligned}$$

Therefore

$$\begin{aligned} E_1E_{2*}(Z_i) &= p\left(\overline{Y} + v_1\right) + (1-p)\left(\overline{X} + v_2\right) \\ E_1Var_{2*}(Z_i) &= p\xi_1^2 + (1-p)\xi_2^2 + p(1-p)E_1((y_i + v_1) - (x_i + v_2))^2 \\ &= p\xi_1^2 + (1-p)\xi_2^2 \\ &\quad + p(1-p)Var_1[(y_i + v_1) - (x_i + v_2)] \\ &\quad + p(1-p)[E_1\{(y_i + v_1) - (x_i + v_2)\}]^2 \\ &= p\xi_1^2 + (1-p)\xi_2^2 \\ &\quad + p(1-p)\left[\left(\sigma_y^2 + \sigma_x^2 - 2\sigma_{xy}\right) + \left\{\left(\overline{Y} + v_1\right) - \left(\overline{X} + v_2\right)\right\}^2\right] \end{aligned}$$

$$\begin{aligned} \text{and } Var_1E_{2*}(Z_i) &= Var_1[py_i + (1-p)x_i + pv_1 + (1-p)v_2] \\ &= Var_1[py_i + (1-p)x_i] \\ &= p^2\sigma_y^2 + (1-p)^2\sigma_x^2 + 2p(1-p)\sigma_{xy}. \end{aligned}$$

Now using the identities

$$E(Z_i) = E_1 E_{2*}(Z_i)$$
$$Var(Z_i) = Var_1 E_{2*}(Z_i) + E_1 Var_{2*}(Z_i)$$

and the fact that under *SRSWR*

$$E(\overline{Z}) = E(Z_i)$$
$$\text{and } Var(\overline{Z}) = \frac{Var(Z_i)}{n}$$

we get

$$E(\overline{Z}) = p\overline{Y} + (1-p)\overline{X} + pv_1 + (1-p)v_2$$
$$Var(\overline{Z}) = \frac{1}{n}\left\{\xi^2 + p\sigma_y^2 + (1-p)\sigma_x^2 + p(1-p)(\overline{Y} - \overline{X})^2\right\}.$$

Therefore for $p \neq 0$

$$\hat{\overline{Y}} = \frac{\overline{Z} - [(1-p)\overline{X} + pv_1 + (1-p)v_2]}{p}$$

is unbiased for $\overline{Y}$ with variance $\frac{\tau^2}{np^2}$, that is,

$$Var\left(\hat{\overline{Y}}\right) = \frac{\tau^2}{np^2}$$

or

$$Var\left(\hat{\overline{Y}}\right) = \frac{1}{np^2}\begin{pmatrix} -p^2\left((\overline{Y} + v_1) - (\overline{X} + v_2)\right)^2 \\ +p\left[\left(\left(\sigma_y^2 + \xi_1^2\right) - \left(\sigma_x^2 + \xi_2^2\right)\right) + \left((\overline{Y} + v_1) - (\overline{X} + v_2)\right)^2\right] \\ +\left(\xi_2^2 + \sigma_x^2\right) \end{pmatrix} \tag{14}$$

Finally, since sample variance $S_Z^2(6)$ is unbiased for the population variance τ^2, an unbiased estimator for $Var\left(\hat{\overline{Y}}\right)$ is given by

$$\widehat{Var}\left(\hat{\overline{Y}}\right) = \frac{S_Z^2}{np^2}. \tag{15}$$

This completes the proof. □

Corollary 1. *In particular for* $p = 1$, which corresponds to the Padmawar–Vijayan model (Padmawar and Vijayan, 2000), we get

$$E(Z_i) = \mu = \overline{Y} + v_1$$
$$\textit{and } Var(Z_i) = \tau^2 = \xi_1^2 + \sigma_y^2.$$

so that

$$\hat{\overline{Y}} = \overline{Z} - v_1$$

is unbiased for $\overline{Y}$ *with variance*

$$Var\left(\hat{\overline{Y}}\right) = \frac{\xi_1^2 + \sigma_y^2}{n}.$$

Note that the expression for $Var\left(\hat{\overline{Y}}\right)$ in (14) depends on the mixing parameter p. When the mixing parameter p in the given framework is at the disposal of the statistician and does not depend on any unknown quantities we can try to obtain an optimal value of p that minimizes $Var\left(\hat{\overline{Y}}\right)$. We first state and prove the following simple but useful lemma.

Lemma 1. *Let* $g : \mathbb{R} \rightarrow \mathbb{R}$ *be the function given by*

$$g(u) = bu + cu^2$$

where $c > 0$ *and* $b \in \mathbb{R}$. *Let* $u_0 = -\frac{b}{2c}$. *Then* $g(u)$ *attains its minimum on* $[1, \infty)$ *at 1 or* $-\frac{b}{2c}$. *More specifically,*

(i) *if* $u_0 \leq 1$ *then* $g(u)$ *attains its minimum on* $[1, \infty)$ *at 1.*
(ii) *if* $u_0 > 1$ *then* $g(u)$ *attains its minimum on* $[1, \infty)$ *at* $u_0 = -\frac{b}{2c}$.

Proof. Observe that

$$\frac{\partial g(u)}{\partial u} = b + 2cu$$

and

$$\frac{\partial^2 g(u)}{\partial u^2} = 2c > 0.$$

$u_0 = -\frac{b}{2c}$ is a solution to $\frac{\partial g(u)}{\partial u} = 0$. Since $c > 0$ the function $g(u)$ is convex on the entire real line $\mathbb{R}$. Thus

(i) if $u_0 \leq 1$ then $g(u)$ attains its minimum on $[1, \infty)$ at 1.
(ii) if $u_0 > 1$ then $g(u)$ attains its minimum on $[1, \infty)$ at $u_0 = -\frac{b}{2c}$. □

Theorem 2. *Under the proposed model (8),* $Var\left(\hat{\overline{Y}}\right)$ *in (12), as a function of* p, $p \in (0,1]$, *is minimized at* $p = 1$.

Proof. Note that minimizing $Var\left(\hat{\overline{Y}}\right)$ in (14) as a function of p on (0,1] is equivalent to minimizing

$$g(u) = bu + cu^2,$$

as a function of u, on $[1,\infty)$, with $u=\frac{1}{p}$, $c=\xi_2^2+\sigma_x^2$ and $b=\left(\left(\sigma_y^2+\xi_1^2\right)-\left(\sigma_x^2+\xi_2^2\right)\right)+\left(\left(\overline{Y}+\nu_1\right)-\left(\overline{X}+\nu_2\right)\right)^2$.

By Lemma 1, the minimum of $g(u)$ on $[1,\infty)$ is attained at $u=1$ if $u_0=-\frac{b}{2c}\leq 1$ and at $u=u_0$ if $u_0>1$. We now show that $u_0=-\frac{b}{2c}\leq 1$.

Since $c>0$, $-\frac{b}{2c}\leq 1$ is equivalent to $-b\leq 2c$.

$$\Leftrightarrow -\left[\left(\left(\sigma_y^2+\xi_1^2\right)-\left(\sigma_x^2+\xi_2^2\right)\right)+\left[\left(\overline{Y}+\nu_1\right)-\left(\overline{X}+\nu_2\right)\right]^2\right]\leq 2\left(\xi_2^2+\sigma_x^2\right)$$
$$\Leftrightarrow -\left[\left(\sigma_y^2+\xi_1^2\right)+\left[\left(\overline{Y}+\nu_1\right)-\left(\overline{X}+\nu_2\right)\right]^2\right]\leq \left(\xi_2^2+\sigma_x^2\right)$$

which is always true.

Thus $Var\left(\hat{\overline{Y}}\right)$ is minimized at $p=1$. □

Corollary 2. *In particular, if we take* $\nu_2=0$ *and* $\xi_2^2=0$*, that corresponds to the situation when we add noise to the sensitive variable but do not add noise to the innocuous variable.*

Remark 1. The above result tells us that in our framework since we are adding noise to true stigmatizing response there are no additional benefits by opting for *RR* vis-a-vis an innocuous variable with or without noise. The optimal value of $p=1$ corresponds to the Padmawar–Vijayan model (Padmawar and Vijayan, 2000).

Remark 2. The estimator $\hat{\overline{Y}}$ in (11) is an *mom estimator* for the population mean $\overline{Y}$ in (1). Apart from point estimators we may also construct confidence intervals. If we are estimating the unknown population proportion then we have to ensure that the estimators belong to the interval $[0,1]$ and the confidence intervals are also contained in $[0,1]$. One may also use bootstrap techniques to produce bootstrap intervals.

2.2 Maximum Likelihood or Pseudolikelihood Method

When the mixing parameter p is at the statistician's disposal and if we want to use *mom estimator* then it suffices to take $p=1$. However, if we want to explore different methods of estimation, we have to go beyond $p=1$. For instance, we may maximize the *likelihood* or *loglikelihood* when the *pdfs* $f_r(.)$, of W_{ri}, $r=1,2$; are known and the mixing parameter p is known or unknown. We may also want to explore the *Bayesian* framework.

Under model (8) the *pdf* of Z_i is $pf_1(z_i-y_i)+(1-p)f_2(z_i-x_i)$. Therefore the joint *pdf* of the *RR*s $Z_1, Z_2,\ldots,Z_n$ based on our *SRSWR sample* of size n is

$$\prod_{i=1}^{n}[pf_1(z_i - y_i) + (1-p)f_2(z_i - x_i)]. \tag{16}$$

Since f_1, f_2 are specified, we may assume that y_i, $1 \le i \le n$, are the parameters. For the innocuous variable x, it may be reasonable to assume that $\overline{X} = \frac{1}{N}\sum_{i=1}^{N} x_i$ is known. Suppose we approximate the joint *pdf* by

$$\prod_{i=1}^{n}\left[pf_1\left(z_i - \overline{Y}\right) + (1-p)f_2\left(z_i - \overline{X}\right)\right], \tag{17}$$

where $\overline{X} = \frac{1}{N}\sum_{i=1}^{N} x_i$.

We can now maximize the approximate joint *pdf (pseudolikelihood)* as a function of $\overline{Y}$ to get maximum *pseudolikelihood* estimator for $\overline{Y}$. We shall later see that in some specific cases we get exact *maximum likelihood estimator (mle)* for $\overline{Y}$.

Alternatively, we may set $\theta = (y_1, y_2, \ldots, y_n)$ as the unknown parameter. Since p is known and f_1, f_2 are completely specified, the *likelihood* function of θ, given the sample $\mathbf{z} = (z_1, z_2, \ldots, z_n)$, may be approximated as

$$L(\theta|\mathbf{z}) = \prod_{i=1}^{n}\left[pf_1(z_i - y_i) + (1-p)f_2\left(z_i - \overline{X}\right)\right]. \tag{18}$$

We may maximize the approximate *likelihood or loglikelihood* whenever possible. Here, we have to solve $\frac{\partial}{\partial y_i}\left[pf_1(z_i - y_i) + (1-p)f_2\left(z_i - \overline{X}\right)\right] = 0$; $1 \le i \le n$, to get an estimate $\hat{y}_i$ of y_i; $1 \le i \le n$. We may then take $\frac{1}{n}\sum_{i=1}^{n}\hat{y}_i$ as an estimator for $\overline{Y}$.

Invariably these estimators would not be available in closed form and we would have to make do with their numerical values. Sometimes we can use techniques like *EM algorithm* to obtain *mle*.

We now work on an example to demonstrate how to obtain *mle*. Consider the following model (19), with the mixing parameter unknown, which is a particular case of our model (8).

Example 1. Suppose we want to estimate the unknown proportion π of a sensitive attribute $\mathcal{A}$. In view of the sensitive nature of the attribute $\mathcal{A}$, we design the following method of collecting data from the respondents belonging to an *SRSWR* sample of size n. The ith respondent is given a die such that the even faces of it are inscribed with $\mathcal{A}$ and the odd faces of it are inscribed with $\lceil\mathcal{A}$. The respondent is then asked to roll the die till s/he gets the face showing her/his true state. The respondent is asked to report only Z_i, the number of times the die was required to be rolled to get a face inscribed with her/his true state for the first time, $1 \le i \le n$. We assume that this reporting would be done truthfully and correctly. It is clear that the die may be replaced by a deck containing two types of cards or an urn containing two types of balls or a container with a transparent window having two types of beads and so on. This experiment can easily be simulated using a computer as well. Let $Z_1,\ldots,Z_n$ be the *RR*s based on the *SRSWR* sample of size n.

Let η be the probability of getting a face inscribed with $\mathcal{A}$ in a single roll of the die. We assume that η, $(0 < \eta < 1)$, is known.

We model this data collection, in terms of (8). In other words, we show how the model under consideration is a particular case of (8). A similar scenario was considered in Padmawar (2015a).

Let $y_i = 1(0)$ if the ith individual in the population possesses $\mathcal{A}$ (does not possess $\mathcal{A}$). Further let $x_i = 1$ for all individuals i, $1 \le i \le N$, in the population.

In model (8) if we set

$$\begin{aligned} Z_i &= Z_{1i} = y_i + W_{1i} \text{ with probability } \pi \\ &= Z_{2i} = x_i + W_{2i} \text{ with probability } 1 - \pi \end{aligned} \tag{19}$$

where we treat W_{1i} and W_{2i} as the noise variables. W_{1i} are geometric random variables taking values 0,1,2,..., with parameter η and the noise W_{2i} are geometric random variables taking values 0,1,2,..., with parameter $1 - \eta$.

We now move on to obtain *mle* for π.

Let $f_{Z_i}(z|\pi)$ denote the probability mass function *(pmf)* of the random variable Z_i, $1 \le i \le n$, then

$$\begin{aligned} f_{Z_i}(z|\pi) &= \pi f_{W_{1i}}(z-1|\eta) + (1-\pi) f_{W_{2i}}(z-1|(1-\eta)) \\ &= \pi(1-\eta)^{z-1}\eta + (1-\pi)\eta^{z-1}(1-\eta) \end{aligned} \tag{20}$$

where $f_{W_{1i}}(z|\eta) = (1-\eta)^{z-1}\eta$, $z = 0,1,2,\ldots$, and $f_{W_{2i}}(z|(1-\eta)) = \eta^{z-1}(1-\eta)$, $z = 0,1,2,\ldots$, are respectively the *pmf* of the random variables W_{1i} and W_{2i}, $1 \le i \le n$.

If $f_{Z_1\cdots Z_n}(z_1,\ldots,z_n|\pi)$ denotes the joint probability mass function of the random variables Z_i, $1 \le i \le n$, then

$$\begin{aligned} f_{Z_1\cdots Z_n}(z_1,\ldots,z_n|\pi) = f_Z(z|\pi) &= \prod_{i=1}^{n}\{\pi(1-\eta)^{z_i-1}\eta + (1-\pi)\eta^{z_i-1}(1-\eta)\} \\ &= \prod_{i=1}^{n}\left[\eta^{z_i-1}(1-\eta) + \pi\left\{(1-\eta)^{z_i-1}\eta - \eta^{z_i-1}(1-\eta)\right\}\right] \\ &= \prod_{i=1}^{n}\eta^{z_i-1}(1-\eta)\left[1 + \frac{(1-\eta)^{z_i-1}\eta - \eta^{z_i-1}(1-\eta)}{\eta^{z_i-1}(1-\eta)}\pi\right] \\ &= \prod_{i=1}^{n}\eta^{z_i-1}(1-\eta) \\ &\quad \times \prod_{i=1}^{n}\left[1 + \frac{(1-\eta)^{z_i-2} - \eta^{z_i-2}}{\eta^{z_i-2}}\pi\right] \\ &= \left(\frac{1}{\eta} - 1\right)^n \eta^{\Sigma z_i} \prod_{i=1}^{n}\left[1 + \left[\left(\frac{1}{\eta} - 1\right)^{z_i-2} - 1\right]\pi\right] \end{aligned}$$

Thus

$$\begin{aligned} f_{Z_1\cdots Z_n}(z_1,\ldots,z_n|\pi) &= f_{\mathbf{Z}}(\mathbf{z}|\pi) \\ &= \left(\frac{1}{\eta}-1\right)^n \eta^{\Sigma z_i} \\ &\quad \times \prod_{i=1}^n \left[1+\left[\left(\frac{1}{\eta}-1\right)^{z_i-2}-1\right]\pi\right] \end{aligned} \tag{21}$$

(21) viewed as likelihood function $L(\pi|\mathbf{z})$ we have

$$L(\pi|\mathbf{z}) = \left(\frac{1}{\eta}-1\right)^n \eta^{\Sigma z_i} \times \prod_{i=1}^n \left[1+\left[\left(\frac{1}{\eta}-1\right)^{z_i-2}-1\right]\pi\right]. \tag{22}$$

One can now obtain $\hat{\pi}_{mle}$, the maximum likelihood estimator(*mle*) for π using (22).

Note that $\prod_{i=1}^n (1+c_i x)$ may be expressed as

$$\prod_{i=1}^n (1+c_i x) = \sum_{k=0}^n S_k x^k, \tag{23}$$

where $S_0 = 1$ and for $1 \le k \le n$; S_k is the sum of $\binom{n}{k}$ terms of the type $c_{i_1} c_{i_2} \cdots c_{i_k}$ where $1 \le i_1 < i_2 < \cdots < i_k \le n$, that is,

$$\begin{aligned} S_0 &= 1 \\ \text{and } S_k &= \sum_{1 \le i_1 < i_2 < \cdots < i_k \le n} c_{i_1} c_{i_2} \cdots c_{i_k}. \end{aligned} \tag{24}$$

Now maximizing (22), as a function of π, is equivalent to maximizing

$$\begin{aligned} \Lambda(\pi) &= \prod_{i=1}^n \left[1+\left[\left(\frac{1}{\eta}-1\right)^{z_i-2}-1\right]\pi\right] \\ \Lambda(\pi) &= \sum_{k=0}^n S_k \pi^k \end{aligned} \tag{25}$$

where S_k are given by (24), with

$$c_i = \left[\left(\frac{1}{\eta}-1\right)^{z_i-2}-1\right];\ 1 \le i \le n. \tag{26}$$

Now

$$\begin{aligned} \frac{\partial \Lambda(\pi)}{\partial \pi} &= \sum_{k=1}^n S_k k \pi^{k-1} \\ \text{and } \frac{\partial^2 \Lambda(\pi)}{\partial \pi^2} &= \sum_{k=2}^n S_k k(k-1) \pi^{k-2} \end{aligned}$$

Since $\frac{\partial \Lambda(\pi)}{\partial \pi}$ is a polynomial of degree $n - 1$, the roots correspond to the optima of $\Lambda(\pi)$. We can thus obtain *mle* for π.

Alternatively, let $\mathcal{L}(\pi)$ denote the loglikelihood, ie,

$$\begin{aligned}\mathcal{L}(\pi) &= \log \prod_{i=1}^{n} \{\pi(1-\eta)^{z_i-1}\eta + (1-\pi)\eta^{z_i-1}(1-\eta)\} \\ &= \sum_{i=1}^{n} \log\{\pi(1-\eta)^{z_i-1}\eta + (1-\pi)\eta^{z_i-1}(1-\eta)\}\end{aligned} \tag{27}$$

Differentiating (27) w.r.t. π we get,

$$\begin{aligned}\frac{\partial \mathcal{L}(\pi)}{\partial \pi} &= \frac{\partial}{\partial \pi}\sum_{i=1}^{n} \log\{\pi(1-\eta)^{z_i-1}\eta + (1-\pi)\eta^{z_i-1}(1-\eta)\} \\ &= \sum_{i=1}^{n} \frac{((1-\eta)^{z_i-1}\eta - \eta^{z_i-1}(1-\eta))}{(\pi(1-\eta)^{z_i-1}\eta + (1-\pi)\eta^{z_i-1}(1-\eta))}\end{aligned} \tag{28}$$

It remains to solve $\frac{\partial \mathcal{L}(\pi)}{\partial \pi} = 0$ to get $\hat{\pi}_{mle}$, the maximum likelihood estimator for π. Note that

$$\frac{\partial^2 \mathcal{L}(\pi)}{\partial \pi^2} = -\sum_{i=1}^{n} \frac{((1-\eta)^{z_i-1}\eta - \eta^{z_i-1}(1-\eta))^2}{\left\{(\pi(1-\eta)^{z_i-1}\eta + (1-\pi)\eta^{z_i-1}(1-\eta))\right\}^2} < 0.$$

Therefore $\mathcal{L}(\pi)$ is a concave function.

In this example, we can also use the method of obtaining *mle* for mixture of densities using *EM algorithm*, as described in Bilmes (1998). They have worked in a very general framework of mixture of densities. In our example, we have a mixture of just two, that too completely known, densities and only the mixing parameter π is unknown. This is, thus, a very simple particular case of their framework. It is no surprise, therefore, that their formidable recursive formula, assumes a simple form while giving *mle* for π in our framework.

The repeated use of the following recursive formula, gives the *mle* for π. For $r = 0,1,\ldots,$

$$\hat{\pi}(r+1) = \frac{\hat{\pi}(r)}{n}\sum_{i=1}^{n} \frac{(1-\eta)^{z_i-1}\eta}{\hat{\pi}(r)(1-\eta)^{z_i-1}\eta + (1-\hat{\pi}(r))\eta^{z_i-1}(1-\eta)},$$

$\hat{\pi}(0)$ being the initial value. $\hat{\pi}(r+1)$ converges to the *mle* of π, as $r \to \infty$.

Remark 3. It may be noted that Example 1 above can be extended to the mixture of more general densities, say $f_1(.)$ and $f_2(.)$, for the noise variables. It may also be noted that Example 1 may be perceived as a *Warner-like* model (Warner, 1965); that incorporates noise and randomizes with unknown π. Here π, the unknown proportion, the parameter to be estimated, also doubles for the mixing or randomizing parameter.

2.3 Bayesian Framework

If the *likelihood function*, given the sample response $z=(z_1,z_2,\ldots,z_n)$, depends on the single parameter $\overline{Y}$, which may be assumed to have a suitable prior, then one can set up the usual *Bayesian* framework to obtain posterior distribution for $\overline{Y}$ and propose posterior mean as an estimator for unknown $\overline{Y}$. We shall later see that in some specific cases we get proper *Bayes* estimator for $\overline{Y}$. Invariably these estimators would not be available in closed form and we would have to again make do with their numerical values. We also explore the possibility of using *pseudolikelihood function*.

In this section, let us use the notation γ for $\overline{Y}$, that is $\overline{Y}=\gamma$. Let us assume that $\overline{X}$ is known. Let us assume a prior $\zeta(\gamma)$ for γ. Again if z is the response vector based on our *SRSWR* sample of size n, let us approximate the conditional joint *pdf* of z given γ as

$$fz(z|\gamma)=\prod_{i=1}^{n}\left[pf_1(z_i-\gamma)+(1-p)f_2\left(z_i-\overline{X}\right)\right]$$

then the joint *pdf* of z,γ is given by

$$fz_\gamma(z,\gamma)=\left\{\prod_{i=1}^{n}\left[pf_1(z_i-\gamma)+(1-p)f_2\left(z_i-\overline{X}\right)\right]\right\}\times\zeta(\gamma)$$

the marginal of z is given by

$$fz(z)=\int_0^{\infty}\left\{\prod_{i=1}^{n}\left[pf_1(z_i-\gamma)+(1-p)f_2\left(z_i-\overline{X}\right)\right]\right\}\times\zeta(\gamma)d\gamma$$

and $\zeta(\gamma|z)$, the posterior of γ given z, is now given by

$$\begin{aligned}\zeta(\gamma|z)&=\frac{fz_\gamma(z,\gamma)}{fz(z)}\\&=\frac{\left\{\prod_{i=1}^{n}\left[pf_1(z_i-\gamma)+(1-p)f_2\left(z_i-\overline{X}\right)\right]\right\}\times\zeta(\gamma)}{\int_0^{\infty}\left\{\prod_{i=1}^{n}\left[pf_1(z_i-\gamma)+(1-p)f_2\left(z_i-\overline{X}\right)\right]\right\}\times\zeta(\gamma)d\gamma}.\end{aligned}$$

We now propose the posterior mean as an estimator for $\overline{Y}=\gamma$.

Remark 4. If the population mean $\overline{X}$ is unknown, it may be replaced by the sample mean $\overline{x}$, if $\overline{x}$ is known, otherwise by some suitable estimator.

We finally give an example where we deal with proper likelihood function so as to get the proper *Bayes estimator*. This example is a generalization of the example considered in Padmawar (2015b).

Example 2. This is also a generalization of Example 1. Let $y_i = 1(0)$ if the individual in the population possesses $\mathcal{A}$ (does not possess $\mathcal{A}$). Further let $x_i = 1$ for all individuals in the population. A respondent is supposed to report a value $y_i + W_{1i}$ if s/he possesses the sensitive attribute $\mathcal{A}$ and a value $x_i + W_{2i}$ if s/he does not possess the sensitive attribute $\mathcal{A}$.

The above scenario corresponds to the model (8) if we set

$$\begin{aligned} Z_i &= Z_{1i} = y_i + W_{1i} \text{ with probability } \pi \\ &= Z_{2i} = x_i + W_{2i} \text{ with probability } 1-\pi \end{aligned} \tag{29}$$

where we treat W_{1i} and W_{2i} as the noise variables with densities $f_1(.)$ and $f_2(.)$ on $(-\infty, \infty)$, say.

Note that (19) is a particular case of (29).

We now move on to obtain Bayes estimator for π. The response Z follows the following mixture density with π as the mixing parameter.

$$f_Z(z|\pi) = \pi f_1(z-1) + (1-\pi) f_2(z-1). \tag{30}$$

We now assume that we can assign an appropriate *prior* to the proportion π. Let us, for instance, consider *Beta prior* $\beta(a,b)$, beta distribution with parameters a and b, where the parameters a and b are assumed to be known.

$$f_\pi(\pi) = \frac{\pi^{a-1}(1-\pi)^{b-1}}{\beta(a,b)}; \ 0<\pi<1, \ a>0, \ b>0. \tag{31}$$

The joint distribution of Z and π may be given by

$$f_{Z\pi}(z,\pi) = \frac{\pi^{a-1}(1-\pi)^{b-1}[\pi f_1(z-1) + (1-\pi) f_2(z-1)]}{\beta(a,b)}.$$

Therefore the marginal distribution of Z is given by

$$f_Z(z) = \frac{a}{a+b} f_1(z-1) + \frac{b}{a+b} f_2(z-1). \tag{32}$$

Note that the mean of the prior distribution $f_\pi(\pi)$, given in (31), also called prior mean, is given by $\frac{a}{a+b}$. Now if we replace π by $\frac{a}{a+b}$ in $f_Z(z|\pi)$ given in (30) then we get the marginal $f_{Z_i}(z)$ given in (32).

We can now find the posterior distribution of π given $Z = z$.

$$\begin{aligned} f_{\pi|Z_i}(\pi|Z=z) &= \frac{f_{Z\pi}(z,\pi)}{f_Z(z)} \\ &= \frac{\frac{1}{\beta(a,b)}\pi^{a-1}(1-\pi)^{b-1}[\pi f_1(z-1) + (1-\pi) f_2(z-1)]}{\frac{a}{a+b} f_1(z-1) + \frac{b}{a+b} f_2(z-1)} \end{aligned}$$

$$= \frac{1}{\beta(a,b)} \times \frac{1}{\frac{a}{a+b}f_1(z-1) + \frac{b}{a+b}f_2(z-1)}$$

$$\times \pi^{a-1}(1-\pi)^{b-1}[f_2(z-1) + \pi(f_1(z-1) - f_2(z-1))]$$

Thus

$$f_{\pi|Z_i}(\pi|Z=z) = \frac{1}{\beta(a,b)} \times \frac{1}{\frac{a}{a+b}f_1(z-1) + \frac{b}{a+b}f_2(z-1)} \times \pi^{a-1}(1-\pi)^{b-1}[f_2(z-1) + \pi(f_1(z-1) - f_2(z-1))]. \tag{33}$$

Using the properties of *beta integral*, it may be verified that the integral of the posterior density (33) is indeed one, that is, $\int_0^1 f_{\pi|Z_i}(\pi|Z_i = z)d\pi = 1$. Further, the mean of the posterior distribution (33) may be shown to be

$$\frac{a}{(a+b+1)} \frac{(a+1)f_1(z-1) + bf_2(z-1)}{af_1(z-1) + bf_2(z-1)}. \tag{34}$$

The posterior mean (34) may be used as an estimator for π.

We can similarly find the posterior distribution of π given $Z_1, Z_2, \ldots, Z_n$, all n responses. Note that the joint distribution $f_{Z_1 \cdots Z_n}(z_1, \ldots, z_n|\pi) = f_{\mathbf{Z}}(\mathbf{z}|\pi)$ can be written down using (30). The joint distribution of $\mathbf{Z} = (Z_1, Z_2, \ldots, Z_n)$ and π is thus given by

$$f_{\mathbf{Z}\pi}(\mathbf{z}, \pi) = \frac{\pi^{a-1}(1-\pi)^{b-1}\prod_{i=1}^{n}[\pi f_1(z_i-1) + (1-\pi)f_2(z_i-1)]}{\beta(a,b)}$$

$$= \frac{\pi^{a-1}(1-\pi)^{b-1}\prod_{i=1}^{n}[(f_2(z_i-1) + \pi(f_1(z_i-1) - f_2(z_i-1))]}{\beta(a,b)}$$

$$= \frac{\left(\prod_{i=1}^{n} f_2(z_i-1)\right) \times \pi^{a-1}(1-\pi)^{b-1}}{\beta(a,b)} \times \prod_{i=1}^{n}\left[1 + \pi\left(\frac{f_1(z_i-1)}{f_2(z_i-1)} - 1\right)\right]$$

We can rewrite the above joint distribution as

$$f_{\mathbf{Z}\pi}(\mathbf{z}, \pi) = \frac{\prod_{i=1}^{n} f_2(z_i-1)}{\beta(a,b)} \sum_{k=0}^{n} S_k \pi^{a+k-1}(1-\pi)^{b-1}, \tag{35}$$

where S_k are given by (24) with

$$c_i = \left(\frac{f_1(z_i-1)}{f_2(z_i-1)} - 1\right); \; 1 \le i \le n. \tag{36}$$

Therefore the marginal distribution of $\mathbf{Z}$ may be obtained as

$$f_{\mathbf{Z}}(z) = \frac{\prod_{i=1}^{n} f_2(z_i - 1)}{\beta(a,b)} \sum_{k=0}^{n} S_k \beta(a+k,b) \tag{37}$$

We can now find the posterior distribution of π given $\mathbf{Z} = z$.

$$f_{\pi|\mathbf{Z}}(\pi|\mathbf{Z}=z) = \frac{f_{\mathbf{Z}\pi}(z,\pi)}{f_{\mathbf{Z}}(z)}, \ i.e.,$$

$$f_{\pi|\mathbf{Z}}(\pi|\mathbf{Z}=z) = \frac{\sum_{k=0}^{n} S_k \pi^{a+k-1}(1-\pi)^{b-1}}{\sum_{k=0}^{n} S_k \beta(a+k,b)}. \tag{38}$$

Note that the posterior density (38) is indeed a density as it integrates to 1. It is now easy to see that the mean of the posterior density (38) may be derived as

$$\int_0^1 \pi \frac{\sum_{k=0}^{n} S_k \pi^{a+k-1}(1-\pi)^{b-1}}{\sum_{k=0}^{n} S_k \beta(a+k,b)} d\pi$$

$$= \frac{1}{\sum_{k=0}^{n} S_k \beta(a+k,b)} \sum_{k=0}^{n} S_k \int_0^1 \pi^{(a+k+1)-1}(1-\pi)^{b-1} d\pi$$

$$= \frac{\sum_{k=0}^{n} S_k \beta(a+k+1,b)}{\sum_{k=0}^{n} S_k \beta(a+k,b)}$$

Thus the posterior mean is given by

$$\frac{\sum_{k=0}^{n} S_k \beta(a+k+1,b)}{\sum_{k=0}^{n} S_k \beta(a+k,b)} \tag{39}$$

Again the posterior mean (39) may be used as an estimator for π. It is a matter of verification that for $n = 1$, (39) reduces to (34).

In conclusion, we have thus proposed and studied an indirect response model. We then obtained estimators for the population mean using various methods of estimation. In the context of *mom*, we established that there is no need to use randomization on noise added responses. Finally, it should be noted that the proposed model is actually a class of indirect response models.

ACKNOWLEDGMENTS

We would like to thank the editors for providing the opportunity to write the present article for Handbook of Statistics on *RRT*. We would also like to thank the referee for her/his comprehensive review.

REFERENCES

Bilmes, J.A., 1998. A Gentle Tutorial of the EM Algorithm and its Application to Parameter Estimation for Gaussian Mixture and Hidden Markov Models. University of California, Berkeley, CA.

Chaudhuri, A., 2010. Randomized Response and Indirect Questioning Techniques in Surveys. Chapman and Hall/CRC Press, Boca Raton, FL.

Chaudhuri, A., Mukerjee, R., 1987. Randomized Response—Theory and Techniques. Marcel Dekker, New York, NY.

Greenberg, B.G., Abdel-Latif, A., Simmons, W.R., Horvitz, D.G., 1969. The unrelated question randomized response model for human surveys. J. Amer. Statist. Assoc. 64, 520–539.

Kim, J.M., Warde, W.D., 2005. A mixed randomized response model. J. Stat. Plan. Inference 133, 211–221.

Padmawar, V.R., 2012. Eliciting information on sensitive matters without inviting respondents' ire: randomized response techniques. Resonance 17, 44–60.

Padmawar, V.R., 2014. On a New Randomized Response Model. Indian Statistical Institute.

Padmawar, V.R., 2015a. On an example of a randomized response model. Model. Assist. Stat. Appl. 10 (4), 431–440.

Padmawar, V.R., 2015b. On a unifying randomized response model. J. Ind. Statist. Assoc. Submitted.

Padmawar, V.R., Vijayan, K., 2000. Randomized response revisited. J. Stat. Plan. Inference 90, 293–304.

Warner, S.L., 1965. Randomized response: a survey technique for eliminating evasive answer bias. J. Am. Stat. Assoc. 60, 63–69.

Yu, J.W., Tian, G.L., Tang, M.L., 2008. Two new models for survey sampling with sensitive characteristic: design and analysis. Metrika 67, 251–263.

Index

Note: Page numbers followed by "*f*" indicate figures, "*t*" indicate tables, and "*b*" indicate boxes.

H

I

Q

R

S

T

Y

FSC
www.fsc.org
MIX
Papier | Fördert
gute Waldnutzung
FSC® C083411

Kolibri 360 GmbH
Ferdinand-Jühlke-Straße 7
99095 Erfurt, Deutschland
produktsicherheit@kolibri360.de